EXPLORING THE URBAN COMMUNITY

**Prentice-Hall Series
in Geographic
Information Science**

KEITH C. CLARKE,
Series Advisor

EXPLORING THE URBAN COMMUNITY
A GIS APPROACH

RICHARD P. GREENE
Northern Illinois University

JAMES B. PICK
University of Redlands

PEARSON
Prentice
Hall

Upper Saddle River, NJ 07458

Library of Congress Cataloging-in-Publication Data

Greene, Richard P.
  Exploring the urban environment through GIS / Richard P. Greene and James B. Pick.
     p. cm.--(Prentice Hall series in geographic information science)
  Includes bibliographical references and index.
  ISBN 0-13-017576-5 (alk. paper)
  1. Urban geography. 2. Geographic information systems. 3. Urban geography--United States.
4. Cities and towns--United States. 5. City planning--United States. I. Pick, James B.
II. Title. III. Series.
  GF125.G74 2006
  307.76--dc22

                                                                    2005004926

Executive Editor: *Daniel Kaveney*
Associate Editor: *Amanda Griffith*
Executive Managing Editor: *Kathleen Schiaparelli*
Assistant Managing Editor: *Beth Sweeten*
Editorial Assistant: *Margaret Ziegler*
Production Editor: *Laserwords/nSight*
Art Editor: *Greg Dulles*
Manufacturing Buyer: *Alan Fischer*
Art Director: *Jayne Conte*
Cover Photo: Bettmann/CORBIS; Chicago's Millennium Park courtesy of Richard P. Greene
All photos and illustrations by the authors, except where noted otherwise.

© 2006 Pearson Education, Inc.
Pearson Prentice Hall
Pearson Education, Inc.
Upper Saddle River, NJ 07458

Pearson Prentice Hall™ is a trademark of Pearson Education, Inc.

Printed in the United States of America

10 9 8 7 6 5 4 3 2 1

ISBN 0-13-017576-5

Pearson Education Ltd., *London*
Pearson Education Australia Pty. Ltd., *Sydney*
Pearson Education Singapore, Pte. Ltd.
Pearson Education North Asia Ltd., *Hong Kong*
Pearson Education Canada, Inc., *Toronto*
Pearson Educación de Mexico, S.A. de C.V.
Pearson Education—Japan, *Tokyo*
Pearson Education Malaysia, Pte. Ltd.

Brief Contents

Contents

Preface

This is a comprehensive textbook on urban geography that applies the tools of geographic information systems (GIS) to the study of the internal structure of cities and urban systems. The need for this book arose from changes in urban geography and the study of cities over the past decade. The field has broadened to include multiple perspectives and interdisciplinary content areas.

This book covers the traditional content of urban geography updated to include contemporary literature and data as well as newer topics. The major topical areas are urban geography as a discipline, city dynamics, urban and metropolitan concepts, city internal structure, classic and postmodern models, systems of cities, central place theory, land use, neighborhoods, population density, urban migration and mobility, transnationalism, race, ethnicity, gender, poverty, industrial location, central city decline, edge cities and urban sprawl, housing, environmental problems, and urban and regional planning. Each chapter includes segments on concepts and theories, applications, examples and cases, and a GIS laboratory exercise.

Topics of recent interest in urban geography such as global cities, activism, technology, postmodernism, sexuality, transnationalism, exurbia, among others, enrich the study of the subject and complement as well as strengthen the traditional concepts and content. The book bridges these content areas and incorporates urban geography concepts and methods, cases and examples, GIS techniques, and spatial analysis to interpret spatial patterns and trends within and across urban areas.

The textbook applies GIS to the analysis of growth and change for three examples of the world's megacities: Chicago, Los Angeles, and Mexico City. The GIS approach enhances traditional spatial analysis and provides students with a modern approach to the study of prominent cities that can be applied to other urban areas.

The three cities were selected because they are contrasting in culture, topography, history of city growth, and urban problems and prospects. At the same time, the book discusses and gives cases, examples, and labs on many other cities and metropolitan areas in the United States and worldwide, including St. Louis; El Paso, Texas; New York; Irvine, California; Portland, Oregon; Las Vegas; the urban system of Snohomish County, Washington; Springfield, Missouri; Shanghai; Tokyo; Nairobi; Lagos; Central American cities; Australian cities, and London.

The book offers a hands-on GIS approach to the study of the city. Urban geography research and practice have been revolutionized by GIS. Computerized spatial analysis can show spatial arrangements and patterns, demonstrate trends, compare different variables, and function in combination with photographs and digital positioning. Urban and regional planners in the United States and many other countries have adopted GIS as an essential, standard planning tool. It is increasingly used by businesses for planning and projects in cities. The combination of GIS technology and the substantive material of urban geography will enhance the analytical skills, as well as the job readiness, of students.

The book provides hands-on demonstrations of how GIS can assist in the urban decision-making process typical of urban and regional planning agencies. It incorporates GIS throughout, including in maps, displays, special chapter sections, and a GIS exercise that accompanies each chapter. In each chapter, the special section, called "Analyzing an Urban Issue," presents essential conceptual background on the spatial analysis of real-world problems, as well as the associated GIS concepts employed. This section supports the GIS exercise at the end of the chapter.

A big plus of this approach to teaching urban geography is that students "get their feet wet" in analyzing real-world urban problems and data. The new approach provides user-friendly software tools, so the students can focus on what is really important: understanding better the spatial structure and processes of cities and urban areas.

The exercise files are contained on the book's CD. They run under the ArcGIS 9 software from ESRI, Inc., and are compatible with earlier ArcGIS 8 software. To run the labs, students need to first have access to an installed copy of ArcGIS 8 or 9. Students should copy the exercise files from the book's CD onto their hard disk, following the directions on the CD. They then follow the instructions for each of the 12 exercises contained in this text. Most urban geography courses have access to a GIS laboratory that has ArcGIS software installed. In many cases, the software is under a campus site license that allows a teaching copy to be provided to a student to put on his or her computer (check with your campus's information services office). Student copies of the ArcGIS software can also be obtained at reduced cost from ESRI, Inc., by contacting the company. A faxed copy of a student ID is required. For short periods of use, demonstration versions of the software are available from the ESRI, Inc., Web site, *www.esri.com*. Some instructors might prefer to give demonstrations of certain exercises, in place of the hands-on lab practice.

The book takes full advantage of the current geographically referenced data that are available in the public domain, including for the three megacities covered. The book emphasizes contemporary data, which draw on the U.S. Census and other government data that were produced in the 2000 censuses and subsequent publications. Some of these data became available up to 2004. For instance, the revised U.S. data with metropolitan definitions published in 2003 are utilized. Megacity data are drawn from United Nations revised publications of 2003. In other instances, a longer time horizon is needed for learning, and in certain sections the book examines historical data and longitudinal time series of up to a century or more.

Although the book is written to fully satisfy urban geography courses, its subject matter is broad and it is also appropriate for courses in urban sociology, urban and regional planning, urban studies, urban policy, and public administration. The authors have drawn material not only from the geography literature, but often from closely allied disciplines as well. In all cases, the subject matter is current and the literature extends up to the most recent references. The resulting up-to-date quality should be an advantage to students using the book in a world that has changed a lot in the past decade.

The authors themselves draw on a variety of backgrounds. Both authors have done extensive research on cities and urban problems. Richard Greene's background is in urban geography and GIS, whereas James Pick's is in urban studies, GIS, demography, environmental studies, and information systems. The book's material and exercises have been tested extensively in the authors' undergraduate classes at Northern Illinois University and University of Redlands, respectively.

It is the hope of the authors that students and professors will find this book contemporary, readable, informative, and knowledgeable; hands-on with spatial problems; and inclusive of the wide range of the body of knowledge and concerns of urban geography.

ACKNOWLEDGMENTS

No undertaking of this magnitude would be possible without the assistance of many individuals. The authors are indebted to many persons who contributed to this project, only some of whom are mentioned here.

Two people stand out as worthy of special acknowledgment. John C. Stager of Claremont Graduate University and University of Redlands provided unflagging consultation, support, motivation, and superb technical expertise for the whole length of the project. We acknowledge his excellent contribution on originally developing several of the labs and improving the others. Richard Forstall, formerly of the U.S. Census, expertly and thoroughly critiqued the entire manuscript sentence by sentence, making numerous perceptive suggestions, small and large, throughout. Both were extraordinary in their contributions and dedication.

We wish to acknowledge the comments and critiques for parts of the book by Rob Burke, Donald Dahmann, Melissa Hyams, Michael Conzen, and Terry Clark. The Prentice Hall reviewers were most thorough and helpful, including Thomas L. Bell, University of Tennessee; Chris Benner, Pennsylvania State University; Fernando Bosco, San Diego State University; Sarah Elwood, University of Arizona; Sallie Ives, University of North Carolina Charlotte; Bruce Pigozzi, Michigan State University; Paul C. Sutton, University of Denver; and David Wong, George Mason University. We thank Keith Clarke for several helpful suggestions. At Prentice Hall, we thank Dan Kaveney for his expertise and thorough attention and support to the project from start to finish, Margaret Ziegler for helpful support, and Jessica Einsig, Kathryn Anderson, and Chris Rapp for their assistance and expertise. We thank Mark Corsey of nSight for his interest and production expertise.

The authors acknowledge ESRI Inc. for its helpful support and assistance, and for the particular data for Figures 1.6, 1.11, 2.20, 2.25, 4.5, 5.5, 5.7, 7.15, and 9.10.

At University of Redlands, James Pick extends thanks to School of Business faculty support staff Joanie James, Susan Griffin, and especially Heather Sarrail for her fine work on permissions, to Dean Jerry Platt for focus and interest in GIS, and to campus Information Technology Services. He thanks all the students in his GIS courses over the past three years, many of whom provided useful comments and critiques of the draft book materials. Particular thanks for lab critiques go to Lea Deesing, Tom Dunn, and Jody Neerman. James Pick acknowledges the support of the University of Redlands Armacost Library and the University of California Irvine Libraries for access to reference information.

At Northern Illinois University, Richard Greene extends his thanks to faculty support staff Barbara Voga, Dawn Sibley, Phil Young (Director of Advanced Geospatial Laboratory), Lenny Walther (Director of Cartography Laboratory), Rick Schwantes, and Jodi Harlan, to Dean Fred Kitterle of Liberal Arts & Sciences and Geography Chairman Andrew Krmenec for the investments they have made in GIS. He thanks all of his students from his urban geography and GIS courses over the years. He gives special thanks to Jamie Benge of ESRI Inc., who was instrumental in taking many of the maps produced in the GIS into final cartographic production while a student in the geography department at Northern Illinois University. Thanks also to Peter Piet, Doug Jones, Soomee Cha Kong, Todd Schuble, and Peter Scizowicz for collaborating in urban geography research during various phases of the book project. Reviews of the book's GIS exercises were aided by former students Adam Aull, Peter Piet, Ray Ulreich, Marek Dudka, Melanie Buell, Tomasz Szczuka, Martin Pinnau, Fred Weiss, and Karen Russ. He thanks the libraries of Northern Illinois University and the University of Chicago, which accommodated much of his research on urban geography.

The authors again wish to acknowledge all these people and many others not mentioned who provided so much assistance and backing.

Finally, James Pick wishes to express special appreciation to his wife, Dr. Rosalyn M. Laudati, for her understanding, interest, and support during the many long hours, some taken away from family time. Richard Greene wishes to express special appreciation to his wife, Anna E. Adamik, for her support during the many long hours of book writing and for engaging in interesting discussions and excursions related to the book. The book is dedicated to Rosalyn and Anna.

1

The Spatial Display of Urban Environments

Urban geography is the study of urban areas and cities based on geographical principles. It includes multiple dimensions of their spatial distributions, activities, and functioning. The study of urban geography is important because 47 percent of the world's population lived in urban areas in 2000 (see Figure 1.1) and that share is expected to increase to 60 percent by 2030 (United Nations, 2003). During that period, the U.S. urban population share is anticipated to grow from 77 percent to 84 percent (United Nations, 2003). Urban areas are the centers of government, economic activity, and business decision making, as well as the primary markets of culture and intellectual accomplishment. Urban geography is aligned with sociology, economics, and urban planning, from which it draws and gives back ideas and knowledge. A geographic information system (GIS) is a modern tool that helps in understanding and analyzing urban spatial distributions and their change over time.

This chapter sets the stage for the book by defining and justifying the importance of urban geography and pointing out why applying spatial analysis to the study of urban geography is a useful approach. It discusses several other geographical approaches that inform the study of the city, including physical, cultural, political, and economic geography. It examines the international view of urban geography, why cities have become more international, and the growth of megacities. An example of an urban issue is presented—the pattern of growth and change of the Sunbelt in the United States—for urban areas in the South and West. The principles and contributions of the Chicago, Los Angeles, and New York City schools of thought are explained. Quantitative techniques, spatial analysis, and the computer, which have

supported more systematic and data-based study of cities and urban areas, have all influenced urban geography. Besides the spatial view, other views of urban geography—political economic, behavioral, urban economic, and urban historical—are introduced. The chapter turns to defining and explaining GIS, how it functions, and especially how it helps in the study of urban geography. The approach of the textbook is explained by a summary of its content, chapter organization, and the role of the GIS practice exercises.

1.1 THE STUDY OF URBAN GEOGRAPHY

Urban areas represent and support most of human activity. They not only are the main focus of civilizations, but they are also dynamic and vibrant forms of community. In their long history, they have been the primary locus for development of social movements, intellectual ferment and discoveries, and the rise and fall of nations and civilizations. For instance, Roman civilization is tied to Rome; jazz to New Orleans, Chicago, and New York; and the Midwest to Chicago, Milwaukee, and Detroit.

Cities exist to form places for economic, social, and political interchange, control, and innovations. Over time, these reasons have changed because of alterations in underlying societal processes that include demographic, economic, political, cultural, and technological changes. For some larger cities, global exchange has been a key factor in their rationale and functioning. The processes that produce and continue to change urban agglomerations are covered in Chapter 2. An example is the influence of the industrial age in the 19th century on cities. Cities grew and changed as manufacturing zones

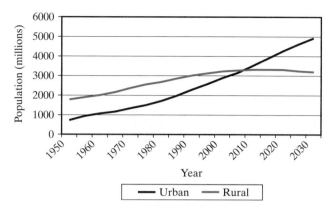

FIGURE 1.1 Urban and rural populations of the world: 1950–2030. Source: United Nations. (2003) *World Urbanization Prospects: The 2003 Revision*. New York: Author.

developed, new forms of transportation such as railroads were introduced, new places were established for residence of industrial workers, and manufacturing cities became interconnected because of economic dependencies.

The study of urban areas and cities is essential. Fortunately, much is known about their social, economic, historical, and geographical evolution. Through this knowledge, real-world practitioners can perform their jobs better. Urban geography can help city planners to understand city sectors and neighborhood change patterns, politicians can learn about neighborhood attributes and the attitudes of their residents, and public health officials can unravel the spatial patterns of human contacts and the spread of disease.

Urban geography depends on the arrangement of phenomena in space. Space is important in the city because it determines the relationships of people, transport, neighborhoods, industry, communications, and environment. How do apartment rentals change from the city center in the westward gradient? How do two neighborhoods change in their attributes and interrelationship with each other over time? How is urban ethnic diversity changing throughout the city? How long does it take citizens in different areas to reach the nearest hospital? How do the city's residents relax and engage in leisure activities? (See Post Alley near Pike Place Market in Seattle in Figure 1.2.) These and other crucial questions can be answered through geographical analysis. The concepts of space have changed as the cities' underlying processes have altered. For instance, as more automobiles were introduced in the 20th century, American cities spread out over the possible car commuting range, a tendency that later led to their sprawl over large spaces.

Within geography, urban geography is linked to *cultural and human geography* because cities have cultural, societal, and human aspects; *physical geography* because they have terrain, rivers, lakes, and floodplains;

economic geography because economic activities are constantly occurring; and *environmental geography* because there is a complex urban environment.

Unlike many subdisciplines of geography, urban geography is relatively new and was not actually offered as a specialization in universities prior to World War II. This is not to say that some courses were not offered or that geographers were not looking at cities—in fact they were—but not with as large a knowledge framework as today.

By the Second World War, therefore, the situation had been reached where a rapid growth of urban geography was inevitable. Preliminary foundations had been laid and many of the basic ideas had been propounded, although in isolation. In many ways the recent expansion in urban geography had been mainly concerned with exploitation of ideas already in existence in the 1930s. (Carter, 1972, p. 5)

Geographers Harold Mayer and Clyde Kohn recognized this evolving synergy for urban geography and compiled what can be considered the first American textbook in urban geography (Mayer & Kohn, 1959). Their collection of readings focused on urban functions, forms, distributions, and growth, the primary urban topics geographers had examined up to that time.

FIGURE 1.2 Post Alley near Pike Place Market in Seattle.

Even at the time that Mayer and Kohn had edited their collection of readings, two contrasting approaches in urban geography were becoming well defined. These were the *urban systems approach* and the *internal structure of cities* approach. The urban systems topics ranged from central-place theory, formulated first by Christaller in the 1930s, to the study of cities as nodes with an emphasis on modeling the interaction among the nodes. Textbooks eventually appeared that broke up the two approaches into major sections. Two books devoting substantial coverage to the urban system approach were *City and Region* by Dickinson (1964) and *Geographic Perspectives on Urban Systems* by Berry and Horton (1970). A key principle of the urban systems approach is that there are underlying economic, social, and political processes that account for spatial variation in phenomena throughout the urban system. These processes and their interaction with urban systems are of major concern to the urban systems approach, as are central place theory and others discussed in Chapter 5. Coverage is also given to this approach in Chapter 7, particularly in the section on migration. The section of this chapter titled "Analyzing an Urban Issue: Sunbelt Growth," and the coverage in Chapter 5 on central place theory are examples of the urban systems approach.

The internal structure of cities approach examines the same economic, social, and political processes, but focuses on how the spatial outcomes play out in the city and urban area. An urban geography textbook in the 1980s bearing the same name, *Internal Structure of the City,* was a volume of readings edited by Larry Bourne (1982) that articulated the approach with examples from 64 authors. Bourne introduced three terms to describe the focus of the internal structure approach: urban form, urban interaction, and urban spatial structure. *Urban form* is the spatial pattern of features in a city, for instance, a working-class Latino neighborhood and an upper-income White neighborhood. *Urban interaction* refers to the linkages and flows that integrate the various features of the city. For example, interaction could be represented by the contrasting routes taken by the two types of households to work. In the Latino case, women domestic workers would move daily overwhelmingly into the upper middle-income neighborhood; in the White case, many workers would commute into a nearby office park or to the city's main downtown area. Flows of buyers and sellers interact daily in outdoor markets in Mexico City (see Figure 1.3). *Urban spatial structure,* according to Bourne, is the highest level of analysis, as it represents the organizational linkages that connect urban interaction and urban form into a cohesive system. In the neighborhood example with its associated workers commuting to their jobs, variation in wage rates among local employment centers might be taken into

FIGURE 1.3   A city market in Mexico City.

account to explain the different patterns of movements and how they are tied to the occupational skill levels of the neighborhoods themselves.

The internal structure of cities approach occupies much of this book. We consider important social and cultural processes that impact the city and its institutions; how people behave, cooperate, and clash; and how governmental institutions grow, change, and affect the city's urban form. The patterns and processes of the natural environment are considered in terms of urban form and urban interaction, because terrain influences city form and contaminants influence its quality of life. However, not all groups in the city are affected by pollution to the same degree. The field known as *environmental justice* seeks to understand these inequities. It requires an understanding of urban spatial structure. Transportation facilitates urban interaction through the movement of people, goods, and services, both within and between cities, and is often used as an example throughout the book.

Urban geography is closely aligned with some other academic disciplines. There are close links with urban sociology, environmental economics, political science, urban and regional planning, and demography. None of these urban disciplines has the last word on urban processes, but often content is shared or overlapping. Urban sociology focuses particularly on the sociological forces and processes in cities, topics such as cities' social structure, social stratification, power structures, the family, crime, and disorganization. Environmental economics includes analyses of economic cost and benefits and trade-offs of pollution. Political science includes urban politics, political geography, and political change, and urban and regional planning is concerned with the theory and practice of planning city development. In this book, we refer to some concepts and ideas from these and other disciplines.

1.2 URBAN GEOGRAPHY AS A DISCIPLINE: ITS IMPORTANCE

For a preliminary glimpse at urban geography, consider a map (see Figure 1.4) showing areas of 50 percent or greater Latino population in Los Angeles in 1970 and 2000. It is clear that the Latino population spread vastly. In 1970, the Latino population was concentrated in the East Los Angeles area, with some small concentrations in San Fernando well to the northwest of East Los Angeles,

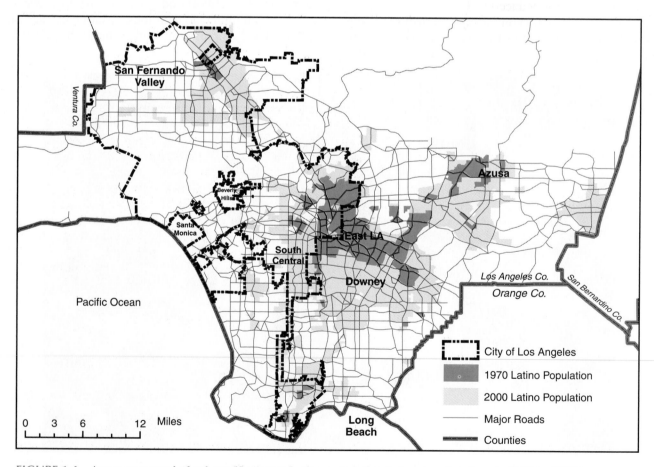

FIGURE 1.4   Areas composed of at least 50 percent Latino population in Los Angeles County, 1970 and 2000.

Azusa to the east, and Downey to the south. By 2000 it had become widely spread out in a much larger area occupying a large part of the central-western section of Los Angeles, a large section of the northern San Fernando Valley, and some areas in the South Bay area northwest of Long Beach. During the same time span, the area of Black population concentrated in south-central Los Angeles was reduced in size. As will be seen in later chapters and especially in a section of the exercise for Chapter 4, the entire Los Angeles metropolitan area expanded over the three decades. Another feature discussed more is that the unusual shape of the Los Angeles city boundaries leaves pockets within it for other cities, such as Beverly Hills and Santa Monica (see Figure 1.4). This stems from the city's rapid growth, as formerly outlying cities were enveloped by it.

Figure 1.5 shows the ethnic components of Los Angeles County from 1940 to 2000. In 1940, the county was 93 percent non-Hispanic White and only 2.2 percent Latino. Latino population grew steadily over 60 years, and by 2000 it comprised 45 percent of Los Angeles County. From 1970 to 2000, the Latino percentage of the county increased by 17 percentage points, or 2,953,597 more Latino persons. By contrast, the proportion of non-Hispanic Whites was reduced to less than half and the share of Black population remained steady at 10 percent. Not only are nearly half of Los Angeles County's persons Latino, but they are now widely distributed, so that few parts of Los Angeles did not have a Latino population in 2000.

The ramifications of this urban change are large. What has happened to the buying habits of consumers? What are the changes to the perceived importance of Latin American and Mexican cultures? Given that the average Latino household tends to be larger than non-Latino white households, how is housing affected? Where are the new migrants to Los Angeles coming from? What are the educational profiles of new migrants arriving into these Latino areas from Mexico and other countries? This small example exemplifies the important contributions stemming from urban geography, illuminating the effects of major urban change on how people live and work, and how they interact with each other.

Geographical analysis can also be *comparative*. This means comparing several cities, comparing a city to its surrounding region or hinterland, or comparing different parts of a single city. If we look at the downtowns and transportation matrix of Los Angeles compared to St. Louis in 1990, we observe very different patterns (see Figure 1.6). St. Louis contains a long-established central downtown, which is located at the central hub of the city's transportation network. By contrast, the main downtown of Los Angeles has many rivals, and none of them dominates as the transportation hub. This example shows that the comparative approach can yield insights beyond what is available in just one city or the other.

1.3 AN INTERNATIONAL VIEW OF URBAN GEOGRAPHY

Another goal of this book is to provide an international perspective. This is accomplished by including international urban examples and information. Foreign cities are important to examine in an urban geography text. The world is shrinking because of continual advances in transport and communications, growing trade, and language diffusions (especially English). For instance, the fastest travel time for passengers to go from Chicago to New York was eight hours in 1900. Today, it is two hours. English is widely dispersed around the world and its use encourages communications. Presently 5.7 percent of the world's population have English as their primary speaking language, second only to Chinese at 20.4 percent (Grimes, 2000). Moreover, 8.5 percent of the world's population speak English when their second language is also counted (Grimes, 2000).

Most of the world's population growth in the 21st century will occur in the developing world, and that will be predominantly in cities. The world's count of megacities of 10 million or more persons has grown from 3 in 1950 (New York, London, and Tokyo) to 20 in 2003 (see Table 1.1). Of the 20 cities, 16 are in developing nations and only 4 in the more developed nations. The seven biggest megacities in 2003 were Tokyo, Mexico City, New York, Sao Paulo, Mumbai (Bombay), Delhi, and

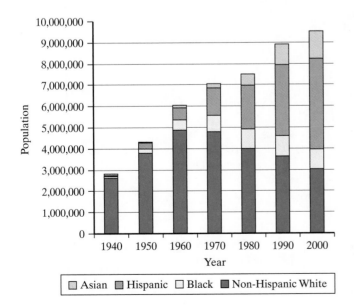

FIGURE 1.5   Ethnic components of population of Los Angeles County. Source: U.S. Census, 1940 to 2000.

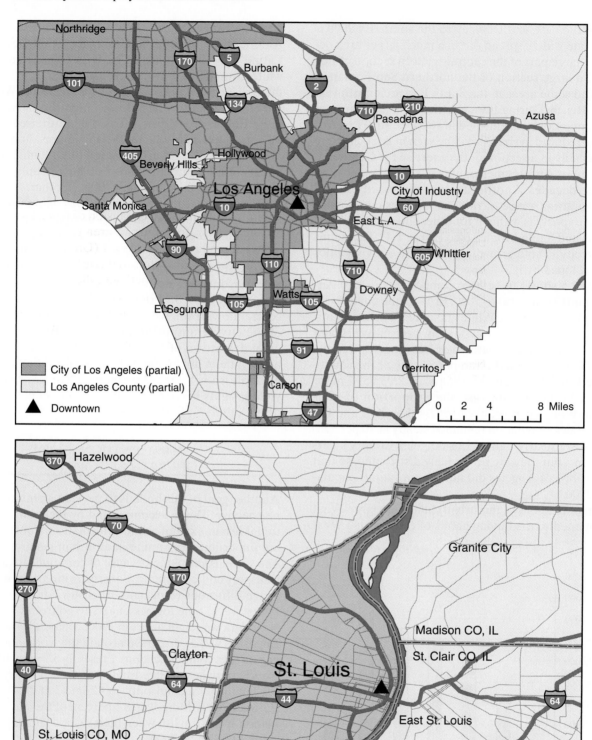

FIGURE 1.6   City center comparisons for Los Angeles and St. Louis.

TABLE 1.1 The world's megacities of more than 10 million population, plus Chicago, 2000

Rank	City	Country	Population (millions)
1	Tokyo	Japan	35.0
2	Mexico City	Mexico	18.7
3	New York	United States	17.9
4	Sao Paulo	Brazil	18.3
5	Mumbai (Bombay)	India	17.4
6	Delhi	India	12.0
7	Calcutta	India	13.8
8	Buenos Aires	Argentina	12.8
9	Shanghai	China	11.6
10	Jakarta	Indonesia	14.1
11	Los Angeles	United States	13.0
12	Dhaka	Bangladesh	12.3
13	Osaka	Japan	11.2
14	Rio de Janeiro	Brazil	10.8
15	Karachi	Pakistan	11.2
16	Beijing	China	11.1
17	Cairo	Egypt	10.8
18	Moscow	Russian Federation	10.5
19	Metro Manila	Philippines	10.4
20	Lagos	Nigeria	10.1
	*Chicago	United States	9.2

*Based on U.S. Census Bureau Metropolitan Definition, 2003.
Source: United Nations. (2003) *World Urbanization Prospects: The 2003 Revision*. New York: Author.

Calcutta. Five of these are in the developing world. Tokyo in this chapter section and its tables refers to the Tokyo greater metropolitan area, which consists of Tokyo and Yokohama. The United Nations counted just 16 world megacities in 2000, and the 4 new ones (Cairo, Moscow, Manila, and Lagos) added by 2003 were all located in the developing world. The United Nations predicts that the ranking will change by 2015 with 8 of the 10 largest urban megacities found in the less developed counties (Table 1.2). In 2015, Tokyo will still be the biggest megacity at 36.2 million inhabitants, followed by Mumbai, India, at 22.6 million, and Delhi, India, at 20.9 million. Continuing an expansion of more than a century, the greater Tokyo urban area grew moderately, by about 1.4 million, in the 1990s (United Nations, 2003). The United Nations refers to this urban agglomeration as Tokyo, although it is sometimes called Tokyo-Yokohama. A comparison of the top megacities in 1970 and those projected in 2015 shows that Mexico City remains in the top 10, whereas Los Angeles, Paris, London, and Osaka will have all dropped out of the top 10.

If the current trends evident on this time-line persist, the advanced world cities that constituted about half of the leading cities in 1970 will have been superseded in 2015 by numerous megacities in the developing world, especially Asia, Africa, and South America, and located

TABLE 1.2 The world's largest megacities, 1970 and 2015

	1970				2015		
Rank	City	Country	Population (in 1000s)	Rank	City	Country	Population (millions)
1	Tokyo	Japan	16,498	1	Tokyo	Japan	36.2
2	New York	United States	16,191	2	Mumbai (Bombay)	India	22.6
3	Shanghai	China	11,154	3	Delhi	India	20.9
4	Osaka	Japan	9,409	4	Mexico City	Mexico	20.6
5	Mexico City	Mexico	8,769	5	Sao Paulo	Brazil	20.0
6	London	United Kingdom	8,594	6	New York	United States	19.7
7	Paris	France	8,498	7	Dhaka	Bangladesh	17.9
8	Buenos Aires	Argentina	8,417	8	Jakarta	Indonesia	17.5
9	Los Angeles	United States	8,378	9	Lagos	Nigeria	17.0
10	Sao Paulo	Brazil	8,308	10	Calcutta	India	16.8
11	Beijing	China	8,087	11	Karachi	Pakistan	16.2
12	Rio de Janeiro	Brazil	7,155	12	Buenos Aires	Argentina	14.6
13	Moscow	Russian Federation	7,107	13	Cairo	Egypt	13.1
14	Calcutta	India	6,912	14	Los Angeles	United States	12.9
15	Chicago	United States	6,716	15	Shanghai	China	12.7
				16	Metro Manila	Philippines	12.6
				17	Rio de Janeiro	Brazil	12.4
				18	Osaka-Kobe	Japan	11.4
				19	Istanbul	Turkey	11.3
				20	Beijing	China	11.1
				21	Moscow	Russian Federation	10.9
				22	Paris	France	10.0

Source: United Nations. (2003) *World Urbanization Prospects: The 2003 Revision*. New York: Author.

at more southerly latitudes. The implications for the study of urban geography are that its models and methods will need to adapt to include these new giant cities in less developed regions. It points to the importance of understanding cultural and economic development aspects of these megacities. Having Mexico City as an example will help us keep in touch with this trend and broaden our cultural reach.

As one instance of an attribute that differs internationally, consider poverty as measured by percentage of low- or no-income inhabitants in greater Mexico City in 2000 (see Figure 1.7). Low income is defined as income less than the minimum wage. The map shows the two parts of Mexico City, the central part located in the acorn-shaped Federal District, and the more outlying part in the State of Mexico that surrounds the Federal District. The base maps for Mexico City are discussed further in Chapter 2. It can be seen that the percentage of workforce with no or low income is moderate to low in the downtown city center and wealthier residential areas to the west. The downtown is located in the upper midcenter of the Federal District. Poverty is very high

in areas way out at the urban fringe to the northeast and southeast, which might stem from their being more rural. Also, in Latin American cities, the poor arriving migrants tend to flow to available "squatter" land in the periphery (Elbow & Pulsipher, 2003; Griffin & Ford, 1980). This differs from U.S. cities, in which the poor tend to be located in the inner part of the city near the downtown, whereas the suburbs and fringe areas have fewer pockets of poverty.

Although having prosperous business and commerce, Mexico City's downtown city center also has a population of poor street vendors and other informal labor forces, some of whom reside in the city center. As in many large Latin American cities, smaller pockets of very poor population are interspersed within mostly wealthy zones. This pattern has other ramifications for working patterns and government planning. For instance, poor inhabitants on the periphery cannot afford automotive transport to commute to jobs in or adjoining the city center, which underscores the need for a cheap and massive public transport system. An area of poverty near the Mexico City periphery is seen in Figure 1.8. On another note,

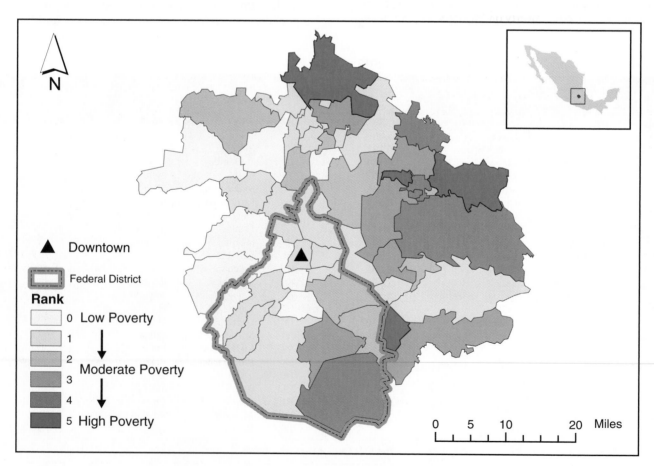

FIGURE 1.7   Low or no-income ranks within Mexico City metro regions (2000 Metro Definition).

FIGURE 1.8 Poverty on Mexico City's periphery.

urban planning in Mexico City, as it is in most U.S. cities, is based less on a zoned approach and is more political and unregulated in nature. Poor squatters are able to arrive and claim land without violating zoning laws and usually will not be evicted.

Thus, studying developing world cities teaches us that there is not one way for urban processes to occur, but alternative ways culturally and societally. We return later in the book to the question of international differences among cities.

In summary, urban geography examines the development, growth, and processes in cities and urban areas in a geographical context. It reflects the many dimensions of the city, including economic, social, human, political, environmental, and physical. Urban geography interacts with many neighboring disciplines, so it is not surprising that a variety of perspectives have emerged within the field itself.

To demonstrate the urban geography approach, the next section examines an urban issue—the growth of Sunbelt cities—from a geographical perspective. The expansion of Sunbelt cities serves as the book's first in-depth example of change in cities and some of the processes underlying this change, especially migration flows. It is an influential issue that has involved the work and lifestyle of millions of Americans. This urban

issue is later pursued in the chapter's GIS laboratory exercise.

1.4 ANALYZING AN URBAN ISSUE: SUNBELT GROWTH

Conventionally, U.S. Sunbelt cities, as shown in Figure 1.9, have been defined as those fast-growing cities that are located below the 37th parallel (Bernard & Rice, 1983). This definition reflects the southward and westward movement of people from the cities of the Northeast and the Midwest to the many cities of the South and West. The Sunbelt city of Miami, Florida, hot and humid in the summer, is shown in Figure 1.10. Mention of the Sunbelt often conjures up an image of destination cities with amenity-rich coastal settings and abundant job generation. However, drawing the line in the sand to demarcate the Sunbelt is difficult. In fact, there might not be a distinct line demarcating the Sunbelt, based on the characteristics most associated with it. If Sunbelt cities are defined solely in terms of population growth, then the Sunbelt might include not only the fast-growing cities of the South and West, but also some cities along the entire Pacific Northwest coast and in the Mountain West.

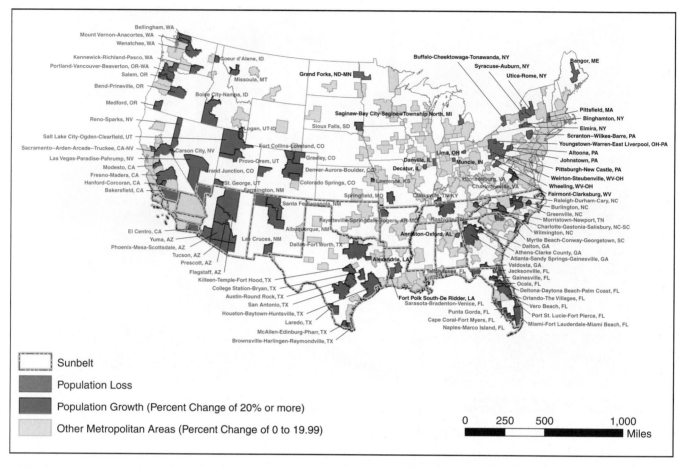

FIGURE 1.9   Metropolitan growth in the 48 conterminous United States: 1990 to 2000.

How might an urban geographer explain these trends? One historic reason cited for the growth of Sunbelt cities was defense and other federal spending that was biased toward Southern and Western cities (Tabb & Sawers, 1984). A boom in high-technology industries and growing affluence, as well as an expanded service and information economy, have also added to Sunbelt city growth. Another factor is the introduction of moderate-priced air conditioning after World War II. This added to the attractiveness of seasonally hot and sometimes humid locations in the South and West. Although these general trends have been true for the United States as a whole, it has been well documented that relatively higher availability of jobs and competitive wages have been strong attractive forces for the South and West (Greenwood & Hunt, 1989; Piet, 1998). Before the 1990s in Figure 1.9, oil and gas exploration and development in Texas, Oklahoma, and Louisiana in the 1970s and early 1980s had helped spur growth in those regions of the Sunbelt.

Florida and Arizona can attribute much of their early and continued expansion to the growth of retirement communities. It will likely be augmented by the retirement of those in the baby boomer generation, which is expected to begin around 2005. On the other hand, these regions could lose some population as retirees reach their 80s and 90s and begin to return from Sunbelt retirement areas to their places of previous residence in the North. In support of this, Manson and Groop (1996) observed an actual decline in interstate movers to the South and West during the early 1990s, whereas interior areas outside of the South and West showed increases. Their study also noted that the impressive overall migration into California in the 1980s turned around to substantial migration losses between 1980 and 1990. Nevertheless, the heavy concentration of recreation and the tourism industry in Florida, southern California, and Las Vegas is an obvious magnet for longer term sustained growth in those regions.

According to a more recent study, city growth in the 1990s followed the same basic patterns found in the earlier Sunbelt literature. In particular, the half-century-long trend of people moving to places with good weather, low population density, and skilled inhabitants continued (Glaeser & Shapiro, 2001). The presence of skilled

FIGURE 1.10   Miami skyline.

inhabitants implies that there are also jobs for migrants. However, there was no general rebirth of high-density cities typical of the Northeast and Midwest in the Sunbelt. Glaeser and Shapiro (2001) also pointed out that some cities experiencing rapid growth outside of the defined Sunbelt were unusually well endowed with college graduates, reflecting the attraction of universities. For instance, the university cities of Lawrence, Kansas, and Charlottesville, Virginia, are both outside of the Sunbelt but are identified in the GIS database of the lab exercise as rapidly growing. Another factor continuing to attract migrants to the Sunbelt is the convenience and comfort associated with air conditioning.

Prior to the 1970s the largest metropolitan areas grew at a faster pace than the United States as a whole, but this trend was reversed in the 1970s when population densities declined across the entire urban hierarchy and internal migration flowed from the Northeast and Midwest to the South and West (Greenwood, 1988; McHugh, 1988; Plane, 1992). Technological changes had allowed

the full diffusion of urban amenities to all parts of the country and increased the importance of environmental, lifestyle, and location amenities in the residential decision-making process (Frey, 1993). People moving to less dense places often cite quality-of-life reasons for their decisions (Fuguitt & Brown, 1990).

Others have pointed to the importance of the age of migrants as a factor in Sunbelt flows. They argue, for instance, that the largest numbers of baby boomers were born in the Northeast and Midwest, the dominant urban employment locations of their parents' generation. This very large cohort of young persons born from 1946 to 1964 entered the age bracket of maximum personal mobility in the 1970s (Borchert, 1983). However, the economic downturn in much of the Northeast and Midwest in the 1970s meant that these areas could not absorb this large cohort of new workers into their labor force. Consequently, net migration accelerated from the Northeast and Midwest to urban destinations in the South and West (Plane, 1992).

Immigration is another factor cited for Sunbelt growth. International immigration flows to key urban areas of the United States have become quite large, often overshadowing earlier internal migration patterns (Champion, 1994; Frey, 1994; Walker, Ellis, & Barff, 1992). As seen in Chapter 8, from the Civil War to World War I, most immigrants came from Europe. For instance, at the turn of the last century 80 to 90 percent of the immigrants were European. As a result, East Coast port cities such as Boston, New York, and Philadelphia were the principal recipients of these immigrants. Waves of immigrants continued to arrive, replacing those who had moved on to the cities of the Midwest and elsewhere.

The Immigration and Nationality Act Amendments of 1965 altered the limits on the number of immigrants legally allowed from each country, which resulted in much higher levels of the "new immigrants," a phrase used to designate immigrants from Latin America and Asia (see Chapter 8). By the 1980s and 1990s, 80 percent of the immigrants were from these regions. The shift in point of origin for most immigrants has been reflected in changes in the main immigration ports of entry. Borchert (1983) observed that the 1970s foreign-born stream settled largely in Southern and Southwestern metropolitan areas that had reached population sizes comparable to the 19th-century ports of entry for the manufacturing belt. The manufacturing belt refers to manufacturing concentrated in the industrial states stretching across the Northeastern and Middle Atlantic states into the Midwest. All this has fueled metropolitan growth rates in the South and West, as many immigrants typically settle, at least initially, near their points of entry (Carlson, 1994; Champion, 1992, 1994; Frey, 1993, 1995). McHugh (1989) noted that the heavy influx of immigrants in the 1980s allowed metropolitan areas in the South and West to continue to grow rapidly despite a decline in internal migration.

In summary, there are many factors offered to explain the spatial variation in the growth of cities in the South and West. Many of the explanations proposed are from urban geographers or persons from related disciplines employing the spatial perspective. In this chapter's GIS lab exercise, you will be able to analyze Sunbelt growth trends for U.S. metropolitan areas in the 1990s to build on these many observations.

The next two sections turn to introducing two influential schools of thought in urban geography over the past 80 years. The Chicago School set a foundation for understanding the spatial arrangement and ecological functioning of cities. It was based on American cities, especially Chicago in the 1920s, but the concepts have been influential in shaping thought on subsequent models of space. The recent L.A. School has adopted postmodernist thought and built a theory that includes the impact of urban sprawl, urban design, culture, and globalization. That school based some of its thinking on Los Angeles as a prototype, but its thought extends to a range of contemporary cities.

1.5 THE CHICAGO SCHOOL AND ITS LINK TO URBAN GEOGRAPHY

A neighboring discipline to urban geography is urban sociology, particularly the urban sociology that developed in Chicago in the early 1900s, which had a significant effect on the spatial perspective of urban geography. This was due in large part to Robert Park (1864–1944) and Ernest Burgess (1886–1966), sociologists at the University of Chicago who guided generations of sociology students doing research on the city of Chicago, many of them on spatial topics. The theories of Park and his associates became known as the human ecology approach to the study of urban development, which was put forth in his classic essay of 1925, "The City: Suggestions for the Investigation of Human Behavior in an Urban Environment" (Chapter 1 in Park, Burgess, & McKenzie, 1925). Park's paper had a profound impact on the development of both sociology and geography in the United States. Many of the students of the city and urban conditions in the early 20th century were at the University of Chicago, which had been established in 1892. In addition, Chicago at the turn of the century was one of the fastest growing major cities in the world, owing in large part to immigration. The Chicago School, as the sociologists Park, Burgess, and McKenzie became known, wanted to learn how immigrants were assimilated into the city; Chicago became a laboratory for this endeavor. As a result, many leading urban concepts were originated through the study of Chicago, and other American cities were less fully investigated. A genealogy of North American urban literature today would reveal that much of it has roots going back to Chicago.

Park believed that the community operated on two levels. The first was the *biotic level,* which was based on competition and division of labor, a concept that was derived from Darwin's "The Web of Life." The second was the *cultural level,* with a focus on communication and consensus on the moral order. Park argued that the proper focus for human ecology was the biotic level, whereas analysis of the cultural level was considered a problem for social psychology. Thus the human ecology approach was to view the human community as a dynamic adaptive system; competition served as the primary organizing agent. Under the pressure of competition, each individual and group was seen as carving out in the city both residential and functional niches in which they could best serve and prosper. The eventual effect of competition was to segregate people and their businesses into relatively homogeneous residential and functional subareas within the community.

Park, Burgess, and McKenzie published Park's original essay along with others in the book *The City* in 1925 (Park et al., 1925). Burgess wrote the book's second essay and developed the diagram, now classic, that illustrates the concentric zone model of urban development. In this model, the inner ring or zone consisted of the central business district (CBD). It was encircled by successive zones of transition, working-class housing, middle-income residential housing, and commuters. The concept was both spatial and temporal. As time passed, the zonal structure continued, but the entire system expanded; that is, the zones expanded outward, consuming additional territory. Burgess contended that the system changed because of the processes of neighborhood change expressed in Park's first essay in the book. The topic of zonal patterns of urban development is reviewed in greater detail at the beginning of Chapter 4 along with other subsequent theories of urban spatial structure.

The third essay, by McKenzie, also had a strong spatial component, as demonstrated by the following excerpt:

> [S]ociety is made up of individuals spatially separated, territorially distributed, and capable of independent locomotion. These spatial relationships of human beings are continuously in process of change as new factors enter to disturb the competitive relations or to facilitate mobility. (Park et al., 1925, p. 64)

The Chicago School's use of social Darwinism to explain the dynamics of city growth caused many scholars of the city to oppose their ideas. Nevertheless, their literature provided much of the foundation for future spatial investigations of the city. In addition, they invented or made familiar many of the spatial terms that we still use today to describe urban settlements. Consider, for instance, Burgess's 1925 description of the physical expansion of cities:

> In Europe and America the tendency of the great city to expand has been recognized in the term "the metropolitan area of the city," which far overruns its political limits, and in the case of New York and Chicago, even state lines. The metropolitan area may be taken to include urban territory that is physically contiguous, but it is coming to be defined by that facility of transportation that enables a business man to live in a suburb of Chicago and to work in the loop, and his wife to shop at Marshall Field's and attend grand opera in the Auditorium. (Park et al., 1925, pp. 49–50).

We turn to the subject of *metropolitan area* definition in Chapter 3, but it is worth noting that this terminology was in use by the human ecology school as far back as 1925. Finally, the human ecology school produced very practical spatial surveys of the city, which helped explain the location of the city's various activities. It is one of the contributions of the Chicago School that would become a major wellspring of the spatial perspective within urban geography. Yeates (2001) also refers to a 1960s Chicago School within The University of Chicago's Department of Geography.

1.6 THE L.A. SCHOOL OF URBAN STUDIES

Recently, a group of urban geographers and affiliated scholars have defined a new urban studies school called the L.A. School (Dear, 2002). Led by Los Angeles area geographers Michael Dear, Edward Soja, and Allen Scott, these scholars use their urban region as a laboratory for many different types of urban investigations. Like their Chicago School predecessors, they argue that their laboratory is the most dynamic metropolis and very reflective of urban processes underway globally (Soja & Scott, 1996). They contend, furthermore, that many of the principles developed by the Chicago School do not pertain to Los Angeles or to the modern-day metropolis in general. Thus, they offer up Los Angeles as the prototype for examining new urban processes, and particularly the impact of globalization.

Among their theoretical points are the following. Postmodern thinking leads to a wide range of methods of gathering knowledge. This includes feminism, architectural design theory, postcolonial concepts, and others. In regards to the study of the city, they contend that rather than having a city with a centralized core, many core functions have moved to the periphery. Accordingly they see an "undifferentiated, centralized grid" instead of the concentric zones proposed by Burgess. As an explanatory framework, the L.A. school draws on the concept of *keno capitalism,* alluding to the patchwork layout of a keno game board (Dear & Flusty, 2002). The urban zones on the keno board are seemingly random, with a lack of relationship between squares. Both Los Angeles and Las Vegas serve as examples of this keno layout (Dear & Flusty, 2002). The L.A. School has been expounded on at length by its adherents (Dear, Schockman, & Hise, 1996) as well as by its critics. The debate also prompted a response to consider a New York school of urban studies, one that emphasizes compact urban forms (Halle, 2003).

Recent debate between advocates of the Chicago, L.A., and New York City schools is taken up in much greater detail with comparisons of the principles and processes espoused by the two schools in Chapter 4.

1.7 DATA AND SPATIAL ANALYSIS IN URBAN GEOGRAPHY

Spatial analysis in urban geography emphasizes the spatial relationships of urban phenomena. This perspective gained ground in the 1950s as the social sciences were becoming more quantitative, in a time sometimes termed the *quantitative revolution* (Adams, 2001). The perspective reached full maturity in the early 1970s, as exemplified by the popular geography textbook of the time, *Spatial Organization: The Geographer's View of the World* (Abler, Adams, & Gould, 1971), which laid out many of the principles and methodologies for measuring urban spatial distributions. Increased computer-processing capabilities coupled with access to census data in electronically readable form allowed researchers to do more systematic and comparative analyses of city distributions. Today the spatial analysis approach to urban geography can be assisted by the more powerful contemporary tool of GIS, which is described in the next section.

As seen in Table 1.3, the spatial analysis view emphasizes the geographical locations of entities in the city and their relationship to each other. The spatial concepts include size, shape, distance, direction, proximity, intervening opportunity, spatial location, connectivity, optimal pathway, buffering, statistical analysis, and modeling. Buffering refers to putting a buffer zone around a point, line, or polygon. For instance, an urban geography study might be analyzing a major highway, represented by a line, which has a 100-yard "buffer zone" on either side that represents the limit of tolerable residential noise impacts. Spatial analysis compares the shapes, sizes, and juxtaposition of points, lines, and polygons, often overlaying layers of spatial features. Statistical analysis examines how correlated one distribution of one characteristic is with another, or applies cluster analysis techniques to group together spatial locations having similar characteristics.

There are today vast amounts of spatially keyed data available and inexpensive GIS tools to analyze the data. Spatial data are increasingly gathered and utilized by national censuses; other agencies of national, state, and local governments; international organizations such as the United Nations; and market research and other information-gathering corporations. Because the trend in government and industry is to gather growing amounts of spatial data, the spatial approach has great promise for the future.

An example of the approach is a city feature map of El Paso, Texas, in 2000 (see Figure 1.11). El Paso is a large city in the southwest corner of Texas just north of the Rio Grande River. As seen in the figure, the city has a diagonal urban footprint that runs from northwest to southeast, mostly along the U.S.–Mexican border (shape). The figure shows the three bridges that connect El Paso with the larger city to the south (direction, proximity) of Ciudad Juarez, Mexico. The CBD is four miles from the Stanton Bridge border entry to Mexico (i.e., this demonstrates distance). The U.S. Interstate Highway 10 is the thick line that runs just north of the CBD (proximity). El Paso Street runs from the downtown to the most easterly bridge, whereas Stanton Street runs from the CBD into the next Stanton Bridge to the

TABLE 1.3   Spatial analysis components. Map features consist of points, lines, or polygons

Spatial concept	Meaning	Example of its use
Size	Spatial size of a feature	Land area of the city of Chicago
Shape	Shape of a feature	The shape of the Tokyo metropolitan area
Distance	Distance between features	Distance between the city centers of St. Louis and Pittsburgh
Direction	Direction of the processes between features	What was the direction of suburban growth in the 1990s in Seattle
Scale	How big or small is a map relative to the real world	One inch on the map of Los Angeles represents 50,000 inches on the ground
Contiguity	Closeness of one feature(s) to other feature(s)	How close is the 10 Freeway in the city of Los Angeles to primary schools
Intervening opportunity	Map feature(s) that affects processes between a sending and receiving feature	The flow of migrants between southern Mexico and the United States is reduced by employment opportunities in the Mexican cities in between
Optimal pathway	The optimal pathway between two or more features	What is the shortest route in distance for a truck going between Long Beach, California, and Bakersfield, California
Buffering	A zone at a certain distance surrounding a feature	The zone of 2,000 feet surrounding Central Park in New York City
Statistical analysis	Statistical analysis of attributes of a feature	What is the correlation of the total population to the unemployment rate for cities in England
Modeling	Modeling of the attributes of features	A model to predict the ethnic changes from 2000 to 2010 in the neighborhoods of Houston, Texas

Source: Modified partly from Clarke, 2003.

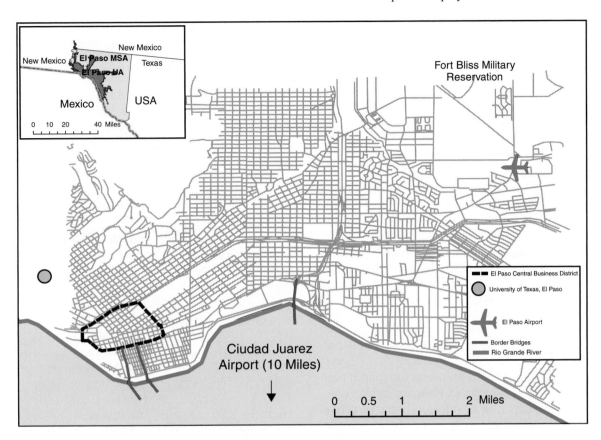

FIGURE 1.11   El Paso urbanized area, metropolitan area, and city features.

east. The University of Texas El Paso campus is located to the northwest, nearby the CBD (i.e., proximity). Military installations are to the north and northeast of the CBD (direction).

As shown in the figure inset, the urbanized area of El Paso (i.e., the area within the urban boundary, as defined by the Census Bureau) is 300 square miles, but the El Paso metropolitan statistical area corresponds to El Paso County, which constitutes 1,013 square miles (i.e., size). Both sides of the border are interconnected through border transport and commerce (connectivity). Although not illustrated on the map, the fastest vehicular route at a particular hour between the El Paso airport and the Ciudad Juarez airport can be identified and its transit time estimated. Although simplified here, this approach provides insight into the urban context of this border city.

1.8 ALTERNATIVE APPROACHES TO URBAN GEOGRAPHY

The spatial analysis view is not the only approach in use today for urban geography. Another is the *political economy* approach. It stresses the importance of political and

economic institutions and processes in shaping city form and urban experience. Examples are how the locations of corporate headquarters, factories, and distribution centers impact local economies; or how changes in political processes and structure of local government affect city planning. The political economy view is emphasized in this book especially in Chapter 9 on industrial location, Chapter 10 on urban fringe growth pressures, and Chapter 12 on urban planning. Although less frequently employed for applications in GIS, political and economic institutions and processes can be modeled using GIS. An example is the widespread use of GIS analysis for political campaigning in the United States. The political economy approach can also come into play in interpreting the urban and metropolitan spatial patterns modeled in a GIS.

Other approaches to urban geography include the behavioral, urban economics, and urban history perspectives. The *behavioral approach* emphasizes urban decision making in a social psychology framework. For instance, the stress levels of residents of downtown Cleveland can be compared to those levels in suburban residents. Cadwallader (1996) introduced the application of the behavioral approach in urban geography with the concept of the *city of the mind*, which is defined as how residents of the city perceive the structure of the city. Cognitive maps or mental maps are an added feature to the approach, as Duncan (1987) argued that people frequently call on their mental maps of the city, which in turn influence their behavior in the city. The ramifications of this approach are far reaching as there are many perceptions of the city contained within a single city and these perceptions serve as the basis for behavior. Cadwallader's (1975) study of consumer behavior in West Los Angeles is illustrative of the approach. The study addressed a classic assumption of retail geography, which is that consumers are rational in their patronage of retail outlets that are nearest to them. Cadwallader found that consumers thought they were rational with respect to the generally accepted retail law, yet in practice they did not always observe it. The conclusion, then, was that these consumers obeyed the law in terms of cognitive distance but not in terms of physical distance.

In the *urban economic approach,* economic concepts are applied as a way of explaining spatial variation in economic phenomena such as land rents and industrial location. An example is the effect of the 2000–2002 economic recession in the United States on the spatial pattern of corporate bankruptcies in Phoenix. In addition, many *urban historians* find a welcome home in urban geography, especially those addressing changes in the built environment or in industrial locations. An example is the history of spatial patterns over the past 50 years of ethnic groups in the Los Angeles urban area. An urban historian might consider ethnic distributions in the context of the California economy's ups and downs during the past half-century, U.S. immigration regulations and policies, the politics of neighborhoods, neighborhood changes, housing, and jobs.

More perspectives of urban geography were developed in the 1980s. Knox (2003) noted that increasing globalization of the economy during the 1980s expressed itself spatially through networks of cities. Knox observed that the response by geographers was toward broad frameworks of inquiry such as structuralist theory exemplified by Harvey's (1989) *The Condition of Postmodernity* and Soja's (1989) writings in social theory. Today, postmodernist geographers emphasize the complexity and ambiguity found in urban land use patterns (Hutton, 2004). Additionally, Knox (2003) suggested that urban geographers were engaged with other social science disciplines in the development of social theory, citing specifically sociology's Anthony Giddens (1984), who also developed a structuration theory centering on time and space.

All of urban geography's traditions flourished in the 1990s (Hanson, 2003). The spatial analysis tradition continued alongside further advancements in social theory (see Table 1.4). Hanson (2003) analyzed urban geography

TABLE 1.4   Themes in urban geography literature, 1990s

Category	Percent
Segregation (including housing, immigration, labor markets, gentrification)	26
Overviews and theoretical perspectives	10
Flows (e.g., airline flows, commuter flows, migration, business linkages)	9
Gender	9
Politics (e.g., NIMBYism, activism, gerrymandering)	9
Globalization (e.g., of labor, capital)	8
Location of economic activity (retail, manufacturing, services)	8
Urban spatial structure (e.g., rural–urban boundary, suburbanization)	8
Urban image (representations of specific cities)	4
Nature/urban environment (water, chemical hazards, topography)	3
Policy evaluations (e.g., banking deregulation, protection of agricultural land)	3
Culture (e.g., cultural practices, understandings of place)	3
Third World cities/megacities	1

Source: Hanson, Susan (2003). "The Weight of Tradition, The Springboard of Tradition: Let's Move Beyond the 1990s." *Urban Geography* 24:465–478 p. 467. Table 1. "Themes in urban geography literature, 1990s: proportion of articles in each category."

articles published in the 1990s in three of the discipline's major journals and found that tradition prevailed:

A substantial proportion of articles take up topics with lineages that stretch back to the Chicago School of urban sociology in the 1920s: immigration, residential segregation by race and class, urban spatial structure, the impact of neighborhood context on households and individuals. Another sizeable group of papers emphasize flows (of people, goods, ideas, and above all money) both within settlements and between them and the ways in which longer-distance flows embed a place in a network of linkages and shape the urban system. Interest in such flows in urban geography also dates to the 1920s (e.g., Platt, 1928). At the intra-urban scale, many articles continue to focus on a particular type of flow—the journey to work, which owes its debut as a star in urban geography to neoclassical urban economists Alonso (1964) and Muth (1969). (Hanson, 2003, pp. 467–468).

Hanson (2003) concluded that important topics such as urban environment, sustainability, identity, and megacities of the developing world deserve greater attention in the future, principally because such research could yield practical solutions for our increasingly urbanizing world.

1.9 GEOGRAPHIC INFORMATION SYSTEMS

GIS is a method that utilizes specialized software to analyze problems spatially. It is difficult to say when the first GIS was introduced. One could argue that the manual map overlay procedure for land suitability studies advocated by Ian McHarg (1969) in *Design with Nature* was GIS in principle (Clarke, 2002). Early computer programs for GIS were introduced in the United States in the late 1950s by the U.S. Census (Clarke, 2002), and in Canada in the 1960s to map natural resources. Since the 1960s, GIS has grown as a field, at first expanding in the government sectors in the United States and Canada, especially local and regional government (Huxhold, 1991). Today, GIS is a standard technique applied by city, county, and state governments in the United States, Canada, Western Europe, and other advanced nations. It has started to catch on with governments in the developing world but might need another decade to be widely utilized there. GIS also is spreading rapidly in the business world, which uses it for marketing, logistics, transportation, siting, real estate, and natural resource exploration (Grimshaw, 2000; Pick, 2005).

The components of a GIS are the *database* containing the attribute data, *boundary files* that contain the geographic coordinates for *digital map layers, analysis tools,* and a *user interface* that enables the user to utilize the software readily.

The database contains attribute information, which is associated with geographical features that are points, lines, or polygons. An example of a polygon feature is Census Tract 10001 in Louisville, shown in Figure 1.12. A *census tract* is a small area of a county defined for publishing census data. Tracts average about 4,000 inhabitants and are fairly consistent internally in their demographic characteristics. Examples of census tracts are shown for Louisville with the following characteristics given: numbers of Asians, average household size (AVE_HH_SZ), number of Black residents, number of Hispanics, and average median age. Census Tract 21111002700 has 2 Asian, 3,020 Black, and 30 Hispanic residents. The tract's average household size is 2.4 and median age is 33.7. An arrow on Figure 1.12 points to that tract, which is part of a digital map layer, consisting of dozens of census tract boundaries. The layer can be regarded as a set of boundaries, without the attribute information, whereas the attribute information is contained in an associated table. Census tracts are small units that together constitute a county, metropolitan area, or state.

One basic purpose of a GIS is to present the data in understandable maps. However, an even more important

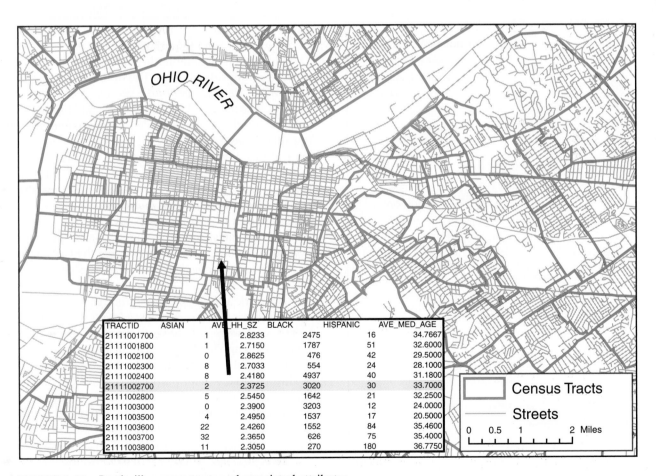

TRACTID	ASIAN	AVE_HH_SZ	BLACK	HISPANIC	AVE_MED_AGE
21111001700	1	2.8233	2475	16	34.7667
21111001800	1	2.7150	1787	51	32.6000
21111002100	0	2.8625	476	42	29.5000
21111002300	8	2.7033	554	24	28.1000
21111002400	8	2.4180	4937	40	31.1800
21111002700	2	2.3725	3020	30	33.7000
21111002800	5	2.5450	1642	21	32.2500
21111003000	0	2.3900	3203	12	24.0000
21111003500	4	2.4950	1537	17	20.5000
21111003600	22	2.4260	1552	84	35.4600
21111003700	32	2.3650	626	75	35.4000
21111003800	11	2.3050	270	180	36.7750

Census Tracts

Streets

0   0.5   1   2 Miles

FIGURE 1.12   Louisville census tracts and associated attributes.

purpose is to perform types of spatial analysis based on the map data. The analysis might consist of comparing two maps, measuring the areas of ZIP codes, determining the fastest transportation route between two points, analyzing neighborhood change, or graphing political opinion by city ward. GIS spatial analysis tools include distance measurement, buffering, statistical, overlay, drawing, and 3-D visualization. A simple measurement tool is the ruler that enables measuring the distance between two points. The buffering tool allows you to measure how many mapping items are within a "buffer" distance from a line or polygon. Statistical tools allow you to calculate means, standard deviations, correlations, regressions, cluster analyses, and other functions. For example, if you have a section of Cincinnati containing 50 census tracts, you can employ this tool to calculate the average income for all the tracts, or the maximum percentage of the workforce that are employed among all 50 tracts. You can also correlate crime and retirement population for the tracts. The overlay tool allows you to superimpose one digital map layer over another. In the example shown in Figure 1.13, Chicago's commuter rail system (layer 1) is superimposed on its population density (layer 2). Overlaying gives

insights that neither layer by itself provides. The map shows a strong relationship between the commuter rail lines and high population density.

There are numerous other tools available. For instance, ArcGIS, the software package utilized for this book's exercises, includes more than three dozen analysis tools.

Today's GIS desktop software is user friendly and visually appealing. Modern user interfaces offer user-friendly features with buttons, menus, and click options to manipulate the maps and launch the spatial analysis tools discussed in the chapter (see Figure 1.14 with the interface for ArcGIS).

The section immediately following this chapter and before the first lab gives a more in-depth introduction to GIS principles.

In summary, GIS is an exciting contemporary tool for spatial analysis and mapping. It is now in widespread use for urban analysis and a multitude of other applications in advanced nations and is expanding in developing countries. It offers digital map layers, databases, and a variety of tools to manipulate them.

1.10 URBAN ANALYSIS USING GEOGRAPHIC INFORMATION SYSTEMS

GIS is a helpful technique to apply in urban geography. Data are widely available through governmental, business, and nonprofit sources. The data range from broad macrolevel data to small-area data and point-location data, such as data for U.S. Census tract and block units or for individual survey respondents. Digital mapping files are readily accessible for states, counties, metropolitan areas, cities, and special regions such as watersheds, pollution control zones, voting districts, school districts, and transportation corridors. These are sometimes also referred to as *boundary files, shapefiles, coverages,* and *geodatabases,* or in ArcGIS as *map layers.*

Web sites can yield a large variety of attribute information and mapping layers. Examples are the U.S. Census at *http://www.census.gov;* Environmental Systems Research Institute (ESRI), Inc. at *http://www.esri.com;* the Geography Network at *http://www.geographynetwork.com;* and national and international maps at *http://seamless.usgs.gov.*

GIS can be applied to solve urban problems. In government it is applied for population analysis, public health planning, analysis of taxation, zoning, tax records, public health, and many other purposes. Businesses solve urban problems through analyzing urban markets, transportation, product distribution networks, site locations, and utility flows (Pick, 2005). Likewise for students, desktop GIS provides an excellent way to practice solving urban problems.

The advantages of GIS for urban problem solving include improved and more efficient financial management,

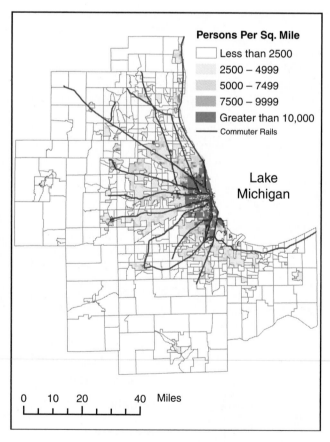

Persons Per Sq. Mile

- Less than 2500
- 2500 – 4999
- 5000 – 7499
- 7500 – 9999
- Greater than 10,000
- —— Commuter Rails

Lake Michigan

0   10   20   40   Miles

FIGURE 1.13 Overlay of Chicago's commuter rail system with year 2000 population density.

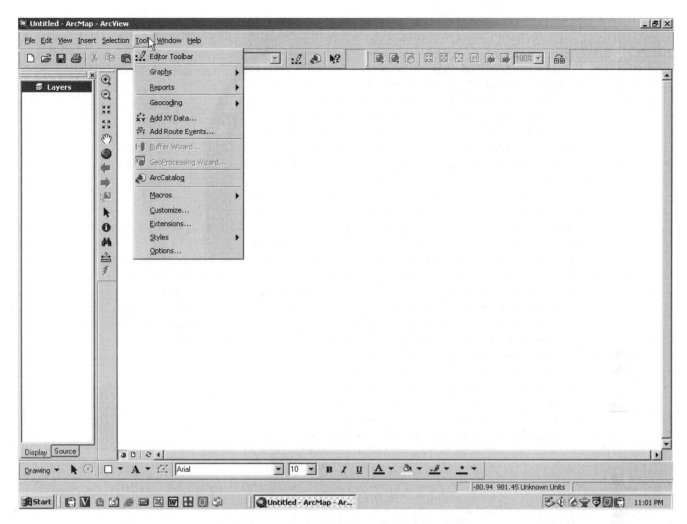

FIGURE 1.14   ArcGIS Software Interface.

welfare services, public health, revenue collection, vehicle fleet management, fire and police services, election redistricting, and city and regional planning (Huxhold, 1991). The GIS can view spatial change within cities, which is important because they are dynamic entities. It can support very large and complex data sets easily. This enables such features as small-area analysis. It can also impose modeling features for urban problems. For instance, GIS can support a density-gradient model and map the results.

GIS is becoming more widely available to urban practitioners. For instance, most urban U.S. counties have GIS capabilities. The GIS unit might be located in planning, economic development, taxation, finance, environment, health, or other offices. GIS is also used by nonprofit organizations concerned with urban problems, such as New Jersey Geographic Information Network, which is a gateway to GIS information for New Jersey including many urban dimensions. In the business sector,

varied firms dealing with urban problems use GIS. For instance, CH2M Hill, a multifaceted environmental consulting firm in Englewood, Colorado, has developed powerful GISs for city and county sewage waste management planning in Los Angeles, Honolulu, and elsewhere.

The use of GIS by the Village of Hoffman Estates, Illinois, in metropolitan Chicago, illustrates how GIS gets introduced to local government and evolves into a fully integrated information system employed by all departments throughout the large suburb. Hoffman Estates is an *edge city*, a type of office-based city located on the outskirts of metropolitan areas, as discussed in Chapter 10. In the first phase of its GIS development, the Village of Hoffman Estates decided to digitize (i.e. convert to computerized form) its property parcel map and to link the county's tax assessment file, containing property-owner information such as the street address, to the digital property map (see Figure 1.15). The property map is registered to a highly accurate aerial photo for the region, which

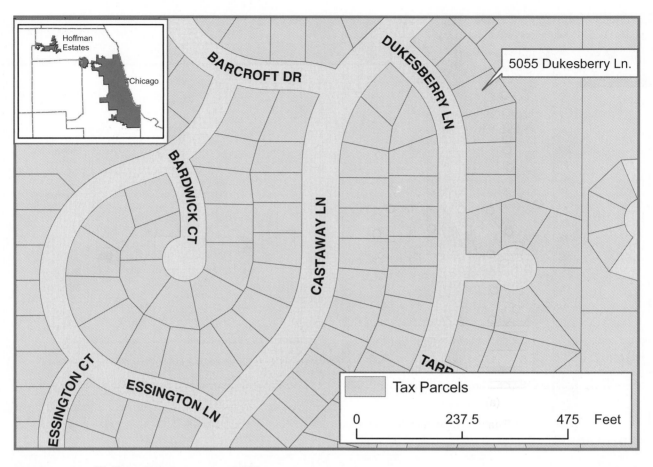

FIGURE 1.15   Hoffman Estates tax parcel GIS.

serves as the base map to which future map layers will be registered (i.e., fitted exactly when placed one on top of another).

In the second phase of its GIS development, the village implemented a notification application allowing village officials to use the GIS to notify residents of important decisions impacting specific areas within the village boundary. For instance, the engineering department wished to notify residents along a particular street that construction of a road improvement would begin within the next week (Figure 1.16a). The engineering department decided to notify all residents along the affected street segment including residents up to 250 feet beyond the affected segment. The engineering department utilized a buffer analysis and intersection overlay of the buffer with the property map to accomplish their goal. Once the residential addresses of affected residents were identified, they were sent directly into the village's mailing address software for immediate mailing of notices about the construction.

In the third phase of its GIS development, the village's uses of GIS mushroomed. Applications developed for many village functions included crime analysis by the

police department, zoning by the planning department, and utility infrastructure by the public works department (Figure 1.16b). By linking a list of addresses representing recent crimes, the police department is able to produce a map with crime symbols, which can assist them in their investigation.

Hoffman Estates was a pioneer in the use of municipal GIS for the northwest suburbs of Chicago. Many other neighboring municipalities have since adopted GIS for their day-to-day business. As seen in Figure 1.17a, the planning department of the nearby city of Crystal Lake, Illinois, northwest of Hoffman Estates, maintains by GIS an updated zoning map that visually displays multiple special use permits for each residential property. The six categories on the maps (RE-1, RE-2, etc.) represent all types of residential zoning. For Crystal Lake, the zoning data are maintained in a relational database that is linked to the GIS. Crystal Lake's public works department can register its utility maps and other public works layers to the same GIS base map (see Figure 1.17b, which shows water lines, water valves, and fire hydrants). The layers can be overlaid on top of each other, resulting in a multilayered GIS map that allows the Crystal Lake planners

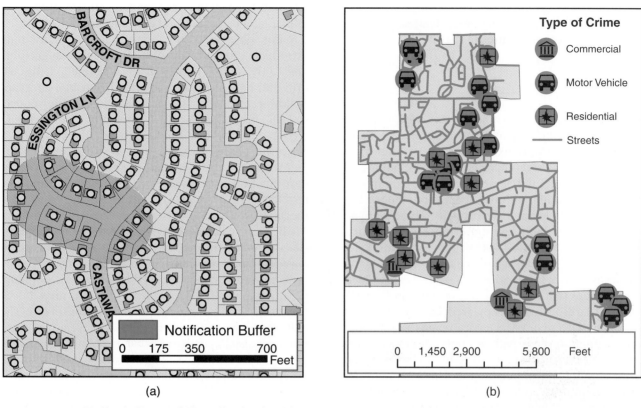

(a)   (b)

FIGURE 1.16   Hoffman Estates GIS application in address notification and crime analysis. (Note: crime locations are illustrative and not actual ones.)

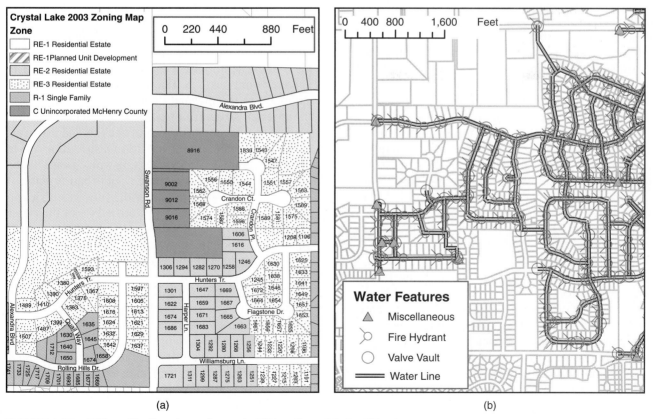

(a)   (b)

FIGURE 1.17   GIS applications in zoning and utilities, Crystal Lake, Illinois.

to model complex flow patterns, such as their water distribution system in relation to water demands, population density, and other features in the city.

1.11 THE APPROACH OF THE BOOK

This book provides a contemporary overview of urban geography including a mixture of traditional and contemporary topics. The book emphasizes the important concepts and theories of urban geography, including contemporary problems and new concepts. The examples in the book are frequently drawn from three megacities, Chicago, Los Angeles, and Mexico City. These were selected because of their key importance in the North American setting, as well as their national and worldwide importance. The three megacities also provide valuable contrasts as well as similarities. Los Angeles and Chicago are old cities that were originally founded in 1776 and around 1812, respectively. They are located in very different industrial and physical environments. Los Angeles is a very spread-out city with multiple business centers, and problems of urban sprawl. It is diverse, with rapidly growing ethnic group populations and high immigration. Chicago, by contrast, is a city of former heavy industry that has been shedding this facet and becoming more of a service city. It has spread out increasingly across the flat former farmlands to the west. It has substantial ethnic diversity, although not as high as that of Los Angeles. Mexico City contrasts with the other two in its giant megacity status, severe environmental degradation, overcentralization of population and industry compared to other developing countries, and strong federal mechanisms of political control.

The book is divided into several sections. The first two chapters are introductory and include fundamentals of the discipline, approaches, and concepts of urban geography. Chapters 3, 4, and 5 define the city, urban area, and metropolis; examine the internal structure of cities; and then consider differences among systems of cities. Chapters 6, 7, and 8 cover interrelated topics of neighborhoods, residential mobility, migration, race, ethnicity, and poverty. Linkages of these topics include, for instance, the influence of residential mobility on changes in neighborhoods, migration and ethnicity, and ethnic neighborhoods. Chapters 9 and 10 have an economic focus on industrial locations in various parts of cities, the urban core, and edge cities as more peripheral service centers. The last two chapters discuss the related topics of the urban environment and urban and regional planning. Environmental issues can often be mitigated through improved planning and regulatory processes. Chapter 12 on planning also serves as a capstone, bringing together concepts from all the prior chapters and integrating them in planning principles and case studies.

In its more detailed contents, the book starts out by introducing the student to the field of urban geography and to the spatial approach to the subject. It discusses GIS and how GIS can be applied to urban geography. Following Chapter 1 is a special section, "An Introduction to GIS and Spatial Analysis Principles," which provides a concise foundation for the lab exercises. Chapter 2 is concerned with the divergences in socioeconomic levels and living patterns, introducing the concepts of social stratification, racial and ethnic segregation, and social conflict. This chapter also discusses the topic of the megacity (i.e., giant cities such as Tokyo, Sao Paulo, and Mexico City). A *megacity* is defined by the United Nations as an urban agglomeration with a population of at least 10 million persons (United Nations, 2003, p. 93). This is an important concept keyed to worldwide population growth that increased exponentially in the mid-20th century and has only recently begun to decline in rate. These cities are all world financial, political, and cultural centers, but most also have urgent urban problems ranging from large poverty zones to clogged transportation to severe air pollution. In this book, the megacities used as examples are Mexico City and Los Angeles, with Chicago very close to the category, as the population of its combined metropolitan statistical area in the 2000 U.S. Census was 9.2 million. Based on 1990s growth rates, we project the Chicago–Naperville–Joliet combined metropolitan statistical area will reach 10 million in the year 2008. For this reason, we refer to it as a megacity in this textbook.

Chapter 3 introduces simple city concepts and then looks at the definitions of city, urban area, urbanized area, and metropolitan area. The metropolitan definitions are extended to the megacity. Chapter 4 is concerned with the internal structure of cities, in particular the demographic and economic dynamics of cities and urban areas inside their boundaries. Population distribution and land use patterns are described. Starting with classical theories of population density and models of urban structure, the chapter considers the major models of concentric zones, sectors, and multiple nuclei. Alternatives of the classical models are introduced with an expanded section on the new L.A. School. In analyzing an urban issue, the population density distributions of Los Angeles and Chicago are compared over a 40-year period, and density gradient models are constructed for Mexico City. An example of urban structure is the downtown of Barcelona, Spain, shown in Figure 1.18. Chapter 5 reviews the systems-of-cities approach to urban geography including central place theory, the evolution of the American urban system from 1790 to today, and an example of central place theory in the state of Washington.

FIGURE 1.18   Overview of Downtown, Barcelona, Spain.

Chapter 6 introduces neighborhood concepts such as definition, change, and composition. Neighborhood topics include gated communities, gentrification, historic preservation, streetcar suburbs, and neighborhood racial composition. The urban issue analyzed is neighborhood transitions in the city of St. Louis. Chapter 7 examines concepts and models of migration and residential mobility. This includes factors in migration, rural-to-urban migration, and residential mobility and its causes. An example of the huge impact of in-migration on urban areas is the vast migration to Florida in the 1990s. From July 1, 2000 to June 30, 2001, Florida's population of 16,054,328 was estimated to have increased by 342,187. Besides migration, the rest of the change was births minus deaths. This has put pressure on urban systems from water supply to administration to voting. Clearly, the 2000 U.S. presidential election raised voting system problems within counties and statewide in Florida. The chapter also covers directional bias (i.e., tendencies of people in cities to move residentially in certain directions) and international immigration.

Chapter 8 introduces the topics of gender, race, ethnicity, and poverty as common characteristics of cities in the United States and worldwide. For instance, in Los Angeles recent data show that Hispanic and Asian ethnic distributions grew and shifted in a major way in the 1980s and 1990s, whereas non-Hispanic Whites decreased, and Blacks were unchanged. Further, poverty is high in the periphery of Latin American cities, but people in these zones can escape poverty through better education, jobs, and home improvements. The chapter also addresses how to measure ethnicity and poverty, ethnic hot spots in the suburbs, racial segregation, the urban ghetto and underclass, and colonias in the Mexican border region.

Chapter 9 concerns industrial location. Today, businesses have a wide choice of places in which to locate, from the classical CBD to shopping malls and industrial parks, some located on the urban fringe. The reasons why industries locate in particular places are discussed. The chapter covers the concepts of industrial classification, location theories, location quotient, and the geography of consumer and producer services. Mexico's *maquiladora* industry illustrates binational industrial location, and the United States–Mexico border cities demonstrate border consumer interactions.

Chapter 10 looks at the issue of urban core versus edge cities. As the urban cores of many U.S. cities have deconcentrated, more population has settled in often peripheral communities of metropolitan areas, referred to as edge cities. For instance, west of Chicago, Hoffman Estates, Schaumburg, and Naperville are examples of edge cities. They are tied to the metropolitan area, and yet sometimes occupy land in the hinterland. This leads to the pressure on the urban fringe and "leapfrog" development, in which formerly rural locations on the metropolitan periphery become urbanized. Fringe growth is not only taking place in every U.S. metropolitan area, but also in all developing world cities, such as Mexico City. Other chapter topics are urban sprawl and smart growth, which is a planning approach to reducing the adverse effects of sprawl.

Chapter 11 emphasizes that urban areas often grow beyond their capacity to control or even monitor environmental degradation. There are many horror stories of environmental neglect ranging from the notorious Love Canal case of toxic waste emission to Mexico City's air pollution. With a geographical approach, urban environmental problems such as land use, water resources, flooding, solid wastes, energy supply, and air pollution are examined. Environmental justice perceives unevenness and often unfairness in the patterns of places and groups impacted by the environment.

Chapter 12, on urban and regional planning, examines the history of urban planning, starting in London in the 17th century and extending to today's planning challenges in the United States and China. It looks at how planning is achieved or not achieved, how to renew or improve central downtowns, and new types of planning including more on smart growth, restoration of downtowns, and open

spaces. Planning in the developing world often relies more on politics and influence and less on objective methods. Several case studies are presented.

Accompanying each chapter in the text is a GIS lab exercise. The exercises cover such topics as density mapping and density gradients (Chapter 4), centrographic analysis of segregation (Chapter 8), and analysis of land use conversion (Chapter 10). The examples are drawn both from the three major megacities that are a focus of the book and from other large and medium sized cities, such as St. Louis and Springfield, Missouri.

The intent of the lab exercises is for students to reinforce their learning of urban concepts and principles by practicing with real-world data and problem solving. The exercises are available on the CD-ROM accompanying this volume and also on the Prentice Hall Web site for this textbook. The GIS software is ArcGIS from ESRI, Inc. For one exercise, Microsoft Excel is also utilized.

The exercise data are drawn mostly from 1990 and 2000 censuses and other government sources. The initial knowledge of computing software called for is rudimentary, but students will be expected to learn some GIS and spreadsheet skills. The emphasis of the labs is not on mastering all the package commands. Students who would like to master the software packages are advised to obtain additional training seminars, courses, and online tutorials.

The labs complement the textbook readings and lectures, leading to a balanced and contemporary learning approach. Each chapter has a section titled "Analyzing an Urban Issue" that will be tied to the lab exercise topic for that chapter. These issue discussions will provide ample background for students to answer questions and write short essays after completing the exercise. In this chapter we have explored the issue of Sunbelt city growth, and that topic is the focus of the GIS exercise. We advocate an integrated learning approach, which might also include field trips to accessible urban areas and work on the World Wide Web to acquaint students with local, regional, national, and global urban geography data, reports, issues, problems, and solutions.

SUMMARY

Urban geography is important because it involves concepts and issues of profound significance for business, commerce, transportation, the environment, and daily life. Its field is cities and urban areas and how they are organized and function. The major perspectives in urban geography are economic, social, cultural, human, environmental, and physical. The allied disciplines of urban sociology, political economy, and others are often useful. This book emphasizes the spatial approach. The modern tool of GIS encompasses the geographic boundaries and multiple dimensions of attributes of cities and urban areas.

Students using this book will learn about diverse aspects of urban geography, including population distribution and land use, neighborhoods, race and ethnicity, gender, industrial location, migration, the internal structure of the metropolis, edge cities, environment, and urban planning. The main text provides the concepts, discusses examples and applications, and considers the implications for business and society. The GIS exercises offer hands-on practice in ArcGIS that helps the student to analyze city processes and see how their problems can be solved.

1.12 AN INTRODUCTION TO GIS AND SPATIAL ANALYSIS PRINCIPLES

This section covers the rudimentary principles of geographic information systems (GIS). Its purpose is to enhance your understanding of GIS principles, which will be helpful in performing the lab exercises. If you already have had other coursework or experience in GIS, you can skip this section.

Common to most definitions of GIS is the manipulation of geographically referenced information with computer hardware and software. GIS is in use today in geographical analysis and in many other fields including business marketing, environmental analysis, tax collection, public health, the military, and transportation (Longley, Goodchild, Maguire, & Rhind, 1999).

A GIS consists of boundary files of graphical data, combined with a database of information associated with the *digital map layer*. Map layers are also sometimes called *coverages* or *boundary files*. We refer to them as map layers. Each map layer contains certain geographical features. A GIS might have a single map layer, but more commonly include two or more layers (Clarke, 2002). For instance, one map layer contains the ZIP code boundaries for the county, a second gives the rivers in a county, and a third gives the point locations of large pollution sources.

As seen in Figure 1.19, the numerical and text data for each map layer are stored in an associated *attribute table* (Clarke, 2002). For example, the attribute table on rivers contains the following attributes for each river: geographical coordinates, name, average river flow rate, and average river water temperature. All the attribute tables together constitute the *geodatabase*.

The GIS performs analyses and the results are output in a variety of ways, including printed output, screen output, and finished plotted output.

A simple example of GIS use would be to map and analyze the road system in a city. This GIS would have

two boundary layers: one for the city boundaries and a second showing the highway transportation network. As shown in Figure 1.20, the boundary layer for the city limits has the following attribute data associated with it: the number of city employees, gross city revenues, and total tax revenues. The map layer for the highway transportation network has the following data associated with each highway: highway width, speed limit, daily vehicle traffic, and number of gasoline stations on each highway.

This application is designed to perform two types of analysis and modeling. First the density of highway transport can be compared to the population density of the city. This allows planners to see how much future need there is for increased highway capacity. Second, the map of the city limits can be compared to the map of the highway transportation network to determine exactly how much of the highway grid is located inside and how much outside the city limits. This is important because it affects the amount of highway tax revenue that the city receives each year.

The outputs from this simple GIS example include the following:

1. Map of the city limits and a map of the highway transportation network
2. Combined map of the city limits and highway transportation network, superimposed on top of each other. This is referred to in GIS terminology as an *overlay*. An overlay can have two or more layers superimposed.
3. Comparison of highway density to density of population. This is presented in the form of a graph and a table.
4. Analysis of how much of the highway grid is inside versus outside the city limits, and a further study of the effect of this distribution on city and state gas taxes. This can be presented in report format as graphs and tables.
5. Analysis of the total length of all highways within the city limits. This GIS analysis might be useful to the city government in estimating total highway maintenance costs.

This example helps in understanding the fundamental capabilities of GIS. Among the capabilities of GIS are the following (Clarke, 2002; Longley, Goodchild, Maguire, & Rhind, 2003):

- Retrieving information based on the spatial location of an item on a map.
- Identifying a spatial location or locations that meet particular criteria. This is sometimes referred to as a query.

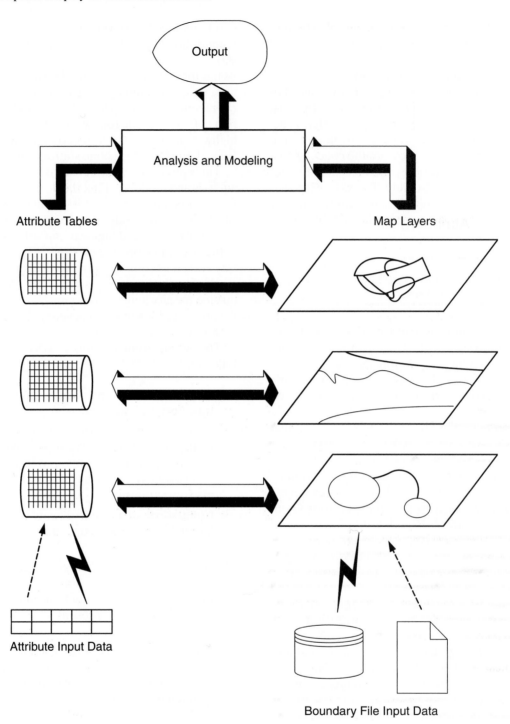

FIGURE 1.19   Design Elements of a GIS.

- Measuring distances and other spatial relationships. GIS has the tools to make such measurements much easier.

- Performing analysis or modeling of spatial problems and expressing the results in map or graph form.

- Selecting certain data for map display and displaying the results.

- Identifying selected data to be transported to another software package outside of the GIS for further processing, or, vice versa, importing into a GIS data from another software package. The ability to transport data into and out of a GIS expands the processing and analysis capability to include many specialized types of computer models and data sources that would not otherwise be available.

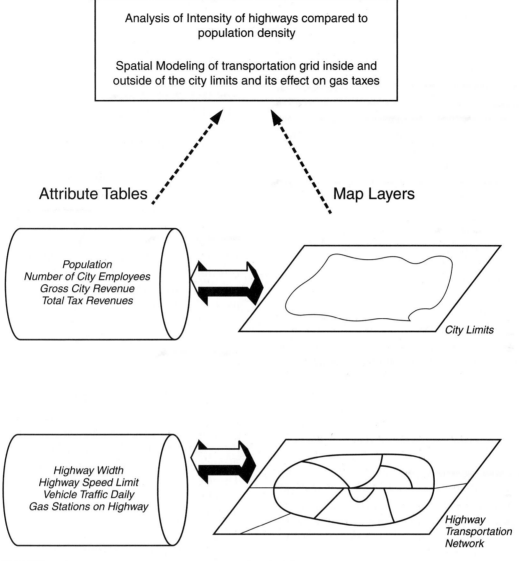

FIGURE 1.20 Example of Processing with a Simple GIS.

As seen by this simple example, to make use of a GIS, you need to have the following components (Huxhold, 1991; Huxhold & Levinsohn, 1995; Longley et al., 1999):

- GIS software.
- Hardware on which to run the GIS software.
- Data, both boundary file inputs in digitized form and attribute input data associated with the boundary files.
- People who understand GIS principles and software and are able to run the applications.
- People with the project management skills to conduct a GIS project successfully from start to finish.
- Geographers, planners, and other specialists studying or managing cities, who can express the relevant questions for a GIS to answer.

GIS Software

For this course, you will be utilizing the GIS software ArcGIS, which is sometimes referenced by its version number, 9.x.

GIS Hardware

GIS software packages run on most personal computers and laptops in use today. They run in reduced form on personal digital assistants (PDAs), such as Palm devices.

GIS hardware also includes larger server-based networks and even mainframes. For this course, it is likely you will be running GIS from a laptop or desktop computer, which will possibly be networked. GIS professionals often have large, high-resolution monitors, because in GIS you will often have many windows open at the same time, which is easier to see on a larger screen.

GIS Data

GIS data are critical to the success of the project. The data can come from a variety of sources, including government, private companies, web resources such as The Geography Network, Geodata.gov, and universities. Gathering the data has often been shown to be the most expensive and time-consuming part of a GIS project (Huxhold, 1991; Huxhold & Levinsohn, 1995).

The following list gives several convenient sources of data:

Census Bureau, Tiger Files
The U.S. Census at *http://www.census.gov* provides boundary files corresponding to its geographic units, such as states, counties, cities, and ZIP codes. Along with the boundary files is a wealth of attribute information from the U.S. censuses of population, housing, economics, and agriculture, as well as specialized surveys that are georeferenced. Quite a few user-friendly mapping features are available.

The most comprehensive boundary files from the Census are called Tiger Files. They give all the geographic boundaries at all levels of census geography from individual city blocks and small streets up to state boundaries and the nation. As a beginner, you are not likely to directly utilize these files.

USGS, Digital Elevation Models
The U.S. Geological Survey (USGS) has been utilizing GIS for many years, and has produced particularly good maps showing digital elevations and topography. These maps are available for the entire United States (*http://www.usgs.gov*). The Web site includes the USGS Geospatial Data Clearinghouse, which contains large amounts of mapping and remote sensing data for the whole nation, and national and international maps at *http://seamless.usgs.gov*.

GIS Data Clearinghouses
These are sources of map layers; attribute data; and satellite, photographic, and remote sensing data. The most prominent data clearinghouses are from the federal and state governments, as well as from overseas. Several examples are the New York State GIS Clearinghouse (*http://www.nysgis.state.ny.us*), the New Jersey Node of the National Spatial Data Infrastructure (*http://deathstar.rutgers.edu/clear/clear.htm*), and the Georgia GIS Data Clearinghouse (*http://www.uga.edu/ucgis/gadata.html*). Some of the state centers are managed by universities. A user needs to be rather experienced to use the clearinghouses.

The Geography Network
This contains a large inventory of GIS map layers and associated data. It is located at *http://www.geographynetwork.com*. It is fairly easy to access and many free data can be loaded into ArcGIS.

The GIS Data Depot
The GIS Data Depot (*http://www.gisdatadepot.com*) is a commercial source of extensive GIS data. It includes census data and map layers, Federal Emergency Management Agency (FEMA) flood data, state-specific geospatial data, and many other sources. This site is for proficient GIS users. Some data are free and others involve a fee.

ESRI Web Site
ESRI, Inc.'s Web site contains a large library of geospatial data, much of it free. For ArcGIS users, the data sets are fully compatible with the software. This site is user friendly and suitable for the beginner and intermediate user.

Geodata.Gov
This is the federal government's Geospatial One-Stop Portal (*http://www.geodata.gov*). It includes mapping services related to all aspects of the government, such as environment, agriculture, politics, culture, demographics, education, health, oceans and estuaries, utilities, and transportation. It includes an information center on standards and tools, as well as simple mapping capabilities. There is a large and growing library of spatial data and map boundary files. Any user can publish data and maps on this site.

Geographical Background for GIS

This section gives a brief introduction to some of the most important geographical concepts for GIS, many of which you will be using in the book's exercises. More complete introductions are available in GIS textbooks (Clarke, 2002; Lo & Yeung, 2002; Longley et al., 2003).

Importance of Projections and Coordinate Systems
The most important component in the GIS definition is *geographically referenced information* (Clarke, 2002; Longley et al., 2003). This means that the information is tied to a geographic coordinate system whereby each observation in the database can be precisely located on the Earth. Locations are represented in GIS with the use of x and y coordinates, which represent real-world coordinates (in three dimensions) that have been projected onto a two-dimensional surface. The geographic grid of *latitude and longitude* is a commonly used geographic reference system. It is not a true coordinate system,

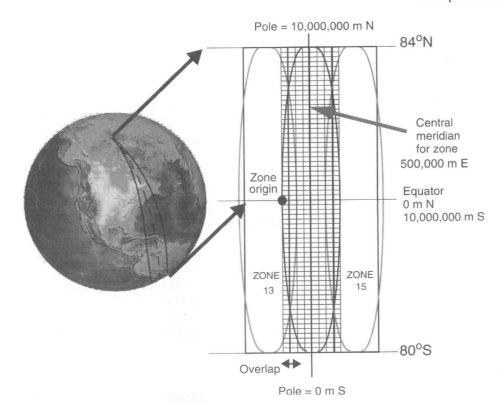

FIGURE 1.21 The Universal Transverse Mercator (UTM) Coordinate System. Clarke, Keith (2000). Source: Clarke, Keith (2000). *Getting Started with Geographic Information Systems.* 4th ed. Prentice Hall. Fig 2.14 p. 50. "The universal transverse Mercator coordinate system".

however, because it measures angles from the center of the Earth in degrees, minutes, and seconds rather than distances on the Earth's surface. Thus, real-world coordinates are typically projected and transformed to a coordinate system; for instance, many of the map layers used in this book are projected to the *Universal Transverse Mercator (UTM) coordinate system*, for which the units are in meters (see Figure 1.21). Another coordinate system that is customized to project to each of the 50 states in the United States is the *State Plane Coordinate System*.

GIS projections transform the three-dimensional (3-D) Earth to a two-dimensional (2-D) flat plane. Some common projections include *Lambert Conic Conformal* and *Albers Conic Equal-Area*, with *x,y* coordinates measured in feet or meters (see Figure 1.22).

Scale is the ratio of the size of the object on the map to the size of the object on the Earth. For example, a scale of 1:100,000 means that an object that is 100,000 feet on the Earth is 1 foot on the map. A scale is a *small scale* if the fraction 1/100,000 is small and *large scale* if the fraction is large. For instance, a map in 1/100 is large scale, because 1/100 or 0.01 is considered very large compared to most scales' fractions of 1/10,000 or smaller. GIS software often adjusts the scale automatically for the user as he or she is engaged in a session. However, it also allows the user to adjust the scale of maps, as he or she is performing tasks.

Often the scale will be adjusted many times during a user session. A tool for automatically adjusting the scale is called a *zoom feature*, meaning that you can zoom in to make things appear larger in size or zoom out to make them appear smaller.

Methods of Data Capture

This course and its exercises are not intended to give you practice with capturing GIS input maps and data. In the exercises, the data and map layers have already been entered on the text's Exercise CD. However, data are a very important aspect of any GIS, and usually the most costly one (Huxhold, 1991; Pick, 2005, Chapter 4). Common ways that GIS data are entered are by digitizing, scanning, and GPS.

GPS

Global positioning system (GPS) devices connect to a multisatellite system, and can sense the device's position within several meters (Clarke, 2002). For instance, if you take a handheld GPS unit on a hike, you can record your exact Earth surface location at all points of interest (called *waypoints*). When you come back from the hike, you can download all the waypoint locations into your computer. They can be further processed and then entered into a GIS map layer.

GPS is also built into many other devices today, from cars to airplanes to military weapons. This topic is beyond

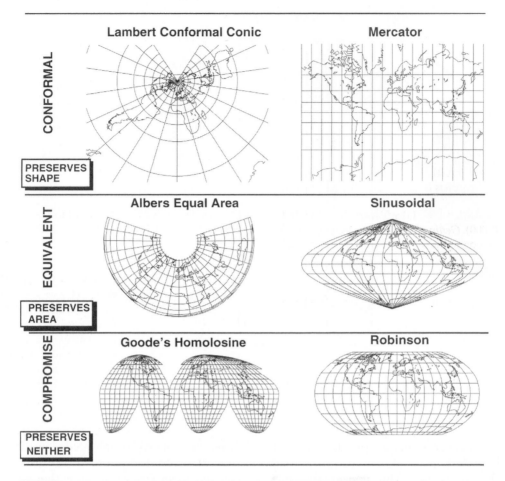

FIGURE 1.22   Examples of projections classified by their distortions. Source: Clarke, Keith (2000). *Getting Started with Geographic Information Systems*. 4th ed. Prentice Hall. Fig 2.10 p. 44. "Examples of projections classified by their distortions."

the scope of this text, but good references are available to explain it (French, 1996).

Digitizing Boards
A digitizing apparatus consists of a digitizing mouse with a crosshair to locate an exact location. The digitizing tablet is a special electronic surface that senses the exact location of the digitizing mouse. Tablets can be small (i.e., several feet square) or quite large (more than six feet on a side). The user moves the mouse from point to point on the digitizing tablet, recording positions from a hard-copy map or photograph. At points of interest, the user clicks the mouse in certain sequences to record point locations.

Scanning
A hard-copy map or photographic surface can be scanned. A crucial aspect of scanning is the need to run vectorizing software on the scanned data, which allows the user to

identify the points, lines, and polygons of interest from the scanned document.

Representing Map Features in Computers
Representing spatial data on a Cartesian (x,y) coordinate system is one of the crucial functions of a GIS. Once this is done, the rest is easy because features in a vector GIS are either represented as points, lines, or polygons. Any map can be built up by numerous combinations of these three elements. This approach to representing a GIS is called a *vector GIS* (Clarke, 2002; Longley et al., 2003). *Raster GIS* is an alternative data model that represents features by recording values in a rectangular grid of tiny cells to represent features. Although an important data structure, raster is not reviewed here because the lab exercises in this book rely only on the vector model (see Clarke, 2002). Raster applies best to continuous data, such as temperature variations in the atmosphere or shapes of the land surface. Thus the exercises do

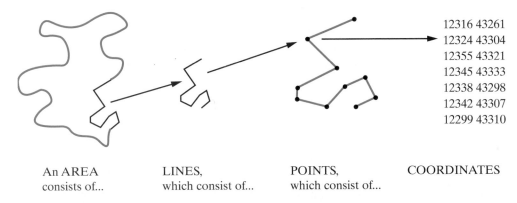

An AREA
consists of...

LINES,
which consist of...

POINTS,
which consist of...

COORDINATES

FIGURE 1.23   GIS features represented as points, lines, and polygons. Source: Clarke, Keith (2000). *Getting Started with Geographic Information Systems.* 4th ed. Prentice Hall. Fig. 2.19 p. 55. "Geographic information has dimension."

not support as well continuous features, such as environmental ones.

So how are points, lines, and polygons represented in a vector GIS?

- A *point* is recorded as a single *x,y* location. For example the manhole point in Figure 1.23 is represented by the coordinate pair 9,10.
- A *line* or *arc* is recorded as a series of ordered *x,y* coordinates. The water line in Figure 1.23 is represented by the coordinate pairs of 2,6, 3,6, 5,7, 9,10, 13,12, and 15,12.
- A *polygon* or *area* is recorded as a series of *x,y* coordinates defining arcs that enclose the area. The coordinate pairs 6,5, 7,4, 8,4, 9,3, 11,3, 10,2, 8,2, 7,1, 5,3, 5,4, and 6,5 define a census tract (Figure 1.23). Notice how the first and last coordinates of a polygon are the same; a polygon always closes.

Common GIS Analytical Methodologies

The last section of this introduction to GIS principles briefly mentions GIS spatial analysis techniques (Getis, 1999; Longley et al., 2003; Mitchell, 1999). This is a collection of methods that is used in the analysis and modeling function of a GIS (see the upper box in Figure 1.19). We have already seen several simple examples of applying these analysis and modeling functions. For instance, an overlay consists of exactly superimposing two or more map layers in a GIS to give a combined layer. The full set of today's analysis and modeling techniques is too numerous to possibly cover in this short introduction. A full treatment is available in specialized sources (Getis, 1999; Mitchell, 1999).

This section, instead, describes only a few important analysis techniques, some of which are utilized in the lab exercises in this book. This short sampling is intended to give you an idea of the range of sophisticated analyses that can be conducted using GIS.

Buffer Analysis

In buffering, GIS software forms bands on either side of a point, line, or polygon to perform analysis within the bands. A simple example would be to assign half-mile buffers on both sides of a highway, and ask how many service stations are within the buffer.

Map Overlay Functions

As already mentioned, overlay consists of superimposing two or more map layers.

Common Analytical Inquiries in Urban Geography Assisted by GIS

Spatial Analysis

These are techniques that compare spatial features (Lo & Yeung, 2002; Longley et al., 2003). Spatial analysis can, for example, determine how many points are inside a polygon, how many line segments cross a polygon boundary, or how much polygons overlap each other. A practical example would be to ask how many times rivers (lines) cross a toxic waste area (polygon).

Proximity Analysis

Proximity analysis assesses how close certain map objects are to other map objects. For instance, it can determine how close the population residing in a census tract is to grocery stores.

Change Detection

This type of analysis seeks to compare maps over time and assess what significant spatial changes have taken place.

Modeling and Forecasting

Forecasting and simulation models can be built with spatial data, and the results can be displayed in map form. An example is a model that projects population distribution within a county, based on starting year data. The future population distributions can be mapped to inform planners and the public.

Statistical Analysis

Statistical models are often utilized to study the relationships of certain attributes to other attributes (Getis, 1999). These techniques include correlation, regression, analysis of variance, cluster analysis, and *t* tests. The input data, as well as the results, of many of these models can be represented as spatial displays. This allows enhanced understanding of the geographical effects and influences. Although beyond the scope of this course, a specialized part of statistics, called geostatistics, takes into account spatial effects and interactions, such as spatial autocorrelation. An example of using regression in GIS is part of the exercise for Chapter 4.

People and their Project Management Skills

Essential for any GIS project are trained people who have the skills to complete a GIS project (Huxhold & Levinsohn, 1995; Sugarbaker, 1999; Tomlinson, 2003). GIS training is available through textbooks, software tutorials, university courses, training courses from vendor companies, and Web-based training such as ESRI's Virtual Campus. Training to a high level of GIS expertise could take a year or more. Besides training, to succeed in GIS, people need good project management skills. They need to be able to formulate the GIS problem, decide what steps are needed to solve it, plan out a schedule of tasks, have a way to get feedback from the GIS users, and proceed to accomplish the project on schedule (Huxhold & Levinsohn, 1995; Sugarbaker, 1999; Tomlinson, 2003). Project management becomes even more important for large GIS projects that involve dozens of specialists.

REFERENCES

Abler, R., Adams, J. S., and Gould, P. (1971 *Spatial Organization: The Geographer's View of the World*. Englewood Cliffs, NJ: PrenticeHall.

Adams, J. S. (2001) "The quantitative revolution in urban geography." *Urban Geography* 22:530–539.

Alonso, W. (1964) *Location and Land Use*. Cambridge, MA: Harvard University Press.

Bernard, R. M., and Rice, B. (eds.). (1983) *Sunbelt Cities*. Austin: University of Texas Press.

Berry, B. J. L., and Horton, F. E. (1970) *Geographic Perspectives on Urban Systems*. Englewood Cliffs, NJ: Prentice Hall.

Borchert, J. R. (1983) "Instability in American metropolitan growth." *The Geographical Review* 73:127–149.

Bourne, L. S. (1982) *Internal Structure of the City*. Oxford, England: Oxford University Press.

Cadwallader, M. T. (1975) "A behavioral model of consumer spatial decision making." *Economic Geography* 51:339–349.

Cadwallader, M. T. (1996) *Urban Geography: An Analytical Approach*. Englewood Cliffs, NJ: Prentice Hall.

Carlson, A. W. (1994) "America's new immigration: Characteristics, destinations, and impacts, 1970–1989." *Social Science Journal* 31:213–236.

Carter, H. (1972) *The Study of Urban Geography*. London: Edward Arnold.

Champion, A. G. (1992) "Urban and regional demographic trends in the developed world." *Urban Studies* 29:461–482.

Champion, A. G. (1994) "International migration and demographic change in the developed world." *Urban Studies* 31:653–678.

Clarke, K. C. (2002) *Getting Started with Geographic Information Systems* (3rd edition). Upper Saddle River, NJ: Prentice Hall.

Dear, M. J., Schockman, H. E., and Hise, G. (1996) *Rethinking Los Angeles*. Thousand Oaks, CA: Sage.

Dear, M.J.(Ed.). (2002) *From Chicago to L.A.* Thousand Oaks, CA: Sage Publications.

Dear, M.J. and Flusty, S. (2002) "Los Angeles as Postmodern Urbanism," in *From Chicago to LA*, ed by M.J. Dear, pp. 61–84. Thousand Oaks, CA: Sage Publications.

Dickinson, R. E. (1964) *City and Region: A Geographical Interpretation*. London: Routledge & Kegan Paul.

Duncan, J. (1987) "Review of urban imagery: Cognitive mapping." *Urban Geography* 8:264–272.

Elbow, G. S., and Pulsipher, L. M. (2003) "Cities of Middle American and the Caribbean," in *Cities of the World: World Regional Urban Development* (3rd edition). ed. by S. D. Brunn, J. F. Williams, and D. J. Zeigler, pp. 93–121. Lanham, MD: Rowman & Littlefield.)

French, G. T. (1996) *Understanding the GPS: An Introduction to the Global Positioning System*. Bethesda, MD: GeoResearch, Inc.

Frey, W. H. (1993) "The new urban revival in the United States." *Urban Studies* 30:741–774.

Frey, W. H. (1994) "The new white flight." *American Demographics* April: 40–48.

Frey, W. H. (1995) "Immigration and internal migration 'flight' from U.S. metropolitan areas: Toward a new demographic Balkanization." *Urban Studies* 32:733–757.

Fuguitt, G. V., and Brown, D. L. (1990) Residential preferences and population redistribution: 1972–1988. *Demography* 27:589–600.

Getis, A. (1999) "Spatial statistics," in *Geographical Information Systems*. ed. by P. A. Longley, M. F. Goodchild, D. J. Maguire, and D. W. Rhind, pp. 239–251. New York: Wiley.

Giddens, A. (1984) *The Constitution of Society: Outline of the Theory of Structuration*. Cambridge, England: Polity.

Glaeser, E. L., and Shapiro, J. (2001) *Is There a New Urbanism? The Growth of U.S. Cities in the 1990s*. Discussion Paper 1925, Harvard Institute of Economic Research. Cambridge, MA: Harvard University.

Greenwood, M. J. (1988) "Changing patterns of migration and regional economic growth in the U.S.: A demographic perspective." *Growth and Change* 19(4):68–87.

Greenwood, M. J., and Hunt, G. L. (1989) "Jobs versus amenities in the analysis of metropolitan migration." *Journal of Urban Economics* 25:1–16.

Griffin, E., and Ford, L. (1980) "A model of Latin American urban land use." *Geographical Review* 70:397–422.

Grimes, B. F. (2000) *Ethnologue* (14th edition). Dallas, TX: SIL International.

Grimshaw, D. J. (2000) *Bringing Geographical Information Systems into Business* (2nd edition). New York: Wiley.

Halle, D. (Ed.). (2003) New York & Los Angeles: Politics, Society, and Culture: A Comparative View. Chicago: University of Chicago Press.

Hanson, S. (2003) "The weight of tradition, the springboard of tradition: Let's move beyond the 1990s." *Urban Geography* 24:465–478.

Harvey, D. (1989) *The Condition of Postmodernity.* Oxford, England: Blackwell.

Hutton, T. A. (2004) "Post-industrialism, post-modernism and the reproduction of Vancouver's central area: retheorizing the 21st-century city." Urban Studies 41:1953–1982.

Huxhold, W. E. (1991) *An Introduction to Urban Geographic Information Systems.* New York: Oxford University Press.

Huxhold W. E. and Levinsohn, A. G. (1995) *Managing Geographic Information System Projects.* New York: Oxford University Press.

Knox, P. L. (2003) "The sea change of the 1980s: Urban geography as if people and places matter." *Urban Geography* 24:273–278.

Lo, C. P., and Yeung, A. K. W. (2002) *Concepts and Techniques of Geographic Information Systems.* Upper Saddle River, NJ: Prentice Hall.

Longley, P. A., Goodchild, M. F., Maguire, D. J., and Rhind, D. W. (1999) "Introduction," in *Geographical Information Systems.* ed. by P. A. Longley, M. F. Goodchild, D. J. Maguire, and D. W. Rhind, pp. 1–20. New York: Wiley.

Longley, P. A., Goodchild, M. F., Maguire, D. J., and Rhind, D. W. (2003) *Geographic Information Systems and Science.* New York: Wiley.

Manson, G. A., and Groop, R. E. (1996) "Ebbs and flows in recent U.S. interstate migration." *Professional Geographer* 48:156–166.

Mayer, H. M., and Kohn, C. F. (1959) *Readings in Urban Geography.* Chicago: University of Chicago Press.

McHarg, I. L. (1969) *Design with Nature.* Garden City, NY: Natural History Press.

McHugh, K. E. (1988) "Determinants of Black interstate migration, 1965–70 and 1975–80." *Annals of Regional Science* 22:36–48.

McHugh, K. E. (1989) "Hispanic migration and population redistribution in the United States." *Professional Geographer* 41:429–439.

Mitchell, A. (1999) *The ESRI Guide to GIS Analysis.* Redlands, CA: Environmental Systems Research Institute.

Muth, R. (1969) *Cities and Housing: The Spatial Pattern of Urban Residential Land Use.* Chicago: University of Chicago Press.

Park, R. E., Burgess, E. W., and McKenzie, R. D. (1925) *The City: Suggestions for the Investigation of Human Behavior in the Urban Environment.* Chicago: University of Chicago Press.

Pick, J. B. (Ed.). (2005) *Geographic Information Systems in Business.* Hershey, PA: Idea Group Publishing.

Piet, P. (1998) *The Determinants of Metropolitan Growth* (Master's thesis). Northern Illinois University, Department of Geography.

Plane, D. A. (1992) "Age-composition change and the geographical dynamics of interregional migration in the U.S." *Annals of the Association of American Geographers* 82: 64–85.

Platt, R. S. (1928) "A detail of regional geography: Ellison Bay community as an industrial organism." *Annals of the Association of American Geographers* 18:81–126.

Scott, A. J., and Soja, E. W. (Eds.) (1996) *The City: Los Angeles and Urban Theory at the End of the Twentieth Century.* Berkeley, CA: University of California Press.

Soja, E. (1989) *Postmodern Geographies: The Reassertion of Space in Critical Social Theory.* London: Verso.

Sugarbaker, L. J. (1999) "Spatial statistics," in *Geographical Information Systems.* ed by P. A. Longley, M. F. Goodchild, D. J. Maguire, and D. W. Rhind, pp. 611–620. New York: Wiley.

Tabb, W. K., and Sawers, L. (Eds.). (1984) *Sunbelt/Snowbelt: Urban Development and Regional Restructuring.* New York: Oxford University Press.

Tomlinson, R. (2003) *Thinking About GIS.* Redlands, CA: ESRI Press.

United Nations. (2003) *World Urbanization Prospects: The 2003 Revision.* New York: United Nations.

Walker, R., Ellis, M., and Barff, R. (1992) "Linked migration systems: Immigration and internal labor flows in the United States." *Economic Geography* 68:234–248.

Yeates, M. (2001) "Yesterday as tomorrow's song: The contribution of the 1960S Chicago School to urban geography." *Urban Geography* 22:514–529.

SUGGESTED READINGS

Brunn, S. D., and Williams, J. F. (1993) *Cities of the World* (2nd edition). New York: HarperCollins.

Chen, N. Y., and Heligman, L. (1994) "Growth of the world's megalopolises." In *Mega-city Growth and the Future.* ed. by R. Fuchs, E. Brennan, F.-C. Lo, and J. I. Uitto, pp. 17–31. Tokyo: United National University Press.

Clarke, K. (2002) *Getting Started with Geographic Information Systems.* Upper Saddle River, NJ: Prentice Hall.

Fuchs, R. J., Brennan, E., Chamie, J., Lo, F.-C., and Uitto, J. I. (Eds.). (1994) *Mega-city Growth and the Future.* Tokyo: United National University Press.

Grimes, B. F. (2000) *Ethnologue* (14th edition). Dallas, TX: SIL International.

Hesp, P., et al. (2000) *Geographica's World Reference.* San Diego, CA: Laurel Glen Publishing.

Pick, J. B., and Butler, E. W. (1997) *Mexico Megacity.* Boulder, CO: Westview.

EXERCISE

Exercise Description

Mapping and interpreting metropolitan growth and decline patterns between 1990 and 2000.

Course Concepts Presented

Sunbelt region, population redistribution, and metropolitan areas.

GIS Concepts Utilized

Thematic mapping and database queries.

Skills Required

ArcGIS Version 9.x

GIS Platform

ArcGIS Version 9.x

Estimated Time Required

1 hour

Exercise CD-ROM Location

ArcGIS_9.x\ Chapter_01\

Filenames Required

Metro48.*
States48.*

Background Information

In this lab, you examine the process of geographical redistribution of metropolitan population and the Sunbelt concept. The Sunbelt concept is 40 years old, and the exercise allows you to examine what Sunbelt trends were taking place for metro areas in the 1990s.

The exercise utilizes ArcGIS software. You will be introduced to map layers and attribute data connected to them. You will conduct query functions, which can select certain geographical entities on the basis of their attributes.

Note: *For all of the exercises associated with this textbook, please copy the files from the CD that contains the work files to your hard disk. After you have copied the files to the hard disk, you need to clear Read Only. For any folder on the hard disk that contains files, you right-click the folder and select Properties. Under Properties, clear the Read Only check box.*

Further Readings

(1) Bernard, R. M., and Rice, B. (Eds.). (1983) *Sunbelt Cities*. Austin, TX: University of Texas Press.

(2) Borchert, J. R. (1983) "Instability in American metropolitan growth." *The Geographical Review* 73:127–149.

(3) Frey, W. H. (1993) "The new urban revival in the United States." *Urban Studies* 30:741–774.

(4) Greene, R. P., and Pick, J. B. *Exploring the Urban Community: A GIS Approach,* Chapter 1.

(5) Manson, G. A., and Groop, R. E. (1996) "Ebbs and flows in recent U.S. interstate migration." The *Professional Geographer* 48:156–166.

Exercise Procedure Flowchart

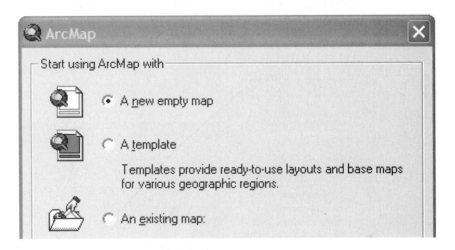

Exercise Procedure

Part 1

1. Start ArcMap with Start Using ArcMap With A New Empty Map selected. Click OK.

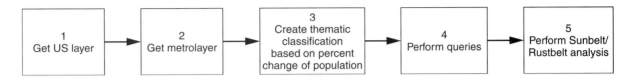

2. Click the Add Data button and navigate to the exercise data folder.

Note: *If you do not see the folder where you placed the exercise data, you will need to make a new connection to the root directory that holds the data.*

3. Add metro48.shp and states48.shp from the Chapter 1 folder. You can add both by clicking on one of the file names, then clicking on the other one while holding down the Ctrl key on the keyboard and click Add.

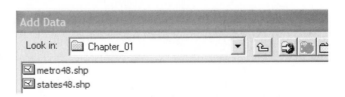

4. Ensure that the metro48 layer appears first in the list under the Layers data frame. If not, you can click and drag layers to a new position.

5. Click the colored box below the states48 under Layers.

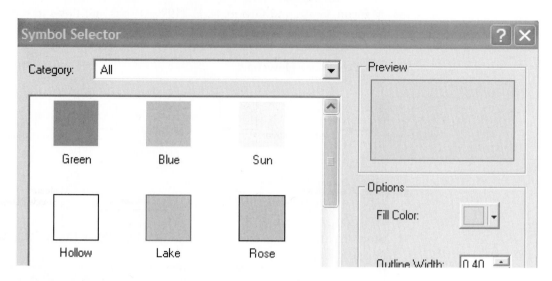

6. Click the Hollow box and click OK.
7. Right-click metro48.
8. Click Properties.
9. Click the Symbology tab.

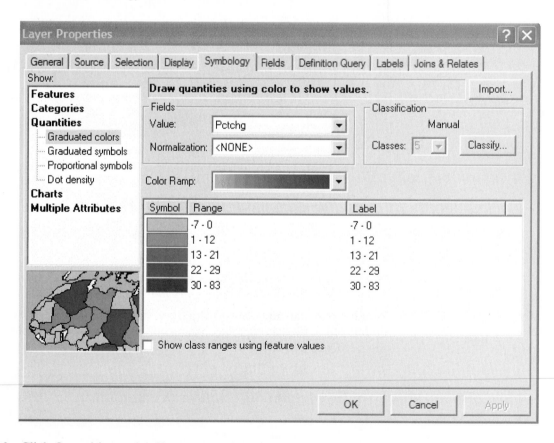

10. Click Quantities under Show.
11. Pick the Yellow to Dark Red Color Ramp setting or a Color Ramp setting of your choice.
12. Pick Pctchg under Value in the Fields box.
    (Pctchg represents the percentage change in population between 1990 and 2000.)
13. Click Classify.

14. Change the break values manually or by sliding the blue bars to the following:
    a. 0
    b. 12
    c. 20
    d. 32
    e. 86 (already set correctly)

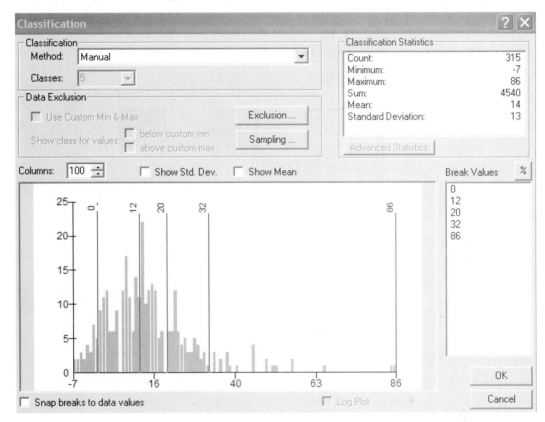

15. Click OK.
16. Click Apply and then click OK to see your new classification applied to the metro48 layer.

Part 2: Sunbelt Queries

Considering only the metro areas with a growth rate of 20 percent or higher, calculate the percentage located in the Sunbelt.

17. Click on Selection from the menu bar and click Select By Attributes.
18. Type the following query statement:

$$\text{"Pctchg"} >= 20$$

After double-clicking Pctchg under Fields, Click >=, and type in 20.

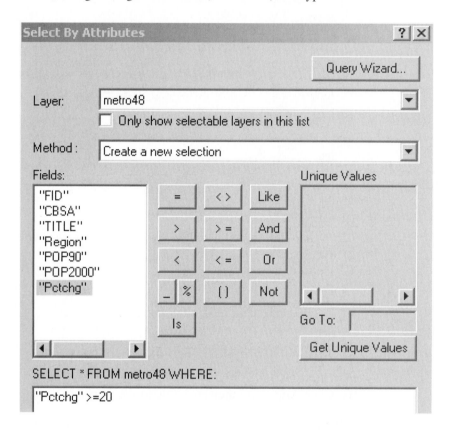

19. Click Apply.
20. Click Close.
21. Right-click metro48 and click Open Attribute Table. At the bottom of the table you see the number of metro areas selected (83 out of 315).

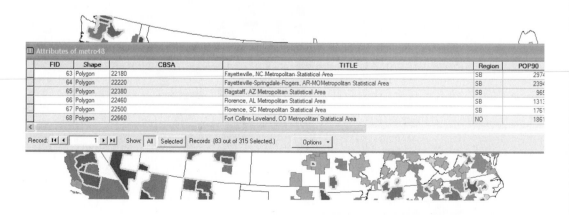

22. Close the table and go back to Selection on the menu bar and click Select By Attributes.
23. Enter the following logical statement to only select these fast-growing metro areas located in the Sunbelt (note: "SB" refers to "Sunbelt" and "no" means "not in Sunbelt"):

$$\text{"Pctchg"} >= 20 \text{ AND "Region"} = \text{'SB'}$$

To see the list of available values for Region, click Get Unique Values.
24. Click Apply and then click Close. Next, open the attribute table by right-clicking metro48 and choosing Open Attribute Table. You see the number of metro areas selected (50 out of 315). You can now compute the percentage located in the Sunbelt: [(50/83)*100]. Use your pocket calculator or the calculator on your computer under the Accessories menu to compute the percentage of fast growing metro areas located in the Sunbelt.
Another question is, are declining metro areas inside or outside of the Sunbelt?
25. Considering only the metro areas that lost population, calculate the percentage of those that are located outside the Sunbelt ("Region" = "NO"). This can be done by returning to Selection, and Select by Attributes and enter the two following queries:

$$\text{"Pctchg"} < 0$$

The above query selects 23 out of 315.

$$\text{"Pctchg"} < 0 \text{ AND "Region"} = \text{'NO'}$$

The above query selects 20.
    You can now compute the percentage of declining metro areas that are located outside the Sunbelt: [(20/23)*100].
26. Considering only the metro areas with a year 2000 population of 1 million or more (2000 population is represented by "POP2000") and that lost population between 1990 and 2000, calculate the percentage of those metro areas that are located outside the Sunbelt.

$$\text{"POP2000"} >= 1000000 \text{ AND "Pctchg"} < 0$$

The above query selects 2 out of 315.

$$\text{"POP2000"} >= 1000000 \text{ AND "Pctchg"} < 0 \text{ AND "Region"} = \text{'NO'}$$

The above query selects 2.
    You can now compute the percentage of large declining metro areas that are located outside the Sunbelt: [(2/2)*100].

Part 3: Compose Map

27. Click Selection and click Clear Selected Features
28. Select Layout View under View to compose a map for printing.
29. Click File and click Page And Print Setup. Change Orientation from Portrait to Landscape and click OK. Resize the data frame to fit the landscape page by clicking a corner and dragging it to fit the page.
30. Click Insert to insert a legend. In the first dialog box, make sure that the only legend item is metro48. You can remove states48 by selecting it in the right column and clicking the single arrow symbol to move it to the left column.
31. Click Next and erase the word Legend by backspacing.
32. Click Next on the next several dialog boxes and then click Finish.
33. Now double-click the selected legend. Click the Items tab and in the right column double-click metro48, click Properties, and click General. Clear Show Layer Name and Show

Heading. Click Apply and then click OK. Click OK as you exit back. Reposition the legend to a suitable spot.
34. Click Insert At The Top and insert a title.
35. Save the file if you wish and Exit ArcMap.

Part 4: Short Essay

Using the Sunbelt metro map you created in Part 3 and the query statistics you computed in Part 2, write a short essay describing the spatial patterns of growing metro areas and declining areas relative to the Sunbelt. Your essay should address whether there are distinct geographic patterns to the growth and decline and offer an explanation for why these 1990s patterns exist. Did Sunbelt metro areas grow more or less than non-Sunbelt ones? Your essay should also address whether the concept of the Sunbelt was still as valid in the 1990s as it was in the period from 1960 to 1990. Please refer to the Chapter 1 section "Analyzing and Urban Issue: Sunbelt Growth" and other literature references listed at the beginning of the exercise or ones you have found.

2

The Dynamics of Cities

This chapter further establishes the foundation of urban geography by emphasizing the dynamics that underpins change in cities. Cities are not static entities, but instead change from year to year and even day to day. Signs of change are evident in every city—rerouting of streets, new high-rise buildings, businesses starting, businesses failing, new corporate parks, air pollution episodes, a new stadium. Over time, the cumulative changes are often huge. Consider that New York City in the mid-19th century was physically much smaller—it roughly corresponded to all of Manhattan Island, did not have motorized transport, and lacked Central Park. Besides physical development, social and cultural changes are continually taking place. These changes are sometimes subtle, but grow profound over time. For instance, with the arrival of a new immigrant group, a city may undergo a language shift that impacts interpersonal communications, theater and the arts, education, and business marketing.

We do not attempt to cover all the possible changes and their causes. Rather, we focus on three areas of change in cities: economic, demographic, and social. The chapter first examines economic change in cities—shifts in the factors of production, industries, and occupations; transitions in the workforce; and deindustrialization and its consequences. Economic change in cities is further examined in later chapters that discuss economic push and pull in urban migration (Chapter 7), industrial location, consumer and producer services (Chapter 9), and economic development (Chapter 12).

Population change in cities depends on births, deaths, migration, and annexation, and a basic model of demographic change for cities is given. An important conceptual base for understanding the population growth of cities is demographic transition theory, which postulates why economically backward nations first undergo reduction in mortality, followed by a later drop in fertility, resulting in a period of very rapid population growth. In the countryside, this growth spurt puts pressure on people to migrate from rural areas to cities, in what is often termed *rural-to-urban migration*. Among the many urban demographic characteristics, age is briefly discussed. Age is important today because the people in cities are from many age groups, which impact the city in different ways.

Social change involves changes in such dimensions as education, social status, language, race, ethnicity, marriage and the family, leisure activities, gender, and lifestyles. These factors are discussed throughout the book, especially in Chapters 6 and 8. Sometimes social change leads to socioeconomic polarization, which refers to the division of a society into groups that are distinctly different from each other in their socioeconomic characteristics. For instance, in some U.S. cities, African-Americans and Whites are socioeconomically polarized, and in Mexican cities, people of Indian background (indigenous people) can be polarized from the majority *mestizo* (mixed blood) population. Socioeconomic polarization stems from historical and contemporary cultural, political, and economic factors, and is expressed in such phenomena as social class distinctions, ethnic tensions, income polarities, segregation, and separation. Often these differences can be overcome in ways that add positive synergy to cities.

After the section discussing social change and socioeconomic polarization, two examples are given to illustrate polarization. The first example of the Hmong population in Minneapolis-St. Paul illustrates a history of polarization for a small ethnic population, both while in Asia following the Vietnam War and later after migrating to the United States. The second example, the city of Nairobi, Kenya, demonstrates sociospatial polarization, with the wealthy living in the advanced center of the city and poorer residents located on the periphery. Chapters 6 and 8 examine social change and polarization further and include more case examples.

The second half of the chapter presents megacity profiles for Chicago, Los Angeles, and Mexico City. As already mentioned, this textbook gives added emphasis to these three cities, although discussing many others as well. In emphasizing the historical growth and change of the cities, the profiles illustrate the urban change concepts from the first half of the chapter.

The profiles provide a base for contemporary knowledge of these cities, which are referred to many times throughout the book. Like other northern cities, the industrial city of Chicago, which prospered by its midcontinent location and diversified industries, has suffered from deindustrialization in recent years, as well as longer term racial tensions, especially within its city limits. *Deindustrialization* refers to the relative decline in manufacturing in advanced cities or sections of cities. The spread-out and low-rise Los Angeles megacity that reaches its fingerlike extensions ever further has experienced huge changes since 1940, including World War II and postwar industrialization, racial tensions, water shortages, fame as the world's entertainment headquarters, massive entry of immigrants from less developed nations (especially since 1975), and the post–Cold War economic downturn in the aerospace industry. Over the past two decades, the influx of millions of immigrants has left Los Angeles without sufficient low-cost housing to accommodate them.

Mexico City, one of the world's largest megacities, expanded vastly in area and size during the 20th century. It serves as the primate city of Mexico, housing its federal government, leading businesses, and a massive industrial complex. The city also provides leadership in Mexico and Latin America in culture and the arts. At the same time, it has had to deal with environmental degradation, poverty, deteriorating infrastructure, and governance.

In the final section on analyzing an urban issue, the rank mobility index is presented as a method to understand how sections of a city change in population rank. This method is explained and applied to change in the population rank of suburbs in Chicago and Los Angeles in the 1990s. It builds on the knowledge of the Chicago and Los Angeles megacities from earlier in the chapter.

2.1 ECONOMIC CHANGE

The growth of cities depends chiefly on their economic resources, organization, markets, and activities. For instance, the economy of a city located in northern Canada, such as Calgary, might emphasize food processing related to its short-season agriculture and natural resources nearby. Flint, Michigan, or Akron, Ohio, are involved principally in supply industries for automobile

manufacturing. By contrast, the city of Cancun, Mexico, is economically driven by its tourism industry, which is tied to hotel, service, construction, and transportation sectors. The economic processes in cities take place over time and space. For instance, trade flows of import and export of goods and services take place over time—time to assemble, transport, receive, and distribute goods and to arrange and make available services. Other processes occur over decades or longer, such as the structural change in Cleveland from manufacturing to services industries (Warf & Holly, 1997). Those processes might be concentrated in certain cities, or within a city in certain locations. In Los Angeles, for example, many import and export activities are concentrated in the Port of Los Angeles. Container cargo in the port is shown in Figure 2.1.

Economic organization refers to how productive units fit together and how the inputs and outputs of these units flow from one to another. An example of this is the U.S. automobile industry, concentrated in Detroit and surrounding cities, along with supplier and service firms; specialized consulting and business services; and related transport, warehousing, and distribution enterprises. This auto manufacturing complex is organized into productive units, connected by relationships and flows. Within the larger region, Toledo, Ohio, has glass manufacturing, and Akron, Ohio, is the center for the tire industry.

As seen in Figure 2.2, the economic model, modified from Knox (1994) and Knox and Agnew (1998), is based on the linkage between economic change and urban change. Economic change influences urban change by capital investment, expenditures, labor markets, real estate transactions, banking, and trade. As also seen in Figure 2.2, outcomes from urban change include land use, urban systems, and social processes. They in turn can lead to social problems and political and government conflict, resulting in policy responses and planning.

Many aspects of this economic change–urban change model influence the spatial distribution of cities. For

FIGURE 2.1   Port of Los Angeles: Container Cargo Dock Area.

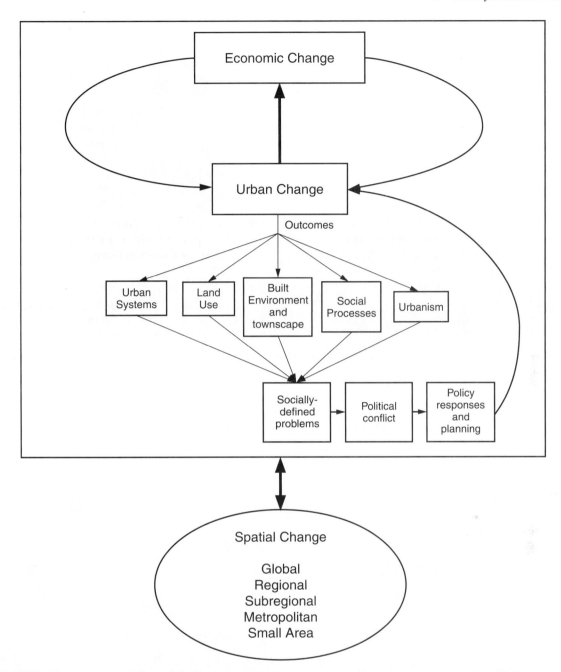

FIGURE 2.2 Economic change and its relationship to urban change and its outcomes. Source: Knox, Paul L. 1994. *Urbanization: An Introduction to Urban Geography*. Prentice Hall. Fig. 1.4, p. 8. "A framework for the study of urban geography: urbanization as a process."

example, the economic demand for tourism leads many major hotel chains to locate in southern Florida. As this builds up into a major tourist complex with more and more amenities, the complex tourism attracts more tourists. Another example of this is the location and relocation of government energy distribution facilities in cities, as shown in Figure 2.3. Economic demands initially call for particular placement of the facilities; however, that placement might stimulate development of new businesses, residences, and transport, which in turn encourages changes in spatial location, and so on. The implications of this model for spatial

change are more fully addressed in Chapter 9, which covers industrial location.

The model applies to different scales, as also seen in Figure 2.2. So far, we have considered the metropolitan scale. However, changes also take place at the global level. Multinational corporations have sites in urban locations in dozens of countries around the world, with flows between these sites of shipments of goods, service provisions, communications, money, and workforce. Consider cities with major trading functions such as Amsterdam; Rotterdam; Hong Kong; Mumbai (formerly

FIGURE 2.3   Energy distribution facilities in Mexico City.

Bombay), India; and Yokohama, Japan. Hong Kong is shown in Figure 2.4. Globalization means that the model of economic change and spatial change can apply at the international level. It can also be brought down to the scale of ZIP codes, city wards, and water districts.

Additional important foundations for economic change in urban geography are factors of production and their spatial arrangement. *Factors of production* are the basic elements leading to production from any economic system, and consist of land, labor, capital, and technology.

FIGURE 2.4   Hong Kong: A global trading and financial city. Source: Omni-Photo Communications, Inc.

In cities it is the combination of these elements that leads to economic production. *Land* includes territory on the earth, as well as the land's soils and natural resources (Knox & Agnew, 1998). *Labor* refers to the number of workers, their skills, experience level, and work quality. *Capital* consists of money, as well as the physical plant for production, such as factories, transport vehicles, and service facilities. *Technology* is the capability for the practical application of knowledge that makes something function or work.

All four factors of production are keys to the development and growth of cities. Urban land is essential to physical changes. Although the market determines its value, its uses can be planned and regulated by government. Labor depends on the city's labor markets, some local and others national or worldwide. Capital is available through the monies accumulated by businesses, governments, and individuals. The patterns of its investment can be influential in a city's development. Businesses make capital investments in such items as industrial plants and other facilities, machinery, buildings, technology and communications hardware and facilities, shopping centers, and sports complexes. Individuals usually allocate their largest capital investment to their home, but might invest in other real estate, financial instruments, and capital goods. As discussed in Chapter 12, governments also invest in capital aspects of cities, sometimes in partnerships with businesses. For example, a mixture of government and private capital might fund a convention–hotel complex.

Technology is an old factor of production in cities that continues to move forward rapidly. Historically, cities have been influenced by such technological developments as the steam engine, railroads, the internal combustion engine, air transport, electronics, telecommunications, robotics, and the Internet (Knox, 1994). Technology factors tend to increase overall productivity. They also "give shape and directions . . . to the pace and character of urbanization" (Knox, 1994, p.11). Technology and the city are emphasized in the last chapter.

An example of the four factors of production, for a steel mill in Seoul, South Korea, are the siting for the mill (land); steel managers and workers (labor); plant, facilities, and money to operate the facility (capital); and the scientific and technological knowledge of the mill and its processes (technology). The factors can also apply to government production. For instance, a water supply office in Sacramento produces management decisions on water system flows. It depends on government water specialists, support staff, managers, and water board members (labor); state funds for budget, office buildings, and office equipment (capital); advanced modeling techniques (technology); and the building site (land). For contemporary Internet-based applications, all these factors also apply, although they were often misapplied in the turn-of-the-century dot-com boom and bust.

Factors of production are arranged spatially. At the global level, a manufacturing firm's capital, consisting of factories, transport equipment, and money, might be spread out around the world. On a local scale, city government staff might be located at six city offices. A university's technology resources of a specific kind could be centralized in a single research and development location or dispersed around to regional sites. The spatial arrangement of factors of production is covered in more detail in Chapter 9.

Industries encompass areas of production. An *industry* is a distinctive group of productive enterprises, for example, the mining or retail industry. Major industry categories include agriculture, construction, manufacturing, retail, information and communications services, financial services, retail and wholesale, and education/health/social services, and other services (U.S. Census, 2002). The industrial distributions of individual cities vary widely. Industrial distributions for the three megacities of Chicago, Los Angeles, and Mexico City are presented in Figure 2.5. The data for Chicago and Los Angeles are for the Combined Metropolitan Statistical Area (CMSA), which is the U.S. Census metropolitan concept that covers the greatest extent of land area among its metropolitan definitions (Office of Management and Budget, 2003; U.S. Census, 2004a, 2004b). The data for Mexico City are compiled for the metropolitan area, using the Instituto Nacional de Estadísticas, Geográfia, e Informática (INEGI) definition that is covered later in the chapter. Figure 2.5 shows the percent distribution of workers employed in each of a dozen industry categories. The Mexico City categories are divided into the central city (Federal District) and the outlying areas (the Mexico City portion of the State of Mexico).

For all three cities, the biggest industry groups are manufacturing, retail/wholesale, and education/health/social services. These three industries together account for almost 50 percent of the workforce in each case. They are of concern in this textbook, particularly Chapter 9, which pays attention to manufacturing, professional services, and retail locations. For these three leading industry sectors, it is important to note that both manufacturing and retail/wholesale services are a larger share in Mexico City than in Chicago or Los Angeles, whereas education/health/social services are a lower share in Mexico City. For manufacturing, this is partly explained by the trend in large companies in advanced nations to transfer their manufacturing to offshore locations. For example, companies with manufacturing formerly located in Chicago have transferred their manufacturing to China, Malaysia, Mexico, and other offshore locations. The reason is simple: Offshore costs are often lower. The higher proportions in the education/health/social services industry in Chicago and Los Angeles stem from the cities' well-developed public education and health services. By contrast, Mexico City has a relatively low percentage of its residents with access to health services.

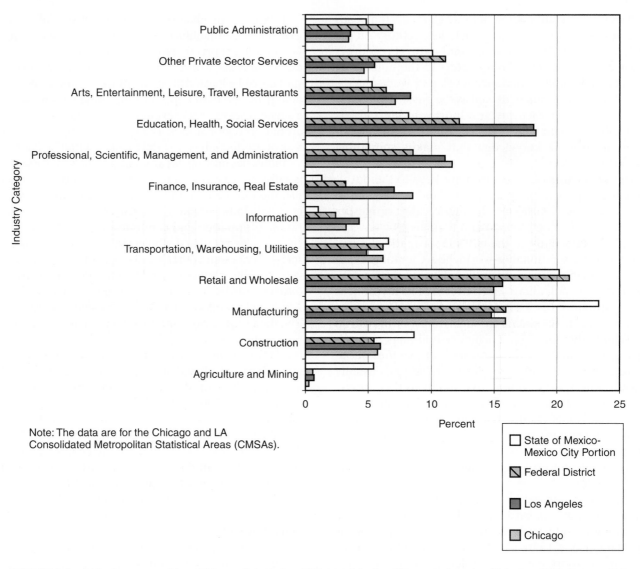

FIGURE 2.5   Industry composition, Chicago, LA, Federal District, Mexico City part of State of Mexico, 2000. Source: U.S. Census 2000 and INGEI, 2000.

This reflects a society and economy that stresses health care less, as well as a younger age structure, compared to the United States.

Over time, the industrial and occupational compositions of cities, as well as nations, change. The U.S. Census aggregates occupations into broad categories such as farming, white-collar, and professional. In the United States, for instance, over the past century the occupation of farming has greatly diminished, whereas white-collar occupations have expanded (see Figure 2.6). Compositional changes in industry and workforce reflect long-term economic change. Agriculture has become much more mechanized so fewer farmers are needed. On the other hand, since 1950, the contemporary service-based and knowledge-based economy in the United States has led to expansion in the white-collar workforce. Cities are

impacted, as the loss of agricultural jobs has pushed displaced farm workers to seek jobs in cities. This was a significant cause of U.S. rural-to-urban migration in the first half of the 20th century, leading to urban growth and causing cities to adjust to the migrants. The new information and knowledge workforce, a growing segment of the white-collar category, is also impacting the structure and functions of cities. Many knowledge workers can be located flexibly at the edges of or even outside cities, because they can communicate anytime from any place with new computer technologies.

Losses of factories, jobs, and workers stemming from deindustrialization can have a very hard impact on a city, unless some other economic sectors compensate. In the case of the Pittsburgh metropolitan area, huge steel and associated plants were operated as its central industry

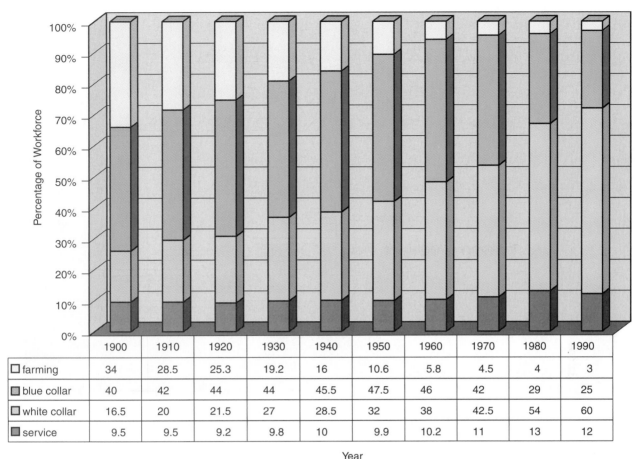

	1900	1910	1920	1930	1940	1950	1960	1970	1980	1990
☐ farming	34	28.5	25.3	19.2	16	10.6	5.8	4.5	4	3
▨ blue collar	40	42	44	44	45.5	47.5	46	42	29	25
☐ white collar	16.5	20	21.5	27	28.5	32	38	42.5	54	60
■ service	9.5	9.5	9.2	9.8	10	9.9	10.2	11	13	12

Year

FIGURE 2.6   Workforce composition in U.S., 1900–1999. Source: U.S. Census 1990 to 1999.

from the late 19th century through the 1930s. A contemporary Spanish steel plant is shown in Figure 2.7. Steel had given the metropolitan area a reputation as a manufacturing, polluting, blue-collar, and somewhat unattractive area. The steel industry began to shut down in Pittsburgh before World War II, due to the lack of expansion space from the city's narrow valleys, the movement of steel demand following the auto industry westward, and later low-cost, efficient foreign competition. As a result, the Pittsburgh area suffered painful job and business losses, along with all the other dependent industries, such as suppliers. The loss was compensated for by the industry moving to less costly U.S. sites, including other Pennsylvania locations, Chicago, and Gary, Indiana. Although the steel industry eventually left the city, Pittsburgh did remain the headquarters, although not the manufacturing center, of Alcoa, the world's largest producer of aluminum products. After the steel companies departed, the good news was that the Pittsburgh metropolitan area was able to effect remarkable environmental improvements, so today the city has a more balanced economic base and is attracting new jobs. London, New York,

and Tokyo all have also lost large amounts of manufacturing, but have compensated by emphasizing even more their world-dominant positions in finance, management control, and high-end services (Sassen, 2001). New York in the year 2000 is shown in Figure 2.8.

In summary, urban economic change can be analyzed and understood by applying economic principles and concepts to geographic phenomena. Economic change and urban change are tightly linked in a two-way feedback relationship. Economic change in cities can often impact urban planning and policy, and also influences spatial change.

2.2 DEMOGRAPHIC CHANGE

Demographic change is important for urban geography because it determines the size, structure, and distribution of cities' populations. It accounts for births and deaths, flows of migrations into and out of cities, spatial arrangements of population, age distributions and patterns, and shifts and changes in demographic-related characteristics, such as morbidity, marriage, poverty, and labor force.

FIGURE 2.7   Steel Plant in the industrial zone of Bilboa, Basque region of Spain. Source: Aurora & Quanta Productions Inc.

The population aspects of cities are given limited attention in this textbook, with some coverage in Chapters 7 and 8. These topics are covered in greater depth in courses on population geography. This section explains the basic determinants of population change in cities, examines the population sizes of U.S. cities in 2000 and their rates of change during the 1990s, discusses the demographic transition and how it relates to cities and urban geography, and discusses population age structure as an illustration of a demographic variable important for urban geography. Age structure influences many other aspects of cities, including schooling, labor force, business marketing, political issues, types of housing, and family structure.

A city's population grows and changes through fundamental components of change that include births, deaths, net migration, and annexation, all of which occur spatially. Annexation is the legal incorporation of additional land area by a city. Annexation applies to cities (i.e., the political/administrative units), but not to urban areas or metropolitan areas, broader concepts of urban agglomeration that extend beyond typical city limits. Those units are formally defined and their extent is not influenced by a city's annexing.

Other population characteristics, such as income, education, and migration, are distributed in sections of the city, its periphery, and hinterlands. Besides growth components, demographic change also relates to urbanization, health and mortality, and the family life cycle.

The principles of population change are expressed in the following equation:

$$POP(t + x) - POP(t) = \Delta POP$$
$$= BIRTHS - DEATHS + MIG + ANNEX$$

FIGURE 2.8   New York City, 2000.

where

$POP(t)$ = population at time t
$\Delta POP(x)$ = change in population over time period x
$BIRTHS$ = births during time period x
$DEATHS$ = deaths during time period x
MIG = net migration during time period x. Net migration refers to the migration into an area (in-migration) minus the migration out of an area (out-migration)
$ANNEX$ = population of territory annexed by the geographical unit during time period x (applies to cities only)

This formula states that the difference in population from time t to time $t + x$ depends on births minus deaths plus net migration plus the population within annexed territory. The number of births minus deaths is often referred to as *natural increase*. The formula can

be applied for a metropolitan area, city, or zone of a city. To apply the formula, the first step is to count the current population. That can be done using the national census or local estimates. The next step is to measure or estimate births and deaths during the period. Birth and death data are collected by local health agencies and are fed into the national statistical system. Net migration can often be estimated through county and state administrative records, such as numbers of drivers licenses issued. The population of territory annexed by cities is not usually available from the U.S. Census, except by special request, and other countries vary on their publication of annexation data. Metropolitan and urban areas do not annex, but can change in territory through census re-classification.

The U.S. metropolitan areas with the largest and fastest growing populations are available from the 2000 U.S. Census. As seen in Table 2.1, the largest metropolitan

TABLE 2.1 Ranking of U.S. metropolitan areas, 2000, based on 2003 metro definitions.

Rank	Combined statistical areas ranked by population, 2000	Census population April 1, 2000	April 1, 1990	Change, 1990 to 2000 Number	Percent
1	New York-Newark-Bridgeport, NY-NJ-CT-PA	21,361,797	19,710,239	1,651,558	8.4
2	Los Angeles-Long Beach-Riverside, CA	16,373,645	14,531,529	1,842,116	12.7
3	Chicago-Naperville-Michigan City, IL-IN-WI	9,312,255	8,385,397	926,858	11.0
4	Washington-Baltimore-Northern Virginia, DC-MD-VA-WV	7,538,385	6,665,228	873,157	13.1
5	San Jose-San Francisco-Oakland, CA	7,092,596	6,290,008	802,588	12.8
6	Philadelphia-Camden-Vineland, PA-NJ-DE-MD	5,833,585	5,573,521	260,064	4.7
7	Boston-Worcester-Manchester, MA-NH	5,715,698	5,348,894	366,804	6.9
8	Detroit-Warren-Flint, MI	5,357,538	5,095,695	261,843	5.1
9	Dallas-Fort Worth, TX	5,346,119	4,138,010	1,208,109	29.2
10	Houston-Baytown-Huntsville, TX	4,815,122	3,855,180	959,942	24.9
11	Atlanta-Sandy Springs-Gainesville, GA-AL	4,548,344	3,317,380	1,230,964	37.1
12	Seattle-Tacoma-Olympia, WA	3,604,165	3,008,669	595,496	19.8
13	Minneapolis-St. Paul-St. Cloud, MN-WI	3,271,888	2,809,713	462,175	16.4
14	Cleveland-Akron-Elyria, OH	2,945,831	2,859,644	86,187	3.0
15	St. Louis-St. Charles-Farmington, MO-IL	2,754,328	2,629,801	124,527	4.7

Rank	Metropolitan statistical areas ranked by % population change: 1990 to 2000	Census population April 1, 2000	April 1, 1990	Change, 1990 to 2000 Number	Percent
1	St. George, UT	90,354	48,560	41,794	86.1
2	Las Vegas-Paradise, NV	1,375,765	741,459	634,306	85.5
3	Naples-Marco Island, FL	251,377	152,099	99,278	65.3
4	Coeur d'Alene, ID	108,685	69,795	38,890	55.7
5	Prescott, AZ	167,517	107,714	59,803	55.5
6	Bend, OR	115,367	74,958	40,409	53.9
7	Yuma, AZ	160,026	106,895	53,131	49.7
8	McAllen-Edinburg-Pharr, TX	569,463	383,545	185,918	48.5
9	Austin-Round Rock, TX	1,249,763	846,227	403,536	47.7
10	Raleigh-Cary, NC	797,071	541,100	255,971	47.3
11	Gainesville, GA	139,277	95,428	43,849	45.9
12	Boise City-Nampa, ID	464,840	319,596	145,244	45.4
13	Phoenix-Mesa-Scottsdale, AZ	3,251,876	2,238,480	1,013,396	45.3
14	Laredo, TX	193,117	133,239	59,878	44.9
15	Fayetteville-Springdale-Rogers, AR-MO	347,045	239,464	107,581	44.9

Source: U.S. Census Bureau, 2004b.
Note: Based on the Metropolitan definitions of June 2003. Phoenix-Mesa-Scottsdale is not a combined statistical area, so does not appear in upper list.

area is the New York–Newark–Bridgeport Combined Statistical Area with a population of 21.3 million. The Combined Statistical Area concept combines several metropolitan areas and is explained more thoroughly in Chapter 3. The Combined New York Statistical Area includes areas in New York, New Jersey, Connecticut, and Pennsylvania. The New York metropolitan area has a larger population, but is smaller in geographical extent, than some well-known nations, such as Chile at 15.6 million, the Netherlands at 16.1 million, and Australia at 19.9 million (Population Reference Bureau, 2003).

In second and third place are the Los Angeles–Long Beach–Riverside Combined Statistical Area (CSA) at 16.4 million and the Chicago–Naperville–Michigan City Combined Statistical Area at 9.3 million. They are also huge, with the Los Angeles metropolitan population approximating the size of the nations listed previously, whereas the Chicago metropolitan population is about the size of Sweden (8.9 million) and Hungary (10.1 million). For Chicago and Los Angeles, the component of change that accounted for most of their growth of 11 to 12 percent in the 1990s was net migration. Among the largest metropolitan areas, the fastest growing were the Phoenix-Mesa Metropolitan Statistical Area (MSA) and the Atlanta CSA at 45.3 and 37.1 percent growth, respectively (see Table 2.1).

Out of the hundreds of U.S. MSAs, the fastest growing were the St. George MSA and its larger neighbor the Las Vegas MSA, which grew by 85.5 percent in population during the 1990s to reach 1.4 million (U.S. Census, 2004b). Las Vegas, seen in Figure 2.9, is a fascinating case of a city located in an inhospitable desert. Its growth has been driven by gambling, a huge convention and entertainment industry, and retirement destination communities (Gottdiener, Collins, & Dickens, 1999). Recently more economic diversification has occurred, particularly by growth in manufacturing, retailing, catalog sales, and credit operations. Las Vegas's demographic increase is due to the growth components of in-migration and annexation (Gottdiener et al., 1999). Much of the migrant stream consists of ex-Californians, who are drawn by lower housing and living costs, an improved environment, and lower taxes. The other rapidly growing metropolitan areas in the U.S. are mostly located in the Sunbelt, including several in the Carolinas. The 15 fastest growing areas all expanded by at least 44 percent. The population growth of these metropolitan areas leads to both positive and negative impacts, which are discussed as growth management in Chapter 12. The next section turns to a major broad and long-term theory of demographic change that might impact the growth of cities over extended periods.

2.2.1 The Theory of Demographic Transition

The theory of *demographic transition* attempts to explain the transformation in population growth from preindustrialized to industrialized societies. The course of the demographic transition is shown in Figure 2.10. A country

FIGURE 2.9   Las Vegas: the fastest growing MSA in the 1990s.

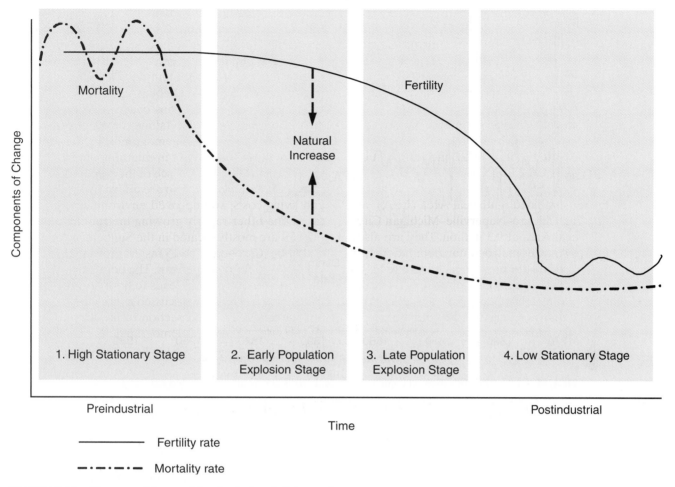

FIGURE 2.10   Demographic transition for industrializing nations.

starts out in a preindustrial status, with high fertility and high mortality. High fertility usually means three or more births per mother, which is a total fertility rate approaching three. Total fertility rate (TFR) is the average number of children a woman would have if birth rates stay constant during her childbearing years, usually taken as 15 to 49. High mortality implies a life expectancy of less than 60 and a crude death rate (deaths per 1,000 population) of 20 or higher (Population Reference Bureau, 2003). This is typical of the developing-world societies today, and was true of many countries historically. Rates are at a high level because these countries do not utilize modern contraception and because health and environment are not at modern standards; that is, there are low levels of sanitation, nutrition, public health services, and medical care. Because the birth and death rates are initially about equal, the population level over time is relatively steady. The first growth component to drop is mortality, which is reduced due to environmental factors such as clean water, sanitation, cleaner food production, preventative public health measures, and bountiful agricultural yield, but rises from diseases such as plague

(Jones, 1990; Weeks, 2001). Historically, for nations that entered the demographic transition 200 years ago, it took 100 years or so for the mortality decline to occur.

Sweden is unusual in having a consistent and reliable series of demographic data that goes back for more than 200 years. The course of its demographic transition from 1770 to 1986 is shown in Figure 2.11 (Jones, 1990). Here, death rates declined quite steadily during the entire transition of more than 200 years.

The theory also postulates that, as a result of the early population boom, fertility rates declined in a delayed pattern after mortality rates, resulting in a period of population expansion. In the first 90 years, because deaths decreased while births remained steady before dropping after 1860, Sweden's population started to increase rapidly (refer back to the population change equation). Fertility drops because it serves as a way for households to cope with the mounting economic pressure from larger families. Families can earn more income in an industrializing society because there are fewer but more educated children. During the whole time portrayed in Figure 2.11, Sweden was industrializing.

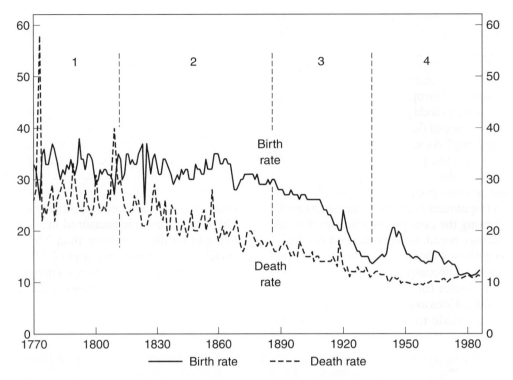

FIGURE 2.11   Sweden's demographic transition, 1770–1986. Source: Jones, Huw. 1990. *Population Geography*. 2nd Ed. Paul Chapman Publishing Ltd. Figure 2.9, p. 19. "Crude birth rates and death rates in Sweden, 1770–1886, and stages of the demographic cycle."

Eventually, the country reaches low fertility and low mortality, and demographic growth slows (see Figure 2.11). By 1980, Sweden's growth from natural increase, which is births minus deaths, had virtually halted. The sequence of demographic transition is often classified into four stages, which are shown in Figure 2.10, as follows:

1. *High Stationary Stage*   Both birth and death rates are high. The economy is preindustrial.

2. *Early Stage of Population Expansion*   Death rates drop but birth rates remain steady. Consequently the population starts to expand rapidly.

3. *Late Stage of Population Expansion*   Birth rates have dropped considerably. The decline in death rates is beginning to slow. The population growth is still rapid, but starting to decelerate.

4. *Low Stationary Stage*   Birth and death rates reach low plateaus. The net population growth rate declines to reach a fluctuating level, averaging zero population growth.

The nations that are advanced today underwent historical demographic transitions lasting one to two centuries (Weeks, 2001). An example is the United States, which in 1790 had a high birth rate of over 50 births per 1,000 population and high mortality, so that life expectancy was

only 35 years, versus 77 years, as it is today (Nam, 1994; Population Reference Bureau, 2003). The difference between births and deaths, or natural increase, was substantial, so the nation grew in population (Nam, 1994). At the same time, around 1790, America increasingly felt the benefit of the economic development coming from Europe, which contributed to a long decline in mortality rates. Among other things, that economic development favored better public health, sanitation, nutrition, and medical care. By the mid-19th century, fertility began to fall, from 52 per 1000 in 1840 to 32 by 1900 and 25 by 1925 (Kurian, 1994; Nam, 1994).

During the first half of the 19th century, when mortality and fertility were both falling, the natural increase continued to be substantial and the large population spurt predicted by demographic transition occurred, further reinforced by immigration. The U.S. population expanded from 5.3 million in 1800 to 23.2 million in 1850, growing at the highest rate in its history of 3 percent yearly. Since 1925, the United States has been in the end stages of the demographic transition. In 1925 the death rate was 12 per 1,000 (Kurian, 1994) and has continued at this low level for the past 80 years, and its fertility rate has remained low, as predicted, although fluctuating with a gradual reducing trend. In 2003, the U.S. birth rate was 14 per 1,000 and its death rate was 9 per 1,000 (Population Reference Bureau, 2003).

Contemporary nations are undergoing demographic transitions that are often faster, as economic changes are more accelerated in today's world of rapid trade and globalization than was the case a century or two ago (Weeks, 2001). The demographic transition theory, which was formulated based mainly on European experience, might or might not be replicated in today's developing nations.

What is the importance of demographic transition theory to urban geography? As seen in Figure 2.12, by the second transition stage, the proportion of high-fertility rural or semirural population is shrinking in favor of a higher proportion of urban population. One consequence is that some rural population is displaced and must move to urban areas. Among the precipitating factors in rural-to-urban migration are population pressure on rural land and economic conditions that include the modernization of agriculture and accompanying rural job losses. An economic "pull" factor for rural workers is the larger number of low-skilled jobs available in cities. The movement from the countryside to the city in demographic

transitions stems from the interaction of the whole set of economic and population phenomena, both at origin and destination.

In Mexico's demographic transition, its mortality rate began to drop significantly in 1907 and continued to drop until it reached a low plateau in 1965 (INEGI, 1999), whereas its birth rate dropped from 1930 to 2000 and has not yet plateaued (INEGI, 1999). This corresponds to a shortened demographic transition expected of a developing nation. In Mexico, and generally in the world, rural fertility is higher than urban fertility. As seen in Figure 2.13, the proportion of rural population began to drop significantly around 1940, and is still continuing to fall today. Urban population is measured in Mexico as population living in localities of more than 2,500 persons, whereas rural population lives in places of 2,500 or less. The losses in rural population percentage corresponded to gains in the cities. You will notice in Figure 2.13 that Mexico's total population grew tenfold during the 20th century. Thus the growth in urban population expanded a remarkable

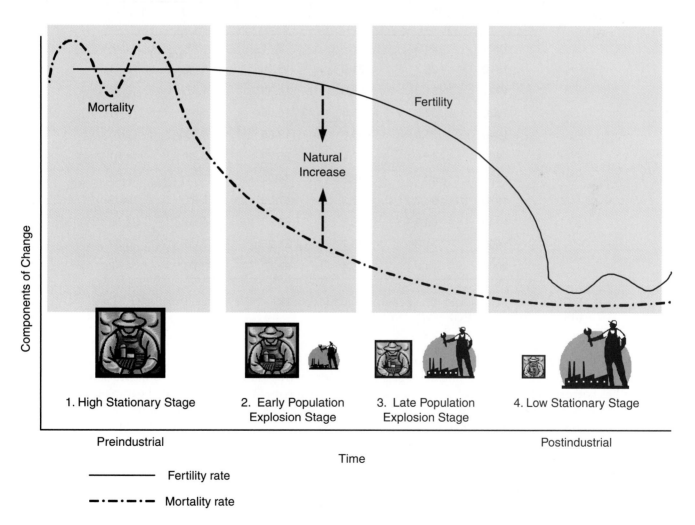

FIGURE 2.12 Demographic transition for industrializing nations, showing rural-to-urban transition.

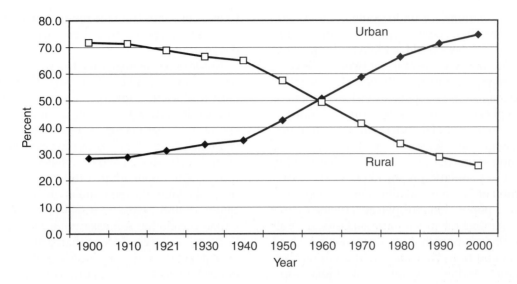

Supporting Data

Year	Urban	Rural	Total Population (in millions)	Urban Population (in millions)	Rural Population (in millions)
1900	28.3	71.7	9.77	NA	NA
1910	28.7	71.3	15.16	4.35	10.81
1921	31.2	68.8	14.33	4.47	9.87
1930	33.5	66.5	16.55	5.54	11.01
1940	35.0	65.0	19.65	6.90	12.76
1950	42.6	57.4	25.79	10.98	14.81
1960	50.7	49.3	34.92	17.71	17.22
1970	58.7	41.3	47.23	27.31	19.92
1980	66.3	33.7	66.85	44.30	22.55
1990	71.3	28.7	81.25	57.96	23.39
2000	74.6	25.4	97.48	72.76	24.72

FIGURE 2.13   Urban and rural population in Mexico: 1900–2000. Source: Estadisticas Historicas de Mexico; INEGI, 1992a; INEGI 1998.

17-fold from 1910 to 2000, whereas rural population grew by a mere twofold. Hence, in Mexico the demographic transition has corresponded to a rural-to-urban transition. For a century, millions of rural migrants made the decision to leave rural homes and jobs to migrate to the supposedly more attractive city. Small farmers, such as the ones shown in Figure 2.14, are susceptible to migration pressures as farming becomes more mechanized. In response to the population increase of the demographic transition, there has been increased immigration to another country, especially because the economy cannot sustain such an enlarged population. This was one of the push factors for Mexican migration to the United States,

starting in 1980. From the standpoint of studying a city or metropolitan area, the case of Mexico demonstrates that it is useful to know what demographic transition stage a city is in, as well as the transition stage for the city's hinterlands, the rest of the country, and important sending or receiving regions.

2.2.2 Population Age Structure

The age structure of a population refers to the distribution of the population by age. It constitutes an example of a demographic characteristic influential for urban geography. It can be expressed by average age, or

FIGURE 2.14   Mexican small farmers: susceptible to migration. Source: Demetrio Carrasco © Dorling Kindersley.

by the proportion of population in certain age groups, for instance, younger than 15 or older than 65. The age structure influences urban geography through the differing impacts of particular age groups. For example, a high proportion of children in a suburb points to the need for K–12 schools, day care programs, and pediatricians. A city neighborhood that is full of elderly people over 65 implies the need for public transport services, elder care facilities, and nursing. The long-term, worldwide trend toward aging accentuates the needs of the elderly.

Contemporary consumer businesses that market to cities know the spatial age structure patterns, and key their programs appropriately, often utilizing GIS. Age can be important on other scales as well. For instance, a national economy might be affected by its social security system, which depends on accurate information about age structure. The reason is that the productive working population of people aged about 20 to 64 contributes social security salary savings into the social security fund, whereas children and adolescents age 0 to 19 are neutral as contributors or beneficiaries of the fund. The elderly population over 65 draws pension payouts from the fund. In modeling U.S. social security demands, projections of the future age structure are among the most important factors.

Age structure can also be measured in smaller gradations, particularly by five-year or one-year age groups, information that is available in most censuses. A useful diagram to display age structure is the *population pyramid*, which typically shows the number or percentage of male population in each age group on the left of the center vertical axis and the number or percentage of female population on the right of the axis. The pyramids for the Chicago and Los Angeles CMSAs are shown in Figure 2.15. The two megacities appear remarkably alike in age structure. For both metropolitan areas, population aged 5 to 14 forms a small "boomlet" that was known in these cities in 2000 to be overloading many K–12 school districts. It portends university enrollments peaking in 2010 to 2015, when 3- to 12-year-olds in the year 2000 are in college. There is a population narrowing for those aged 15 to 25, young people born from 1975 to 1985, a time of reduced immigration affecting children and somewhat lower fertility. The "baby boomers" are seen in the bulge of ages 25 to 49.

In summary, population change is a key factor in the dynamics of cities. Particular attention is given to migration in Chapter 7. Demographic transition theory informs urban geography in a long-term perspective. Age structure influences the economic productivity, demand for services, and educational needs of a city. It also influences

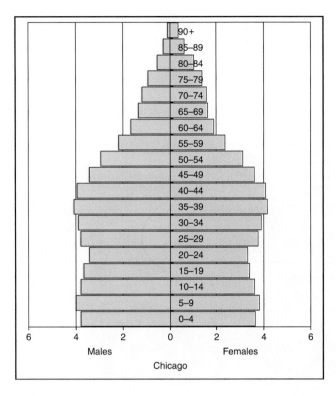 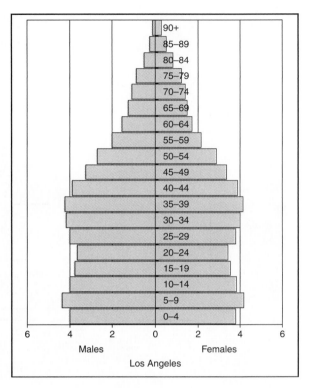

FIGURE 2.15   Population age distributions for Chicago (on left) and Los Angeles (on right) CMSAS, 2000, by percent.

the social character of cities and their neighborhoods and zones, as the next section discusses.

2.3 SOCIAL CHANGE AND SOCIOECONOMIC POLARIZATION

Social change and socioeconomic differences provide another cornerstone in the dynamics of cities. Cities consist of people and groups having different social characteristics and social interactions with each other. Sometimes those interactions result in a positive, cooperative force, but other times, they can lead to tensions or even hostilities. This section covers some significant social characteristics of cities, as well as causes and outcomes of socioeconomic polarization. Polarization is illustrated by two case studies, one of the Hmong population in Minneapolis-St. Paul and the other of Nairobi, Kenya.

Social change is influenced by socioeconomic characteristics, as well as other forces such as economic, political, and demographic change (see Figure 2.16). Social changes can lead to socioeconomic polarization. For instance, the proportion of elderly population (over 65) rises in a suburb, leading to political tensions with the younger population of childbearing age. In this suburb, the older population has tended to live in a section with fewer children and schools, but more amenities for older

people, such as stores and entertainment. The young population has located itself in residential areas closer to the schools. Thus, the social change might in turn cause spatial changes to take place. Such a confrontation has occurred in Youngstown, Arizona, just north of Phoenix (Moehringer, 2003). A former retirement community has received a large influx of young families starting in 1998. Now there are concerns from the retirees about the city's rising crime, drug problems, and domestic violence. This has led to political turmoil between young and old that is unresolved. Besides this example, the changes can occur on different geographic scales. For example, some social changes (e.g., large-scale immigration) influence the population and spatial distribution of the entire United States. There can even be social changes at the global scale, such as the tendency toward increased leisure activities, that influence worldwide spatial arrangements.

Socioeconomic characteristics are important factors in social change in cities. Data are available for nations, regions, and cities, and vary between nations. For instance, marital status varies culturally in how it is measured. Mexico has three marriage measures, religious, civil, and free union (*union libre* in Spanish), the latter two of which are available from its census (INEGI, 2002). Civil marriage is required to legally formalize the arrangement, whereas religious marriage brings families together with ceremony.

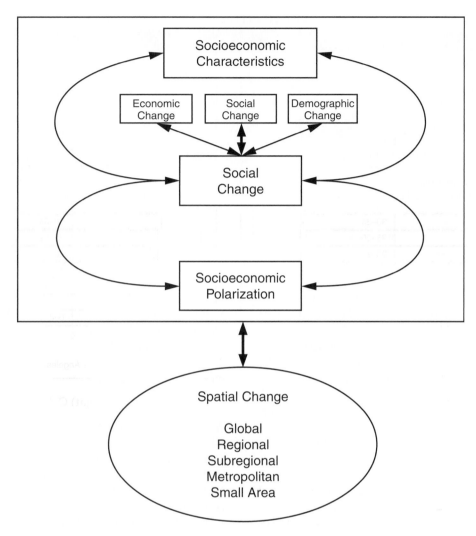

FIGURE 2.16 Social change and socioeconomic polarization in cities. Source: Knox, Paul L. and John Agnew. 1998. *The Geography of the World Economy*. Arnold Publishers (Arnold is part of Hodder Headline Group). Figure 1.1, p. 6. "The inter-relationships surrounding economic organization and spatial change."

Free union marriage avoids the need for either civil or religious marriage (Quilodrán, 1991). A census respondent declares himself or herself to be in a free union. Likewise, disability status can depend on how a particular nation defines disabilities. For instance, is a person with limited walking capability classified as disabled?

Examples of social characteristics for the Chicago and Los Angeles metropolitan areas, as well as the United States, are given in Table 2.2. Each metropolitan area is distinctive in certain respects. For example, the Los Angeles metropolitan area has a higher proportion of married couples with children as a percentage of all families than does Chicago or the nation. This reflects child members of immigrant families that arrived in the Los Angeles metropolitan area over the past decades, as well as the high average fertility of those families. Chicago

demonstrates some of this effect also, compared to the nation. Another set of distinctive characteristics is ethnicity. Chicago has 19 percent Black, 16 percent Hispanic, and 4 percent Asian population, versus 8 percent Black, 40 percent Hispanic, and 11 percent Asian for Los Angeles. On almost all percentages except Los Angeles's Black proportion, the United States has less ethnic diversity. The presence of more ethnic diversity in the biggest cities is not surprising—this characterizes megacities around the world.

Related to ethnicity is the proportion of population that speaks a non-English language at home. Although for the United States this proportion is only 18 percent, it is a remarkable 47 percent in Los Angeles. It is also higher in Chicago at 25 percent, compared to the country. Los Angeles is somewhat more college-educated at

TABLE 2.2   Social characteristics of Chicago and Los Angeles CMSAs and United States, 2000.

	Chicago CMSA	LA CMSA	United States
African-American, Percent	19.1	8.0	12.6
Asian, Percent	4.3	10.9	3.7
Hispanic, Percent	16.4	40.3	12.2
Married Couples with Children, Percent of Families	36.8	39.0	34.6
College Education, 25+ Yrs., Percent	15.6	18.2	15.9
Never Married, 15+ Yrs., Percent	27.1	30.9	31.3
Divorced, 15+ Yrs., Percent	9.7	8.4	8.9
At Least One Disability, 65+ Yrs., Percent	40.7	43.1	41.9
Owner-Occupied Housing, Percent	65.2	54.8	66.2
Native to State, Percent	63.1	47.9	60.0
English Ancestry, Percent	5.1	6.1	8.7
Polish Ancestry, Percent	10.0	1.5	3.2
Greek Ancestry, Percent	1.0	0.3	0.4
Non-English Speaking at Home, 5+ Yrs., Percent	24.9	46.8	17.9

Source: U.S. Census, 2003.

18 percent than Chicago or the nation, which are at 15 to 16 percent. Chicago has been influenced culturally by its large Polish population, as evident by its 10 percent Polish ancestry, compared to only 1.5 percent in Los Angeles. It is also not surprising that Chicago has two to three times the percentage of citizens with Greek ancestry, because it is known for its Greek Town and cultural presence. This small sampling of differences emphasizes that cities have distinctive social characteristics that influence neighborhoods, city politics, urban activities, and cultural traditions, and can sometimes lead to polarization. Social characteristics are discussed more in this textbook, related to a number of topics including neighborhoods, ethnicity and poverty, and environmental justice in Chapters 6, 8, and 11.

Socioeconomic polarization occurs when social groups or individuals separate themselves from one another, or when tensions and conflicts occur. Examples abound. At several times in the 20th century and especially in the late 1960s in the United States, many major cities, including Detroit, Chicago, and Los Angeles, erupted in racial riots reflecting enormous stress and conflict between Blacks and Whites. Israeli cities have often reflected ethnic divisiveness, segregation, and hostility. Belfast in Northern Ireland and Baghdad in Iraq have experienced bloody conflict between religious groups. The historical, cultural, and economic causes of these differences are beyond the scope of this chapter, but are considered in other parts of the book.

Two short case examples are presented of polarization that has influenced cities: the Hmong ethnic group in Minneapolis-St. Paul, and ethnic separation in Nairobi, Kenya.

2.3.1 Hmong in Minneapolis-St. Paul

The Hmong are members of a Laotian ethnic group that was concentrated for centuries in the highlands of Laos. The group had its own language and largely lived a nomadic

life (Lee, 2003; Strohl, 2000). During the 1950s French-Vietnamese conflict in Indochina and the 1964–1975 Vietnam War, the Hmong sided with the French and then the Americans, against the Viet Cong communists. Laos was drawn into the Vietnam War when the United States attacked and bombed Laos, because the North Vietnamese were in Laos, working together with the communist Pathet Lao against other Laotian troops aligned with the Americans. The Hmong were sustained during both periods by special American airlifts of essential supplies to their highland home areas. Following the fall of Vietnam to the Viet Cong communists, the Hmong entered a devastating period, in which the Pathet Lao, now the ruling communists, decided to try to exterminate the Hmong (Lee, 2003; Strohl, 2000). The clannish and traditional families had no choice except to pack up and attempt the hazardous trip to Thailand, which provided refugee camps. Many did not survive (Lee, 2003; Strohl, 2000).

Most of the Hmong who made it to Thailand went on to other countries, and more than 200,000 migrated to the United States (Lee, 2003). After originally settling in rural northern California, the Hmong migrated mostly to Minneapolis-St. Paul and surrounding areas, so today there are about 60,000 Hmong in Minnesota (Strohl, 2000), with Wisconsin as the second most important location for them nationwide.

The Hmong have had mixed success in the United States and many of them have faced continuing problems of poverty, illiteracy, and lack of assimilation (Strohl, 2000; Hughes, 2003). Some of the challenge in adjustment stems from their original culture, which was based on primitive agriculture and a barter economy. Their language does not have vocabulary for modern business practices; for instance, there is no word for *bank*. They have been beset by other problems, including pregnancy among unmarried teens, delinquency in gangs, and a low starting educational level. In spite of this, a small but growing proportion of Hmong have achieved middle-class status, including higher education, middle to upper middle incomes, and business success, often through small family enterprises. The states of Minnesota and Wisconsin have paid particular attention to helping the Hmong transitioning from poverty and the welfare rolls to low-level work, with some success (Strohl, 2000).

The plight of the Hmong was reflected in the decision by Minneapolis-St. Paul Metro Transit on the languages to use in the everyday customer signage and materials for the Hiawatha Light Rail line that opened in 2004. At issue was the choice of a third language, besides the givens of English and Spanish. The Metro Transit finally decided on Hmong, after comparing it to Somali, particularly. The reason was that the agency felt the Hmong had a much lower level of assimilation than Somalis or others, and could hence benefit the most from including their language in the Hiawatha's signage and materials (Coleman, 2002).

The Hmong experience reflects many of the elements of social change, socioeconomic polarization, and spatial change discussed earlier (see Figure 2.17). They started out radically polarized from the main Laotian society, followed by the social change of refugee semi-isolation in Thailand. These elements came into play in their subsequent experience in the United States, where the Hmong have not assimilated quickly, and hence have remained rather isolated and polarized through language and educational barriers from the mainstream of the Twin Cities' life and economy.

A second example of socioeconomic polarization is that of Nairobi, Kenya. It also illustrates some of the challenges for African cities (Stren and Halfani, 2001). Nairobi is the capital and dominant city in Kenya, and one of the largest cities in Africa, with a 2000 population of 2.23 million (United Nations, 1995, 2002a). It is the most important industrial, business, and tourism center of Eastern Africa (Mehretu & Mutambirwa, 2003). The United Nations (2002b) estimates that Nairobi is growing at the rapid rate of 4.7 percent yearly. The population growth is driven chiefly by migration from rural areas of Kenya. Africa's biggest cities are not large by world standards—mostly in the 1 to 3 million range—but appear large to the citizens (Potter & Lloyd-Evans, 1998). As in many developing countries, Africa's large cities tend to be national capitals and primate cities that dominate nations in population size.

Nairobi's rapid population growth rate over the past 50 years has brought with it the following problems (Mehretu & Mutambirwa, 2003; United Nations, 1995):

- Native Kenyans who live in Nairobi tend to regard the city as an alien enclave located in a predominantly rural nation, and usually identify with their home village or town rather than the city in which they are living.

FIGURE 2.17   Hmong teen marchers at Asian American Festival, Minnesota. Source: Skjold Photographs.

- Asians and Europeans living in Nairobi identify themselves primarily with their home nations, whereas Kenyan natives identify with Kenya.

- The three main ethnic groups—Africans, Asians, and Europeans—live mostly in separate residential areas and are stratified by separation of lifestyle and jobs.

The urban morphology of Nairobi reveals a modern central business district (CBD) that is a continental center of commerce, education, trade, and business (see Figure 2.18). The urban poor are economically marginalized in separated zones (Mehrutu & Mutambirwa, 2003). This example demonstrates polarization of the major racial groups leading to spatial separation and lack of common identity. Rural-to-urban migration keeps funneling poor people into this city, and these rural newcomers differ greatly from the affluent international, government, and business population in the city's center. They end up in squatter settlements on the urban periphery (see Figure 2.19), largely cut off from the wealth and power of the CBD.

Megacity Profiles

This second part of the chapter presents short profiles of the three megacities of Chicago, Los Angeles, and Mexico City, demonstrating the change concepts in the first part of the chapter and setting a foundation to examine the three cities throughout the book.

2.3.2 Chicago

Chicago, Illinois, was founded in 1833, and grew rapidly with the U.S. frontier expansion of the 19th century, the expansion of railroads, and then location of major industry in the city. The key features of the Chicago metropolitan area today are shown in Figure 2.20. This metropolitan area, which borders on Lake Michigan to the east, consists of mostly level topography. It is supported by major arterial highways that focus on the CBD, known as the Chicago Loop (from the "loop" of elevated train tracks that goes through it). The Chicago Loop is shown in Figure 2.21. Figure 2.20 displays the 2003 Census metropolitan concept of the Chicago–Naperville–Michigan City, IL–IN–WI Combined Statistical Area, which extends north along Lake Michigan into Wisconsin and east along the south end of Lake Michigan to include Gary, Michigan City, and other parts of northwestern Indiana (see Figure 2.21 for an aerial view). A Combined Statistical Area is a group of counties containing large metropolitan areas, termed metropolitan statistical areas (U.S. Census, 2004a). In the past several decades, the Chicago metropolitan statistical area grew from a population of 7.25 million in 1980 to 8.27 million in 2000 (U.S. Census, 2002). In 2000, the larger Chicago–Naperville–Michigan City, IL–IN–WI Combined Statistical Area had a population of 9.31 million (U.S. Census, 2002).

FIGURE 2.18   Nairobi's Downtown. Source: Peter Arnold, Inc.

FIGURE 2.19 Squatter settlement in Nairobi. Source: © David Turnley/CORBIS.

Chicago was incorporated as a city in 1837, in the era of rapid westward expansion and settlement in what was then the western reach of urban development. Chicago was located with key water access to Lake Michigan, the Chicago River, and after 1848 to the Mississippi River system by canal. In 1850, Chicago was a settlement with 30,000 people (U.S. Census, cited by Forstall, 2003). The city grew rapidly and by 1870, it had 310,000 in population (U.S. Census, cited by Forstall, 2003). The rapid city growth had been helped by annexation of surrounding communities. For instance, a group of area communities in 1889 voted, and agreed to be annexed (Lafrenz, 2001). Just before the world's Columbian Exhibition of 1893, the city in 1890 had a then very large population of 1.10 million (U.S. Census, cited by Forstall, 2003) and covered 125 square miles (Lafrenz, 2001). The Columbian Exhibition, celebrating the 400th anniversary of the discovery of America, added prominence to the growing city. It included the famous "white city" buildings, some designed by Daniel Burnham, an architect who later made contributions to planning the city, and is discussed in the final chapter.

Throughout Chicago's 19th-century growth, a large European immigration arrived, first from northern and western Europe and later from the south and east of Europe. They formed neighborhoods and brought along their cultural traditions, which gave varied character to

the city that still remains partly in place today, even after a different wave of late-20th-century immigration. By 1900, Chicago, with a population of 1.72 million, had reached the rank of fifth among the world's cities and was a newcomer on the list (Chandler, 1987).

An important 20th-century trend was large-scale migration of Blacks from the South, starting during World War I. They found the appeal of manufacturing employment in Chicago far outweighed their mostly rural jobs in the South. Black population in the city grew from 30,000 in 1900 to 234,000 in 1930, and concentrated on the South Side of the city, south of the downtown Loop area (Lafrenz, 2001). As discussed in Chapter 8, the growing Black population found itself increasingly isolated, leading to segregation and eventually some areas of ghettoization, features that today, many decades later, are present in Chicago's South Side. Other Black areas grew on the West Side and later in areas further south, as well as other parts of the city.

Other crucial factors in the growth of the city were growth in manufacturing and the advent of automobile commuting transport. Chicago grew in the first half of the 20th century as a manufacturing center, with particular emphasis on steel and related industries. World-class steel plants were constructed to the far south of the city along Lake Michigan and further south in Indiana. Like Detroit, Chicago became known for its traditional,

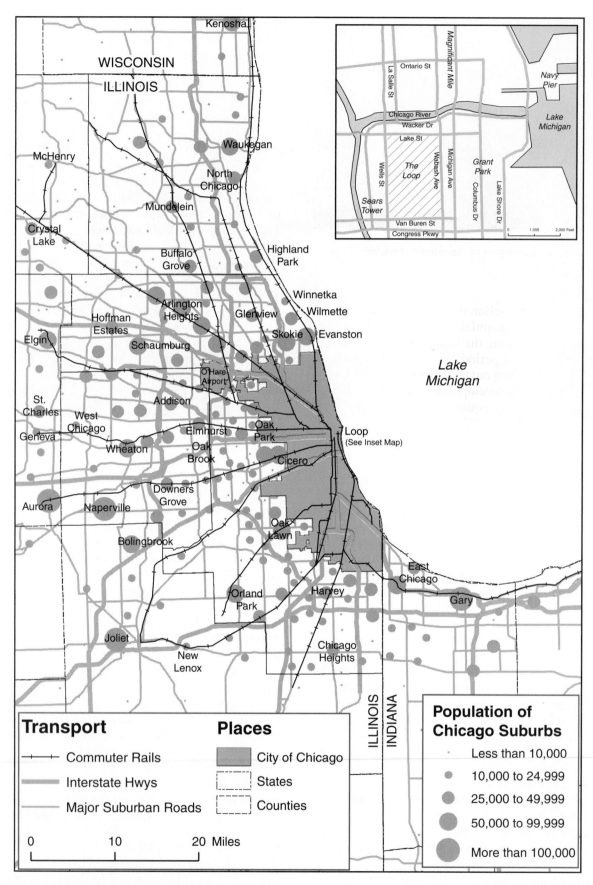

FIGURE 2.20   Chicago metropolitan area reference map.

FIGURE 2.21   Chicago: part of downtown stretching north.

large-scale mass production that had originated in Detroit with automobile manufacturing and the first Henry Ford especially. Later on, the term *Fordist* was given to this type of city. In the period after World War I, motor vehicle use grew, pushing out from the metropolis, a trend in the city's structural development covered more in the next chapter. As a consequence of many factors, the urban area population by 1930 had reached 4.26 million (U.S. Census, 2000).

From the end of the 1960s, manufacturing waned. The urban periphery expanded, Hispanic immigrants arrived in volume, and some of the city's traditional ethnic tensions grew. In the 1960s and 1970s, the decline in manufacturing impacted Chicago and other "Fordist" cities such as Detroit, Pittsburgh, and Cleveland. Foreign low-cost competition was taking its toll on large-scale, integrated manufacturing. Countries such as Japan and Korea, with newer technologies and lower costs, became competitive with the United States in many heavy production markets. It is estimated that the Chicago metropolitan area lost more than 1 million industrial jobs from 1967 to 1982 (Lafrenz, 2001).

In the year 2000, the Chicago CMSA had 15.9 percent of jobs in manufacturing, which is slightly more than the Los Angeles CMSA at 14.8 percent, and more than the 14.1 percent proportion for the United States overall (U.S. Census, 2004b). The former leader in manufacturing employment had fallen to an almost average proportion. Nevertheless, falling manufacturing employment was offset by increased productivity per worker through automation and technology. Today's manufacturers, such as the metal manufacturer seen in Figure 2.22, tend to emphasize technology to boost productivity (Stein, 2002). Furthermore, after 1970, Chicago switched from its manufacturing emphasis to trade and service, and the downtown was developed more (Lafrenz, 2001). These

advances added to its longer term strengths in insurance, retail/wholesale, and corporate headquarters management. A similar pattern was reported for Cleveland, which underwent severe deindustrialization in the 1970s and 1980s, to be followed by a "renaissance" in services, smaller scale advanced manufacturing, and cultural attractions (Warf & Holly, 1997).

The Chicago urban periphery grew, as did other metropolitan areas, so that by 2000, two thirds of the Chicago metropolitan area population lived outside of the Chicago city limits (U.S. Census, 2004b). Chicago spread out in all directions, although not much to the east. Yet, on a historical note, it had even spread east in the early 1900s through in-fill of Lake Michigan, mostly for parks, but also for a part of the downtown north of the Chicago River and east of Michigan Avenue. As will be seen in the next chapter, its highway systems expanded, and much residential and business development took place along or near these suburban arteries. The built-up areas in this much larger territory are variously termed fringe development, edge cities, and extraurbia, topics that are taken up in other chapters, especially Chapter 10.

Another factor has been the large immigration since World War II of Hispanics. Today the metropolitan area has a Hispanic population of 1.5 million, constituting a sixth of the total (U.S. Census, 2002). Most of the Hispanics came from Mexico, bringing with them many traditions, customs, and characteristics, in an echo of the European immigration nearly a century earlier. However, Hispanics today are about half located outside the city limits of Chicago in the remainder of the metropolitan area (U.S. Census, 2002), a lesser share than for total population.

As Chicago enters the 21st century, it has maintained its role as a global city, but faces a continuing deep problem: racial tension. Although its manufacturing role is reduced, Chicago is second in the nation in corporate headquarters and finance, a leader in science and higher education, and one of the largest U.S. transportation hubs and convention centers (Lafrenz, 2001). Its downtown Loop area gleams with modern skyscrapers, many of architectural note. At the same time, the ethnic tensions between the Black and White populations persist. As seen in Figure 2.23, the Black concentration in the south of the city has grown; in 2000, two thirds of the metropolitan area's Blacks lived in the city limits, mostly on the South and West Sides. On average, Blacks have been disadvantaged by lower educational opportunities and elevated poverty. Political tensions, which had exploded in city riots of Blacks in 1968, remain tense (Abu-Lughod, 1999). The key issue is whether the city's African-Americans will in the future experience less racism and receive more education, employment, and other opportunities.

FIGURE 2.22   Small manufacturer in Chicago.

2.3.3 Los Angeles

Los Angeles contrasts with Chicago in many ways. The city was founded by Spaniards as the Mission San Gabriel Arcángel in 1781, located in the present-day city of San Gabriel, seven miles northeast of the present downtown Los Angeles. In 1781, 11 families departed from the mission to found El Pueblo de la Reyna de Los Angeles de Porciuncula (The Town of the Queen of the Angels of Porciuncula), near today's downtown (City of San Gabriel, 2003). Spain controlled the territory until 1822, when the newly independent Mexico claimed it. In 1847, Los Angeles became a U.S. city, and was incorporated in 1850. At that time, it was a rowdy cowtown, feeling the overflow from California's Gold Rush. Today, 150 years later, the Los Angeles urban agglomeration is the world's eleventh-largest megacity (United Nations, 2003), with 13.0 million people. Its skyline appears in Figure 2.24. Its urban area spreads over 4,320 square miles, which is somewhat smaller than the 5,498 square miles for the Chicago urban area (U.S. Census, 2002). The population of the Los Angeles-Long Beach-Riverside, CA Combined Statistical Area in 2003 is 16.4 million (U.S. Census, 2004b). As seen in Table 1.2, the United Nations projects

that Los Angeles will have a population in 2015 of 12.9 million (United Nations, 2003). The Los Angeles metropolitan area will have an estimated 18.9 million.

Adjacent to the Pacific Ocean, Los Angeles spreads over varied topography, encompassing a large coastal plain, mountain ranges, and deserts (see Figure 2.25). The Los Angeles-Long Beach-Riverside, CA Combined Statistical Area, shown in Figure 2.25, contains the five counties of Los Angeles, Orange, Riverside, San Bernardino, and Ventura. This region encompasses 33,000 square miles of both urban and rural area, with more than 140 incorporated cities (Abu-Lughod, 1999). However, the Los Angeles urban area occupies only one eighth of it (Forstall & Greene, 1997). It represents a polycentric metropolis, a topic discussed in the following chapter. As seen in the figures, Los Angeles is criss-crossed by its famous freeway system, which continues to expand into outlying areas.

Los Angeles has been built up by migration, like Chicago. The migration has been both domestic and foreign, and has varied for different groups at various times. As with nearly all large cities, a common theme has been the dominance of economic motivation for migration; in other words, Los Angeles's job pull looked better for

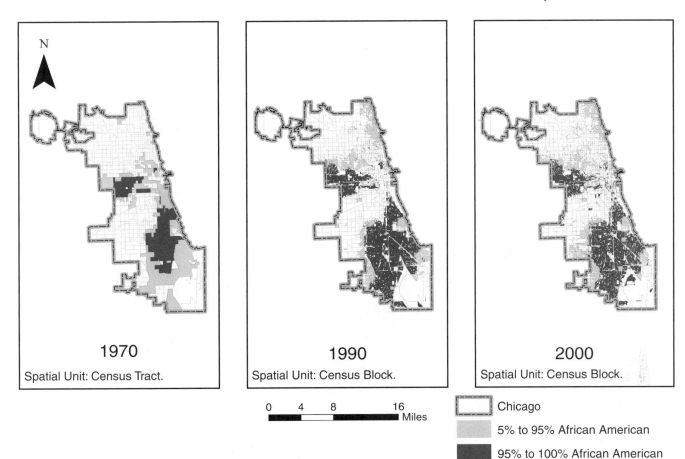

FIGURE 2.23 Chicago's African-American population: spatial changes 1970–2000. Source: Abu-Lughod, Janet L. 1999. New York, Chicago, Los Angeles: America's Global Cities. University of Minnesota Press, Minneapolis, MN. p. 336 Map 11.2a.

people contemplating migration. In the late 19th century, after the completion of the transcontinental railroad, California's citrus industry boomed, which drew population. The movie industry, which became large in the 1930s, continued to expand into today's entertainment industry. Over the last century, southern California has drawn

FIGURE 2.24 Los Angeles skyline.

many retirees attracted by the sunny climate and amenities. Yet some of them have found the cost of living to be high and have returned to the workforce. A non-job-related reason for migration before the 1930s was health, as the climate was so favorable (United Nations, 1995). During and after World War II, there was a huge manufacturing and military-related influx of migrants (United Nations, 1995).

Manufacturing in Los Angeles grew in the second half of the 20th century. Following World II, war-related manufacturing continued to develop and contributed to a diversified manufacturing sector. This diversification has enabled Los Angeles not to suffer the large Fordist-type decline of Chicago. In fact, Los Angeles avoided deindustrializing in the 1970s and 1980s, but rather had net gains in manufacturing jobs (Abu-Lughod, 1999). Although there was a decline in aerospace employment after the Cold War in the 1990s, that did not equal the proportionate losses in Chicago and other earlier manufacturing cities.

Although some of its industries are world-class, high-tech, such as aerospace, electronics, and biotechnology, other manufacturing has developed that draws on

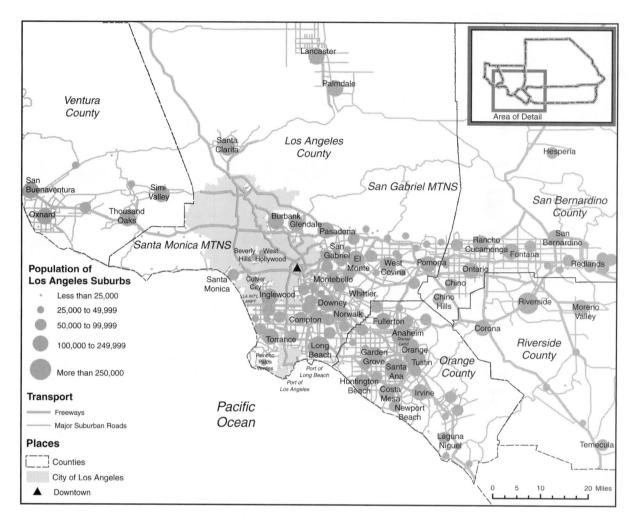

FIGURE 2.25   Los Angeles metropolitan area reference map.

the low-cost workforce, especially the large-scale garment industry. The latter is located centrally and draws disproportionately on immigrant populations (Abu-Lughod, 1999).

Los Angeles is the world center of the entertainment industry, a vast source of investment, employment, and excitement. This aspect of the metropolis has affected its urban image. Movie studios, Hollywood, Mann's Chinese Theatre, and Disneyland are all associated with the image of Los Angeles (see Figure 2.26). The Los Angeles metropolitan area also has some reputation as a tourist center that focuses on its gentle climate and proximity to beaches, surfing, and boating. The locus of tourism over the decades has shifted over time from the City of Los Angeles to outlying tourist centers such as Palm Springs, Laguna Beach, and Santa Barbara.

Los Angeles has had a major multi-ethnic aspect that has been influenced in recent decades by massive foreign immigration. This can be seen by its components of population growth in the 1980s and 1990s (see Figure 2.27).

For each of these decades, the components represent the change, in millions of persons, in natural increase (births minus deaths), net domestic migration (net flow of migrants into the area from other parts of the country), and

FIGURE 2.26   Los Angeles entertainment.

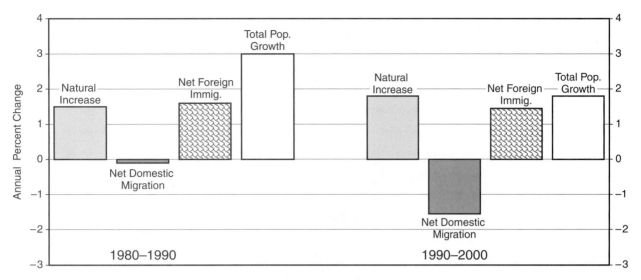

FIGURE 2.27 Components of change, greater Los Angeles metropolitan area, 1980s and 1990s. Source: Chang, Ping. 2002. The State of the Region 2002. Los Angeles: Southern California Association of Governments. Figure 5.2, p. 52. "Port Cargo in Los Angeles and Long Beach."

net foreign immigration (net flow of migrants from foreign countries). The total population growth is also shown. It is clear that in the 1980s, the Los Angeles metropolitan region had high rates of foreign immigration, combined with no net domestic migration. By the 1990s, the continuing positive foreign immigration was offset by equally negative flows of domestic migrants away from the Los Angeles metro region (Chang, 2002; Southern California Association of Governments, 2000). However, natural increase, consisting three quarters of fertility rather than mortality, grew moderately, the result partly of large Mexican and Hispanic immigration, which brought along higher fertility rates (California Department of Finance, 2003). The 1990s domestic movement away from the Los Angeles metropolitan area might stem from more job competition, shortages of affordable housing, dissatisfaction with the declining environment, and flight of White population to areas outside the metro region, some as far as Las Vegas (Pulido, 2000).

The increase in the number and percentage of foreign-born population in the five-county Los Angeles metropolitan area is graphed in Figure 2.28. The percentage of foreign-born residents grew from 18 percent in 1980 to over 30 percent in 2000, as the foreign-born population increased from about 2 million in 1980 to 5 million in 2000. The foreign-born residents are located in huge areas in the center and center west of the City of Los Angeles, as well as in the eastern San Fernando Valley and central Orange County, especially around Santa Ana, plus smaller areas near San Bernardino and Riverside (see Figure 2.29).

Hispanic population has grown especially, so that today, the City of Los Angeles has 1.7 million Hispanic people, or 46.5 percent of the total population; the

metropolitan area has 6.6 million, or 40 percent. Asians are also numerous due to immigration; the metropolitan area has 1.9 million Asians, comprising 11.5 percent. By contrast, African-Americans constitute only 8.3 percent of the metropolitan area. Los Angeles has racial tensions, as seen by L.A. Police Department controversies that recurred throughout much of the 1990s. This reached its peak in the Rodney King controversy, in which police officers were videotaped beating a suspect. The police involved were acquitted from the beating, leading to riots in Los Angeles in 1992. Later two of the officers were convicted of violating King's civil rights. Earlier, in the 1960s, the much larger Watts riots occurred, an episode that left

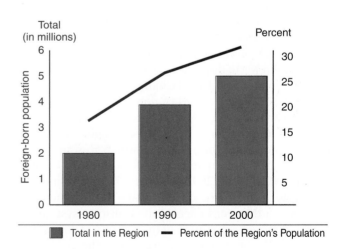

FIGURE 2.28 Total and percent foreign-born population in the Los Angeles metropolitan area, 1980–2000. Source: U.S. Census, various years.

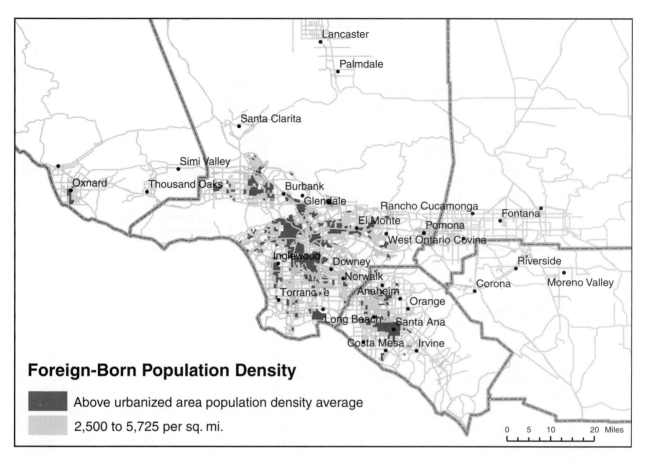

Foreign-Born Population Density

◼ Above urbanized area population density average

▨ 2,500 to 5,725 per sq. mi.

0 5 10 20 Miles

FIGURE 2.29 Foreign-born population density in Los Angeles, 2000.

major physical damage in central Los Angeles, as well as a decade or more of social impacts. However, other tensions have been more multiethnic than in Chicago, such as youth gangs of different ethnicities. As with other urban areas, Los Angeles also evidences large divisions between the rich and poor, which are often associated with race.

The Los Angeles megacity has experienced failure with transportation and environmental problems, as well as some successes. Among its environmental challenges are adequacy of water supply, energy, air pollution, inadequate solid waste disposal, as well as the natural hazards of earthquakes and fires (Chang, 2002; Southern California Association of Governments, 2000). A trend that has exacerbated the transport load is the tremendous growth in the sea cargo ports of Los Angeles and Long Beach, among the world's largest, which grew in volume from about 75 million tons in 1990 to about 150 million tons in 2001 (Chang, 2002; see Figure 2.30). This increase brings along with it growth in trucking through the urban area, which threatens parts of the transport system and points to more air pollution in the future. Yet, the air pollution in the metropolitan area over the past three decades has dropped considerably (see Figure 2.31) due to more stringent vehicle and industrial restrictions.

With such substantial population growth, the metropolitan area's transport situation has worsened. Metropolitan area commute times grew by 12 percent from 1990 to 2000 to average 29 minutes one way. In Los Angeles County, about 30 percent of one-way trips exceed one hour (Chang, 2002). There is a limited system of public commuter rail, urban rail, and rapid buses (Southern California Association of Governments, 2000; see Figure 2.32). Many observers feel that the rail systems are outdated due to their inability to serve much of the vast dispersion of the Los Angeles urban area.

In summary, Los Angeles metro region has moved from a tiny mission to a global megacity. Along with the growth have come opportunities, world-class business leadership in some sectors, and strong science, education and technology, while on the other hand, the city faces crime, ethnic tensions, and problems in providing water and facilitating transport, as well as the sheer ability to manage such a vast geographic area.

2.3.4 Mexico City

Mexico City exceeds Los Angeles in size as the world's second-largest megacity with a population of 18.7 million

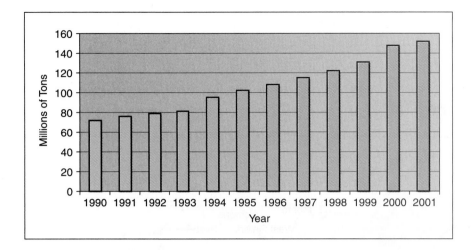

FIGURE 2.30 Port Cargo at Los Angeles and Long Beach Seaports, 1990–2001, in millions of tons. Source: Chang, Ping. 2002. The State of the Region 2002. Los Angeles: Southern California Association of Governments. Figure 5.2, p. 52. "Port Cargo in Los Angeles and Long Beach."

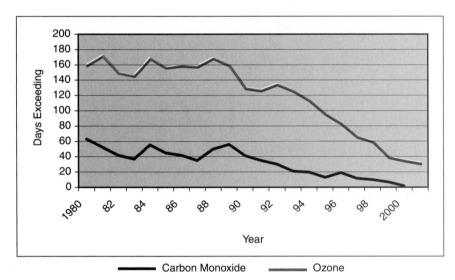

FIGURE 2.31 Number of days exceeding federal air pollution standards, South Coast Air Basin, 1980–2001. Source: Chang, Ping. 2002. The State of the Region 2002. Los Angeles: Southern California Association of Governments. Figure 5.3, p. 58. "Number of days exceeding federal standards in the SCAG region."

Note: Ozone data represents the total number of days the Federal 1-hour standard was exceeded at all monitoring stations in the South Coast Air Basin

in 2000. It is exceeded only by Tokyo at 35.0 million (United Nations, 2003). It is located in a valley at 7,000 feet on an ancient lake bed that still contains marshy areas (Figure 2.33). It is surrounded to the south by a semicircle of volcanic mountains, with the huge volcanoes of Popocatépetl and Ixtaccihuatl in the southeast, towering at 17,930 and 17,159 feet, respectively. Administratively, the Mexico City urban area falls into two Mexican states. The Federal District contains the original city center called the Zócalo, which housed the centers of the original Aztec as well as Spanish Colonial city. Surrounding the Federal District to the west, north, and east is the State of Mexico (see Figure 2.33). That state contains many *municipios* (i.e., the Mexican equivalent of the U.S. county) that are today part of the metropolitan area of Mexico City. There is no standard federal

government definition of which municipios are included in the Mexico City metropolitan area. We have utilized an INEGI-defined metropolitan area in 2000 that associates 34 municipios with Mexico City (INEGI, 2001). Also shown on the maps are the 16 delegations, or city wards, within the Federal District.

Mexico City is an ancient city. In the Aztec civilization, it was founded in the 13th century as the city of Tenochtitlán, located on a large lake, Lake Texcoco. The city had a well-developed civilization and population of about 50,000 when in 1522, Hernando Cortés conquered it and all of Mexico for Spain. Shortly thereafter, its population was halved, due partly to military losses, but mostly to smallpox brought by the invaders. During the Spanish era, which lasted until Mexican independence in 1823, Mexico City became the most important center of

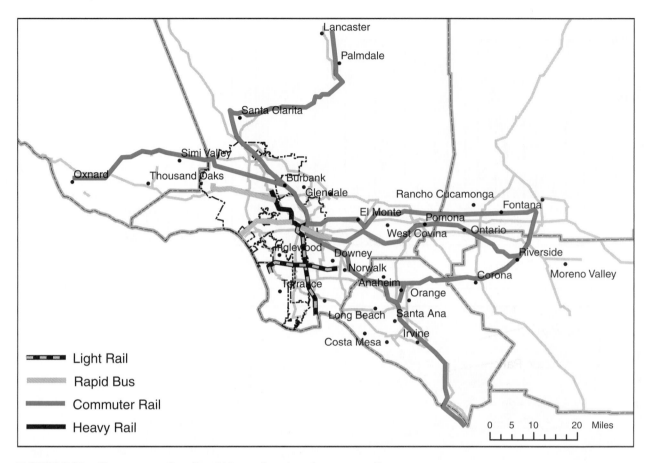

FIGURE 2.32   Commuter rail and rapid buses, Los Angeles metropolitan area.

colonial Spain. Spanish rule brought with it features of European cities of the time, leading to a new design for the city center, based on Spanish architecture. The population grew gradually to reach 56,000 in 1650 and 165,000 in 1823 (Garza, 2001). By 1900, the city had 344,000 residents. Mexico City today remains influenced by its Aztec and earlier roots, and especially by the three centuries of Spanish rule (Garza, 2001; Ward, 1998). These have shaped the culture of the nation and city, in such fundamental aspects as language, architecture, educational traditions, music and dance, culinary arts, street commerce, and other profound ways. Contemporary Mexico City appears in Figure 2.34.

In the 20th century, Mexico City's population surged to 3.1 million in 1950, 8.8 million in 1970, and 18.3 million in 2000 (INEGI, 2002). It has always far surpassed the other leading Mexican cities and is the country's unchallenged *primate city* (see Figure 2.35), a term which refers to a city that dominates a country's economy and population. There are a number of measures of primacy, with the simplest being the primate city's proportion of national population, which in 2000 was very large at 19 percent for Mexico City. The other cities in the urban system of a country are related to the primate city by the rank-size

rule originated by Zipf in 1949. According to this rule, the second largest city is estimated to be half the size of the primate city; the third largest city is one third the size of the primate city, the fourth largest one fourth the size, and so on. In the year 2000, the next three largest Mexican metropolitan areas of Guadalajara (3.669 million), Monterrey (3.237 million), and Puebla (2.343 million) had a combined 2000 population of only 9.249 million, about half of Mexico City's population. These secondary cities are all substantially smaller than the rank-size rule predicts. For instance, the rank-size rule would predict Guadalajara's population at 9.15 million. This implies that Mexico City has greater primacy than is theoretically expected. Because Mexico City has high primacy, its impact on the Mexican economy is large. Mexico's economic leaders are mostly located in the city, and because it is the seat of the federal government, national economic policy is established there. Its dominance is seen by the Federal District accounting for 71 percent of finance and insurance revenues, 60 percent of professional and technical revenues, and 44 percent of real estate revenues (Pick & Butler, 1997, p. 278), and the Mexico City metropolitan area contained 41 percent of the country's major corporate headquarters in 1997

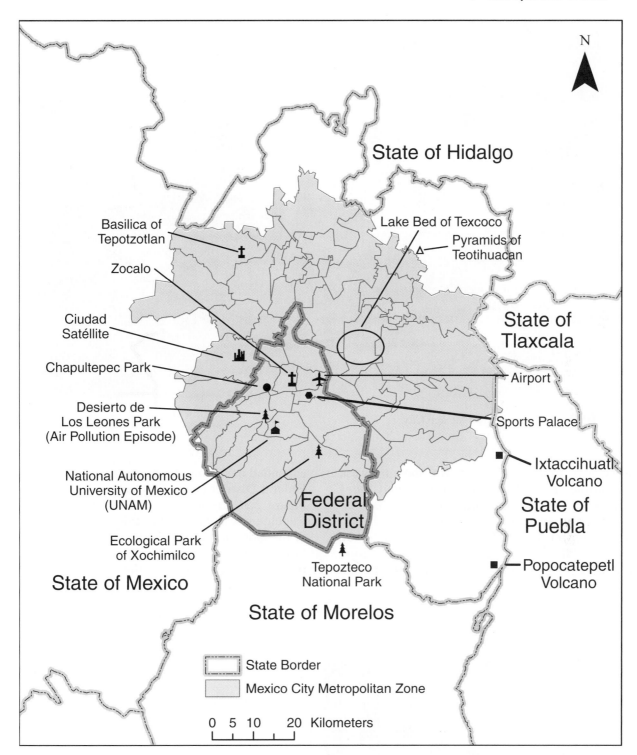

FIGURE 2.33   Mexico City metropolitan area reference map.

(Butler et al., 2001). Mexico City is involved in a significant amount of all the economic transactions and activity in the nation.

The huge population growth is also reflected in the massive growth in metropolitan land area (Ward, 1998). As seen in Figure 2.36, the 1900 land area encompassed

the original Aztec and colonial city, centering on the Zócalo and surrounding areas, and comprising about 5 percent of the northern Federal District. By 1950, this area had grown about sixfold, still located in the northern Federal District. However, 35 years later in 1985, the urban area included more than half of the Federal District and

FIGURE 2.34 Downtown Mexico City.

extensive areas to the northeast, north, and northwest. This territorial growth has continued unabated, today reaching much further to the north and northeast.

Within the urban area, the population in the Federal District did remain stable in the 1980s and 1990s, whereas the State of Mexico portion continued to grow, by around 5 million (see Figure 2.37). As the urban area

expanded in size and land area, the central core of the Federal District began to lose population. As seen in Figure 2.38, from 1960 to 1980, the delegations of Cuauhtemoc and Miguel Hidalgo, which contain much of the modern business center of the Federal District, lost population, whereas a ring of surrounding delegations and municipios were the biggest gainers. In the

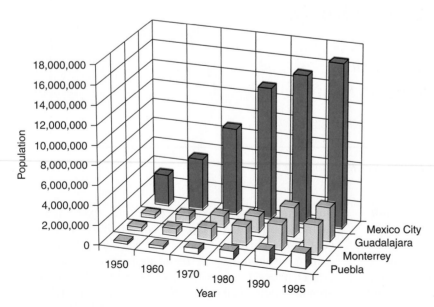

FIGURE 2.35 Population of Mexico's four largest metropolitan areas, 1950–1995. Source: INEGI, 1999; INEGI, 2003.

FIGURE 2.36   The physical expansion of Mexico Megacity, 1900 to 1986. Source: Ward, Peter. 1998. *Mexico City*. John Wiley and Sons. P. 51. Figure 2.2. "The physical expansion of the Metropolitan area 1900–1985."

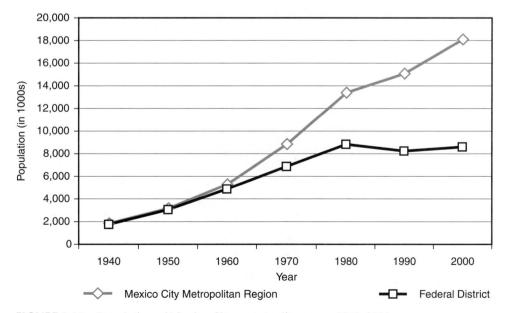

FIGURE 2.37   Population of Mexico City metropolitan area, 1940–2000.

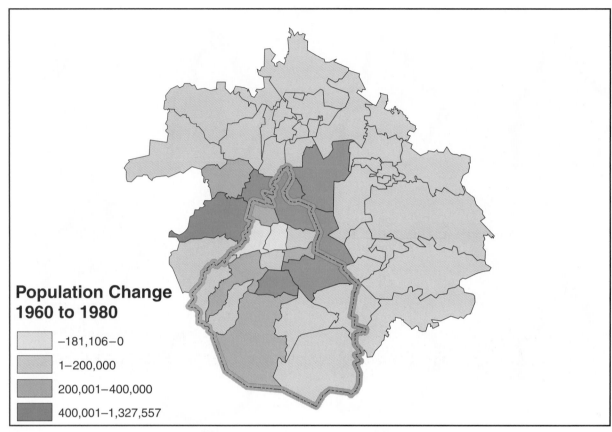

**Population Change
1960 to 1980**

- −181,106−0
- 1−200,000
- 200,001−400,000
- 400,001−1,327,557

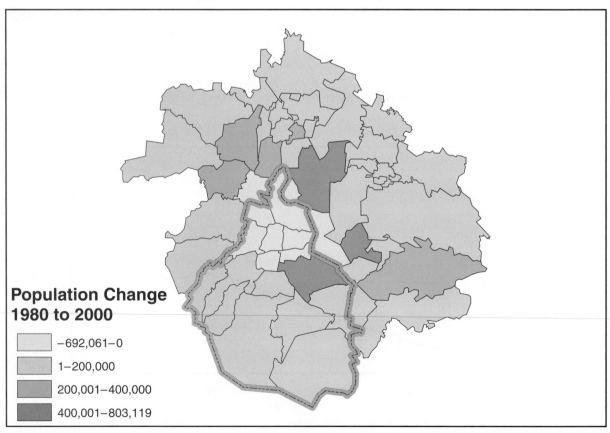

**Population Change
1980 to 2000**

- −692,061−0
- 1−200,000
- 200,001−400,000
- 400,001−803,119

FIGURE 2.38   Spatial population change, Mexico City metropolitan area, 1960–2000.

FIGURE 2.39   Youths in a park in Mexico City's deconcentrating core.

1980 to 2000 period, an even larger deconcentrating core, comprising seven delegations and two municipios, was a source of migrants to the growth ring that had moved outward and more into the State of Mexico. This deconcentrating core and expanding outward rings of growth reflect a 40-year process that is likely to continue (Garza, 2001). In Figure 2.39, youths are seen in Chapultepec Park, located in the deconcentrating core.

By 1980, more and more State of Mexico municipios were part of the expanding urban boundaries of Mexico City (Garza, 2001). During this time, Mexico City moved from the fifth-largest megacity in 1970 up to second in 2000 (United Nations, 2003). Projections by the Mexican

government shown in Figure 2.40 indicate that the Mexico City metropolitan area will grow at a reduced rate to reach a population of 22.9 million in 2030, with all of the approximately 5 million population growth in the period between 2000 and 2030 coming from expansion into the State of Mexico. Garza (2001) went beyond this to foresee the expansion of Mexico City to form a megalopolis extending beyond the State of Mexico to encompass the five peripheral cities of Querétaro in the State of Querétaro, Pachuca in the State of Hidalgo, Tlaxcala in the State of Tlaxcala, Puebla in the State of Puebla, and Cuernavaca in the State of Morelos (see Figure 2.41). This megalopolis will constitute a subsystem of cities that are

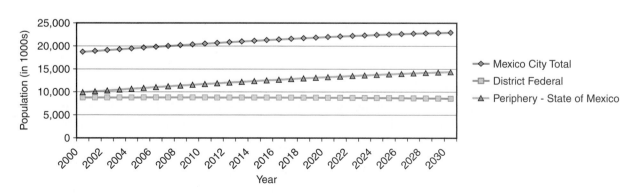

FIGURE 2.40   Projected population of Mexico City metropolitan area, 2000–2030. Source: CONAPO, 2003.

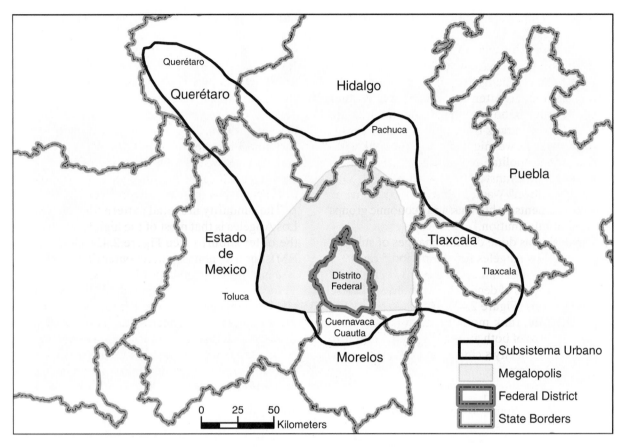

FIGURE 2.41   Mexico City as a megalopolis. Source: Garza, Gustavo. 2001. "La Megalopolis de la Ciudad de Mexico en el Ocaso del Siglo XX," in Gomez de Leon Cruces, Jose and Cecilia Rabell Romero (eds.), La Poblacion de mexico, Mexico City, D.F.: Fondo de Cultura Economica. Carretera Picacho-Ajusco, 227, 14200 Mexico, D.F., Mapa 2, p. 621.

linked together by job commuting and close communications. It would preserve the dominance of Mexico City, while increasing the population of the region to approach 25 to 30 million. It could approach the size of the megalopolis many recognize on the eastern seaboard of the United States (Gottmann, 1957).

Mexico City in the 21st century presents many problems and prospects. Among the prospects are improved transport, better pollution and environmental control, and better municipal management (Pick & Butler, 1997). At the same time, there are huge challenges that include economic weakness, poverty, deteriorating infrastructure, deficits of water, solid waste disposal, and traffic congestion (Garza, 2001; Pick & Butler, 1997; Ward, 1998). One small indication of the planning challenge is the effort of the Federal District government to phase out highly polluting and sometimes criminally dangerous Volkswagen Bugs (Weiner, 2003). They have the old car design that originated in the 1950s (see Figure 2.34). A small percentage of passengers, Mexican and foreign alike, have been assaulted and robbed in the VW taxis for many years. There are currently

70,000 bugs in Mexico City, most without catalytic converters, safety glass, or seat belts, but appealing to taxi drivers since new they only cost $7,500 (Weiner, 2003). This is a small example of tests that will challenge the long-term will of the city government. The answers to this and numerous other problems are unresolved. The bottom line for the megacity's future might reside in its leadership and its will to plan better, and govern strongly with a long-term perspective (Pick & Butler, 1997; Ward, 1998).

2.4 ANALYZING AN URBAN ISSUE: CHANGES IN POPULATION RANKS AMONG AREAS OF CHICAGO AND LOS ANGELES

Between 1990 and 2000, the populations of the Chicago and Los Angeles metropolitan areas grew by 11.0 percent and 12.7 percent, respectively (U.S. Census, 2004b). Most of the growth between 1990 and 2000 in both metropolitan areas occurred in the suburban areas well outside the city limits. In the case of Chicago, the growth occurred mostly in far western and far northern suburbs. For Los Angeles, the most rapid growth took place in dispersed urban areas to the east and northeast.

Although a simple approach would be to just map the percentage population change for small areas within the two metropolitan areas, often city planners and other parties are interested in how suburban communities are moving up and down in rank from year to year or census to census. For instance, we might ask how the rank within the Chicago metropolitan region changed for Chicago's north suburb of Evanston between 1990 and 2000. Did it move up or down? Was it high or low to start with? Often city governments, politicians, and economic groups seek this kind of information.

A GIS analysis was done of rank changes of suburbs for Chicago and Los Angeles for the period from 1990 to 2000 to identify rings and sectors of edge city rapid growth that offset areas of decline in the interior of these metropolitan regions (Figure 2.42). This map shows the spatial pattern of the rank mobility index (RMI) computed for the suburbs of both areas (Marshall, 1989). The RMI is a measure of a city's change in population rank among a group of cities:

$$M = \frac{(R1 - R2)}{(R1 + R2)}$$

where

M = RMI
R1 = city's rank at time 1
R2 = city's rank at time 2

An RMI value can range from −1.0 to +1.0. A negative RMI value indicates a decrease in rank, whereas a positive RMI value reflects an increase in rank. An RMI value of 0 indicates no change in rank.

It is clear from Figure 2.42 that, for the Chicago metropolitan area, RMI has a considerably different pattern than for the Los Angeles metropolitan area. For instance, Chicago's areas of greatest growth in RMI are to the west, whereas Los Angeles's largest RMI growth is in fairly scattered suburbs to the southeast, east, and north.

The use of the RMI is especially appropriate for studying the changing dominance of suburbs because RMI values are larger when higher rank increases are involved. A rank change carries more weight when it is higher on the list of rankings. The larger the city, the more difficult it is to overtake. For example, let us examine the city of Aurora in the Chicago metropolitan area. In 1990, Aurora was the third-ranked city in the metropolitan area (i.e., the second-largest suburb of Chicago). In 2000, it was the second-ranked city (first-ranked suburb of Chicago), resulting in an RMI value of 0.20 for the period from 1990 to 2000. Let's examine another city that also moved up

one position, Naperville. Naperville's rank increased from fourth to third, however, its RMI was only 0.14. An equal-sized change in ranking did not result in an equally large RMI, reflecting that it is more difficult to overtake a larger city. Incidentally, Gary, Indiana, was the second-largest city (i.e., first-ranked suburb of Chicago) in 1990, but as Aurora and Naperville moved up the suburban hierarchy in the 1990s, Gary descended to rank at position five. Not coincidentally, the U.S. Office of Management and Budget (OMB) changed the name of the metropolitan area from Chicago–Gary–Kenosha (as named in 1993) to Chicago–Naperville–Joliet in 2003 (see Chapter 5).

The similarity in spatial pattern for both Chicago and Los Angeles is that most of the high RMIs are located in the outer suburbs (see Figure 2.43). Chicago's highest RMIs form an almost perfect outer ring, whereas the pattern is more interrupted for Los Angeles, for reasons explained in Chapter 3. The growth of edge cities (suburban employment centers), a topic introduced in Chapter 10, helps explain these general patterns. The downtown of Chicago is the largest employment center for its metropolitan area, but the share is a lot less for downtown Los Angeles and the share of jobs in the edge cities of both places is growing.

RMI analysis can be done based on jobs rather than population, which would show these disparities. Thus the spatial concentration of high RMI clusters indirectly reflects the changing patterns of employment growth in outlying edge cities or employment centers. The pattern of declining RMIs might reflect areas of aging residential suburbs. Industrial or manufacturing areas were designed to facilitate industry. Residential suburbs often located themselves close to these industrial centers, but once the industry died so did the residential areas (Harris, 1994). Because suburbs can be ranked on a variety of characteristics, RMI can be applied to any urban characteristic expressed on a numerical scale.

Focusing on the Chicago area alone illustrates the power of the RMI for spatial analysis. Consider, for instance, the half-arc stretching from Waukegan to Joliet, with RMIs that were all greater than the average (Figure 2.43). Many of the cities included in this arc had positive RMI values that were three standard deviations over the average RMI, including Waukegan, Crystal Lake, Elgin, West Chicago, Aurora, Naperville, and Joliet. Reasons for these positive climbs up the suburban hierarchy include proximity to the region's interstate highways (e.g., see Elgin, Aurora, and Naperville). Other variables can also be considered in the interpretation for why certain suburbs overtook others in population during the time period between 1990 and 2000. For instance, immigration played an important role in increasing the RMIs of Elgin, West Chicago, Aurora, and Cicero three standard deviations over the average (Figure 2.43). In all of these

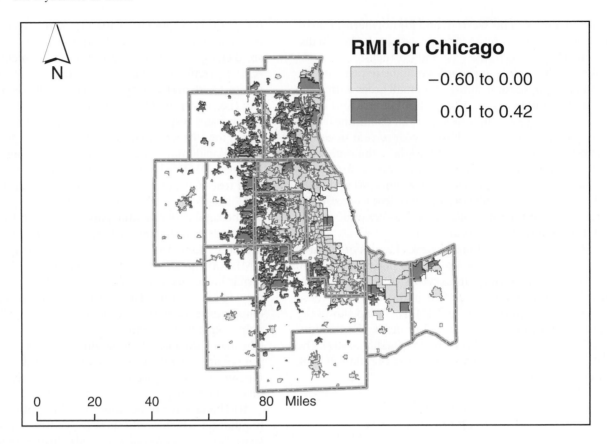

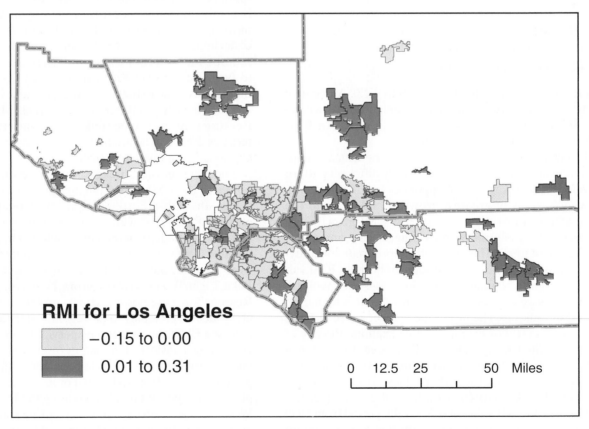

FIGURE 2.42   Rank mobility indexes for Chicago and Los Angeles, 1990–2000.

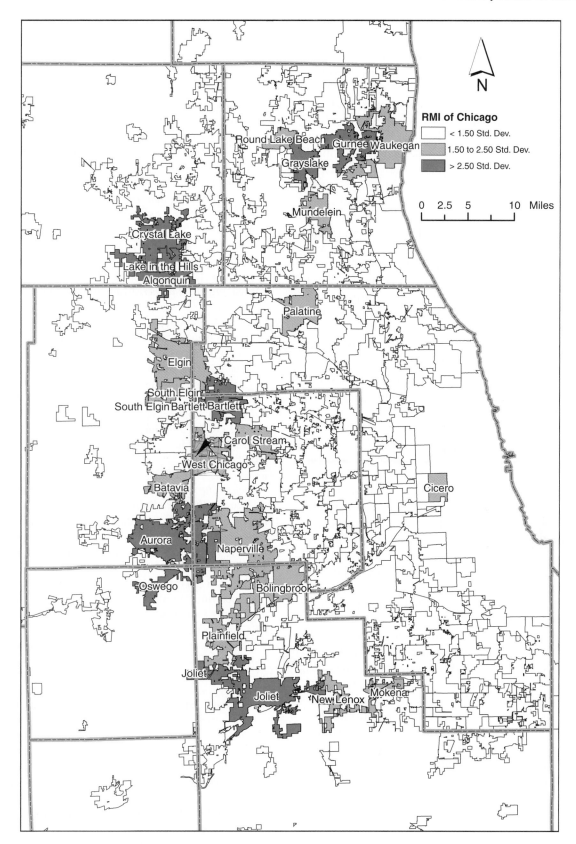

FIGURE 2.43 Chicago communities in 2000 with RMIs 1.5 or more standard deviations above the mean RMI.

suburbs, the increase in foreign-born population was substantial, especially among Latino populations.

The RMI allows for easy comparisons of city dynamics within a metropolitan area. The measure gives insight into the changing suburban hierarchy within a metropolitan area and identifies cities and suburbs that are strengthening or weakening relative to others. For some purposes, this ranking of places is more important than their total percentage gains or losses, such as determining the best place for consumer market gains in which to locate a business, or identifying the place with the worst declines in residential tax base. RMI is a good indicator of population, economic change, or social change, depending on which characteristic is chosen for the index. It constitutes a good measure for this chapter, which was concerned with change in cities.

In this chapter's exercise, students will compute RMI maps for population, to show rank changes during the 1990s for suburbs in the Chicago and Los Angeles metropolitan areas. The focus is on comparing two of the metropolitan areas that were profiled in this chapter, and critiquing the use of the RMI index for urban analysis. The exercise and its suggested essays draw on concepts and facts from the two chapter parts on change in cities and megacity profiles.

SUMMARY

Chapter 2 contributes further to the book's foundation by introducing change in cities, in particular economic change, demographic change, and social change, including the potential for polarization and even confrontation. A number of theories and concepts are presented from economics and demography that are helpful in understanding urban change processes.

The three megacities of Chicago, Los Angeles, and Mexico City constitute useful examples of city growth and change. They are utilized throughout the book to illustrate spatial organization, geographical concepts, urban systems, real events, and case examples, and hence are emphasized in this chapter. The book considers a variety of other cities in the United States and overseas. The three large cities have contrasting geographic settings, histories, and recent growth and change. All are world or at least world-regional leaders for certain urban phenomena. For example, Chicago is a cultural and financial leader, along with its strength in certain business sectors. Los Angeles is a world capital for entertainment, aerospace, and certain high-tech sectors. Mexico City is a leading cultural and intellectual city in Latin America, as well as a major manufacturing center for the western hemisphere. At the same time, each city faces its own particular problems and challenges, many of which constitute topics of discussion throughout this text.

REFERENCES

Abu-Lughod, J. L. (1999) *New York, Chicago, Los Angeles: America's global cities*. Minneapolis, MN: University of Minnesota Press.

Butler, E. W., Pick, J. B., and Hettrick, J. W. (2001) *Mexico and Mexico City in the world economy*. Boulder, CO: Westview Press.

California Department of Finance. (2003) *Live Births by County and Deaths by County*. Sacramento, CA: Author.

Chandler, T. (1987) *Four Thousand Years of Urban Growth: An Historical Census*. Lewiston, NY: St. David's University Press.

Chang, P. (2002) *The State of the Region: Measuring Progress in the 20th Century*. Los Angeles: Southern California Association of Governments.

City of San Gabriel. (2003) "The Mission San Gabriel Arcangel." *http://www.sangabrielcity.com*.

Coleman, T. (2002) "Hiawatha Light Rail: Hmong is rail's third tongue." *Pioneer Press*. *http:/www.twincities.com/mld/pionerpress*.

Forstall, R. L., Greene, R. P. (1997) "Defining job concentrations: The Los Angeles case." *Urban Geography* 18 (8): 705–739.

Forstall, R. L. (2003) Special compilation of U.S. Census data provided to the authors.

Garza, G. (2001) "La megalópolis de la Ciudad de México en el ocaso del siglo XX," in *La Población de México: Tendencias y Perspectivas Sociodemográficas Hacia el Siglo XXI*. ed. by J. Gómez de León Cruces and C. Rabell Romero, pp. 605–632. Mexico City, D.F.: Consejo Nacional de Población.

Gottdiener, M., Collins, C.C., and D.R. Dickens (1999). *Las Vegas: The Social Production of an All-American City*. London: Blackwell Publishers.

Gottmann, J. (1957) "Megalopolis, or the urbanization of the Northeastern seaboard." *Economic Geography* 33:189–200.

Harris, R. (1994) "Chicago's other suburbs." *Geographical Review* 84:394–410.

Hughes, A. (2003) "Hmong in Minnesota still face language and cultural barriers." *Minnesota Public Radio News*. *http://news.mpr.org*

INEGI. (1999) *Estadisticas Historicas de Mexico* (2 volumes). Aguascalientes, Aguascalientes: Instituto Nacional de Estadísticas, Geográfia, e Informática.

INEGI. (2001) *Estadísticas del Medio Ambiente del Distrito Federal y Zona Metropolitana, 2000*. Aguascalientes, Aguascalientes: Instituto Nacional de Estadísticas, Geográfia, e Informática.

INEGI. (2002) *XII Census de Población y Vivienda. Tabulados Básicos* (3 volumes). Aguascalientes, Aguascalientes: Instituto Nacional de Estadísticas, Geográfia, e Informática.

Jones, H. (1990) *Population Geography* (2nd ed.). London: Paul Chapman.

Knox, P. (1994) *Urbanization: An introduction to urban geography*. Englewood Cliffs, NJ: Prentice Hall.

Knox, P., and Agnew, J. (1998) *The Geography of the World Economy* (3rd ed.). London: Wiley.

Kurian, G. T. (1994) *Datapedia of the United States, 1790–2000*. Lanham, MD: Bernan Press.

Lafrenz, J. (2001"Chicago: Transformation processes of the metropolis in the American Midwest." English summary translation in *Hamburg und Seine Partnerstadte Sankt Petersburg, Marseille, Shanghai, Dresden, Osaka, Léon, Prag, Chicago*," pp. 483–502. Hamburg, Germany: Selbstverlag Institut für Geographie der Universität Hamburg.

Lee, G. Y. (2003) *Refugees from Laos: Historical Background and Causes. http://www.hmongnet.orgn/hmong-au/refugee.*

Marshall, J. U. (1989). *The Structure of Urban Systems.* Toronto: University of Toronto Press.

Mehretu, A., and Mutambirwa, C. (2003) "Cities of Sub-Saharan Africa," in *Cities of the World: World Regional Urban Development.* ed. by S. D. Brunn, J. F. Williams, and D. J. Zeigler, pp. 293–330. Lanham, MD: Rowman & Littlefield.

Moehringer, J. R. (2003, July 30) "Life with age-old problems." *LATimes.com.*

Nam, C. B. (1994) *Understanding Population Change.* Itasca, IL: Peacock.

Office of Management and Budget. (2003, June 6) "Revised definitions of metropolitan statistical areas, new definitions of micropolitan statistical areas and combined statistical areas, and guidance on uses of the statistical definitions of these areas." *OMB Bulletin No. 03-04. http://www.whitehouse.gov/omb/bulletins/b03-04.html.*

Pick, J. B., and Butler, E. W. (1997) *Mexico Megacity.* Boulder, CO: Westview.

Population Reference Bureau. (2003) *2002 World Population Data Sheet.* Washington, DC: Author.

Potter, R. B., and Lloyd-Evans, S. (1998) *The City in the Developing World.* Harlow, England: Addison Wesley Longman.

Pulido, L. (2000) "Rethinking environmental racism: White privilege and urban development in southern California." *Annals of American Geographers* 92(1):12–40.

Quilodrán, J. (1991) Niveles de Fecundidad y Patrones de Nupcialidad en Mexico. Mexico City, Mexico: El Colegio de Mexico.

Sassen, S. (2001) *The global city.* Princeton, NJ: Princeton University Press.

Southern California Association of Governments. (2000) *Regional Comprehensive Plan and Guide: Energy.* Los Angeles: Southern California Association of Governments.

Stein, B. (2002, July 29) "Still hot at 50." *Recycling Today. http://www.recyclingtoday.com.*

Stren, R., and Halfani, M. (2001) "The cities of Sub-Saharan Africa: From dependency to marginality," in *Handbook of Urban Studies.* ed. by R. Paddison, pp. 466–485. London: Sage.

Strohl, L. (2000) "Asians ascending." *Horizon Magazine. http://www.horizonmag.com.*

United Nations. (1995) "The challenge of urbanization: The world's large cities." *Population Division Report ST/ESA/SER.A/151.* New York: United Nations.

United Nations. (2002) *2000 Demographic Yearbook* (52nd edition). New York: United Nations.

United Nations. (2003) "World urbanization prospects: The 2003 revision." New York: United Nations.

U.S. Census. (2000) *Historical Statistics of the United States.* Washington, DC: Author.

U.S. Census. (2002) *2000 Census of Population and Housing.* Washington, DC: Author. *http://www.census.gov.*

U.S. Census. (2004a) *Metropolitan and Micropolitan Statistical Area Definitions. http://www.census.gov.*

U.S. Census (2004b) "Ranking Tables for Population of Metropolitan Statistical Areas, Micropolitan Statistical Areas, Combined Statistical Areas, New England City and Town Areas, and Combined New England City and Town Areas: 1990 and 2000." Washington, D.C. U.S. Census Bureau. *http://www.census.gov.*

Ward, P. (1998) *Mexico City* (2nd edition). Chichester, England: Wiley.

Warf, B., and Holly, B. (1997) "The rise and fall and rise of Cleveland." *The Annals of the American Academy of Political and Social Science* 551: 208–221.

Weeks, J. (2001) *Population: An Introduction to Concepts and Issues* (8th edition). Belmont, CA: Wadsworth.

Weiner, T. (2003, January 5) "Mexico City's VW Bugs are headed for extinction." *New York Times. www.nytimes.com.*

Wheeler, J. O. (2002) "From urban economic to social/cultural urban geography." *Urban Geography* 23(2):97–102.

Zukin, S. (2002) "Our World Trade Center," in *After the World Trade Center.* ed by M. Sorkin and S. Zukin, pp. 13–22. New York: Routledge.

EXERCISE

Exercise Description

Mapping and interpreting rank changes for Chicago and Los Angeles Suburbs: 1990 and 2000.

Course Concepts Presented

Rank Mobility Index (RMI)

GIS Concepts Utilized

Table edits, calculations, string manipulations, and queries.

Skills Required

ArcGIS Version 9.x

GIS Platform

ArcGIS Version 9.x

Estimated Time Required

1 hour

Exercise CD-ROM Location

ArcGIS_9.x\Chapter_02\

Filenames Required

1. Los Angeles: La_places.shp, Co_la_sp83.shp, lacity.shp, and LosAngeles_names.mdb
2. Chicago: Ch_places.shp, chcity.shp, Cnty13.shp, and Chicago_names.mdb

Background Information

A comparison of 1990 and 2000 population census numbers shows a general decline in the population ranks of suburbs close to the "inner city," whereas outer suburbs have experienced the greatest increases in population rank. In this exercise you will be working with the population ranks of suburbs in Chicago and Los Angeles.

Rank Mobility Index (RMI): Calculated by using a city's population rank at two particular times.

$$M = \frac{(R1 - R2)}{(R1 + R2)}$$

where

M = RMI

R1 = city's rank at time 1

R2 = city's rank at time 2

An RMI value can range from −1.0 to +1.0. A negative RMI value indicates a decrease in rank and a positive RMI value indicates an increase in rank. An RMI value of 0 indicates no change in rank. The use of the RMI is especially appropriate for studying the changing dominance of suburbs because RMI values are larger when higher rankings are involved. A rank change carries more weight when it is higher on the list of rankings. The larger the city, the more difficult it is to overtake.

Readings:

(1) Forstall, R. L., and Greene, R. P. (1997) "Defining job concentrations: The Los Angeles case." *Urban Geography* 18:705–739.

(2) Getis, A. (1986) "The economic health of municipalities within a metropolitan region: The case of Chicago." *Economic Geography* 62: 52–73.

(3) Greene & Pick, *Exploring the Urban Community: A GIS Approach: Analyzing an Urban Issue,* Chapter 2.

(4) Harris, R. (1994) "Chicago's other suburbs." *Geographical Review* 84:394–410.

(5) Marshall, J. U. (1989) *The Structure of Urban Systems.* Toronto: University of Toronto Press. (Rank mobility index explained on pp. 37–42).

(6) Schiesl, M. J. (1991) "Designing the model community: The Irvine company and suburban development, 1950–1988," in *Postsuburban California: The Transformation of Orange County Since World War II.* ed by R. Kling, S. Olin, and M. Poster, pp. 55–91. Berkeley: University of California Press.

Exercise Procedure Flowchart

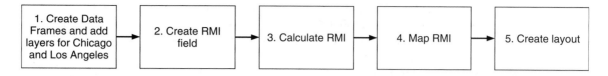

Exercise Procedure

1. Start ArcMap with the "Start Using ArcMap with a new empty map" checked. Click OK.

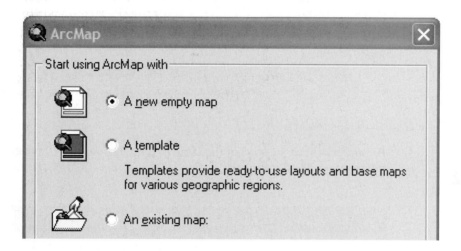

2. Right-click on Layers.

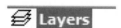

3. Click Properties.
4. Click the General tab, and change the name "Layers" to "Los Angeles."

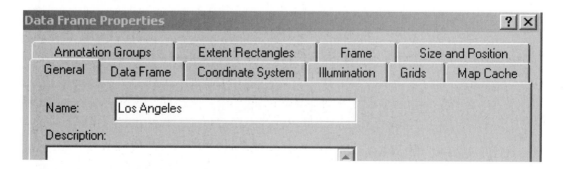

5. Click OK.
6. Click the Add Data button and navigate to the exercise data folder.

__Note:__ If you do not see the folder where you placed the exercise data, you will need to make a new connection to the root directory that holds the data.

7. Add La_places, co_la_sp83, and lacity from the Chapter 2 folder. You can add all three at once by clicking on one of the file names, then click on the others while holding down the Ctrl key on the keyboard and click the Add button.

8. Ensure that the co_la_sp83 layer appears first in the list under the Los Angeles data frame. If it is not then you can click on a layer name in the table of contents and drag the layer to a new position.

9. Right-click on co_la_sp83 in the table of contents and click Properties. Click the Symbology tab and then the Symbol box.

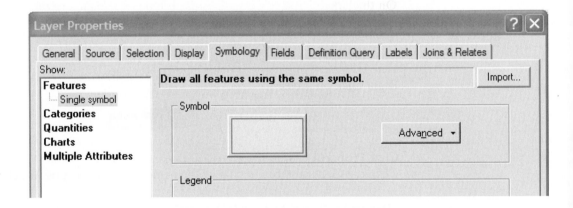

10. On the Symbol Selector dialog box, click the Hollow box. Then click the Properties button.

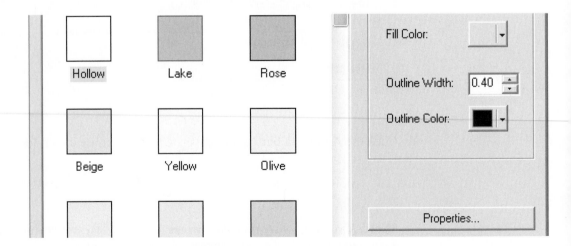

11. On the Symbology Property Editor dialog box, click the Outline button.

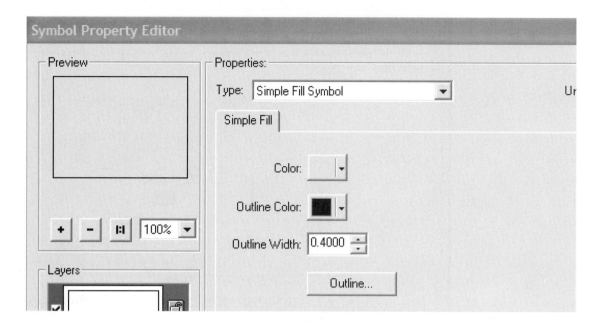

12. On the Symbol Selector dialog box, scroll down and choose the symbol for county boundaries. Click OK on each box as you return to the map.

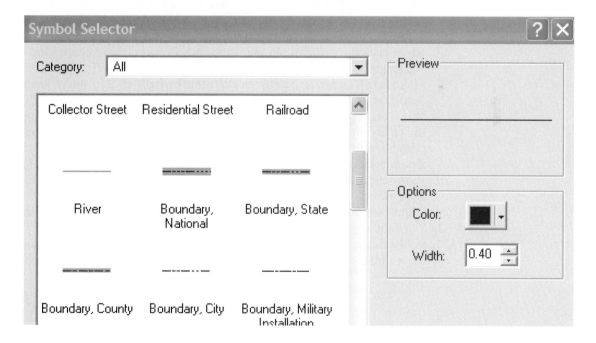

13. Next make lacity the next in order below the county layer (click its name and drag into position). Click the symbol box below the lacity layer and on the Symbol Selector dialog box change the fill color to white (the City of Los Angeles is excluded from the rank analysis; we are only examining suburbs). This will be useful when you make your final map.

14. Click the Add Data button again and double click LosAngeles_names.mdb and add CityNames_LosAngeles (this is an annotation feature class contained within the geodatabase LosAngeles_Names.mdb)

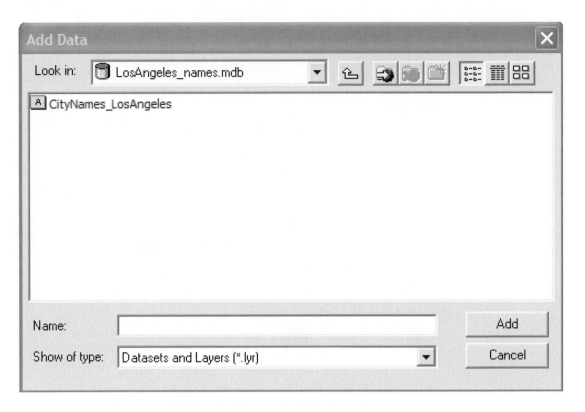

15. Once the City Names are added, zoom in on a portion of the map and pan around the area to get oriented to the map. Zoom back out to the full map extent before moving onto the next step.
16. Select Data Frame under Insert to insert a new data frame.
17. Right-click on New Data Frame, then click Properties and change the data frame's name to "Chicago" and click OK.
18. Make sure Chicago is the active data frame by right-clicking and selecting Activate. Click the Add Data button and add Ch_places.shp, cnty13.shp, and chcity.shp. Repeat the process above that you did for Los Angeles to make a county outline map (cnty13 layer) and change the chcity layer to a solid white fill. Again, add the CityNames_Chicago annotation feature class last and zoom and pan around the map to familiarize yourself with Chicago's suburbs. Zoom back out to the full map extent before moving onto the next step.
19. Right-click on the Los Angeles data frame and select Activate to activate the Los Angeles data frame.
20. Right-click on La_places in the Los Angeles data frame and select Open Attribute Table.
21. Click the Options button and select Add Field.
22. Add a field called RMI with a type of float, a precision of 12, and a scale of 5. Precision gives the number of digits in the number. Scale refers to the number of decimal places.

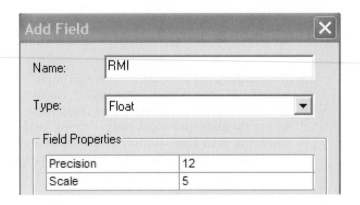

23. Click OK.
24. Click the Options tab below and click Select by Attributes. In the Select by Attributes dialog box, double-click "Rank1990" and single-click "<>" and double-click "Rank2000" (rank in 1990 not equal to rank in 2000) and click Apply.

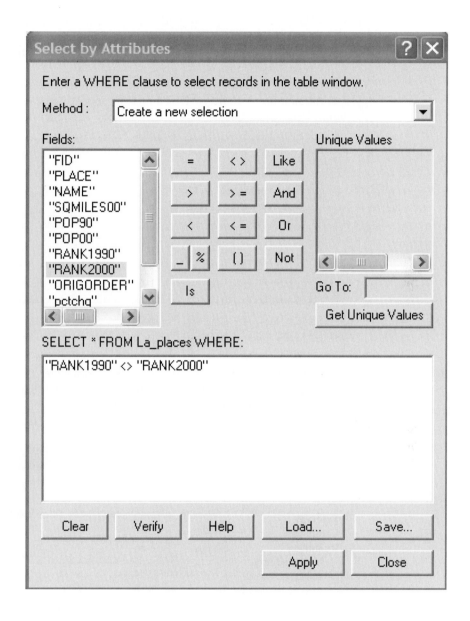

*Note: *If you do not do the above operation to select only suburbs that changed rank between 1990 and 2000 before you run the RMI calculation you will receive an error because you cannot divide into zero. The error occurs because Los Angeles is classified as rank 0 in both 1990 and 2000 leading to a division by 0.*

25. Close the Select by Attributes dialog box.
26. Scroll over to the right in the table and right-click on the field heading RMI and select Calculate Values. If you receive a message box warning asking if you wish to continue, click Yes.
27. Compute the RMI = ([Rank 1990] − [Rank 2000])/([Rank 1990] + [Rank 2000]). Note: "RMI =" is already entered for you in the calculation. You only need to enter the right side of the equation. Also note that you must type in parentheses around the denominator and numerator of the equation. Click OK.

28. Click the Options tab below and click Clear Selection and close the attribute table.
29. Right-click on La_places and select Properties, then select the Symbology tab.
30. Select Quantities. Select Graduated Colors. Select RMI for the Value under Fields.
31. Click the Classify button.
32. Choose 3 for Classes; Select a Method of manual; and under the Break Values column select the first break to be −0.00001 and the second to be 0; keep the third at the current maximum limit. Click OK.

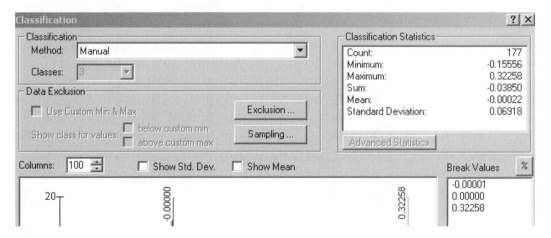

33. Select an appropriate color ramp (i.e., shades of blue).
34. The Label Field gives the text that actually appears in the map legend. Now change the cells in the Label Field to something more meaningful for the map legend, such as "Decline in Rank" for the first, "No Change in Rank" for the second, and "Increase in Rank" for the third class.

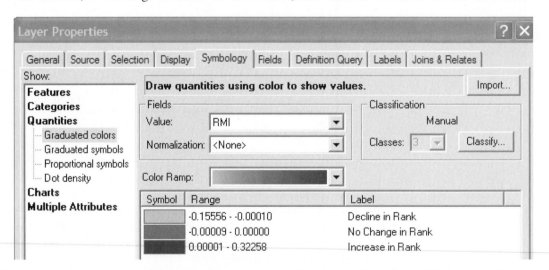

35. Now you want to remove the outlines on the symbol boxes. The map will be more effective without the lines. Double Click on the symbol (light blue) for the first category and set Outline Color to No Color by clicking the dropdown arrow. Click OK.
36. Repeat for the remaining two classes and click Apply and OK to see your results.
37. Turn off the CityNames_LosAngeles city name labels by un-checking its check box.
38. Right-click the Chicago data frame and click Activate.
39. Right-click Ch_places in the Chicago data frame and select Open Attribute Table.
40. Repeat Steps 21 through 37 for Chicago.

Creating a Final Map Layout

41. Prior to leaving the Data View you may want to zoom in on the central area patterns in Chicago and Los Angeles in each of the Data Frames as they include vast land areas, especially Los Angeles.
42. Select Layout View from under View to compose a map for printing.
43. Within the Layout View you can see multiple data frames at a time. Both Data Frames are there, you just need to re-position them in the following format with associated legend and text.

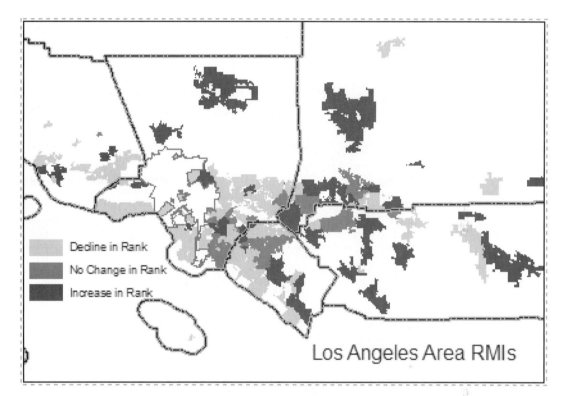

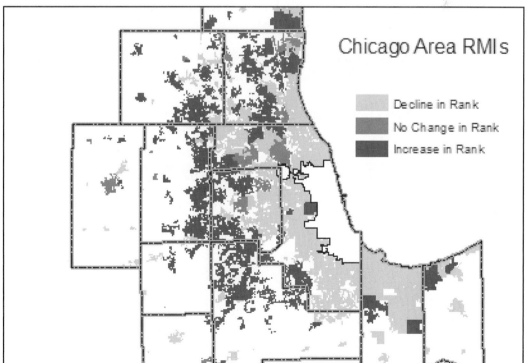

44. In order to achieve a similar format to that above, you can standardize the x, y position of the lower left corner of each Data Frame and assign a standard height and width by right clicking on a Data Frame, selecting Properties, and clicking the Size and Position Tab.

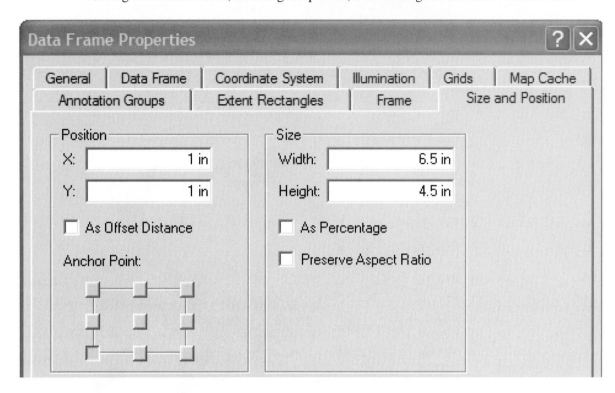

45. In the above example for the Chicago Data Frame, the size included a width of 6.5 inches and a height of 4.5 inches, while the x position was set to 1 inch and the y position to 1 inch. Note that the Anchor Point that is depressed is the lower left, this is the x and y position you are specifying on that corner of the page. The Los Angeles Data Frame should include the same width and height as well as the same x position. However, its y postion should be set to 5.75.
46. Make the Chicago Data Frame active and click the Insert button to insert a legend. In the first dialog box, make sure that the only legend item is Ch_places. For instance, for the

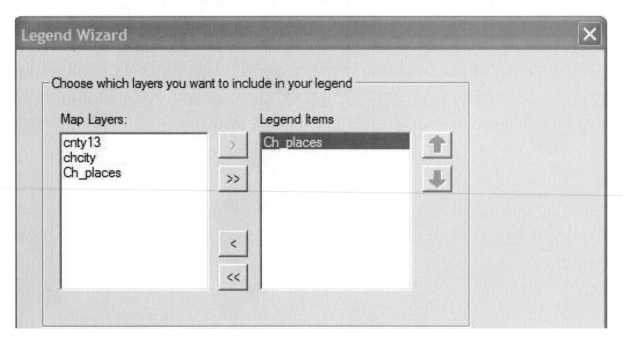

other layers in the box on the right, select and use the arrows to push them over to the left column.

47. Click Next and erase the word "Legend" by backspacing it out.
48. Click Next on the next several dialog boxes and click Finish.
49. Now double click on the selected legend. Click the items tab and in the right hand column double click on Ch_places, click on properties, click the General button, uncheck "show layer name" and uncheck "show heading," click Apply and OK. Click OK as you exit back. Reposition the legend to a suitable spot.
50. Click Insert at the top and insert a Title for each map.
51. Click File and select Print to print the final map (alternatively you could Click File and select Export Map and export to one of your favorite image formats).
52. Exit ArcMap.

Essay Question 1

Write a one- to two-page essay describing and interpreting the geographic distributions of the RMIs. Please compare the Chicago and Los Angeles metro areas. The bands of the mobility index appear more concentric for Chicago than for Los Angeles. Why are the patterns different?

Essay Question 2

Based on the examples of Chicago and Los Angeles, write a one-page essay that discusses the RMI. Explain the index's usefulness. Do you see any drawbacks to it? Please refer to the "Analyzing an Urban Issue" section of Chapter 2 as well as relevant readings that you may have access to.

3

Defining the Metropolis

This chapter introduces you to the definitions and concepts of key elements in the study of the city. It starts with the definitions of city, urbanized area, and metropolitan area. These concepts distinguish urban and rural, urbanized area and urban cluster, metropolitan and nonmetropolitan, and administrative definitions of city. The terms for now should be thought of as follows: The metropolitan area is a population nucleus together with surrounding communities that are closely linked to it economically. The urbanized area is a densely settled area, which is defined by establishing a threshold population density that is reflective of an urban landscape. The city is a legally defined place. Cities emerge and grow through incorporation, addition of unincorporated areas, and annexation. A metropolis or metropolitan area is regional in scope with a dense urban core and a more sparsely populated periphery. The metropolitan area can be more or less integrated socially and economically, but it often has fragmented political and governance issues.

There is a range of sizes of cities. At the giant end are ones of more than 10 million population, which we term megacities. An example of such a giant metropolis already discussed is Mexico City. Its 20th-century growth from city to megacity was fueled by fertility and rural migration to its periphery. It grew over 50 years by expanding growth rings and a diminishing central core. A limiting factor in its growth has been its natural environment.

The study of defining the boundaries, urban entities, and census distinctions for the metropolis is important because it enables students of urban geography to refer to the appropriate concept and then apply it systematically to support other parts of urban geography. For example, in an analysis of urban environmental disputes across the western United States, it is essential to define at the beginning whether the frame of reference is the West's cities, metropolitan areas, or urbanized areas.

Mobility patterns in the metropolis cannot be understood without defining what boundaries movers are crossing, whether they are city, urbanized area, or metropolitan area boundaries. Although this chapter is principally based on U.S. definitions, the concepts of Tokyo and Mexico City are briefly discussed as examples of the variety worldwide. You should understand that other nations not covered in the textbook will have their own distinctive definitions of the metropolis that need to be understood before you can analyze the country's urban geography.

This chapter underpins most other chapters in the book, including the ones on migration, neighborhoods, urban core and edge city contrasts, environmental problems, and urban planning. This chapter's concepts will be used as building blocks to a better understanding of the dimensions of cities and urban areas.

3.1 URBAN AND CITY CONCEPTS AND PROCESSES

The focus of urban studies is on the city, but the most difficult task facing a student of urban studies is defining the city. Often, people loosely use the term in different ways in everyday conversation. How many times have you been visiting a new place and been asked "Where are you from?" The answer to this question at first might seem simple, but depending on how much time you have, the answer can be quite lengthy, especially if you are from a large metropolis. Consider the person from Oak Lawn, Illinois, a suburb of Chicago, who wishes to be brief in responding and replies "Chicago." This answer will be satisfactory to most Americans, as they can easily place the location of Chicago in their mental atlas (see Figure 3.1). On the other hand, to someone knowledgeable of the Chicago area, the response might be, "Do you mean Chicago proper?" Now the

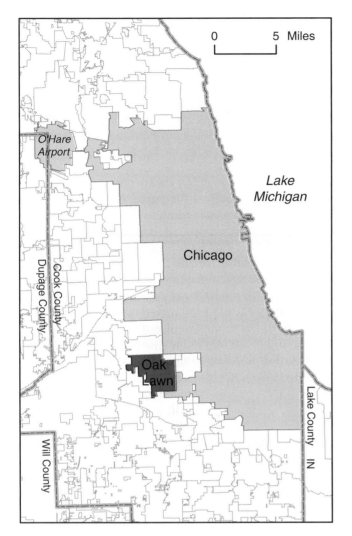

FIGURE 3.1   City concepts: Oak Lawn and Chicago.

When Macaulay sought to instruct his readers on how London circa 1850 differed from what it had been in 1685 this is what he first noted. "The town," he wrote, "did not, as now, fade by imperceptible degrees into the country." Engels found London strange because "a man may wander for hours together without reaching the beginning of the end, without meeting the slightest hint which could lead to the inference that there is open country within reach . . . " Both were concerned to assess the social significance of the spatial reorganization of a city whose wanton spread beyond its previous limits was erasing all discernable boundaries. (pp. 4–5)

3.1.1 Definitions of City, Urban Area, and Metropolitan Area

A *city* is typically defined as an inhabited settlement of contiguous urban area that is larger than a town or village. In the United States, a city is an incorporated place that is contrasted with a small town or village of fewer than 2,500 persons, which is rural.

Identifying urban places and cities depends on the following factors (Frey & Zimmer, 2001):

- Ecological elements
  a. Size of population
  b. Density of population
  c. Space
- Economic elements
  a. Economic function
  b. Labor supply and demand
  c. Transportation
  d. Economic organization
- Social elements
  a. Social organization
  b. Social character
- Administration

A *metropolitan area* is the largest urban concept in area, which consists of a region with a dominant city. For instance, Chicago is a metropolitan area that covers a large area of northern Illinois, and Minneapolis-St. Paul is a large metropolitan area of Minnesota. A metropolitan area might also dominate a small region. For example, Billings is at the center of a fairly small metropolitan area in western Montana.

Urban places, compared to rural places, and *cities*, compared to villages, are more dense in population; more crowded for space; have higher order economies, that is, not just primary production, but secondary and tertiary production; have labor supply and demand over broader regions; have more sophisticated economic organization; possess a social character that includes urbanism and city values; and have stronger and more sophisticated administration.

Chicagoan, or so she claimed, has a new challenge: to identify herself as being from Oak Lawn while holding strong to her original statement. The confusion in this situation arises because there are multiple concepts of the city operating in our day-to-day communication about cities. Yet, geographers and government planners are able to define clearly a variety of concepts of city, metropolitan area, and urbanized area.

The difficulty in defining the city has been a problem for the past two centuries. As early as 1850 in London, the problem of where to draw the boundary between city and countryside existed. This was evidenced by two writers of the time, Thomas Babington Macaulay and Frederick Engels, as cited by Goheen (2000):

Among the generation of writers who instructed their contemporaries on the modern city even as it was taking shape were Thomas Babington Macaulay and Frederick Engels. . . .

The ecological elements imply a dense agglomeration of population that uses space in a different way than rural areas. The built infrastructure can accommodate the higher density. Land is divided into more specialized uses.

The economy of a city has more diverse and specialized functions. The urban place engages in little if any production of agriculture or raw materials, but receives them from outside in exchange for manufacturing and services. There is a larger labor market of supply and demand that extends beyond the city to a wider region, national, or international. The more advanced economy requires a larger and more complex transportation network for movement of people and goods. Flows of money and information are also more complex. The size and scope of enterprises tend to be larger in urban places.

Urban places and cities, compared to rural areas and villages, also might have different values, perceptions of the world, and behavior patterns—a different "way of life" (Frey & Zimmer, 2001; Wirth, 1938). Wirth termed it *urbanism as a way of life*. It refers to a more sophisticated, plural, multicultural, and complex way of living and viewing life. The differences are often portrayed in literature or movies, with images of the "farmer" versus "city slicker."

The administration of cities versus villages is quite different. A *village* is a small settlement of hundreds to 1,000 persons. Often a village or town has small administrative offices, reduced government structure, and limited functions. A *town* is somewhat larger than a village of up to 2,500 persons. On the other hand, a city has a multifunction government administration, elected officials, and a city council or equivalent, with broad government powers to tax, set regulations, control, police, plan, and provide services. In the United States, a village or town sometimes is incorporated .

The governments of cities, as well as many associated metropolitan agencies, perform significant municipal functions. The services they provide are essential to the daily living of residents as well as the functioning of businesses and organizations. Among these services are transportation, public health, environmental monitoring and control, consumer affairs, business development, family services, public education, building and housing monitoring and approvals, municipal courts and legal services, information services, public safety, jails, utilities, sanitation, libraries, cultural institutions, parks and recreation, and planning. In the United States, the services are provided through administrative offices that are directed by elected officials and overseen by public representatives. For a large city, the annual budgets to provide these services can run into billions of dollars. The municipalities also need to generate the revenue to support these functions. All these entities must strike a balance from year to year in the provision of services versus the receipt of revenues through taxes, service charges, revenue sharing, and other sources.

National censuses and statistical agencies have problems measuring all these dimensions, so the main ones used worldwide to officially define urban places are size of population and administration (Frey & Zimmer, 2001). In the United States, urban places are defined as those having a population agglomeration of 2,500 or more. This is the most common way to define urban places, although many nations such as India have different cutoff sizes for urban agglomeration (e.g., for India, it is generally 5,000 or more population).

An *urbanized area* (UA) consists of a central city and its surrounding thickly settled area. On the other hand, a *metropolitan area,* according to the U.S. Census, is defined as a substantial population nucleus, surrounded by adjacent counties that have significant commuting linkages with the central nucleus. This metropolitan definition implies a more spacious region, in which not all land is built up, but has a mixture of open and built-up land on the periphery; the main determining factor for the inclusion of peripheral open land is a certain level of commuting to the central nucleus. Later in the chapter, the concepts of UA and metropolitan area are discussed with respect to the official U.S. definitions.

3.1.2 International Comparisons

Across nations, the definitions of all these concepts—urban, city, UA, and metropolitan area—vary. This becomes a weakness for efforts to study cities worldwide because it is hard to classify them systematically. Even organizations such as the United Nations base their world urban area rankings on urban definitions that do not use consistent geography (see Table 3.1). For

TABLE 3.1   United Nations 2000 population estimates for 10 urban areas, by type of geographic area (population in millions)

Urban area	Urbanized area	Metropolitan area	Municipality (China)
Tokyo	26.4		
Mexico City		18.1	
Sao Paulo		18.0	
New York	16.7		
Mumbai (Bombay)	16.1		
Los Angeles	13.2		
Kolkata	13.1		
Shanghai			12.9
Dhaka		12.5	
Delhi	12.4		

Source: Forstall, R., Greene, R., and Pick, J. (2004) "Which are the largest? Why published populations for major world urban areas vary so greatly." *City Futures Conference Proceedings*, Chicago, IL.

TABLE 3.2 Population, area, growth, and density for Tokyo by different definitions

Urban definition for Tokyo	Population 7/1/2000	Area (km²) 2000 census	Population per (km²) 7/1/2000	Annual percent change 1990–2000
City Proper	8,126,000	621	13,085	0.41
Administrative Area	12,049,000	2,187	5,509	0.49
Urbanized Area	28,228,000	3,084	9,153	0.61
UA (administrative boundaries)	30,360,000	6,657	4,560	0.55
Metropolitan area (MA) (1)	34,542,000	13,565	2,546	0.56
Metropolitan area (MA) (2)	30,681,000	7,631	4,020	0.56

Source: Forstall, R., Greene, R., and Pick, J. (2004) "Which are the largest? Why published populations for major world urban areas vary so greatly." *City Futures Conference Proceedings*, Chicago, IL.

instance, the latest U.N. rankings for Tokyo, New York, and Delhi are based on the UA populations of each, whereas Mexico City and Sao Paulo are ranked on the basis of metropolitan area population. Meanwhile, the ranking of Shanghai is based on a city proper (municipal) definition, although for China these areas typically include substantial rural areas (Forstall, Greene, & Pick, 2004).

The implications of various urban definitions for such measures as population rank, population density, and growth rate are numerous, as demonstrated by Tokyo (see Table 3.2). For Tokyo, the table shows the city proper and the administrative area called Tokyo Metropolitan Government (Tokyo-to), which is larger than the city but includes only a small portion of Tokyo's suburbs (see Figure 3.2). The city's mean population density is quite

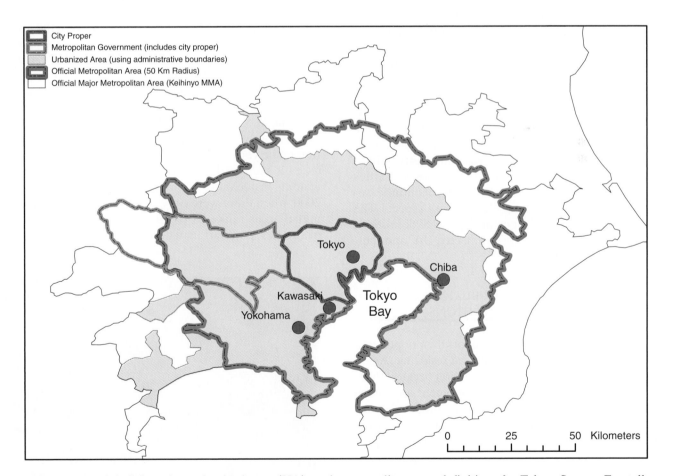

FIGURE 3.2 Administrative, urbanized area (UA), and metropolitan area definitions for Tokyo. Source: Forstall, R., Greene, R., and Pick, J. (2004) "Which are the largest? Why published populations for major world urban areas vary so greatly." *City Futures Conference Proceedings*, Chicago, IL.

high, at more than 13,000 per km^2 or more than 30,000 per square mile.

The Tokyo UA is given both with an exact definition (based on their densely inhabited districts, or DIDs) and with that definition adjusted to administrative boundaries. The exact definition has a density over 9,000 per km^2. Adjusting to administrative boundaries adds less than 10 percent to the exact UA's population but more than doubles the area included and halves the density (see Figure 3.2).

Two official Tokyo metropolitan areas are shown in Table 3.2. The first is the major metropolitan area (MMA), defined very generously to include all communities that have at least 1.5 percent of their population commuting either to work or to school in the MMA's four central cities (Tokyo, Yokohama, and two less familiar places, Kawasaki and Chiba). The MMA definition is as of 1995 because the 2000 definition has not yet been published (Figure 3.2).

The second official metropolitan area is defined as the administrative units within 50 km of central Tokyo. It has a somewhat smaller population and a much smaller area than the MMA.

Tokyo's population is growing slowly according to all of the definitions. The more inclusive definitions all have rather similar rates. The rate for the UA reflects that it expanded geographically during the last intercensal period; for all the other definitions, the growth rates use constant geography.

This text does not have room to delve into all the international differences in these definitions, but the student is referred to Paddison (2001) and Forstall et al. (2004). A unifying theme of all these definitions is to realize that cities, urbanized areas, and metropolitan areas include the elements of urban places and cities described earlier. These are ecological, economic, social, city administration, and city services aspects. The following sections dig deeper into the U.S. definitions and distinctions of urban and rural, city, UA, and metropolitan area.

3.2 CITY SIZE DISTRIBUTIONS AND CENSUS DEFINITIONS

Most national statistical agencies report their population statistics broken out by urban and rural residence. Agencies consider this an important distinction because there is often an assumed economic development advantage to living in urban areas and many of the reported statistics reinforce this idea. Also, it can help differentiate rural versus urban growth patterns. Prior to 2000, the U.S. Census Bureau included places with a minimum population of 2,500 as part of its urban definition, but it now defines as urban all territory, population, and housing units located within a UA or an urban cluster (UC). It delineates

UA and UC boundaries to encompass densely settled territory, which consists of the following:

- Core census block groups or blocks that have a population density of at least 1,000 people per square mile.
- Surrounding census blocks that have an overall density of at least 500 people per square mile.

A *block* is approximately equivalent to a square block in a city, contains a population of about 300 persons, and is the smallest unit for complete census data collection. A *block group* is a group of blocks at about 1,500 population; this is the smallest unit for sample data collection (U.S. Census, 2004).

In addition, under certain conditions, less densely settled territory can be part of a UA or UC. This density-based definition is regarded as an improvement on the earlier definition of all persons living in places of 2,500 or more inhabitants because it is no longer based at all on municipal boundaries. It follows then that *rural* consists of all territory, population, and housing units located outside of UAs and UCs.

Metropolitan areas have a range of sizes because of historical growth patterns. Over time this leads to differences in population and extent that stem from the area's history of economics, jobs, transport, lifestyle, and amenities. In 2003, the U.S. Census Bureau defined 370 metropolitan areas, all having more than 50,000 persons. The largest metropolis was New York City at nearly 21.4 million persons, whereas the smallest was Carson City, Nevada, at 52,457 people. Boston had a metropolitan population of 5.7 million and ranked seventh in the nation in the year 2003 (see Figure 3.3). The importance of cities is seen in the 82 percent of the U.S. population residing inside these 370 metropolitan areas. In the year 2000, when there were only 280 metropolitan areas, the metro areas ranked 1 to 10 accounted for 32 percent of the U.S. population, whereas those ranked 11 to 30

FIGURE 3.3   Boston metro landscape.

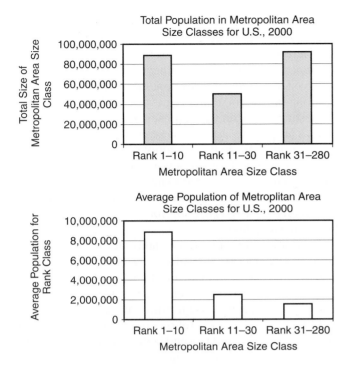

FIGURE 3.4 Varying sizes of metropolitan areas, United States, 2000. Source: U.S. Census, PHC-T-3, 2001.

comprised 18 percent. Metro areas ranked from 31 to 280 accounted for 33 percent of the population (see Figure 3.4). The average sizes for the three classes are 8.9 million, 2.5 million, and 1.5 million. Notice that the difference in average size is less between the second- and third-ranked groups. The third-ranked group has the most cities—197 of them—with populations between 50,000 and 500,000. As seen in Figure 3.4, in the United States, many Americans are likely to live in one of these smaller metropolitan areas. All this demonstrates that people reside in areas across a range of city sizes. Further, the lifestyle, city environment, and complexity of industry and culture varies from giant cities to smaller ones.

3.3 CITY AS MUNICIPALITY

The concept of city most frequently used in everyday life is the city as a self-governing unit, for instance, the city headed by a mayor and city council. This is the *incorporated place,* which is a concentration of population with legally defined boundaries and legally constituted governmental functions. Incorporated places can have a variety of labels including city, town, village, and borough. The reason a group of citizens and landowners desire incorporation includes the desire to establish identity, control development, act as a legally recognized body (including the ability to tax and collect revenues), and gain some autonomy from higher levels of government.

Some places in the United States have an area identity and have locally recognized boundaries but are not legally incorporated. They are referred to by the U.S. Census as *census-designated places.* An example is Mentone, California, a place of 7,803 population in 2000 that is located in San Bernardino County, on the eastern side of the Los Angeles urban area, 86 miles west of the city of Los Angeles and 21 miles northeast of the city of Riverside. Mentone has an elementary school, small business area, restaurants, and below-average income for the region.

Another legal process is *annexation,* which is the inclusion of land territory into an existing municipality through expansion of the city limits. The city of Redlands, California, with a population in 2000 of 67,771, is located in an expanding urban region to the east of Los Angeles, and adjacent to Mentone. It annexed a significant amount of land between 1990 and 2000 (as seen in Figure 3.5). The motivation for a municipality to annex territory is to regulate development, accommodate population growth, and increase the municipal tax base. A property owner's motivation to be annexed by a municipality is often to gain services such as water and sewer hookups. Other services such as fire and police protection and street maintenance might be enhanced if an area is annexed by a city. The likelihood of approvals to

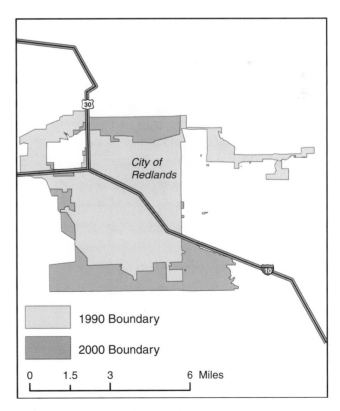

FIGURE 3.5 Annexation in the 1990s for the city of Redlands, California.

develop property and a sense of certainty about how and when development might occur are additional reasons owners might choose to agree to be annexed.

State laws differ with respect to new incorporations, new unincorporated areas, and annexations, and the annexation process differs throughout the United States. Illinois serves as a good example, as it has more municipalities than any other state with a total of 1,282 compared to a U.S. state average of 385 (U.S. Census, 1993). *Incorporation* requires a petition by a certain number of electors residing within the area proposed for incorporation. If a majority of voters approve the incorporation, it passes into law. There is no regional or state oversight, although counties can review incorporation petitions within their borders. The result of this process has been the creation of an especially large number of municipalities (i.e., cities, villages, etc.) in Illinois, a good proportion of which are in the Chicago metropolitan area. The many municipalities in this metropolitan area lead to a heterogeneity of city and town development and governance (they are shown in Figure 3.6). Critics of this fragmented

municipal geography argue that the structure makes it difficult to achieve consensus on many regional issues (Barnett, 1995).

Annexation can occur in either of two forms, voluntary or forced. Through voluntary annexation, a property owner or group of owners submits a petition to the municipality requesting to be annexed. Forced annexation can happen when a property is less than 60 acres and is surrounded by one or more municipalities, a forest preserve, and/or state-owned property. In either case, it is necessary for the parcels to be contiguous to the municipality to which they are to be annexed. Some municipalities have stretched the definition of contiguous. For example, it is not uncommon to see municipalities annexing a thin strip of land along a road for some distance to reach and include in the annexation a larger parcel of some economic interest. See, for instance, the city of Redlands, California, in Figure 3.5, for which some annexation is just south of U.S. Interstate 10 and whose acquisition was for water rights.

Florence, Arizona, is an example of the annexation extremes a municipality can employ to increase its municipal tax base and gain money from state revenue sharing that is based on per-capita population (see Figure 3.7). Florence has annexed to incorporate prisons on its periphery, allowing it to tap into state and federal funding to pay for such projects as a new town hall and senior center, projects that most towns finance through revenue generated by local property and sales tax (Whiteside, 2002).

A city must have sources of revenue. Incorporated places like cities are taxing jurisdictions as well as service providers. In a metropolitan region where there are often multiple incorporated places, the fiscal health of each one is often dictated by its geographic extent, relative location within the region, and whether it has adjacent land to acquire through annexation. Incorporated places that are hemmed in and have no room to expand often find it difficult to balance incoming tax revenue and outgoing payments for services, especially if there is a net movement away among its tax-paying population. In this sense, the landscape of cities that are defined legally almost seems arbitrary in that they take on a variety of shapes and sizes.

Los Angeles in 2000 revealed much variety in its boundaries (see Figure 3.8). Los Angeles is an extreme case. The city was incorporated as a 25-mile-square area in 1850 including the Spanish central plaza, and encompassing the current downtown. After 1900 it took on a linear shape as it sought to acquire and develop its port to the south in San Pedro Harbor. It took on a large mushroom shape with the annexation of the San Fernando Valley in 1905. This was done for the purpose of increasing the city's assessed tax value to secure

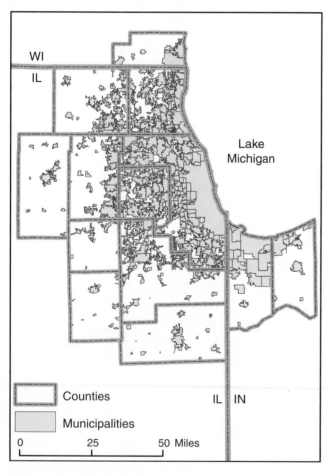

FIGURE 3.6   Municipalities of the Chicago metropolitan area.

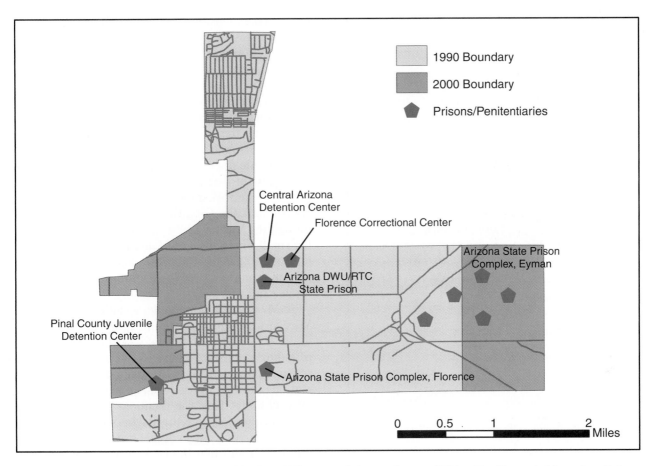

FIGURE 3.7 Annexations in 1990s by the city of Florence, Arizona. Source: Whiteside. Harper's Magazine. February 2002, p. 88.

bonds to pay for a 250-mile aqueduct from Owens Valley in the Sierra Nevada Mountains to Los Angeles (Reisner, 1986).

Ironically, there have been periodic secession movements in the San Fernando Valley to break off from Los Angeles, starting in the 1970s. The most recent attempt for secession was strongly voted down by the Los Angeles city electorate in November 2002. The affluent and mostly White San Fernando Valley voters and others in favor of the valley's independence from the Los Angeles "bureaucracy" were voted down by the city's multiethnic majority, which found the present city services useful and fairly efficient, including those for the valley (Steingold, 2002). Another notable characteristic is the number of municipalities that the city of Los Angeles surrounds, such as Beverly Hills, West Hollywood, Santa Monica, and Culver City. These cities have maintained their independence for more than a century, resisting overtures for annexation. Today, Los Angeles has a large and complex shape, with many arms of development.

The annexation history of Chicago's suburban municipality of Naperville demonstrates that the shape, extent, and spatial orientation of a municipality is often a result of competition with neighboring municipalities for available land for expansion (see Figure 3.9). Prior to 1964, Naperville was fully contained within DuPage County, but as Naperville annexed land south and westward in the mid-20th century, it eventually encountered the city of Aurora at State Route 59. Naperville then turned south into Will County to acquire more territory. In the late 1980s and 1990s, it crept further southward, consuming a lot of land in Will County. Eventually, Naperville will be limited to the south as it converges with the cities of Joliet and Plainfield, which are annexing northward toward Naperville. This century-long history of growth in Naperville demonstrates the powerful force of annexation in converting and filling in metropolitan areas. Los Angeles and Mexico City also represent a long-term process of filling in what were once scattered towns and cities into a closely knit urban complex, a process that has included some annexation.

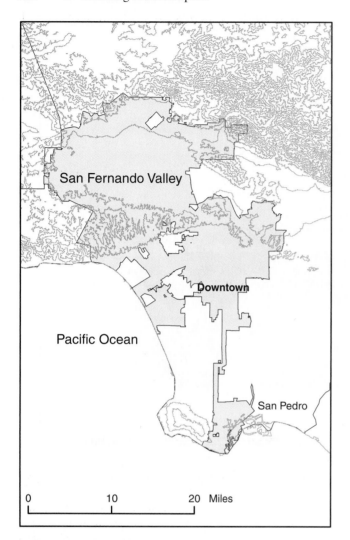

San Fernando Valley

Pacific Ocean

Downtown

San Pedro

0   10   20   Miles

FIGURE 3.8   City boundaries of Los Angeles, with 500-foot elevation contours.

3.4 CITY AS A BUILT-UP AREA

The municipal definition of the city is often arbitrary with respect to the placement of boundaries. Through the way it grew and annexed land, a city might contain considerable undeveloped land within its boundaries (over-bounded) or it might not be fully inclusive of the developed land of an area (underbounded). This can be problematic if the municipal definition of city is used for differentiating between a nation's urban and rural population. Consider the case of the *underbounded city,* which fails to include considerable built-up land within its city limits. Within the United States, the large cities—especially in the Northeast and Midwest—tend to be underbounded, as their annexation histories were of limited success, and as neighboring suburbs incorporated and in effect hemmed them in on all sides, preventing additional

territorial expansion. Underbounded cities such as Buffalo, St. Louis, Boston, Chicago, and Philadelphia contain less than 50 percent of their metropolitan populations, and their suburban areas have been gaining a larger population share since the end of World War II. The latter trend has been referred to as the "doughnut" city phenomenon, where the loss of population in the inner city conjures up the image of an empty hole in the center. It is evident from St. Louis in the late 20th century that the city limits do not reflect a great part of the region's urban development (see Figure 3.10). Underbounded cities in the South and West include Seattle, Portland, San Francisco, Salt Lake City, Denver, and New Orleans. Even though the former 1990s U.S. urban definition included the small, emerging suburban municipalities with populations of 2,500 or more, the underbounded city phenomenon underscores the arbitrariness of municipality shapes and extents.

In Mexico, the same underboundedness occurred for Mexico City, which started to spill over outside its city limits (i.e., those of the Federal District) into a surrounding state starting in the 1960s and 1970s. This has continued to such an extent that today, most of the population of the Mexico City urban area resides outside the District limits. This also points to potential governance problems. For instance, the mayor of Mexico City is elected to govern only the Federal District and has little if any influence regarding the parts of the urban area in the State of Mexico. Thus, to this point, it has been difficult for the mayor to implement public programs that benefit the whole metropolis.

The opposite case of the underbounded city is the *overbounded city,* which contains considerable rural territory and population. The large cities of the Southwestern United States are more often overbounded, due in part to liberal annexation laws in the states of the Southwest. The former mayor of Albuquerque, David Rusk (1993), referred to these Southwestern cities as "elastic" cities because they can expand to contain their sprawling suburban development, whereas "inelastic" cities like Detroit and Cleveland are far removed from the new development of their regions. The overbounded city often annexes in advance or in anticipation of new development on its edge, thus preventing the incorporations that limited territorial expansion in the older metropolitan areas of the Northeast and Midwest. It is usually easier for the Southwestern cities to annex because the surroundings of the city are often thinly settled. As a result, cities such as Albuquerque, San Antonio, and Phoenix contain more than 70 percent of their metropolitan population. There is also a correlation between the fiscal health of cities and the degree to which they are truly bounded; that is, underbounded cities have few options in curtailing middle-class flight to adjacent suburbs,

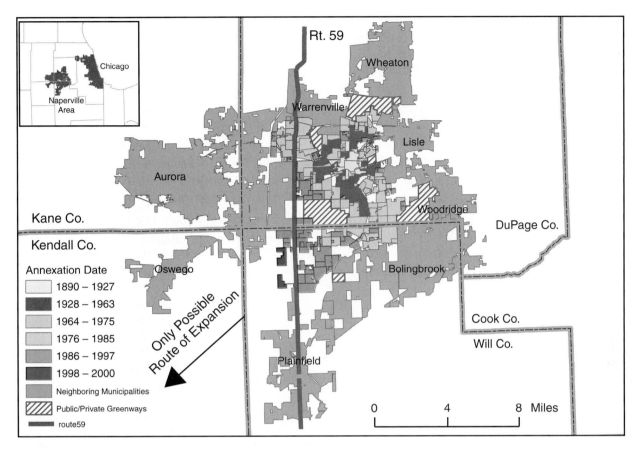

FIGURE 3.9 City of Naperville annexation history, 1890 to 2000.

whereas overbounded cities can annex in the direction of anticipated "hot spots."

However, overbounded and underbounded cities have both advantages and disadvantages in their urban settlement patterns. Underbounded cities have the advantages of fewer resources and efforts being poured into planning and annexing at the edges. On the other hand, they lose control of development of the edge. Overbounded cities are better able to plan and develop the entire urban area in an integrated way. Their governments have more influence on how the city grows in the broad and strategic sense. The downside is that they might have to sink resources into maintaining undeveloped and even useless land.

The concepts of underbounded and overbounded have major influences on the revenues of and service provision by cities. An underbounded city might not gain adequate revenue sources from its own residents. This is because it might be supporting a workforce larger than its resident population can provide for and could be responsible for services for a larger region, extending beyond the city limits. This been recognized in the revenue sharing and grants provided by state and federal

government, but fiscal balance might still not be achieved, as was seen, for instance, in the near bankruptcy of New York City in 1975. On the other hand, an overbounded city can tap into a wide revenue base, including affluent areas located on its periphery. However, because it stretches over larger reaches of territory, it could have problems deploying and managing effective municipal services. It also has more areas of undeveloped land in its outlying reaches that require special types of environmental and natural disaster services, such as fire and flood control. It also faces the challenges of urban planning and development for its outlying territory. Some of these challenges for the two situations can be seen in examples and cases in later chapters.

The complications presented by the underbounded and overbounded city have brought about alternative concepts and definitions to better differentiate urban and rural population. UAs are statistical areas that were first designed by the Census Bureau in the 1950s to provide a more realistic separation of the urban and rural population surrounding large cities. In many respects, it was to mimic the exercise of drawing boundaries around the urban agglomerations you would see on an aerial

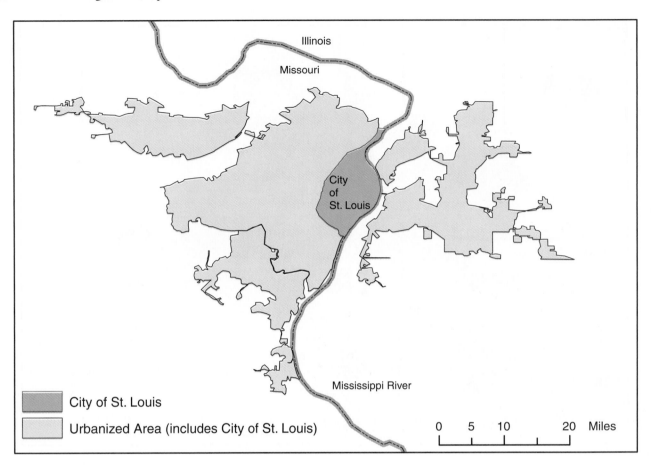

FIGURE 3.10 City of St. Louis relative to built-up area.

photograph of a city while leaving the open spaces between the agglomerations blank. An example of drawing the urban boundaries is seen in Figure 3.11, in which the aerial photograph makes clear how urban and rural territory are interwoven.

A UA consists of a central city and its surrounding thickly settled area. The area boundary is first established by the limits of settlement. Once this area is established, the most important criterion for a UA is that it has a census population of at least 50,000 people. Second, a UA consists of a geographic core area of block groups and blocks with a population density of 1,000 people per square mile, and the adjacent block groups and blocks have at least 500 people per square mile. In cities, a block is often just as it sounds—a city "street block" includes its buildings and residents. However, for the census, it does not always need to be bounded by streets, and could have other limits such as rivers or a lake. It is important that a UA can cross over and does not need to correspond to the boundaries of incorporated places or census-designated places (see, e.g., Figure 3.12, which shows the Chicago UA crossing back and forth across the place boundary

of Wayne, Illinois). This is a major departure from the 1990 UA delineation rules, which in some cases "forgave" areas that were within municipalities but had fewer than 500 persons per square mile.

As illustrated by the village of Wayne, Illinois, the 2000 rules for the Chicago UA ignored the Wayne municipal boundaries, resulting in a more accurate depiction of the truly built-up portion of the region. Extracting the Chicago area territory that was UA in 1990 but not UA in 2000 gives one an understanding of the estimated overbounded area resulting from the UA rules prior to 2000. As seen in Figure 3.13, the 2000 UA differences form a ring around the Chicago area reflecting undeveloped areas of the region's many overbounded urban fringe municipalities. These were counted as urban in 1990, but not in 2000.

A UC is a new census concept for 2000 that can be thought of as a smaller version of the UA. Like a UA, it is defined in terms of core blocks or block groups surrounded by noncore ones; however, a UC has a total population of only between 2,500 and 49,999 persons.

Figure 3.14 demonstrates the concepts of UA and UC for the Chicago metropolitan area. The difference between

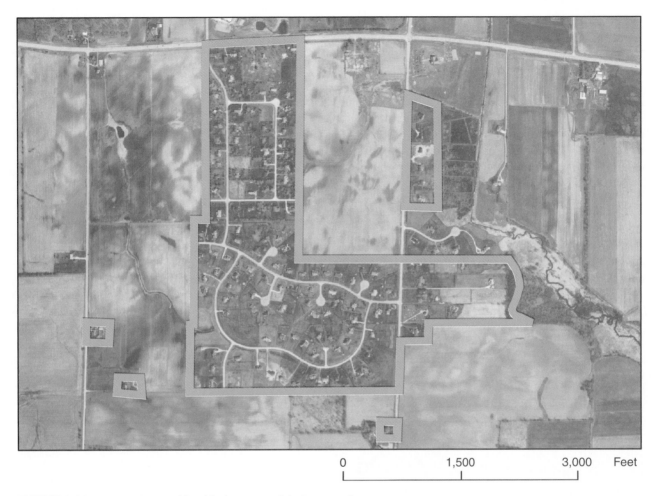

0   1,500   3,000   Feet

FIGURE 3.11   Urban clusters identified on an aerial photograph.

a UA and a UC is that UCs are much smaller in population, and they are separated from the urban territory of UAs as illustrated by the Chicago case. Similar to the historical development of the UA, they lie in proximity to major transportation routes. UAs and UCs together are referred to as *urban areas*. Any area that is not classified as an urban area is *rural*.

Another advance in understanding UAs is the availability of aerial photos. These show a more complete picture of the physically built-up portions of cities. It is possible to superimpose digital photos with UAs defined by the census. With the widespread availability of digital orthophoto quads (DOQs) from the U. S. Geological Survey (USGS), a physical approach to UA boundary delineation is becoming more realistic. Even though aerial photography was available in previous censuses, it was too difficult to apply consistently and to do the necessary processing for the vast extent of the U.S. urban system. The USGS has recently begun a state, local, and federal cost sharing program to produce DOQs nationwide (Robinson et al., 1995). Although not all counties are

available at the current time, many urban counties will eventually sign on and others could find alternative means for financing their own DOQ production. The only caveat to using a DOQ image for a nationwide program, similar to the delineation of UAs by the census, is that the acquisition date of the photo will vary and will often not correspond to census-taking time.

Nonetheless, the use of high-resolution aerial photographs that are georeferenced to other GIS map layers could allow regional governments to overcome any perceived shortcomings in the national UA definition by creating their own local definitions. An example of an aerial photo compared to an urban area in the city of Aurora, Illinois, is seen in Figure 3.15. Note the portion of Aurora, Illinois, that was included in the 1990 UA but excluded in the 2000 UA. The area has a low population density, as it is composed of an industrial park, a school and open land. The local government, on the other hand, would still consider the area to be urban in terms of its built-up character. Nevertheless, the 2000 census UA definition, although still population based, has helped improve the correspondence

Change in U.S. Census Definition of Urbanized Area from 1990 to 2000

In 1990, the U.S. Census used a different definition of UA. The underlying idea of a densely settled contiguous area was similar to that used in 2000, but the 1990 definition was much more complicated because modern geographic information systems (GIS) tools could not be applied to automate and simplify the definition. A UA in 1990 was defined by the following three rules:

1. Territory made up of one or more contiguous census blocks having a population density of at least 1,000 people per square mile, provided that it is (a) contiguous with and directly connected by road to other qualifying territory or (b) noncontiguous with other qualifying territory, provided specified conditions are met.

2. A place containing territory qualifying on the basis of the above criterion will be included in the UA in its entirety, if that qualifying territory includes at least 50 percent of the population of the place. If the place does not contain any territory qualifying on the basis of the above criterion, or if that qualifying territory includes less than 50 percent of the place's population, the place is excluded in its entirety.

3. Other territory with a population density of less than 1,000 people per square mile, provided that it meets specified conditions contained in the guidelines.

This was a complicated definition. Recent advances in GIS and digital aerial photography enabled the improved UA definition for 2000. The 2000 definition does not include rule 2; thus in 2000, all of the block groups and blocks constituting a UA are really built-up ones, avoiding potential problems of over- and under-boundedness.

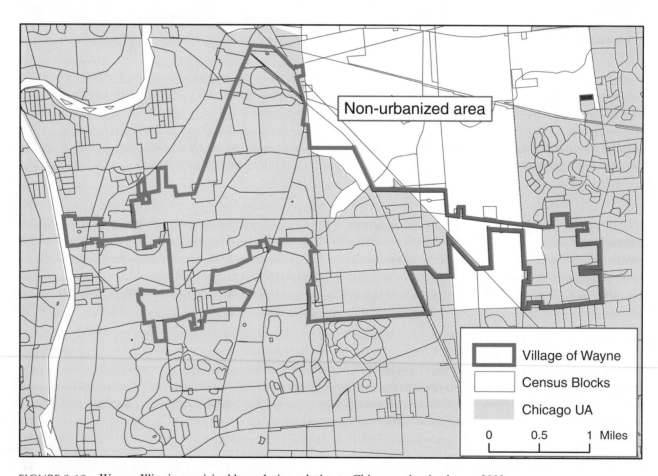

FIGURE 3.12   Wayne, Illinois, municipal boundaries, relative to Chicago urbanized area, 2000.

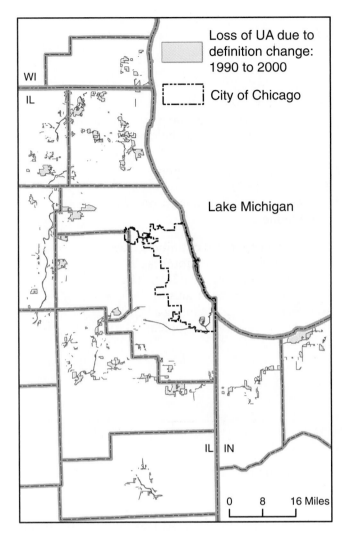

FIGURE 3.13 Loss of Chicago urbanized area (UA) from 1990 to 2000, as a result of UA definition change.

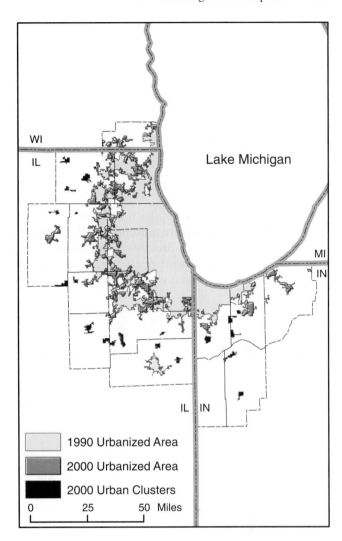

FIGURE 3.14 Chicago urban area in 1990 and 2000 and its urban clusters in 2000.

of UA and physical built-up area in many fringe farming areas. The *fringe* is where urban land ends and rural land begins, although no single line separates the two.

3.5 CITY AS METROPOLIS

As mentioned earlier, the most prominent metropolitan area concept is based on the U.S. Office of Management and Budget's (OMB) functional integration principle. According to this principle, metropolitan areas are defined according to the presence of a nucleus and a zone of influence. In 2000, this principle was translated into rules to define both metropolitan and micropolitan statistical areas that were to be based on the commuting data from the 2000 Census (U.S. Census, 2003). A metropolitan statistical area is therefore a large population

nucleus, coupled with adjacent communities that have a high degree of economic and social interaction with that nucleus. In the official definitions, the population nucleus consists of one or more central counties containing the area's main population concentration. The "zone of influence" component of the metropolitan definition refers to certain standards that outlying counties must meet with respect to their central counties. The 2000 official guidelines require only a specified level of commuting to or from the central counties, whereas prior standards also considered not only commuting, but also population density, urban population share, and county growth rate.

The OMB's new metropolitan guidelines were applied in 2003 to produce a new list of metropolitan statistical area definitions, ones that were based on commuting data from the 2000 Census. There are three primary guidelines

FIGURE 3.15   Aerial photo of section of Aurora, Illinois, included in Chicago's 1990 urbanized area, but excluded in 2000.

to move from concept to definition, by (1) setting a population size requirement for a metropolitan statistical area, (2) establishing boundaries for the urban nucleus in the form of central counties, and (3) by defining the geographic extent of the influence of the nucleus in the form of outlying counties. Considering size, a metropolitan statistical area has at least one UA. To define the nucleus or the core, a county can be considered central if at least 50 percent of the population is located in the UA. This latter rule would tend to bias the selection of the core counties toward the denser portions of the urban region. Finally, to define the area over which the nucleus extends its influence, outlying counties are identified in terms of their economic and social interaction with the central counties. Commuting flows to and from central counties is the empirical measure of the degree to which central and outlying counties are socially and economically integrated.

According to the new rules that took effect in 2003, a county was considered outlying if at least one fourth of the working residents of the county worked in the central counties or at least 25 percent of the employment in the county was accounted for by workers who resided in the

central counties. To illustrate, DeKalb County, Illinois, was once thought too far out from Chicago to be metropolitan, but in the 1990s it was included in the Chicago metropolitan area because of the large number of workers who commuted inward toward Chicago (see Figure 3.16).

It is important to recognize that commuting, the key determinant of metropolitan inclusion, is defined by where workers live relative to their workplace. If all workers lived near their jobs, there would be little commuting flow between the metropolitan nucleus and outlying areas. For all U.S. metropolitan areas and many in Mexico and other developing nations, it has become customary to have extensive commuting networks between where people work and where they live. There are, of course, other reasons that people commute besides their work—they sometimes do it for social reasons. However, the work basis for commuting is utilized because censuses and other agencies concentrate on collecting this type of commuting data. In later chapters in this text, especially those on migration and residential mobility, ethnicity, and urban core and edge city contrasts, the changing residential location of workers versus their job

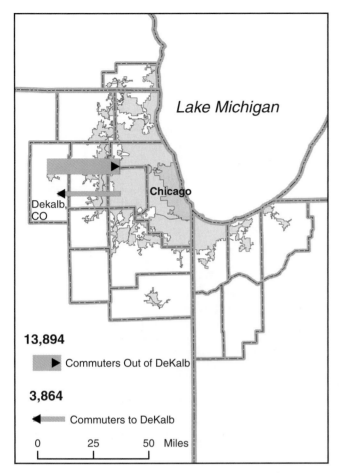

FIGURE 3.16 Commuting flows to and from DeKalb, Illinois, 2000.

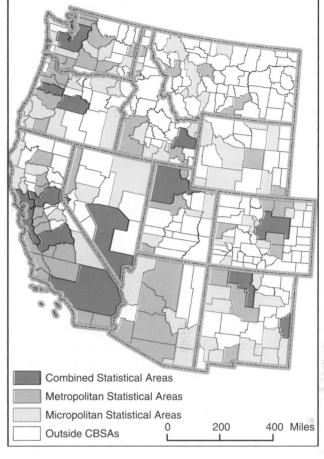

FIGURE 3.17 Core-based statistical areas (CBSAs) for the Western United States, June 2003.

locations will be addressed further. Especially since World War II, there has been a trend for workers to leave the metropolitan nucleus and move out to the suburbs, and in the 1990s even further out to exurbia. *Exurbia* is defined as an area with a high proportion of former urbanites, but far beyond the conventional urban–rural fringe. A key question is to what extent jobs have moved outward to exurbia as well, and whether that has tended to reduce commuting levels. The main point for now is to realize that commuting is based on the balance of residence and employment locations.

The formal term for metropolitan statistical areas and micropolitan statistical areas taken together is *core-based statistical area* (CBSA). A CBSA includes the county or counties that contain either a UA or UC, plus adjacent counties having social and economic linkages with the core as exemplified in commuting ties (U.S. Census, 2003). CBSAs for the western United States are displayed in Figure 3.17. During the development phase for the 2000 Census, the Census Bureau tested the new procedures

throughout the entire United States using the earlier data derived from the 1990 Census. As seen in Figure 3.17, although some areas look similar to the earlier definitions (e.g., Los Angeles) note the three different types of core based statistical areas: (1) metropolitan statistical areas, (2) combined statistical areas, and (3) micropolitan statistical areas. The combined statistical areas result from the joining of adjacent metropolitan statistical areas that have significant commuting ties with each other. The micropolitan statistical areas represent the biggest change from the earlier metropolitan maps that did not recognize areas of that intermediate size. These new areas contain between 10,000 and 49,999 people and have at least one UC.

The micropolitan area played a decisive role in the 2004 Presidential campaign in the United States. Nationally, micropolitan areas went for George Bush over John Kerry by a margin of 61 percent to 39 percent, while the total Bush margin was a thin 51 to 49 percent (Lang et al., 2004). Bush's popular voting edge over Kerry in the micros was 2,403,264 votes, which constituted a good part to his national popular-vote victory edge of 3,429,575. Even more significant was that the "micros" accounted for Bush taking Ohio with its 20 electoral votes, which was the key to his electoral victory. His micropolitan edge over his opponent in Ohio was 154,981 votes, far more than his statewide winning margin of 18,000 votes (Lang et al., 2004). In fact, he took 27 of Ohio's 29 micros. Political journalists reported that Bush's campaign team was aware of the micro vote and consciously cultivated it, placing the candidates George Bush and Dick Cheney in the Ohio micros in the days before the election, whereas the Kerry-Edwards campaign focused on the metropolitan areas of Cleveland, Akron, and Columbus (El Nasser, 2004).

3.6 ANALYZING AN URBAN ISSUE: METROPOLITAN DEFINITION

Since the metropolitan concepts originally appeared in 1910, there have been many changes in their definitions. This makes sense because the nation has changed a great deal over the past 90 years in its cities and its urban system. A significant change in the latest 2003 procedures for defining metropolitan areas concerns inclusion rules for outlying counties that are somewhat more reflective of current-day urban forms. For instance, an outlying county is included in a metropolitan area if it either sends commuters to central counties or if it receives commuters from central counties. In prior guidelines, the bias was on the sending of commuters to central counties, but now they are weighted more equally.

Prior to accepting the new 2003 rules for defining metropolitan areas, the U.S. Census Bureau commissioned four studies to evaluate the former 1990 procedures (Dahmann & Fitzsimmons, 1995). The desire to capture the true extent of a city's influence commenced in 1910 with metropolitan districts, which were created for cities of over 200,000. In 1930 and 1940, metropolitan districts were created for cities of 50,000 or more and generally including adjacent minor civil divisions, such as townships and so on, with population density of 150 persons per square mile.

By the 1940s there was much dissatisfaction with metropolitan district classifications, mainly because the districts were not defined in terms of counties, and that reduced the amount of statistical data available. UAs and standard metropolitan statistical areas (SMSAs) were then conceived for the 1950 Census, whose guidelines were the predecessors of the 2003 guidelines applied to the Springfield, Missouri, metropolitan statistical area in the GIS exercise at the end of the chapter (see Figure 3.18).

In later years, critics continued to argue that the system was not consistent with the rapid restructuring that

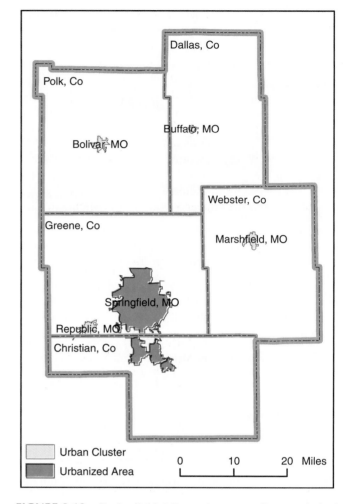

FIGURE 3.18   Springfield, Missouri, metropolitan statistical area, 2000.

was occurring throughout the entire American urban system. One critic went so far as to state that the system was "so arcane, so needlessly complex, so lacking in underlying principle and so afflicted by "ad hoc-ism" in the selection and modification of criteria for inclusion or exclusion of counties in statistical areas as to be ludicrous" (Berry, 1995, pp. 86–87).

Some have even questioned the use of journey to work in measuring interaction, as it might not be the most significant trip in the modern-day metropolis (Adams, 1995). Others have pointed to recent developments in telecommunications whereby economic integration of outlying households with the central core can be maintained without ever leaving home.

Some of these criticisms were taken into consideration when developing the procedures for defining metropolitan statistical areas that were implemented in 2003. Other suggestions were not feasible to implement; for instance, there are no consistent national data on telecommunications links at the city and county levels. Nevertheless, only time will tell if the latest guidelines will hold up to continually changing urban forms or additional criticisms that might arise of the latest definitions.

In 2000, 141.1 million Americans lived in the nation's top 30 metropolitan statistical areas (see Table 3.3). This was half of the U.S. population of 281.4 million. The largest U.S. combined statistical area was New York–Newark–Bridgeport with 21.3 million persons. This is an area so big that it extends across parts of four states: New York, New Jersey, Connecticut, and Pennsylvania. In second place was the Los Angeles–Long Beach–Riverside, California combined statistical area, having 16.4 million people, and the Chicago–Naperville–Michigan City

combined statistical area was third at 9.3 million. New York and Los Angeles are the megacities of the United States, according to a cutoff of 10 million. On the basis of U.S. Census data given here and a continuation of 1990s growth rates, we project Chicago will join them as a megacity in 2009. Thus, we refer to Chicago in this volume as a megacity, with the understanding that it is just on the brink of being one and might become one while this edition of the textbook is in use. We discuss these cities further in this book, particularly Los Angeles and Chicago.

The country's 30 largest metropolitan statistical areas grew in the 1990s at an average rate of 12.1 percent,

TABLE 3.3 Largest 30 metropolitan areas in the United States, 2000

Rank	TITLE	POP1990	POP2000	Change	PctChg 1990–2000
1	New York-Newark-Bridgeport, NY-NJ-CT-PA Combined Statistical Area	19,710,239	21,361,797	1,651,558	8.38
2	Los Angeles-Long Beach-Riverside, CA Combined Statistical Area	14,531,529	16,373,645	1,842,116	12.68
3	Chicago-Naperville-Michigan City, IL-IN-WI Combined Statistical Area	8,385,397	9,312,255	926,858	11.05
4	Washington-Baltimore-Northern Virginia, DC-MD-VA-WV Combined	6,665,228	7,538,385	873,157	13.10
5	San Jose-San Francisco-Oakland, CA Combined Statistical Area	6,290,008	7,092,596	802,588	12.76
6	Philadelphia-Camden-Vineland, PA-NJ-DE-MD Combined Statistical Area	5,573,521	5,833,585	260,064	4.67
7	Boston-Worcester-Manchester, MA-NH Combined Statistical Area	5,348,894	5,715,698	366,804	6.86
8	Detroit-Warren-Flint, MI Combined Statistical Area	5,095,695	5,357,538	261,843	5.14
9	Dallas-Fort Worth, TX Combined Statistical Area	4,138,010	5,346,119	1,208,109	29.20
10	Miami-Fort Lauderdale-Miami Beach, FL Metropolitan Statistical Area	4,056,100	5,007,564	951,464	23.46
11	Houston-Baytown-Huntsville, TX Combined Statistical Area	3,885,180	4,815,122	959,942	24.90
12	Atlanta-Sandy Springs-Gainesville, GA Combined Statistical Area	3,317,380	4,548,344	1,230,964	37.11
13	Seattle-Tacoma-Olympia, WA Combined Statistical Area	3,008,669	3,604,165	595,496	19.79
14	Minneapolis-St. Paul-St. Cloud, MN-WI Combined Statistical Area	2,809,713	3,271,888	462,175	16.45
15	Phoenix-Mesa-Scottsdale, AZ Metropolitan Statistical Area	2,238,480	3,251,876	1,013,396	45.27
16	Cleveland-Akron-Elyria, OH Combined Statistical Area	2,859,644	2,945,831	86,187	3.01
17	San Diego-Carlsbad-San Marcos, CA Metropolitan Statistical Area	2,498,016	2,813,833	315,817	12.64
18	St. Louis-St. Charles-Farmington, MO-IL Combined Statistical Area	2,629,801	2,754,328	124,527	4.74
19	Pittsburgh-New Castle, PA Combined Statistical Area	2,564,535	2,525,730	−38,805	−1.51
20	Denver-Aurora-Boulder, CO Combined Statistical Area	1,875,828	2,449,044	573,216	30.56
21	Tampa-St. Petersburg-Clearwater, FL Metropolitan Statistical Area	2,067,959	2,395,997	328,038	15.86
22	Cincinnati-Middletown-Wilmington, OH-KY-IN Combined Statistical Area	1,880,332	2,050,175	169,843	9.03
23	Sacramento- Arden-Arcade- Truckee, CA-NV Combined Statistical Area	1,587,249	1,930,149	342,900	21.60
24	Portland-Vancouver-Beaverton, OR-WA Metropolitan Statistical Area	1,523,741	1,927,881	404,140	26.52
25	Kansas City-Overland Park-Kansas City, MO-KS Combined Statistical Area	1,695,974	1,901,070	205,096	12.09
26	Charlotte-Gastonia-Salisbury, NC-SC Combined Statistical Area	1,501,663	1,897,034	395,371	26.33
27	Indianapolis-Anderson-Columbus, IN Combined Statistical Area	1,594,779	1,843,588	248,809	15.60
28	Columbus-Marion-Chillicothe, OH Combined Statistical Area	1,613,711	1,835,189	221,478	13.72
29	San Antonio, TX Metropolitan Statistical Area	1,407,745	1,711,703	303,958	21.59
30	Orlando-The Villages, FL Combined Statistical Area	1,256,429	1,697,906	441,477	35.14

Note: This table is based on the metropolitan definitions in effect as of June 2003.

Source: U.S. Census Bureau, 2004.

0.5 percent faster than the United States as a whole. By comparison, Mexico's four largest cities grew during the 1990s at 23.1 percent, 2.9 percent faster than Mexico in total (Instituto Nacional de Estadística, Geografía, e Informática [INEGI], 2001). The higher rate of metropolitan growth in the developing world has many ramifications, some of which are examined in Chapters 11 and 12.

The nation's metropolitan statistical areas in 2000 were concentrated most heavily in the mid-Atlantic and southern New England states, as well as in Florida and California. The top 30 are mostly in the Northeast, California, Texas, and Florida. This helps to understand the major routes for economic flows, transport, communications; and it meant that Florida was a critical battleground for the 2004 presidential campaign, while the other three were big electoral vote contributors.

Understanding the metropolitan concept allows further studies of how metropolitan areas are arranged spatially, what determines their spheres of influence, and what transfers of goods, people, services, and information take place between them.

3.7 OTHER APPROACHES TO DEFINING THE METROPOLIS: THE CASE OF MEXICO

Like the United States, Mexico has major urban population, large cities, and expanding metropolitan areas. The concepts of city and metropolitan zone are similar to those in the United States, yet they are different. For one thing, cities have grown and developed differently in Mexico. A major factor is that new migrants, who are usually poor, tend to arrive at the metropolitan periphery. The more affluent areas of the city are located toward the city center. Yet the cities have persistently grown outward and their geographical areas have enlarged. This stems both from the overall higher growth rate of the Mexican urban population, which is due to higher fertility than in the United States, and from the steady migratory streams of Mexicans from rural to urban areas during the 20th century.

This section looks at Mexico's urban and metropolitan system, with particular attention to Mexico City. After examining the urban and metropolitan definitions and how they differ from those in the United States, the chapter discusses more about the metropolitan area of Mexico City and how the metropolis has increasingly spread out over two states. It expands on the introduction in Chapter 2 to the tremendous expansion of the metropolitan population and land area since the 1950s; the city's continuing deconcentration in the center; some cultural differences that account for these urban patterns; and this growth in terms of the city's natural features, especially its volcanic mountains.

In Mexico, an urban area (*area urbana*) consists of a central city plus a contiguous area that has buildings and inhabitants, with land use being nonagricultural. The urban, built-up area is physically contiguous in all directions, up to limits where it is interrupted by nonurban land use, such as forests, fields, or water bodies. The urban area contains population that is classified as urban from the geographical, social, and economic perspectives. The definition implies tightly built-up urban territory, leading to an urban area that has an irregular form and does not coincide with the political-administrative boundary of the city, such as that defined by INEGI in 1995 (see Figure 3.19). Because the urban area is so closely defined by its built-up quality, it is sometimes perceived by Mexican academic experts as an "urban blot or smudge" (*mancha urbana*).

This urban area concept is similar to the UA or UC in the United States. Like the block groups in the UA or UC, the urban area in Mexico is built up of contiguous areas geográficas estadísticas basicas (AGEBs). An AGEB is a basic urban geographic unit built up from city blocks and includes about 2,500 persons. The approximate 2,500 minimum ensures that the minimum urban size is about 2,500 persons, which is equivalent to that of the United States urban threshold.

In the Mexico City area, there are 37 *municipios* reached by the continuous urban area defined as just described. Together these comprise a definition of the UA adjusted to administrative boundaries and referred to as the continuous UA on Figure 3.19.

A broader urban concept in Mexico is that of *conurbation*. This is the union of two or more urban areas that have grown very close or are touching each other. These urban areas are in different political-administrative units. Often these urban areas formerly were distinct and have grown together, as the Mexico City urban complex has expanded (see Figure 3.20). For instance, an example of this conurbation is the coalescence of the Toluca urban area with the Mexico City urban area.

A metropolitan area (*area metropolitana*) in Mexico is defined as territory composed of complete municipios, or a combination of city wards (*delegations*) and municipios for Mexico City. A *municipio* is similar in size but somewhat different in concept from a U.S. county. In this textbook, we keep this Spanish word, and refer to them as municipios, not translating the term to counties, to distinguish them from the United States. A Mexican municipio is an incorporated unit having a municipio government headed by an elected president, administrative offices, and responsibilities. Although the very largest cities extend over several municipios, most large cities are located in one municipio, which serves as the city government.

The outlying metropolitan municipios maintain an intense, daily social and economic interchange with the central city (Figure 3.19). This concept also resembles that

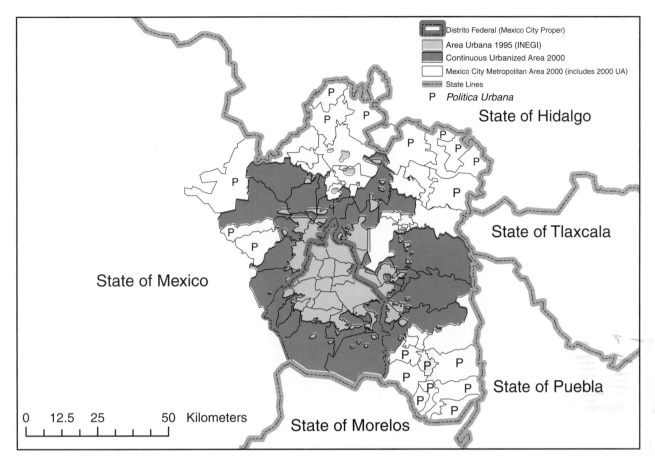

FIGURE 3.19   City, urban, and metropolitan zones of Mexico City. Source: Corbis Digital Stock.

for the United States, although the U.S. CBSA defines its connections with outlying counties strictly by commuting flows, whereas Mexico defines these connections more broadly. The metropolitan zone in Mexico has smoother boundaries because it is built up of large units; however, it also contains a variety of political-administrative units, including those of the central city, municipio or municipios, and the peripheral towns and cities. The broad variety of administrative units might be one reason that cities in Mexico lack clear-cut planning.

INEGI defined the metropolitan zone of Mexico City in 2000 to consist of 16 city wards in the Federal District plus 58 municipios in the surrounding State of Mexico and one in the State of Hidalgo (Figure 3.19). However, 18 of these municipios are included for nonstatistical reasons, for example, for planning purposes or to round out the metropolitan boundary to match an existing state boundary (referred to as *politica urbana* and labeled P in Figure 3.19). The Federal District is one of the 32 Mexican states, but is special in constituting the seat of the federal government, somewhat like Washington, DC. This huge metropolitan zone in 2000 had a population of 18.5 million. As seen in

Table 3.3, this metropolitan zone is larger in population than all the U.S. metropolitan areas in year 2000 except for the New York–Newark–Bridgeport combined statistical area.

The Mexico City metropolis has been growing rapidly, especially since the 1950s (see Figure 3.21). It expanded in the 1990s by 18 percent, with nearly all the growth concentrated in the conurbated municipios in the State of Mexico. This growth also reflects an expansion of territory of the metropolis by inclusion of more municipios. By contrast, in the 1990s, the three largest U.S. combined statistical areas grew at slower rates—New York at 8 percent, Los Angeles at 13 percent, and Chicago at 11 percent.

Mexico City's metropolitan population of only 3.2 million in 1950 increased to 8.8 million in 1970, and to 18.5 million in 2000. At the same time, the older city consisting of the Federal District grew until it peaked at 8.8 million in 1980 and has declined slightly in the two decades since. What is remarkable in the last 40 years has been the explosive growth of the metropolitan municipios in the State of Mexico. With respect to the underbounded old city limits of the Federal District, the metropolitan zone has been overbounded; it has leapt over this boundary.

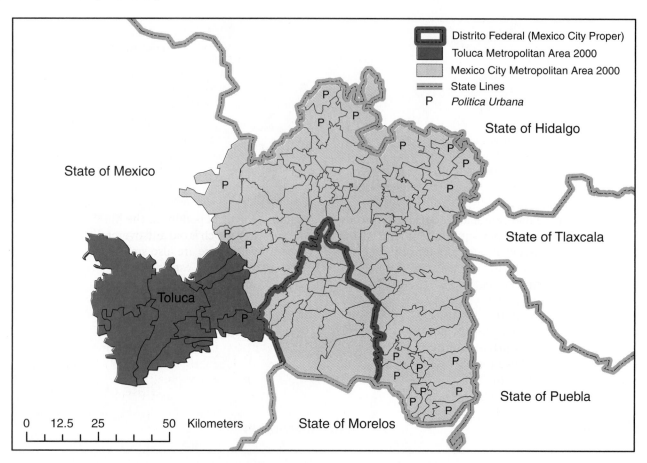

FIGURE 3.20   Mexico City and Toluca conurbation, 2000.

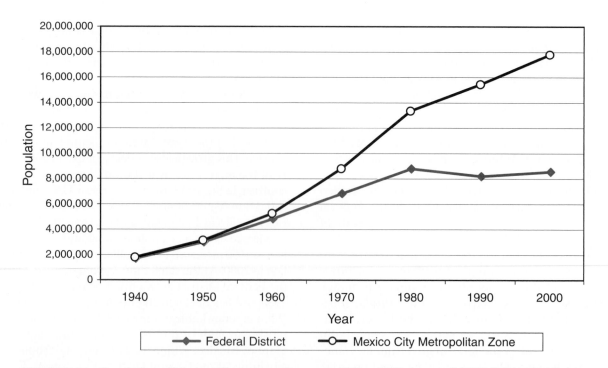

FIGURE 3.21   Population of Mexico City metropolitan zone, 1950 to 2000. Source: INEGI (2001) *Estadísticas del Medio Ambiente del Distrito Federal y Zona Metropolitana 2000*. Aguascalientes, Mexico: Institute Nacional de Estadísticas, Geografía, e Informática.

From a geographical standpoint, as mentioned in Chapter 2, the metropolitan growth occurred each decade in wider and wider rings of growth, while the city core depopulated. Mexico City's core lost increasing amounts of population from 1960 to 2000. The present core includes both high-rise and low-rise buildings, seen in Figure 3.22. What is the lesson of this 50-year scenario of outward-moving growth rings and a depleting core? Much of the explanation is that land opportunities were greater and land prices cheaper in the growth rings, so population moved increasingly outward toward them over the 50 years. On the other hand, during this period, the dense urban core had population densities in some wards over 12,000 persons per square kilometer, rising land prices, greater insecurity, and a deteriorating environment, forming push factors to leave. Mexico City's private urban developers helped in this 50-year process by investing in housing and buildings in these expanding rings to attract and absorb the population.

Another reason for this growth pattern was the attraction of Mexico City's urban periphery for migrants from the rural areas. These migrants moved there because they perceived opportunities to better their life experiences and standard of living. They could not afford to settle closer. In fact, some of them did not purchase the land they settled on, but rather squatted on it. These migrants have been referred to as *parachutists,* meaning that they dropped in, landed, and claimed their land.

An important effect of the expanding rings has been the development and problems of transportation in the metropolis. Mexico City suffers from a deficit in major highways. Its major freeway, termed the *Periferico,* was built in the 1960s when the city was much smaller, and it is inadequate for current commuting flows. There are major traffic stoppages nearly every day during peak commuting times. Recently, the Mexico City government built a second story on parts of the Periferico that offers some relief. Because the major employment centers remain centralized in the urban core (Pick & Butler, 1997), the pressure continues. One response is the large number of commercial minibuses in the metropolis, which carry millions of commuters every day, but are inefficient and serious sources of pollution. The Mexico City *metro* (i.e., the subway), which is old and overcrowded, is heavily utilized as well. There are also small-scale commuting patterns, which apply especially to the lower classes in Mexico City. They do not tend to commute long distances and mostly use public transport, especially inexpensive minivans and the subway.

The Mexico City metropolis also has natural features that have been in interplay with the massive metropolitan growth of the last 50 years. Five hundred years ago, the Aztec city was originally set in a huge shallow lake with rich forest ecosystems surrounding it. Its natural setting then and now is one of towering volcanoes to the south and southwest, mountains to the west, and a high plain stretching to the north across the State of Mexico and beyond. The ancient lake has mostly dried up, but residuals of the lake remain in the northeastern part of the Federal District and in the east including Texcoco.

FIGURE 3.22 Mexico City core.

Metropolitan growth has caused major losses to the forests and other biota of the metropolitan region. The discussion of the natural setting here is in terms of how the natural features have been barriers influencing the metropolitan distributions, shapes, and human densities. This discussion is helpful because any city or metropolitan area interacts and adjusts with its natural features in its development.

The natural regions of Mexico City are shown in Figure 3.23 (also see Figure 2.33 in Chapter 2 for additional detail). The huge snow-capped volcanoes of Popocatepetl and Ixtaccihuatl and surrounding mountainous regions are included in this semicircle extending to southwest, south, and southeast. These areas are not very habitable and even less so because of volcanic activity in recent years in Popocatepetl. In the middle and north of the Federal District and extending northward through the State of Mexico is an area of plains. This area is the most habitable and also forms a corridor for future growth. It is broken up at the tip of the Federal District and to the northwest by several smaller volcanic mountain regions. The large hills to the east and west of the plains are also barriers and have not had much habitation.

The influence of this topography is clear when looking at the urban area of Mexico City. It is clear that the south part of the plains is almost entirely filled with urban area. The northern part of the plains, located in the State of Mexico, has many pockets of urban settlement. These have the potential, as the city expands more, to coalesce through processes we have talked about into a continuous urban area rivaling that already present to the south. Toward the southern volcanic semicircle, the physical barriers are so formidable that it is unlikely in the foreseeable future to have more than minor additions of population. The elevations of Popocatepetl and Ixtaccihuatl are 17,930 and 17,159 feet, respectively. A further constraint to metropolitan expansion in the south is a water supply, which is more restricted because of limited supply from the underground basin. We examine Mexico City's water problems in Chapter 11.

In the southern city wards of Mexico City, the volcanoes have caused overboundedness. These wards are much larger than their habitable portions, which are in the north.

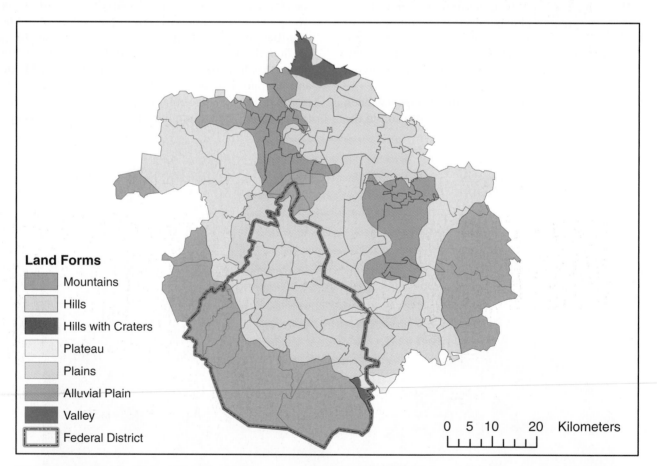

FIGURE 3.23   Mexico City and its natural regions. Source: INEGI. (2001) *Estadísticas del Medio Ambiente del Distrito Federal y Zona Metropolitana 2000*. Aguascalientes, Mexico:Institute Nacional de Estadísticas, Geografía, e Informática.

To the south are large stretches of natural areas that adjoin the lower reaches of the volcanoes with little population. This is a different type of overboundedness, not discussed earlier. It is the result of inclusion in the metropolitan region of relatively uninhabitable land areas.

This example emphasizes that natural barriers can influence urban settlement patterns and can point a metropolis in certain directions for future expansion. It has touched also on environmental losses and carrying-capacity limits that constrain the size and shape of cities.

SUMMARY

Urbanized areas, cities, and metropolitan areas form key elements of urban geography. Most countries define urban places, at least in part, as places larger than a specified minimum size. The metropolitan concept is focused on central areas that are connected through commuting and other social, transport, and economic linkages to peripheral areas. Overall, the metropolis is less tightly defined than UAs. Both urban areas and metropolitan areas are associated with administrative and government units. However, the definitions are not based on government, but rather on agglomerations of people. In many cases, there is a poor fit between the government arrangement in space and the ability to govern a city or metropolitan region well.

Metropolitan areas are not isolated to themselves, but form systems nationally and regionally. The metropolises interact with each other through spheres of influence, transport, trade, migration, and communications. The chapter exercise applies spatial analysis to examine commuting exchanges. Many other models exist to try to account for city and metropolitan interchanges.

REFERENCES

Adams, J. S. (1995) "Reconsidering the conceptual basis of federally defined metropolitan areas," in *Metropolitan and Nonmetropolitan Areas: New Approaches to Geographical Definition*. ed. by D. C. Dahmann and J. D. Fitzsimmons, pp. 9-83. Washington, DC: Working Paper Series of the Population Division, U.S. Bureau of the Census.

Barnett, J. (1995) *The Fractured Metropolis: Improving the New City, Restoring the Old City, Reshaping the Region*. New York: HarperCollins.

Berry, B. J. L. (1995) "Capturing evolving realities: Statistical areas for the American future," in *Metropolitan and Nonmetropolitan Areas: New Approaches to Geographical Definition*. ed. by D. C. Dahmann and J. D. Fitzsimmons, pp. 85-137. Washington, DC: Working Paper Series of the Population Division, U.S. Bureau of the Census.

Dahmann, D. C., and Fitzsimmons, J. D.(eds.). (1995) *Metropolitan and Nonmetropolitan Areas: New Approaches to Geographical Definition*. Washington, DC: Working Paper Series of the Population Division, U.S. Bureau of the Census.

El Nasser, Haya. (2004). "For political trends, think micropolitan." *USA Today*, November 23, page A13.

Forstall, R., Greene, R., and Pick, J. (2004) "Which are the largest? Why published populations for major world urban areas vary so greatly." *City Futures Conference Proceedings*, Chicago.

Frey, W. H., and Zimmer, Z. (2001) "Defining the city," in *Handbook of Urban Studies*... ed. by R. Paddison, pp. 14–35. London: Sage.

Goheen, P. (2000) "Urban edges at the end of the twentieth century: An introduction to the Canadian experience," in *Re-Development at the Urban Edges....* ed. by H. Nicol and G. Halseth, pp. 3-17.. Waterloo, Ontario: University of Waterloo Press.

Institute Nacional de Estadísticas, Geografía, e Informática. (2001) *Estadísticas del Medio Ambiente del Distrito Federal y Zona Metropolitana 2000*. Aguascalientes, Mexico: Author.

Lang, R. E., Dhavale, D., and Haworth, K. (2004). "Micro Politics: The 2004 presidential vote in small-town America. *Metropolitan Institute at Virginia Tech Census Note* 04:03. Blacksburg, Virginia: Virginia Tech.

Meyers, D. (1992) *Analysis with Local Census Data: Portraits of Change*. Boston: Academic Press.

Paddison, R. (ed.). (2001) *Handbook of Urban Studies*. London: Sage.

Pick, J. B., and Butler, E. W. (1997) Mexico Megacity. Boulder, CO: Westview.

Reisner, M. (1986) *Cadillac Desert: The American West and Its Disappearing Water*. New York: Penguin .

Robinson, A., Morrison, J. L., Muehrcke, P. C., Kimerling, A. J., and Guptill, S. C. (1995) *Elements of Cartography (6th Edition)*. New York: Wiley.

Rusk, D. (1993) *Cities without Suburbs*. Baltimore, MD: Johns Hopkins University Press.

Steingold, J. (2002) "Succession skepticism: Residents of San Fernando Valley doubt breakaway city, if OK'd, could provide vital services quickly enough." San Francisco, CA, *Chronicle*, October 30, 2002. Available 10/03 at *www.sfgate.com*.

U.S. Bureau of the Census. (1993) *The 1992 Census of Governments*. Washington, DC: U.S. Government Printing Office.

U.S. Census (2003). "About Metropolitan and Micropolitan Statistical Areas." Washington, DC: U.S. Census. Available 10/03 at *www.census.gov*.

U.S. Census (2004). "Geographic Changes for Census 2000 + Glossary." Washington, DC: U.S. Census. Available 11/04 at *www.census.gov*.

Whiteside, R. (2002) "Sprawling for prisoners." *Harper's Magazine* (February):88.

Wirth, L. (1938) "Urbanism as a way of life." *American Journal of Sociology* 44(1):1–24.

EXERCISE

Exercise Description

To illustrate how the U.S. Census Bureau implements the Office of Management and Budget's (OMB) standards for defining metropolitan statistical areas (MSAs). The exercise uses the Springfield, Missouri, MSA as an example.

Course Concepts Presented

Metropolitan statistical area, urbanized area, urban cluster, core county, outlying county, and commuting.

GIS Concepts Utilized

Spatial queries and overlays.

Skills Required

ArcGIS Version 9.x

GIS Platform

ArcGIS Version 9.x

Estimated Time Required

1 hour

Exercise CD-ROM Location

ArcGIS_9.x\Chapter_03\

Filenames Required

metro_bndry.shp
metro_uauc00.shp
metro_cos.shp
greeneco_blks.shp
chrstnco_blks.shp
metro_micro_procs.pdf (Description of Official Procedures)

Background Information

In this lab, you examine the method used by the U.S. Census Bureau to define a metropolitan area. The metropolitan area concept is an important urban concept because it represents the broad geographic region that is influenced by a city. You will be working with the Springfield, Missouri, MSA to determine how its core and outlying counties were determined.

Readings

(1) Berry, B. J. L. (1995) "Capturing evolving realities: statistical areas for the Amercian future." In *Metropolitan and Nonmetropolitan Areas: New Approaches to Geographical Definition* (Working paper series of the Population Division). ed. by D. C. Dahmann and J. D. Fitzsimmons, pp. 85-137. Washington, DC: U.S. Bureau of the Census.

(2) Greene, R. P., and Benhart, J. E. (1992) "The encroachment of megalopolis into the great valley: Evidence from the Cumberland Valley." *The Pennsylvania Geographer* 30:30–46.

(3) Greene, R. P., and Pick, J. B. *Exploring the Urban Community: A GIS Approach*, Chapter 3.

Exercise Procedure Flowchart

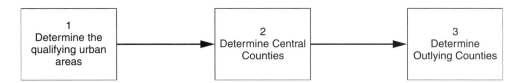

Exercise Procedure

Introduction

The U.S. Census Bureau's 2003 definition of the Springfield, Missouri, MSA is composed of a five-county area centered on the city of Springfield. Using the official procedures for defining MSAs (see metro_micro_procs.pdf in Exercise folder), we will use GIS to replicate their definition. The exercise only completes the first three sections of the procedures that relate to identifying the population requirements for urban areas, establishment of central counties, and establishment of outlying counties.

Part 1: Population Size Requirements for Qualification of Core-Based Statistical Areas (see page 10 of Metro_micro_procs.pdf document)

Each core-based statistical area (CBSA) must have a Census Bureau–defined urbanized area of at least 50,000 population or a Census Bureau–defined urban cluster of at least 10,000 population. (Urbanized areas and urban clusters are collectively referred to as urban areas.)

MSAs are a type of CBSA while micropolitan statistical areas are the second type.

1. Start ArcMap with Start ArcMap With A New Empty Map selected. Click OK.

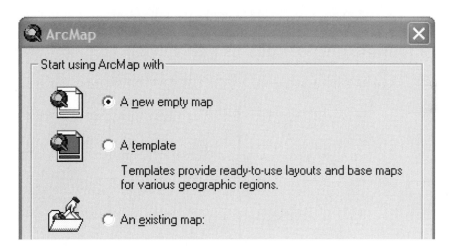

2. Click Add Data and navigate to the exercise data folder.

Note: *If you do not see the folder where you placed the exercise data, you will need to make a new connection to the root directory that holds the data.*

3. Add metro_bndry, metro_uauc00, and metro_cos from the Chapter 3 folder. You can add all three at once by clicking on one of the file names, then clicking the others while holding down the Ctrl key on the keyboard and clicking Add.

4. Ensure that the metro_bndry layer appears first in the list under the Layers data frame followed by metro_cos and metro_uauc00. If they do not, you can click on a layer name in the table of contents and drag the layer to a new position.

5. Click the fill symbol below metro_bndry and click the Hollow box. In the same Symbol Selector dialog box, click Properties and then click Outline. Scroll down and select Boundary, State and change the color to blue. To exit, click OK on each dialog box until the Symbol Selector is dismissed. Your Metro boundary should look like the following:

6. Click the fill symbol below metro_cos and click the Hollow box. In the same Symbol Selector dialog box, click Properties and then click Outline. Scroll down and select Boundary, County. To exit, click OK on each dialog box until the Symbol Selector is dismissed. Your County boundaries should look like the following:

7. Your map should look like the following:

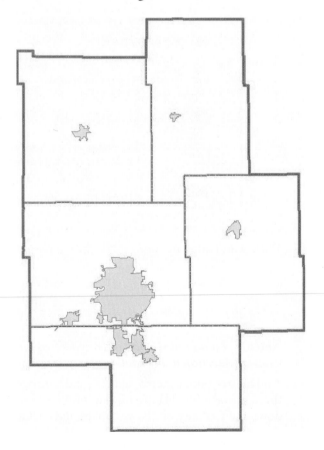

Print the Map

8. Select Layout View under View to compose a map for printing.
9. Click Insert and insert a title (Springfield, MO Metropolitan Statistical Area).
10. Click File and then click Print.
11. Select Data View under View to return to the Data View.

Queries

12. Click the Selection menu tab at the top of the ArcMap screen and choose Select By Attributes. In the dialog box choose metro_uauc00 as the Layer. Enter the following query and click Apply and then click Close:

    'POP2000" > =10000

13. Open the attribute table of metro_uauc00 (right-click the layer name) and note that one of the five urban areas was selected (Springfield, MO Urbanized Area). Thus, for section one of the guidelines, the Springfield, Missouri, MSA qualified initially for consideration based on the fact that it had one qualifying urban area. Because the qualifying urban area was an urbanized area rather than an urban cluster, it would later be identified as an MSA rather than a micropolitan statistical area. Close the table.

	FID	Dissolve	UA	Cou	First	First_LSAD_TRANS	First_NAME	POP2000
▶	0	Polygon	08812	1	76	Urban Cluster	Bolivar, MO	8746
	1	Polygon	11323	1	76	Urban Cluster	Buffalo, MO	2681
	2	Polygon	55117	1	76	Urban Cluster	Marshfield, MO	6032
	3	Polygon	74233	1	76	Urban Cluster	Republic, MO	8846
	4	Polygon	83953	8	75	Urbanized Area	Springfield, MO	215004

Attributes of metro_uauc00

Part 2: Central Counties (see page 10 of metro_micro_procs.pdf document)

The central county or counties of a CBSA are those counties that (a) have at least 50 percent of their population in urban areas of at least 10,000 population; or (b) have within their boundaries a population of at least 5,000 located in a single urban area of at least 10,000 population.

14. Right-click metro_cos and click Label Features. Note that Greene and Christian are the only counties that intersect the qualifying urban area. Accordingly, it is only necessary to test these two counties for central county status.

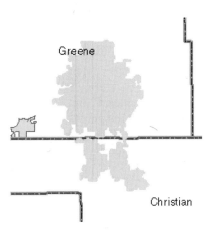

Greene

Christian

15. Use the Identify tool and click Greene County. Make sure to click the Layers drop-down box in the Identify dialog box and select metro_cos as your layer to identify. Once you have the correct identifying layer, click Greene County again. Note that the 2000 population for Greene County is 240,391.

16. Now click Christian County with the Identify tool and note that it has a 2000 population of 54,285. Close the Identify dialog box, as you are now ready to determine the percentage of these counties' population within the Springfield urbanized area.

17. Click Add Data and add greeneco_blks (these are the census blocks for Greene County that contain population counts).

18. Click the Selection menu tab at the top of the ArcMap screen and choose Select By Location. In the dialog box, select greeneco_blks as the layer to select from, change Intersect to Have Their Center In, and click the drop-down box to select metro_uauc00 for the Features In This Layer option (see below for correct settings).

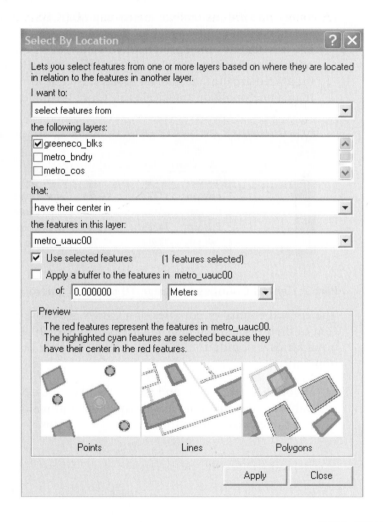

19. Click Apply and then click Close.

20. Right-click the layer name greeneco_blks and select Open Attribute Table. Scroll over and right-click on the POP2000 field and click Statistics. Note a population sum of 196,730, which represents 78 percent of Greene County's population that is within the Springfield

urbanized area. Thus Greene County qualifies as a central county under (a) of Central Counties in the guidelines. Close Statistics Results and close the table.

21. Right-click greeneco_blks and click Remove.

22. Click Add Data and add chrstnco_blks (these are the census blocks for Christian County that contain population counts).

23. Repeat Steps 18 through 21 by substituting greeneco_blks with chrstnco_blks. If you did your work correctly, you should determine a population sum of 26,561 for Christian County, which represents 48 percent of Christian County's population in the Springfield urbanized area. This is short of the 50 percent required to qualify under (a) of Section 2, but because it has more than 5,000 people within a single urbanized area, Christian County qualifies as a central county under (b) of Section 2.

Part 3: Outlying Counties (see page 10 of metro_micro_procs.pdf document)

A county qualifies as an outlying county of a CBSA if it meets the following commuting requirements: (a) at least 25 percent of the employed residents of the county work in the central county or counties of the CBSA; or (b) at least 25 percent of the employment in the county is accounted for by workers who reside in the central county or counties of the CBSA.

24. Click Add Data and add commute_matrix.dbf.

25. Right-click the name commute_matrix and click Open. The table includes fields for where workers live (RESIDENT), where those workers work (WORKPLACE), and the number of those workers with a residential and workplace match (COUNT).

OID	RESIDENT	WORKPLACE	COUNT
0	Dallas Co. MO	Maricopa Co. AZ	12
1	Dallas Co. MO	Benton Co. AR	2
2	Dallas Co. MO	Pope Co. AR	17
3	Dallas Co. MO	Pulaski Co. AR	8
4	Dallas Co. MO	Fresno Co. CA	6
5	Dallas Co. MO	Cook Co. IL	20
6	Dallas Co. MO	Johnson Co. KS	4
7	Dallas Co. MO	Wyandotte Co. KS	2
8	Dallas Co. MO	Itasca Co. MN	3
9	Dallas Co. MO	Benton Co. MO	2

Attributes of commute_matrix

Record: 1   Show: All Selected   Records (0 out of 140 Selected.)   Options ▼

26. Click Options in the lower right corner of the dialog box and click Select By Attributes. Reposition the table and the Select By Attributes dialog box so you can see both.

27. Enter the following query and click Apply:

"RESIDENT" = 'Dallas Co. MO'

28. In the table, right-click Count and select Statistics. Note that the Total Resident Workers value for Dallas County is 6,425.

29. Close the Statistics Results box and change the query to the following, then click Apply:

'RESIDENT" = 'Dallas Co. MO' AND('WORKPLACE" = 'Greene Co. MO'
OR 'WORKPLACE" = 'Christian Co. MO')

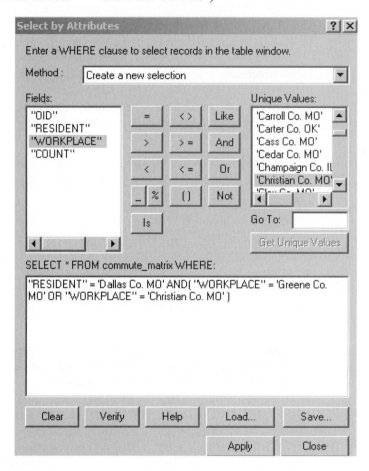

30. In the table, right-click Count and select Statistics. Note that the Total Resident Workers for Dallas County commuting to Greene and Christian Counties is 1,876. Given that the total labor force of Dallas County is 6,425, then 29 percent of them are commuting to the central counties of the Springfield MSA. Accordingly, Dallas County qualified as an outlying county of the MSA under (a) of Section 3 of the guidelines.

31. Repeat Steps 26 through 30 for Polk and Webster Counties to determine under which rule they qualified for inclusion as outlying counties in the Springfield MSA. If you do it correctly, you should find that 30 percent of Polk County's workers and 49 percent of Webster County's workers commute to the central counties of Greene and Christian. Thus all three outlying counties qualify as outlying counties under (a) of Section 3 (Outlying Counties) of the MSA Guidelines. Thus it is not necessary to test the reverse commuting criteria contained within Section 3 (b) of the guidelines.

Short Essay

MSAs are defined by the strength of journey to work flows between outlying counties to central counties or from central counties to outlying counties of the metropolitan area in question. This reflects the motor vehicle way of life in large cities, stemming from the 20th century. However, today there are more telecommuters in cities (i.e., people who work several days of the week at home and are connected to the main office through the Internet and other technologies). Please write a one-page essay on the following issue: Do you consider the telecommuting phenomenon important enough today to change the current MSA inclusion rules? If so, suggest how you would change the inclusion rules to reflect other aspects other than just commuting. Include a map of the Springfield, Missouri, MSA with your essay.

4

The Internal Structure of Cities

The internal structure of cities approach to urban geography considers the patterns of urban distributions within cities and attempts to understand the processes that cause the differentiation of spatial patterns. The extreme opposite of this approach, covered in the next chapter, is an urban systems approach, which considers the relations of cities to each other and to the areas and hinterlands they serve. Some refer to these two approaches to the study of urban geography as intraurban (within cities) investigations versus interurban (among cities) investigations. Not all urban investigations can be neatly placed into one of these two categories, but the designations serve a useful pedagogical tool for discussing urban geography research.

Among the questions the internal structure approach seeks to answer are as follows: Do rich people and poor people appear to reside within different sections of cities? Do new immigrants congregate in certain sections of the city? Is this also true of individual immigrant groups? How does the distribution of population relate to urban transportation corridors? What is the spatial relationship of city workers to their jobs? Do urban minorities live in closer proximity to environmental hazards than their nonminority counterparts?

The chapter starts with a review of the classic models of urban internal structure. First is the concentric zone model of urban development, which postulates that cities tend to grow and develop in concentric rings around their downtowns. The Chicago School of sociology advanced the concentric zone model in the 1920s, a time when immigrants had been the critical stimulant for urban growth in Chicago for several decades. As a result, much attention was paid to the effect immigrant assimilation had on the ring pattern. The second model is the sector model, which suggests that urban growth and development occur in wedges emanating out from a city's downtown. This

model put much emphasis on high-status households as the principal agents of change as opposed to low-status households. The third model is known as the multiple nuclei model of urban development, which hypothesizes a more complex urban area than the concentric zone model, with multiple centers rather than just a single dominant center of activity.

These classic models of urban development have persisted in the urban literature, and they continue to influence some modern interpretations of city growth. Nevertheless, the models have come under heavy criticism lately from the L.A. and New York Schools of urban studies. The L.A. School claim is that these classic models have had a tight grip on how urban analysis has taken place in the last century and do not capture the evolving realities of a postmodern urbanism. The L.A. School draws on processes and trends seen in the Los Angeles region and offers that region as a model for current and future urbanism. Responses by some Chicago School advocates are reviewed, highlighting the issue as one of the more critical debates within urban studies. The New York School also points out limitations in the Chicago School model, but does not agree that Los Angeles is the principal prototype for current and future urbanism. Rather, its proponents identify several characteristics of New York City's urbanism that are found in other urban centers and advocate that those characteristics should be considered alongside the processes being attributed to Los Angeles urbanism.

The "Analyzing an Urban Issue" section is presented in two parts. The first compares and contrasts the historic urban development patterns for both Chicago and Los Angeles to help acquaint the reader with the spatial context of these urban areas. The second part introduces density gradients as a tool for illustrating intraurban structure and contrasts Mexico City's internal density patterns to the two American cases and offers the Latin American city model as an explanation for the differences. A concluding case study of recent growth trends in Chicago shows a city equal in dynamics to that of Los Angeles on several key variables.

4.1 CLASSIC URBAN DEVELOPMENT MODELS

4.1.1 Concentric Zone Model

The *concentric zone model* of urban development is an early depiction of internal urban structure (see Figure 4.1). Ernest Burgess's concentric zone model first appeared in Park, Burgess, and McKenzie's (1925) *The City: Suggestions for the Investigation of Human Behavior in an Urban Environment.* The influence of the Chicago School of sociology on the discipline of urban geography, as discussed in Chapter 1, was much broader than the concentric zone model. However, the model had great appeal because of its explicit spatial aspects. Living and experiencing the tremendous growth of Chicago in the 1920s, Burgess hypothesized that cities grew in concentric rings outward from a *central business district* (CBD). The inner ring consisted of the CBD, which for Chicago was the Loop, the original name for the elevated tracks that formed a loop in the central area of the city. The CBD was encircled by what Burgess called zones in transition, workingmen's homes, residential, and commuters (see Figure 4.2). The

FIGURE 4.2   Modern-day view of Chicago's zone of transition.

zone in transition was a port of entry for poor immigrants who worked in nearby factories and was the most dynamic of all of the rings as the population was always turning over. The next outer ring, *zone of workingmen's homes,* consisted primarily of industrial workers who had moved out from the zone in transition. Made up largely of second-generation immigrants, the residents of this ring had accumulated sufficient wealth to purchase their own homes. The fourth ring was referred to as the residential zone, a zone of better residences. In Chicago, the housing in this ring contained many bungalows, one-story houses on 30-foot lots, as well as three-story apartment buildings more numerous near El lines for middle-income Loop workers mostly without autos. Finally, there was the commuter zone, suburban areas or satellite cities within a 30- to 60-minute trip of the CBD.

The concentric zone concept was both spatial and temporal. The rings would continue to expand through time because there were always new immigrants replacing the residential vacancies created by those exiting the zone in transition. As time passed, the entire structure expanded, and each ring grew at the expense of the one next further out. Burgess argued that these changes came because of the general invasion and succession processes espoused by the Chicago School (see discussion in Chapter 1).

4.1.2 Sector Model

Homer Hoyt's (1939) *sector model* was a second contribution to our understanding of the spatial aspects of city growth (see Figure 4.3). Hoyt proposed the sector model in *The Structure and Growth of Residential Neighborhoods in American Cities,* prepared for the Federal Housing Authority. In that study, the purpose of which was to classify neighborhoods according to their mortgage-lending risk, Hoyt examined 142 American cities. Like Burgess, Hoyt recognized that cities developed a center

I Loop

II Zone of Transition

III Zone of Workingmen's Homes

IV Residential Zone

V Commuters Zone

FIGURE 4.1   The concentric zone model of urban development. Source: Burgess, Ernest. 1925. "The Growth of the City," in R.E. Park, E.W. Burgess, and R.D. McKenzie, The City, University of Chicago Press. P. 51. Chart 1. "The growth of the city."

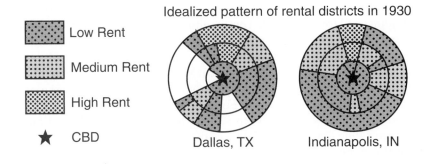

Idealized pattern of rental districts in 1930

Low Rent

Medium Rent

High Rent

★ CBD

Dallas, TX   Indianapolis, IN

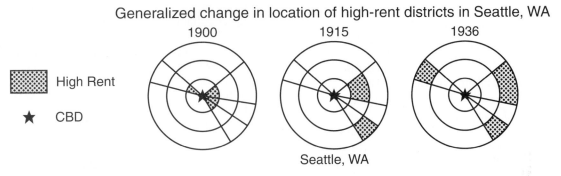

Generalized change in location of high-rent districts in Seattle, WA

1900   1915   1936

High Rent

★ CBD

Seattle, WA

FIGURE 4.3   The sector model of urban development. Source: Hoyt, H. (1939) *The Structure and Growth of Residential Neighborhoods in American Cities.* Washington, DC: Federal Housing Administration.

characterized by business and service activities, but Hoyt's model suggested that residential growth extended outward from the center in wedges or sectors rather than concentric rings. Urban residential growth was therefore seen as axial, associated with streetcar lines and other major transportation radials (see Figure 4.4).

Hoyt believed that high-income households were the engines of growth in that they preempted the most desirable land in the emerging city, mostly well away from industrial activity and often in the direction of natural amenities. Thus, in the sector model, high-rent areas take the form of wedges along axial corridors away from the

FIGURE 4.4   Streetcar sector in Toronto.

center. The medium-rent areas occupy land adjacent to the high-rent households on either side, whereas low-rent wedges protrude in opposite directions from the high-rent sectors, often alongside industrial areas. The notion that high-income households played the most important role in the spatial growth of cities was in stark contrast to Burgess's emphasis on invasion and succession of lower status households within the zone in transition as the driving force for growth and change.

The conclusions of Hoyt's study are summarized here:

1. The highest rental (or price) area is located in one or more specific sectors on one side of the city. These high-rent areas are generally peripheral locations, although some high-rent sectors extend continuously from the center of the city.

2. High-rent areas often take the form of wedges, extending in certain sectors along radial lines outward from the center to the periphery of the city.

3. Middle-range rental areas tend to be located on either side of the highest rental areas.

4. Some cities contain large areas of middle-range rental units, which tend to be found on the peripheries of low- and high-rent areas.

5. All cities have low-rent areas, frequently found opposite high-rent areas and usually in the more central locations.

Some interpretations of contemporary urban landscapes continue to draw linkages to the findings of Hoyt. For instance, Adams (1991) noted persistence in the sector patterns by reviewing a less well known follow-up investigation by Hoyt himself with 1960 census data. Adams (1991) extended the record of observation with his own analyses of 1970 and 1980 census data for 20 metropolitan areas. For the Twin Cities, Adams (1991) argued that builders initiate sector development by concentrating new construction on the suburban edge and setting in motion vacancy chains. A vacancy chain occurs when households move to a new address and leave behind vacancies. Adams demonstrated that the vacancy chains behaved in sectors, at least in the Minneapolis housing market (see Figure 4.5). One sector in particular, Sector B, saw its relative market value of neighborhoods drop as high-priced new housing was built on the suburban edge. The box inset on the figure shows the decline in housing price rank of neighborhoods in Sector B for 1980 compared to 1970. The assumption is that many household moves occurring within sectors originate closer into the city center and have destinations on the outer edge often on the same transport corridor. Such moves in the aggregate might account for the overall decline in the market values in the inner neighborhoods of the sector as wealth is transferred along with the moves to the outer suburb. Gentrification of the inner part of the sector can offset these declining market values, a topic discussed in detail in Chapter 6.

The persistence of sectors in everyday behavior is further demonstrated for Chicago where communities with the highest percentage of workers using public transportation in the year 2000 showed a strong orientation to commuter rail lines (see Figure 4.6). Living close to commuter rail lines, particularly train stations, is one of the best predictors of the use of these lines. The attraction of such access accounts for higher density and higher land values following these transit lines, which all reinforce the sector pattern of urban development that Hoyt had identified in his model. "Kiss and ride" parking lots are often placed close to these train stations resulting in even greater sector flows during peak commuting times.

4.1.3 Multiple Nuclei Model

In contrast to Burgess and Hoyt, who both emphasized growth and development away from a single CBD, Harris and Ullman (1945) developed the multiple nuclei model of urban development, arguing that growth and development occur around multiple centers of economic activity in a metropolitan area (see Figure 4.7). For instance, not only does this hypothetical metropolis have a CBD but it also has a wholesale light manufacturing center, a heavy manufacturing center, an outlying business center, and an industrial suburb. Some of these outlying industrial and business centers are quite compact and look like a downtown, such as Burbank in the Los Angeles area. Wholesale and light manufacturing centers, on the other hand, are spread over a wider territory (see Figure 4.8).

With respect to the origins of the various nuclei, the model recognized differences across cities:

> In some cities these nuclei have existed from the very origin of the city; in others they have developed as the growth of the city stimulated migration and specialization. An example of the first type is Metropolitan London, in which "The City" and Westminster originated as separate points separated by open country, one as the center of finance and commerce, the other as the center of political life. An example of the second type is Chicago, in which heavy industry, at first localized along the Chicago River in the heart of the city, migrated to the Calumet District, where it acted as a nucleus for extensive new urban development. (Harris & Ullman, 1945, p. 283)

This passage points out that a number of major world cities developed around more than one early center, most of which absorbed existing villages, the irregular street plans of which often can still be discerned within the larger city (e.g., Harlem and Greenwich Village in New York City; Forstall, 2003). Cities will have a dominant nucleus, typically around the initial point of settlement near a landing place or waterfront in many American cities. Harris and Ullman (1945) attributed the rise of separate nuclei to a combination of four factors:

1. Certain activities require specialized facilities. The retail district, for example, is attached to the point of greatest intracity accessibility, the port district to suitable waterfront, manufacturing districts to large blocks of land, and water or rail connection, and so on.

2. Certain like activities group together because they profit from cohesion.... Retail districts benefit from grouping, which increases the concentration of potential customers and makes possible comparison shopping. Financial and office-building districts depend upon facility of communication among offices within the district. The Merchandise Mart of Chicago is an example of wholesale clustering.

3. Certain unlike activities are detrimental to each other. The antagonism between factory development and high-class residential development is well known. The heavy concentrations of pedestrians, automobiles, and streetcars in the retail district are antagonistic both to the railroad facilities and the street

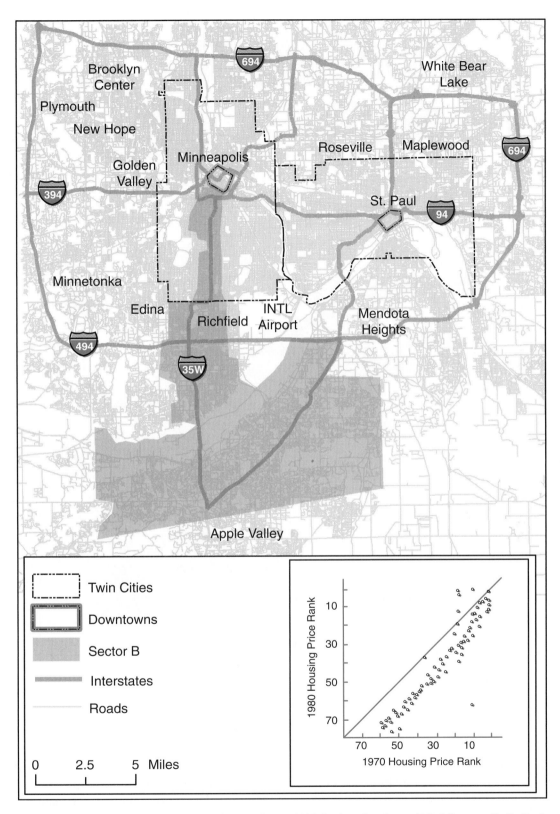

FIGURE 4.5   Changes in housing price ranks, 1970 to 1980, for housing Sector B in Minneapolis-St. Paul. Source: Adams, John S. 1991. Housing Submarkets in an American Metropolis. In J.F. Hart, editor, Our Changing Cities. Johns Hopkins Press, Baltimore, MD. Fig. 7.3 "Changes in Tract Ranks in Housing Submarkets".

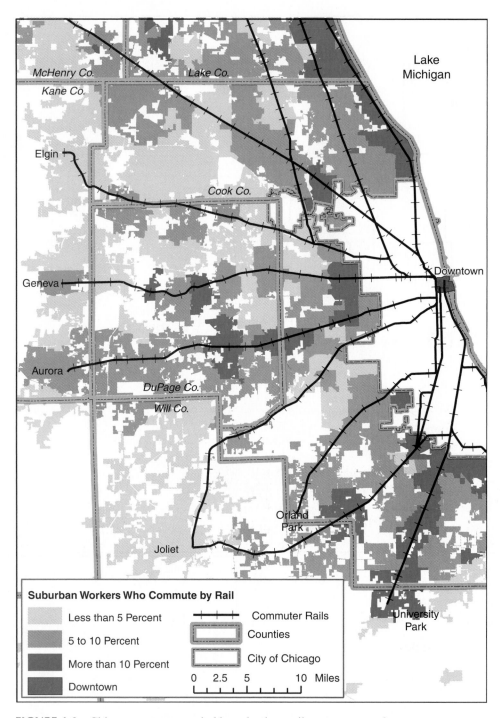

FIGURE 4.6   Chicago sectors revealed by suburban rail commuter preferences.

loading required in the wholesale district and to the rail facilities and space needed by large industrial districts, and vice versa.

4. Certain activities are unable to afford the high rents of the most desirable sites. This factor works in conjunction with the foregoing. Examples are bulk wholesaling and storage activities requiring much room, or low-class housing unable to afford the luxury of high land with a view. (Harris & Ullman, 1945, pp. 283–284)

These combinations of forces have brought about the development of separate nuclei within cities, with the number of nuclei that develop varying mainly with the size of the city.

In addition to establishing the existence of multiple nuclei, Harris and Ullman (1945) identified four urban components or districts that have developed in virtually all large American urban areas. The first and most dominant is the *CBD,* which Harris and Ullman reminded us

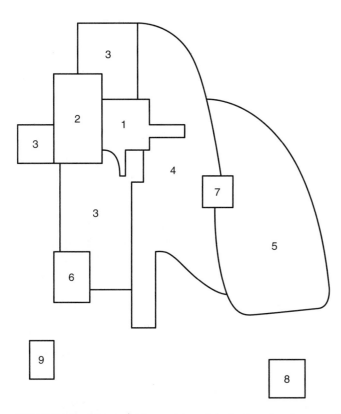

1. Central business district
2. Wholesale light manufacturing
3. Low-class residential
4. Medium-class residential
5. High-class residential
6. Heavy manufacturing
7. Outlying business district
8. Residential suburb
9. Industrial suburb

FIGURE 4.7 The multiple nuclei model of urban development. Source: Harris, C.D. and Ullman, E.L. 1959. "The Nature of Cities," in Mayer, H.M., and Kohn, C.F., Readings in Urban Geography, University of Chicago Press. P. 281. Figure 5. "Generalization of Internal Structure of Cities."

FIGURE 4.8 Wholesale and light manufacturing district in Seattle.

is not necessarily at the city's geographic center, and might even be near one edge if the city is on the sea or a lake. The CBD is the focus of intracity transportation facilities, making it the single most convenient location to access from anywhere in the urban area.

The *wholesale and light manufacturing district* is usually adjacent to but not surrounding the CBD and is accessible to railroads and trucks. This district's activity is primarily manufacturing and its location takes advantage of both transport facilities and access to the urban area's labor. Next, the *heavy industrial district* is on the edge of the city where it could occupy large tracts of bad space with good transport. Although once on the edge of the city, this district appears relatively close in as the urban area has expanded. Chicago offers good examples of such districts:

> In Chicago about a hundred industries are in a belt three miles long, adjacent to the Clearing freight yards on the southwestern edge of the city.... The stockyards of Chicago, in spite of their odors and size, have been engulfed by urban growth and are now far from the edge of the city. They form a nucleus of heavy industry within the city but not near the center, which has blighted the adjacent residential area, the "Back-of-the-Yards" district. (Harris & Ullman, 1945, p. 285)

Next is the *residential district,* differentiated into high-class, middle-class, and low-class segments. The high-class segments are typically on high, well-drained land and away from air and noise pollution, whereas low-class segments are situated close to factories and other nuisances. Other minor nuclei were recognized by Harris and Ullman, which represented cultural centers and parks. Finally, residential and industrial suburbs were distinguished from satellites that were separated by large distances to the central city with little daily commuting to or from it. At the time that Harris and Ullman devised the model, Gary was considered an industrial suburb, whereas Elgin and Joliet were considered industrial satellites.

4.2 STRUCTURALISM AND POSTMODERNISM: ALTERNATIVE PERSPECTIVES

The classic models of urban development were developed early in the 1920s and through the 1950s. As a consequence, a number of researchers have dismissed them for lack of power in explaining contemporary urban change. Beginning in the 1970s, Harvey (1973) offered an alternative perspective that applied Marxian economics to the study of conditions in the city. The approach became known as the structural approach, which emphasized analyses of uneven development and the role of capitalists in fostering it and widening it. Gottdiener (1994) reviewed how Harvey was influenced by the ideas

of the French philosopher Henri Lefebvre, who is credited with reviving the Marxist tradition in social science in the 1970s. Lefebvre's concept of circuits of capital was particularly appealing to Harvey:

> Harvey argued that capitalists involved in the first industrial circuit are principally interested in location within the urban environment and in reducing their costs of manufacturing. Capitalists in the second circuit hold a different set of priorities relating to the flow of investment and the realization of interest on money loaned or rent on property owned. (Gottdiener, 1994, pp. 134–135)

The result of these differences in how capital is invested and reinvested in these circuits is uneven development; for instance, a heavy investment in real estate in the suburbs at the expense of abandoned neighborhoods in the central city.

A more recent extension to Marxist urban geography approach is that advanced by the postmodernists. In rejection of modernist urban theory, which would include the classic urban models, the postmodernists cannot be pigeonholed; rather, they represent varying views of the city:

> [T]here are at least two camps: neo-Marxian urbanists, such as Fredrick Jameson and David Harvey, who have proclaimed the arrival of a "condition of postmodernity," and more self-conscious postmodernists, such as Jean Baudrillard, Michael Dear, and Edward Soja (in his recent work), who claim that the radical critique of modernism carried out by such theorists as Jacques Derrida and Jean-Francois Lyotard should guide a post-Marxian epistemology for interpreting the postmodern condition. (Ethington & Meeker, 2002, p. 414)

4.3 HAVE THE CLASSIC MODELS STOOD THE TEST OF TIME?

In conclusion, a number of spatial models of urban development have been proposed for understanding how urban areas grow and develop. At the same time, alternative perspectives have arisen as a result of changing times and urban processes. Mapping the census year when Chicago census tracts reached 2,000 persons per square mile shows that the region exhibits spatial elements of all three of the classic models (see Figure 4.9). The 1950 high-density tracts form a pattern in between a ring and sectors. However, the 1960 high-density tracts show that sector development patterns were well in place at that time. Also, outlying industrial suburbs can also be discerned from the 1950 and 1960 density patterns. The sector pattern was reinforced by the development trends of the 1970s and 1980s as density increased along former sectors. The 1990s also saw a continuation of growth along these same sectors, although multiple nuclei were

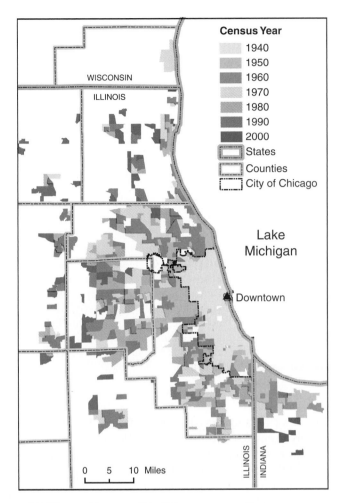

FIGURE 4.9 The year that Chicago metropolitan area census tracts reached 2,000 persons per square mile, 1940 to 2000.

evident in the concentration of new high-density areas around traditional outlying centers as well as in the newly emerging edge cities.

However, could imposing such old models on the urban landscape limit our understanding of current-day urban patterns and processes? Some urban scholars continue to make note of the pattern and legacies, especially for Chicago:

> Located on an essentially featureless plain at the shore of Lake Michigan, Chicago has a physical organization that is stunningly simple. The symmetrical semicircular core expands almost monotonically into the featureless plain from along the lakeshore, a pattern distorted only modestly by the sectoral divisions formed by radial rivers and rail lines. Beyond Cook County are its five collar counties, which contain the overflow in independent suburbs and satellite towns. (Abu-Lughod, 1999, pp. 21–22)

Berry and Kasarda (1977) showed that in spite of debate about the relative merits of the concentric zone and

sector models, there was substantial evidence to support the existence of both:

> A succession of large-scale factor analytic studies conducted since the end of the Second World War now makes it possible to state definitively that the three models are additive contributors to the total socioeconomic structuring of city neighborhoods. In each of the studies the answer is the same: there are three dominant dimensions of variation. These are (1) the axial variation of neighborhoods by socioeconomic rank; (2) the concentric variation of neighborhoods according to family structure; and (3) the localized segregation of particular ethnic groups. (Berry & Kasarda, 1977, p. 89)

Nevertheless, these models have come under intense scrutiny, especially recently by the new postmodernist L.A. School:

> Much of the urban research agenda of the 20th century has been predicated on the precepts of the concentric zone, sector, and multiple-nuclei theories of urban structure. The influences can be seen directly in factorial ecologies of intra-urban structure, land-rent models, studies of urban economies and diseconomies of scale, and designs of ideal cities and neighborhoods. The specific and persistent popularity of the Chicago concentric ring model is harder to explain, however, given the proliferation of evidence in support of alternative theories. The most likely reasons for its endurance (as I have mentioned) are related to its beguiling simplicity and the enormous volume of publications produced by adherents of the Chicago School. (Dear, 2002, p. 16)

In the first issue of *City & Community*, Dear (2002) summarized the major points of the school's doctrine and invited a debate on the topic among contemporaries of the Chicago School of sociology. Dear (2002, p. 24) summarized the major differences between the L.A. School and traditional Chicago School as follows:

1. Traditional concepts of urban form imagine the city organized around a central core; in a revised theory, the urban peripheries are organizing what remains of the center.

2. A global, corporate-dominated connectivity is balancing, even offsetting, individual-centered agency in urban processes.

3. A linear evolutionist urban paradigm has been usurped by a nonlinear, chaotic process that includes pathologic forms such as transnational criminal organizations, common-interest developments (CIDs), and life-threatening environmental degradation (e.g., global warming).

Dear offered an alternative diagram to that by Burgess, which he considered to capture the spatial manifestations

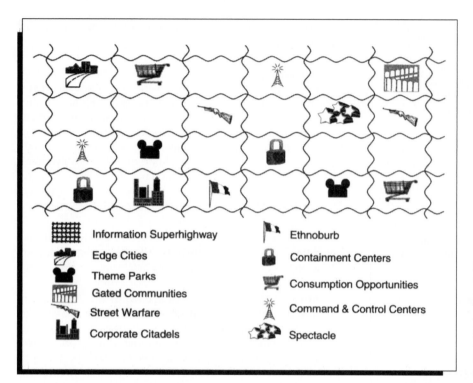

Information Superhighway

Edge Cities

Theme Parks

Gated Communities

Street Warfare

Corporate Citadels

Ethnoburb

Containment Centers

Consumption Opportunities

Command & Control Centers

Spectacle

FIGURE 4.10   Dear and Flusty's L.A. School "gaming board" diagram. Source: Dear, Michael and Steven Flusty. 2002. "Los Angeles as Postmodern Urbanism," in Michael J. Dear (ed.), From Chicago to La, Sage Publications, p. 80, Fig. 3.4. "Keno capitalism: a model of postmodern urban structure."

of these assumptions (see Figure 4.10). Note that the diagram is not spatially referenced because in the words of Dear (2002, pp. 24–25):

> Urbanization is occurring on a quasi-random field of opportunities, in which each space is (in principle) equally available through its connection with the information superhighway. Capital touches down as if by chance on a parcel of land, ignoring the opportunities on intervening lots, thus sparking the development process. The relationship between development of one parcel and non-development of another is a disjointed, seemingly unrelated affair. While not truly a random process, it is evident that the traditional, center-driven agglomeration economies that have guided urban development in the past no longer generally apply. Conventional city form, Chicago Style, is sacrificed in favor of a non-contiguous collage of parcelized, consumption-oriented landscapes devoid of conventional centers yet wired into electronic propinquity and nominally unified by the mythologies of the (dis)information superhighway. Los Angeles may be a mature form of this postmodern metropolis; Las Vegas comes to mind as a youthful example. The consequent urban aggregate is characterized by acute fragmentation and specialization—a partitioned gaming board subject to perverse laws and peculiarly discrete, disjointed urban outcomes. Given the pervasive presence of crime, corruption, and violence in the global city (not to mention geopolitical transitions, as nation-states give way to micro-nationalisms and transnational criminal organizations), the city as gaming board seems an especially appropriate 21st-century successor to the concentrically ringed city of the early 20th.

Thus the diagram offered up by the L.A. School contrasts sharply from the concentric zone, sector, and multiple nuclei models principally with respect to spatial relationships of the urban components. That is, the L.A. School's diagram as represented by Dear appears as nonspatial or as a "gaming board" because of the apparent chaos in Los Angeles development as cited in the preceding passage.

Dear's (2002) essay was a call for debate and a debate is what he got. The following are several excerpts from articles submitted by some Chicago School contemporaries that appeared in the same issue of *City & Community*. Abbott (2002) suggested that the processes being identified by Dear as new in Los Angeles were previously considered by the Chicago School:

> More generally, the Chicago view of the city as a mass of interwoven processes and peoples, a complex ecology of groups and spaces, continually under renegotiation, is precisely the processual vision that Dear thinks so revolutionary in the Los Angeles School. He has mistaken Burgess's little diagram for the whole of an enterprise of which he and his colleagues are, like the rest of us, lucky legatees. (p. 34)

Molotch (2002) also suggested that the L.A. School is identifying nothing new, but he also stated that they are mistaken to attribute what is happening in Los Angeles to local processes rather than national processes:

> The Chicago scholars were, of course, products of their day, excited by the prospect of a science of society and immersed in the unprecedented scale of urban growth surrounding

their campus. They took as general what was specific to their time and place, including its topical agenda. I think Dear and some of his colleagues err in the opposite direction, treating some of what they see in L.A. as more local than it really is. What they take to be a harbinger of the future from Southern California reflects national (at least) dynamics, rather long in evidence. (Molotch, 2002, p. 40)

Sampson (2002) saw Dear's L.A. School proposition as a disguise for nothing more than advancing postmodernism:

My assessment is ultimately optimistic—a new generation of original research is being carried out unencumbered by the theoretical baggage evident in the new L.A. and the old Chicago. The L.A. baggage, of course, is the revealed "a priori" commitment of the Los Angeles School to advance postmodern theory. (p. 45)

Sampson (2002) also suggested that the L.A. school might be misrepresenting both Los Angeles and Chicago; that is, that patterns and processes identified as uniquely Los Angeles are actually found in Chicago:

Consider Latino immigration and the Pacific Rim emphasis said to characterize modern L.A. Metropolitan Chicago now has 1.5 million residents of Latino or Hispanic origin, by some estimates the second largest concentration in the country. The city proper grew during the 1990s and is one of the most ethnically diverse cities in the United States. (p. 46)

Sassen (2002) agreed with Sampson's point on similarities between Chicago and Los Angeles on such patterns as segregation, concentration effects, inequality, and the distribution of crime. On Dear's (2002) claim that what appears as unique in Los Angeles, Sassen (2002) claimed they might be mere spatial outcomes to quite similar processes:

The underlying dynamics may be similar even as their spatial outcomes diverge sharply. Differences in spatial outcomes in this case result from the presence or absence of particular kinds of inherited built environments and metropolitan administrative structure. Being older, much of New York and Chicago's built environment responds to earlier locational logics and their distinct construction and transport options. In the case of Los Angeles, we see the opposite effect: far fewer constraints. But what is feeding spatial outcomes may well be similar. (p. 49)

Sampson defended a more contemporary Chicago School by amplifying Coleman's (1994) earlier articulation. In the words of Sampson, the school is especially concerned with:

1. A relentless focus on context (especially place).
2. A focus on properties of communities and cities as social systems.

3. A relational concern with variability in forms of social organization as opposed to population attributes (or composition).
4. Continual attention to neighborhood change and spatial dynamics (time and space).
5. An eclectic style of data collection that relies on multiple methods but always connects to some form of observation.
6. A concern for public affairs and the improvement of community life.
7. An integrating theme of theoretically interpretive empirical research. (p. 46)

The debate over whether the collection of Los Angeles urban scholars and collective viewpoints signifies a new school of urban studies will no doubt continue. One recent paper identified a Chicago School II (school of quantitative urban geography), which the authors argued has remained resilient to the critiques by the postmodernists (Shearmur & Charron, 2004). Nijman (2000) made the case that "many cities in the U.S. are likely to be affected fundamentally by trends that are found in extreme form in present-day Miami" (p. 144). He proposed the concept of the paradigmatic city as one that best captures the most typical trends and features of an urban system.

More recently, Halle (2003) proposed an alternative New York School of urban studies. Halle did not take issue with the Los Angeles School; rather he suggested that the urban character (a dispersed and centerless urban area) ascribed by L.A. School proponents to the Los Angeles area, does not capture urban settlements with strong centers such as New York City. Unlike the intellectual divide between postmodernism and urban ecology perspectives used as the basis for the establishment of a Los Angeles School, Halle turned to differences in urban structure as the bases of the argument for a New York School.

In contrast to the L.A. School's emphasis on low-density sprawling urban development, Halle suggested that the New York School is characterized by an interest in the central city. Like the L.A. School, the New York School has many voices including William Whyte (1956), Jane Jacobs (1961), Sharon Zukin (1982), Kenneth Jackson (1985), Richard Sennett (1990), and Robert Stern (Stern, Mellins, & Fisman, 1997). These authors have in common "a fascination with contemporary New York City, especially Manhattan, and a belief, in some cases passionate, in the superiority of city life over suburban life" (Halle, 2003, p. 15). In contrasting the New York School with the Chicago School, Halle noted that the Chicago School did not depict the central city as desirable for the middle class or the rich, whereas the New York School sees central New York City "as a place for the middle class and the rich, not just the poor and the working class" (Halle, 2003, p. 15). Halle credited Jane Jacobs as the founding voice for the New York School due to her role in high-profile events such as the defeat of the Lower Manhattan Expressway in the 1960s, as well as her seminal book *The Death and Life of Great American Cities*. Other major contributions to the development of a unique New York School include Zukin's (1982) study of loft living in SoHo, one of the earliest accounts of urban gentrification. Furthermore, Jackson's *Crabgrass Frontier* and Whyte's *The Organization Man* are books that Halle included as essential voices for the New York School, as they provide some of the best critique of suburban living.

To illustrate the difference between the two schools, Szanto (2003) contrasted the visual art worlds of New York and Los Angeles. New York, he argued, had the edge early on due to its connections to Europe and an elite willing to spend large sums of money on art. By the 1960s the visual arts in Los Angeles was taking off, but it remained in the shadows of New York. This was due in part to geography; that is, just like urban development in Los Angeles, the visual arts there sprawled as it expanded. Meanwhile, the density found in New York's art communities provided energy to its participants. The diffuse pattern in Los Angeles, on the other hand, made it difficult for the artists to interact.

In the next section, "Analyzing an Urban Issue," we apply geographic information systems (GIS) to map historic urban development patterns in Chicago and Los Angeles. These GIS maps allow for comparisons to be made between the two cities at the center of the first debate, Chicago and Los Angeles, and might provide insight into the similarities and differences between the two with respect to the spatial pattern of development. A second part of the section analyzes urban density patterns of Mexico City as a global contrast to the American cases. These urban issues are later pursued in the chapter's GIS exercise.

4.4 ANALYZING AN URBAN ISSUE

4.4.1 Chicago and Los Angeles Urban Development Patterns

A GIS analysis of Chicago and Los Angeles using census tract population data for the period from 1960 to 2000 identifies patterns of concentric circles, sectors, multiple nuclei, and edge city development for both urban places. The growth and development of these places during this period shows the importance of transportation technology to sector-oriented growth (Piet & Greene, 2000). The importance of this period is that there was rapid expansion of urbanized areas as well as a shift in the major mode of urban transportation away from mass transit to automobiles. The percentage of the population living within an

easily accessible distance of the major transportation corridors is compared for each census period. Because census tract boundaries changed over time, a GIS methodology was developed to proportionately relate earlier census population figures to current census tract boundaries (Figure 4.11). The concept of census tract was covered in Chapter 1 on page 17. This was accomplished by using the tract comparability tables that the Census Bureau provides for each census to construct a spreadsheet that related the current tracts to their parent tracts. Where possible, the relationships were traced back to the 1940 Census. Alternative methods exist for allocating population based on census tract boundary changes over time (Howenstine, 1993).

Utilizing the comparability tables, the spreadsheet was built backward to incorporate the 1990 Census. Where there was a one-to-one relationship between the tracts for the two censuses, the corresponding 1990 tract number was entered for each 2000 tract. Where a 2000 tract consisted of parts of multiple 1990 tracts, additional rows were added to the spreadsheet to allow the relationship to be expressed. This process was repeated for each census in the study period, ultimately resulting in a spreadsheet that contained more than 3,000 rows for Chicago and more than 5,000 rows for Los Angeles.

Table 4.1 illustrates how the method was employed for one tract within the City of Chicago. The area encompassed by the current tract 1005 has been included within a number of different tracts over time. This resulted in the current tract appearing 10 times in the spreadsheet.

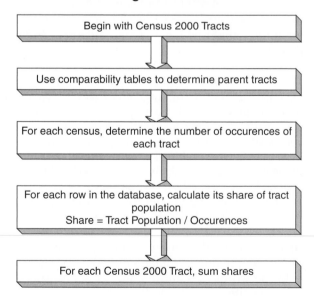

Building the Database

FIGURE 4.11 Census tract comparability procedures. Source: Piet, P., and Greene, R. P. (2000) "Population growth and the transportation link: Large city comparisons between the humid east and arid west." *Forum of the Association of Arid Lands Studies* 16:45–52.

TABLE 4.1 Comparability Tables and Spreadsheet Entries for Chicago Census Tract 1005

2000-1990	1990-1980	1980-1970	1970-1960	1960-1950	1950-1940
1005 ⇒ 1005	1005 ⇒ 1005	1005 ⇒ 1005 / 7603(pt)	1005 ⇒ 142-Z(pt) NPT-85(pt) 7603 ⇒ 142-Z(pt) NPT-85(pt)	142-Z ⇒ 142 141(pt) 143(pt) CC-6(pt) NPT-85 ⇒ CC-6(pt)	141 ⇒ 1 41 142 ⇒ 142 143 ⇒ 143 CC-6 ⇒ Not tracted

			Year				
	2000	1990	1980	1970	1960	1950	1940
	1005	1005	1005	1005	142-Z	141	141
	1005	1005	1005	1005	142-Z	142	142
	1005	1005	1005	1005	142-Z	143	143
	1005	1005	1005	1005	142-Z	CC-6	-
Tract ID	1005	1005	1005	1005	NPT-85	CC-6	-
	1005	1005	1005	7603	142-Z	141	141
	1005	1005	1005	7603	142-Z	142	142
	1005	1005	1005	7603	142-Z	143	143
	1005	1005	1005	7603	142-Z	CC-6	-
	1005	1005	1005	7603	NPT-85	CC-6	-

Once the spreadsheet was filled in, the population allocated to the current tract was calculated for each census. First, the frequencies of each tract were determined for each census. The tract population was divided by the occurrences, giving a share of the population per each occurrence. Finally, the shares were summed for each 2000 tract. In the example of tract 1005 illustrated in Table 4.1, the 10 entries for the 1960 Census contained 8 occurrences of tract 142-Z and 2 of NPT-85. Overall these two tracts appear a total of 16 and 4 times, respectively, in the spreadsheet, thus half of the 1960 population of each tract was assigned to tract 1005. The 1960, 1980, and 2000 population data were extracted as GIS databases to be used in the GIS exercise at the end of this chapter.

The growth of the Chicago metropolitan area is shown through changing density patterns in Figure 4.12. Initially, the suburban population shows a strong sector orientation. Over time, the population disperses, filling in the areas between the transportation routes. The growth pattern in Los Angeles is similar (see Figure 4.13), but the terrain of southern California constrains the dispersion. For Chicago, the suburban population living within 1 mile of the commuter railroads declined steadily over time from 1960 to 1990 (see Figure 4.14). This result would be expected as the automobile has become a more important transportation option, aiding in the dispersion of the population. This period also saw great increases in

suburban employment concentrated in and around edge cities. This served to decrease the importance of the commuter railroads that were constructed to connect downtown businesses with suburban employees, but also resulting in no connectivity from suburb to suburb.

In both Chicago and Los Angeles, the changes in the percentage of people living within 3 miles of a limited-access highway display a similar pattern (see Figure 4.15). As the picture shows, automobile traffic along major highways into cities is fed by limited on- and offramps. In both the Chicago and Los Angeles case, the percentage shows a sharp increase from 1960 to 1970 in people living close to highways built at those times (see Figure 4.16A). After this rise, the percentage remains fairly steady, declining slightly over time. This supports the commonly held idea that highway construction will fuel population growth as people utilize the easier access to relocate farther from their place of employment. It also conforms to the Hoyt sector model, according to which households move in corridors, often transport or amenity corridors. However, if we look at the percentage over time of the suburban population residing within 3 miles of the built-out (1990) highway system, a different story begins to emerge. As illustrated in Figure 4.16B, the share of the population living within this corridor was highest in 1960 and has declined steadily since then. This indicates that the highways could also have been built to

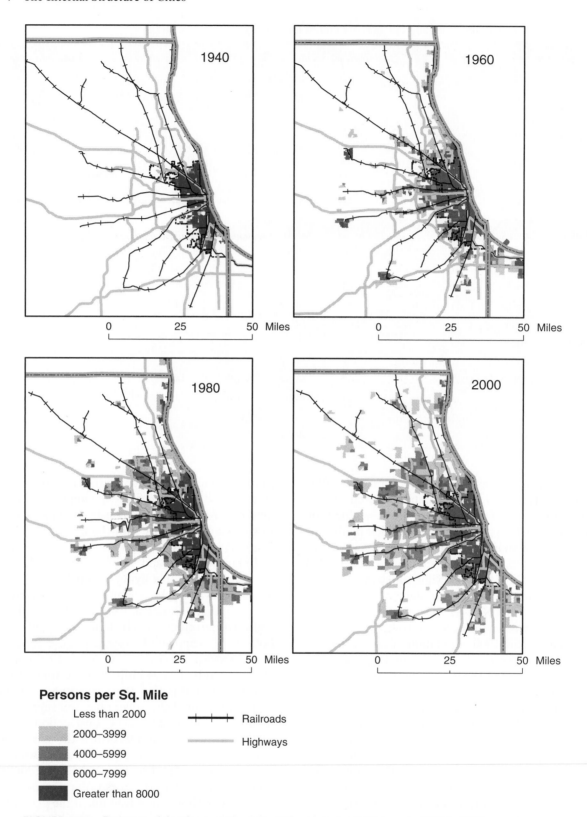

Persons per Sq. Mile

Less than 2000

2000–3999

4000–5999

6000–7999

Greater than 8000

Railroads

Highways

FIGURE 4.12   Patterns of density growth of the Chicago metropolitan area, 1940 to 2000.

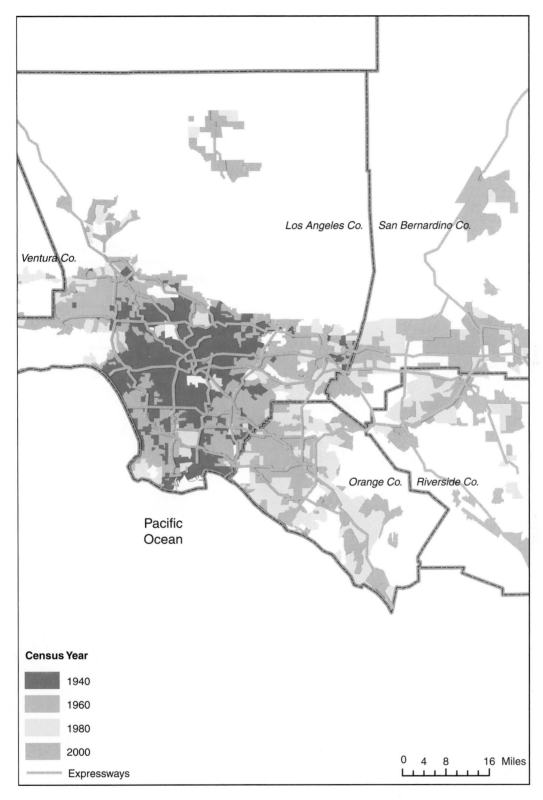

FIGURE 4.13   Patterns of density growth of the Los Angeles metropolitan area, 1940 to 2000.

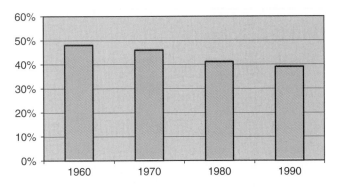

FIGURE 4.14   Share of suburban Chicago metropolitan area population living within 1-mile rail buffer, 1960 to 1990. Source: Piet, P., and Greene, R. P. (2000) "Population growth and the transportation link: Large city comparisons between the humid east and arid west." *Forum of the Association of Arid Lands Studies* 16:45–52.

service an existing population rather than simply serving as the conduit for growth. The story is identical for Chicago and Los Angeles, with Los Angeles exhibiting a consistently higher percentage residing near the highway, quite likely because of the constraints the Los Angeles physical landscape imposes on growth. The topographic obstructions literally force a higher density on all buildable land including that close to the highways.

Transportation has had a profound effect on urban form, from the electric streetcar that resulted in residential development conforming to rail lines all the way to the automobile and its associated freeway network that have resulted in a more dispersed urban form.

The second part of the GIS exercise consists of a density map analysis of Mexico City. Similar to Chicago and Los Angeles, the density pattern of Mexico City conforms to elements of all three of the classic urban development

FIGURE 4.15   Commuting to Chicago.

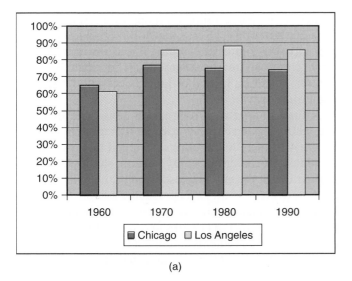

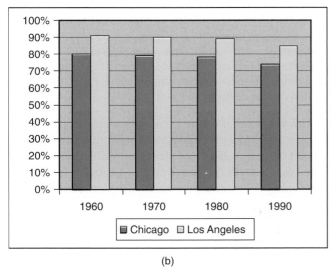

FIGURE 4.16 Suburban population living within 3-mile high-way buffer of (a) built-out highway system as of year shown on bar and (b) the 1990 built-out highway system: Chicago versus Los Angeles metropolitan areas. Source: Piet, P., and Greene, R. P. (2000) "Population growth and the transportation link: Large city comparisons between the humid east and arid west." *Forum of the Association of Arid Lands Studies* 16:45–52.

models: concentric zone, sector, and multiple nuclei. However, Mexico City density exhibits a unique pattern of development on the edge explained by the growth of squatter settlements, which can be illustrated with a density gradient in the next section.

4.4.2 Mexico City Urban Density Patterns: Density Gradients

First it will be useful to review the concept of an urban density gradient and its use in the study of urban structure.

Clark (1951) is credited with bringing urban density gradient models into wide use, models that showed that urban population densities decline with distance from city centers. With 36 cities as examples, Clark showed that urban population densities decline in a negative exponential manner with increasing distance from the city center:

$$D_x = D_0\, e^{-bx}$$

where D_x is the population density at distance x, D_0 is the population density at the center of the city, e is the base of natural logarithms, and b (density gradient) is an exponent expressing the rate of change of density with distance x.

Taking natural logs of both sides of this equation results in

$$\ln D_x = \ln D_0 - bx$$

Clark and others showed that this equation approximated a pattern of urban densities that had been replicated in a large number of cities. Nevertheless, the parameters of the equation, the extrapolated central density and the gradient, do vary across urban systems. One factor that contributes to regional variations in these parameters is the age of the city. Adams (1970) demonstrated for the American urban system that older cities have higher central densities and steeper density gradients than younger cities (see Figure 4.17). For instance, New York City is expected to have a steep density gradient because

FIGURE 4.17 Central density related to city age. Source: Adams, John S. 1970. Residential structure of Midwestern cities. *Annals of the Association of American Geographers* 60: 37–62. Fig. 3 p. 41.

it is in the Northeast, where cities are generally older than in the Midwest, South, and West, all of which developed as the urban settlement frontier advanced westward:

> Yet when age and size data are plotted some general locational regularities can be displayed. The old cities of the East Coast occupy one section and the new cities of the West, Southwest, and the Gulf Coast occupy another cluster. Between the two is the zone occupied by Midwestern cities, founded in the nineteenth century during the middle period of American urban history. (Adams, 1970, p. 41)

Adams noted that the relationship between population densities and age was due in large part to the transportation technology available to urban residents during the period of the city's most rapid population growth. New York City experienced rapid growth early in the preauto and prerail era, which helped account for its high residential density infrastructure. Even today, the urban infrastructure of this period has persisted as evidenced by the population densities that can still support an extensive public transportation system best epitomized by the New York subway network. In contrast, a place like Dallas experienced most of its growth and development well into the auto–air-amenity epoch and would exhibit a shallow pitch in its density gradient. Its public transport system is not as well developed as that of New York.

Another finding regarding urban density gradients is that smaller cities have steeper density gradients because they are more compact than bigger cities (see Figure 4.18).

Yeates (1990) made a similar finding with respect to land consumption rates (LCRs), which he found to decrease with city size. Yeates interpreted this as the efficiency of large cities where the competition for land is much greater, resulting in higher land prices and hence the higher intensity of use.

A final finding regarding urban density gradients is that within a given city they become flatter over time (see Figure 4.19). Clark and others observed that, over time, the density gradient of a city became shallower as the population of the city increased and as the built-up area grew outward. Clark (1957) presented evidence suggesting that this relationship was true for London and Chicago. Berry and Horton (1970) showed that the shallowing trend of density gradients continued for Chicago through the 1960s. Newling (1969) established an axiom for the decline:

> This behavior of the density gradient through time appears as an almost invariable rule (though exceptions have been detected), while the central density seems to behave in a less systematic fashion, sometimes increasing, sometimes remaining constant, and sometimes falling during the associated decline of the gradient. (p. 2)

The application of the density gradient to the developing world has yielded somewhat different results from the developed world examples (Berry & Kasarda, 1977; Wang & Zhou, 1999). In the case of Calcutta, Berry and Kasarda (1977) found that the density gradient did not

FIGURE 4.18   Density gradients related to size of city. Source: Adams, John S. 1970. Residential structure of Midwestern cities. *Annals of the Association of American Geographers* 60: 37–62. Fig. 4 p. 41.

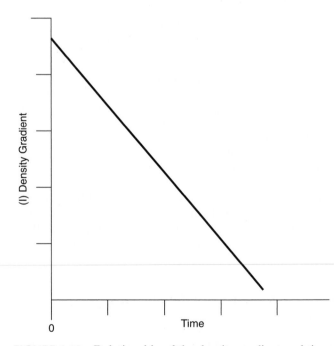

FIGURE 4.19   Relationship of the density gradient and time. Source: Newling, B. E. (1969) "The spatial variation of urban population densities." *Geographical Review* 59:242–252.

tend to flatten out over time as observed for cities of the developed world. By examining the period 1881 to 1951, Berry and Kasarda (1977) observed that although central density increased and the urbanized area expanded, the density gradient remained constant. The flattening out of the density gradient in the cities of the developed world has been explained by the process of redistribution from the center of cities to their edges with resulting decreases in crowding. The results for Calcutta suggest that the city has remained compact and crowded through time, perhaps a result of less expansion on the periphery relative to what has normally occurred in the cities of the developed world. However, the findings for Calcutta might not be generalized to all cities of the developing world, as Wang and Zhou (1999) found that Beijing's density gradient became flatter over time, while its extrapolated central density dropped between 1982 and 1990.

In the second part of this chapter's GIS exercise, a population density gradient is calculated for the Mexico City urban region for the year 2000 to establish whether it conforms to the developed world pattern or the developing world pattern. According to Griffin and Ford (1993), by the end of the 1970s, Latin America was the most urbanized region in the developing world with more than 69 percent of its population living in cities. The Mexico City urban area reached approximately 18 million people in 2000, making it one of the half-dozen largest cities in the world.

Urban life in Latin American cities is generally centered on the main plaza and employment opportunities are concentrated in the city center. Also, in most of Latin America, cities were administrative centers and market towns that did not experience rapid immigration or industrialization during the 19th century. Consequently, they did not experience the same type of urban growth and structural change as the cities of North America. Latin American cities generally remained specialized, traditional, and relatively small. There were other factors limiting their growth. For instance, the urban elite were concentrated in the core and the low-income majority on the periphery. Urban areas remained compact as the spread of the city was often associated with lower standards of living for those living on the periphery (Griffin & Ford, 1993).

Griffin and Ford (1993) developed a model to describe this process and urban structure (see Figure 4.20). The center of the city is occupied by the CBD, similar to the North American city. In contrast to North American cities, the CBD in Latin American cities has remained the center of most employment as well as commercial and entertainment life. Also characteristic of the Latin American city is a commercial spine. The spine is an extension of the CBD and is lined with upper-class and upper–middle-class housing in addition to high-rise office buildings. Socioeconomic levels and housing quality decrease with increasing distance from the CBD in Latin American cities, which is the inverse situation from the typical North American city. Another notable difference between Latin American cities and North American cities is the development of large squatter settlements

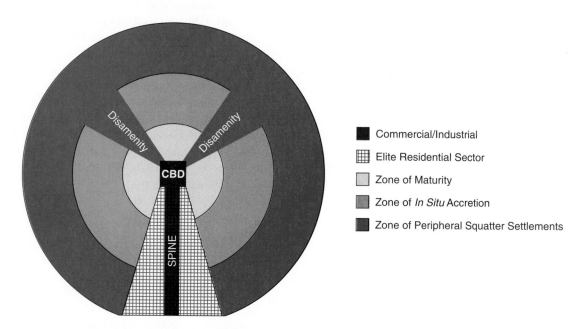

FIGURE 4.20   Latin American city model. Source: Griffin, E. and L. Ford. 1980. "A Model of Latin American Urban Land Use." *Geographical Review* 70:397–422. p. 406 and Figure 4. "A model of Latin American City Structure."

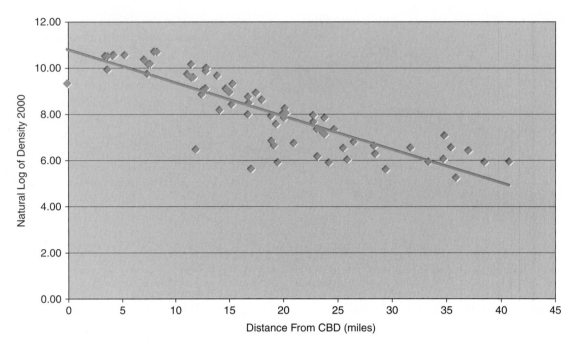

FIGURE 4.21   Mexico City density gradient, 2000.

on the urban periphery that house the poor, many of whom have recently arrived from even poorer rural conditions. In these settlements, most residents live in makeshift housing built illegally and often lacking proper sanitation.

With these structures in mind, we now turn to the analysis of Mexico City's density gradient for 2000 (see Figure 4.21). In some respects, the density gradient pattern is consistent with previous observations made about cities of the developed world. Nevertheless, high-density outliers on the urban area edge appear on the Mexico City density gradient graph confirming elements of the Latin American city model; that is, it can be concluded that these outliers on the edge have the density characteristics of squatter settlements. In the exercise, mapping four of the outliers with especially high densities, given their distance from downtown, will provide evidence of them as potential squatter settlements (see Figure 4.22).

4.5 CONVERGENCE AND DIVERGENCE FROM THE CLASSIC MODELS: THE CASE OF CHICAGO

We now turn specifically to a case study of the city of Chicago, with some comparisons to the city of Los Angeles, to illuminate the urban processes. This section attempts to tease out the many processes identified by

the contemporary adherents of the Chicago School as being on par to those identified by the L.A. School for Los Angeles. The city of Chicago showed remarkable growth during the 1940s when it added almost 225,000 people (see Figure 4.23). It was remarkable because of the extreme stability exhibited in the 1930s, but more important, Chicago's city population would reach its all-time decennial census peak of 3.6 million people by 1950. Subsequently during the 1950s, Chicago would lose about 70,000 people before undergoing a deep descent that would last through 1990, the date that Los Angeles officially replaced Chicago as the nation's second largest city (see Figure 4.23). In the Chicago metropolitan area overall, the city's share of the total metropolitan population had fallen to 43 percent by 1970 (see Figure 4.24). The rate of loss in Chicago accelerated from the 1950s to the 1960s and again in the 1970s, when it reached its all-time-high decennial loss of 362,000 persons. However, in the 1980s, the rate of loss in Chicago slowed with a loss of approximately 220,000 persons.

Reversing this long trend, the central city of Chicago expanded in population between 1990 and 2000, the first time it had done so since the 1950 Census. Chicago's 1990s metropolitan population spurt of 11.1 percent is much higher than the city growth rate of 4.0 percent, indicating that the area outside the central city is continuing its post-1950 trend of rapid growth. In 2000, the city's proportion of the metropolitan area declined to only 32 percent.

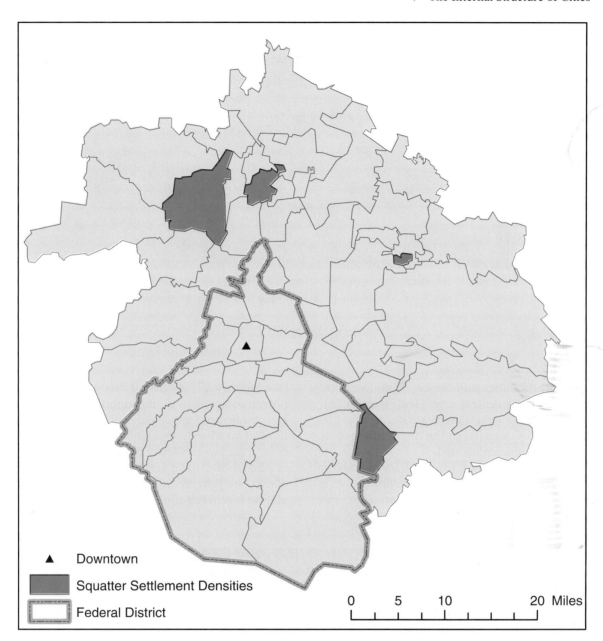

FIGURE 4.22   Peripheral areas of Mexico City with potential squatter settlements.

Plotting density gradients through time is another way to illustrate urban development trends within a metropolitan area (see Figure 4.25). The literature on this subject argues that the gradients should become flatter through time as filling in occurs between the dense core and the less dense fringe. Although that happened for Chicago recently, it was only slight. An especially notable difference between Los Angeles and Chicago is the slope of the density gradient from their downtowns to the edges of their metropolitan areas. Note the steeper gradient for

Chicago caused by more observations in the high-density and low-density categories and Los Angeles with many tracts in the medium-density categories. This is because there is more high-rise and multifamily housing in Chicago, accounting for relatively high densities in the core. The Chicago metropolitan area also includes large zones of low-density areas, providing plenty of room to grow. Los Angeles, on the other hand, has largely run out of a low-medium to low-density urban fringe, due to being hemmed in by mountains, deserts, and ocean as shown by

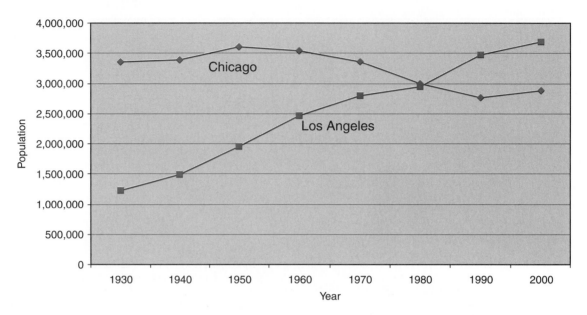

FIGURE 4.23   Population growth of the city of Chicago versus the city of Los Angeles. Source: U.S. Census 1930 to 2000.

this National Aeronautics and Space Administration (NASA) image of the central area (see Figure 4.26). Clearly, competition for land is much greater in a metropolitan area like Los Angeles, so the distribution of its tract densities is more concentrated.

To illuminate the components of growth for Chicago in the 1990s within this first 10-mile ring from downtown, a map shows 1990 and 2000 extreme poverty area patterns

(see Figure 4.27). At first, it might seem odd to consider population growth in the context of poverty, but this is relevant because much of the 1990 research on these areas showed massive depopulation since 1980 and subsequent rises in the poverty level. The addition of the 2000 poverty areas shows that a remarkable contraction in the zone of extreme poverty had occurred in the 1990s while population loss in the zone was only slight.

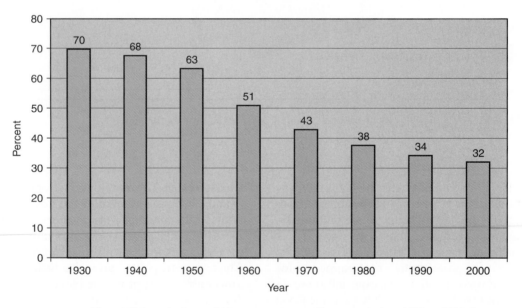

FIGURE 4.24   City of Chicago's share of 13-county metropolitan population, 1930 to 2000. Source: U.S. Census 1930 to 2000.

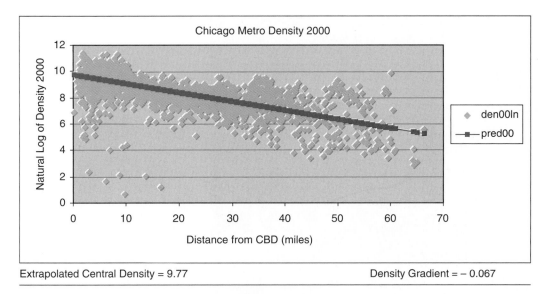

Extrapolated Central Density = 9.77   Density Gradient = – 0.067

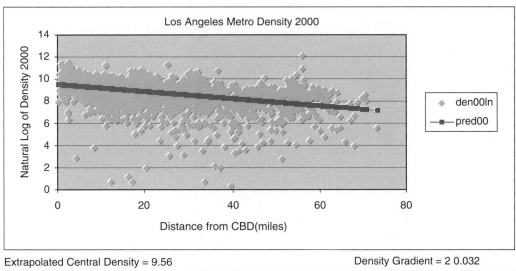

Extrapolated Central Density = 9.56   Density Gradient = 2 0.032

FIGURE 4.25   Chicago and Los Angeles density gradients, 2000.

Analyzing the city's emerging immigrant areas, defined as tracts with a 1990 foreign-born population that had increased by at least 1,000 between 1990 and 2000, helps explain Chicago's rebound. To put this in perspective, consider that the city of Chicago's total population increase was 112,290 between 1990 and 2000. The total population increase of the emerging immigrant areas was 81,000, and their foreign-born population increased by 93,000, suggesting a decline in the area's native-born population. It is also interesting that the Latino increase in these areas was approximately 114,000, which might relate to nearness to employment centers particularly in the northwest. Incidentally, these emerging immigrant areas are mutually exclusive of the extreme poverty areas westward and southward of downtown Chicago.

New immigrants are also locating close to edge cities in the Chicago area. As we saw previously, many of the old models describing the entrance of immigrants into cities recognized that the new arrivals moved into neighborhoods that adjoined either the CBD or close-in industrial districts. In light of recent manufacturing job shifts out of central cities and the growth of employment in outlying edge cities, the change in the location pattern of immigrant settlement is not surprising. An edge city destination is likely to hold more economic promise

Pacific
Ocean

0   2.5   5   10 Miles

FIGURE 4.26   Landsat image of Los Angeles urban area with city boundary superimposed. Source: CalView Landsat Data: http://gis.ca.gov/ (date: 4 March 2002; bands: 4, 5, and 7).

for a new immigrant than a neighborhood closer into the city, which has experienced a decline in its industrial job base.

The following maps illustrate that a significant number of the Chicago area's new immigrants are bypassing the traditional zone in transition for emerging job opportunities in the region's edge cities. In the absence of sufficient demand, some of these bypassed neighborhoods have fallen into a cycle of landscape dereliction, which is characterized by the decay of public infrastructure and large-scale housing abandonment. On the other hand, a number of these neighborhoods were rediscovered by gentrifiers in the 1990s, a topic discussed at greater length in Chapter 6.

During the 1980s and 1990s, the number of legal immigrants entering the United States rose sharply, and in 2000 approximately 12 percent of Americans were foreign born (see Figure 4.28). Ten years ago, this was significant news for Chicago because of the approximately 8 million legal immigrants that had entered the United States during the 1980s, Illinois was one of six states receiving 75 percent of them. Once again, a Los Angeles comparison is in order here. Note that Figure 4.28 shows a downward trend in percentage foreign born for both Chicago and Los Angeles from 1870 to 1960. However, prior to 1960, Chicago's percentage was always higher. Chicago's then continued to fall, whereas Los Angeles's begins to rise and remains above Chicago through 2000.

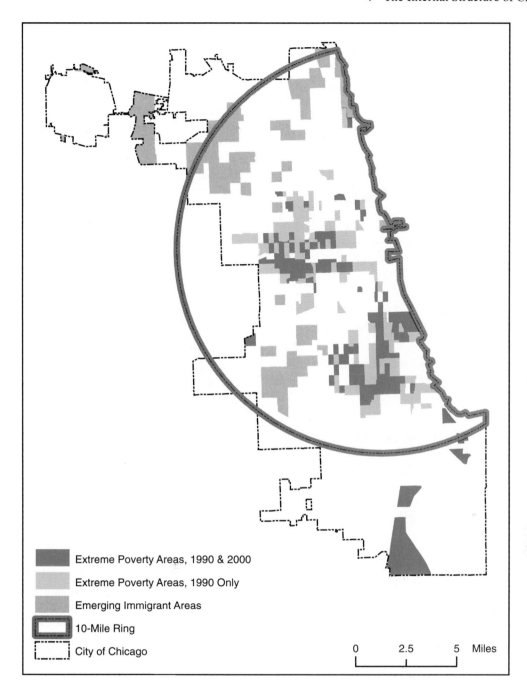

FIGURE 4.27   Chicago poverty and immigrant settlement areas.

Chicago only descended for one more decade and then made sharp ascents from 1970 to 1980 and 1980 to 1990, with its sharpest ascent from 1990 to 2000 (even sharper than Los Angeles for that decade only).

Chicago's suburbs and outlying industrial satellite cities have experienced large increases in Latino population in the last decade or two. The growth in Latino population was especially pronounced in Elgin (see Figure 4.29); confined mainly to the east side of the Fox River in 1990, the distribution had spread to the west side by 2000. In many respects, the spatial pattern exhibited by the growth of Latinos in Elgin, Illinois, is similar to that seen in much larger cities such as the 1970 to 2000 expansion of Latinos in Los Angeles already

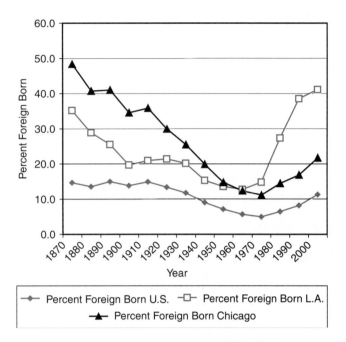

FIGURE 4.28   Percentage foreign born in Los Angeles, Chicago, and United States, 1870 to 2000.

discussed in Chapter 1. In Elgin, the older core of the Latino population was east of the Fox River. However, as middle-class non-Latino. White households on the west side of the river moved out of Elgin and west along an established high-income sector, vacancies and better housing opportunities were created for the city's growing Latino population.

Emerging immigrant concentrations, defined as tracts that experienced increases of 1,000 or more foreign-born persons between 1990 and 2000, emanate outward from the city of Chicago, often adjoining many of the region's employment centers (see Figure 4.30). A substantial increase in immigrants outside the city of Chicago has shifted the geographic center of the region's immigrant population in a northwest direction, which is an orientation that corresponds with the directional shift in the Chicago region's overall population, jobs, and major ethnic groups including Latinos and Asians (see Figure 4.31). The only apparent diverging movement is that exhibited by the African Americans. The African American population of the Chicago region has been suburbanizing as have all groups, however their movement has been biased in the direction of south and southwest sectors.

As with other very large U.S. urban areas, one of the most significant demographic trends for the Chicago region in recent decades has been population redistribution from the core of the region to its periphery, with

the most recent decade showing a slight reversal of that trend. The most noticeable effect of this redistribution has been large-scale housing abandonment or neglect in many of Chicago's poorer neighborhoods. Some of these neighborhoods have experienced gentrification, the effect of which is picked up in the recent city population increases. Meanwhile, the urban fringe on the outskirts of the metropolitan area has been the recipient of large-scale investment in its built environment. Although significant, the number of immigrants entering Chicago and its inner suburbs has not matched the redistributive growth of population to the edge. Also, new immigrants arriving in the Chicago region appear to be orienting themselves in proximity to the region's emerging job centers on the edge, but removed from even farther outlying scattered residential enclaves of the upper class.

In summary, in the Chicago urban area, population and economic decentralization have been long-term trends, with the exception of the most recent growth period of 1990 to 2000 when the core experienced renewed growth. New immigrants are settling in a geographic pattern that is quite different from immigrant groups that entered Chicago in earlier periods when its core was the major location for job opportunities. Today, immigrants entering the Chicago area encounter an industrial geography that is becoming increasingly dispersed. The high employment growth rates in the outlying ring of municipalities surrounding the city of Chicago are an obvious magnet for new immigrants in search of job opportunities. The older and persistently low-rent areas of the west and south, close to the urban center, are not attracting new immigrants at a sufficient rate to offset the recent population declines incurred there by the outward migration of the African American middle class to the suburbs. As a result, these old, bypassed neighborhoods are becoming the locus of the region's extreme poor, who are economically disadvantaged by location. The new immigrants, on the other hand, have been settling on the periphery of the region's traditional immigrant neighborhoods as well as outside Chicago's city limits and are thus more strategically located relative to the region's geographic distribution of jobs. In addition, a counter trend of gentrification by young urban professionals and artists has also contributed to the population turnaround exhibited by Chicago's core in the 1990s, a topic covered in Chapter 6.

SUMMARY

The internal structure of cities has always been a prime topic of interest to urban geographers. In the United States, the first formal model of the internal urban structure

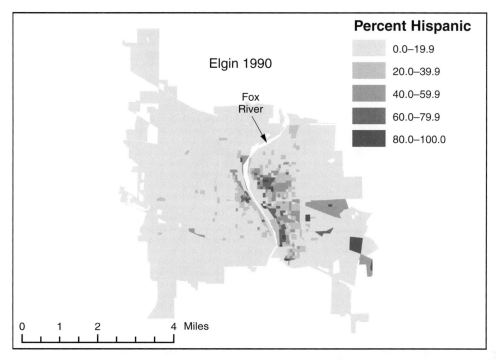

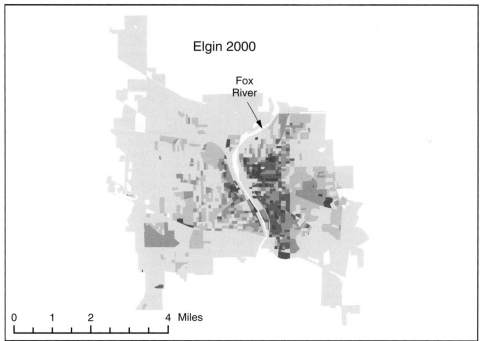

FIGURE 4.29   Change in percentage Latino population in Elgin, Illinois, 1990 to 2000.

of cities can be dated to the Chicago School of sociology in the 1920s. The concentric zone, sector, and multiple nuclei models represent the classic American statements of how city distributions are organized and change. Today, there are both proponents and critics of these classic urban development models. Proponents continue to find evidence to support aspects of one or more of the three models, whereas the critics argue that the models no longer reflect current-day urban processes. Historical development patterns for Los Angeles and Chicago show that the built-up areas advanced in patterns consistent with all three models. A review of recent trends in

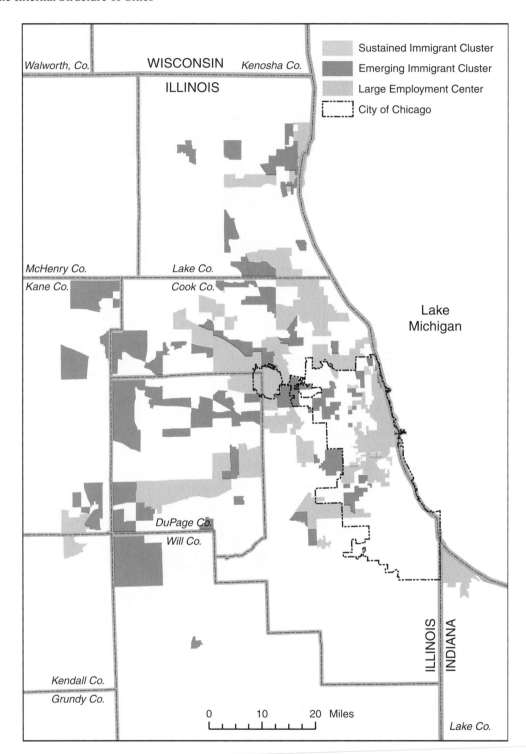

FIGURE 4.30 Immigrant clusters and job centers in the Chicago metropolitan area, 2000.

Chicago illustrates that many of the patterns of settlement identified by the earlier models are similar but that the context has changed. For instance, on the basis of the Chicago analysis, one can argue that the zone in transition, originally located next to factories, has shifted to new and outlying employment centers such as O'Hare Airport, which is a major employment center for workers that include many new immigrants who live nearby and work in the service-related jobs associated with the airport.

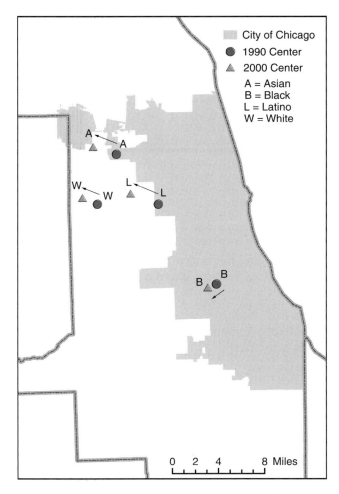

FIGURE 4.31 Shift in the population centers for Chicago metropolitan area, 1990 to 2000.

REFERENCES

Abbott, A. (2002) "Los Angeles and the Chicago School: A comment on Michael Dear." *City & Community* 1:33–38.

Abu-Lughod, J. L. (1999) *New York, Chicago, Los Angeles: America's Global Cities.* Minneapolis: University of Minnesota Press.

Adams, J. S. (1970) "Residential structure of Midwestern cities." *Annals of the Association of American Geographers* 60:37–62.

Adams, J. S. (1991) "Housing submarkets in an American metropolis," in *Our Changing Cities.* ed. by J. F. Hart, 108–126. Baltimore: Johns Hopkins University Press.

Berry, B. J. L., and Horton, F. (1970) *Geographic Perspectives on Urban Systems.* Englewood Cliffs, NJ: Prentice Hall.

Berry, B. J. L., & Kasarda, J. (1977) *Contemporary Urban Ecology.* New York: Macmillan.

Clark, C. (1951) "Urban population densities." *Journal of the Royal Statistical Society Series A* 114:490–496.

Coleman, J. (1994) "A vision for sociology." *Society* 32(1): 29–34.

Dear, M. J. (2002) "Los Angeles and the Chicago School: Invitation to a debate." *City & Community* 1:5–32.

Ethington, P. J., and Meeker, M. (2002) "Saber y Conocer: The metropolis of urban inquiry," in *From Chicago to LA: Making Sense of Urban Theory."* ed. by M. J. Dear, pp. 405–420. Thousand Oaks, CA: Sage.

Forstall, R. L. (2003) Personal communication.

Gottdiener, M. (1994) *The New Urban Sociology.* New York: McGraw-Hill.

Griffin, E., and Ford, L. (1993) "Cities of Latin America," in *Cities of the World.* ed. by S. Brunn, and J. Williams, pp. 225–265. New York: HarperCollins.

Halle, D. (2003) *New York and Los Angeles: Politics, Society, and Culture.* Chicago, IL: University of Chicago Press.

Harris, C. D., and Ullman, E. L. (1945) "The nature of cities." *Annals of the American Academy of Political and Social Science,* 242:7–17 (reprinted in Mayer, H. M., and Kohn, C. F., eds., 1959, *Readings in Urban Geography,* 277–286. Chicago: University of Chicago Press.

Harvey, D. (1973) *Social Justice and the City.* Baltimore: Johns Hopkins University Press.

Howenstine, E. (1993) Measuring demographic change: The split tract problem. *Professional Geographer* 45:425–430.

Hoyt, H. (1939) *The Structure and Growth of Residential Neighborhoods in American Cities.* Washington, DC: Federal Housing Administration.

Jackson, K. T. (1985) *Crabgrass Frontier: The Suburbanization of the United States.* New York: Oxford University Press.

Jacobs, J. (1961) *The Death and Life of Great American Cities.* New York: Random House.

Molotch, H. (2002) "School's out: A response to Michael Dear." *City & Community* 1:39–43.

Newling, B. E. (1969) "The spatial variation of urban population densities." *Geographical Review* 59:242–252.

Nijman, J. (2000) "The paradigmatic city." *Annals of the Association of American Geographers* 90:135–145.

Park, R. E., Burgess, E. W, and McKenzie, R. D. (1925) *The City: Suggestions for the Investigation of Human Behavior in the Urban Environment.* Chicago: University of Chicago Press.

Piet, P., and Greene, R. P. (2000) "Population growth and the transportation link: Large city comparisons between the humid east and arid west." *Forum of the Association of Arid Lands Studies* 16:45–52.

Sampson, R. J. (2002) "Studying modern Chicago." *City & Community* 1:45–48.

Sassen, S. (2002) "Scales and spaces." *City & Community* 1:48–50.

Sennett, R. (1990) *The Conscience of the Eye.* New York: Knopf.

Shearmur, R., and Charron, M. (2004) "From Chicago to L.A. and back again: A Chicago-inspired quantitative analysis of income distribution in Montreal." *Professional Geographer* 56:109–126.

Stern, R., Mellins, T., and Fisman, D. (1997) *Architecture and Urbanism Between the Second World War and the Bicentennial* (2nd edition). New York: Monacelli Press. 393–422

Wang, F., and Zhou, Y. (1999) "Modeling urban population densities in Beijing 1982–90: Suburbanization and its causes." *Urban Studies* 36:271–287.

Whyte, W. H. (1956) *The Organization Man.* New York: Simon & Schuster.

Yeates, M. (1990) *The North American City* (4th edition). New York: Harper & Row.

Zukin, S. (1982) *Loft Living: Culture and Capital in Urban Change.* New Brunswick, NJ: Rutgers University Press.

EXERCISE

Exercise Description

Part 1: Comparing the urban expansion of Chicago and Los Angeles: 1960, 1980, and 2000.
Part 2: Identifying squatter settlements in Mexico City with the use of a density gradient.

Course Concepts Presented

Density maps, population density, density gradients, comparative urban analysis, and urban development models (North American vs. Latin American).

GIS Concepts Utilized

Thematic mapping, data view versus layout view, table relates, and spatial query.

Skills Required

ArcGIS Version 9.x and Excel spreadsheets is an option in Part 2.

GIS Platform

ArcGIS Version 9.x

Estimated Time Required

2 hours

Exercise CD-ROM Location

ArcGIS_9.x\Chapter_04\

Filenames Required

Part 1:
    chic_msa.shp
    Ch_zoom.shp
    losa_msa.shp
    La_zoom.shp

Part 2:
    mx_mun03_final.shp
    fed_dist03.shp
    mexden00.dbf (available if you decide to skip the Excel worksheet step in Part 2)

Background Information

In this lab, you examine the urban expansion of Chicago and Los Angeles from 1960 to 2000 by making time series maps of population density. The maps will be used to illustrate how well the growth experiences of the two metropolitan areas conform to the concentric zone, sector, and multiple nuclei models of urban development. The second part of the lab requires you to make a density gradient for the Mexico City metropolitan area to see if it conforms to the Latin American city model of urban development.

Readings

(1) Abu-Lughod, J. L. (1999). *New York, Chicago, Los Angeles: America's Global Cities.* Minneapolis: University of Minnesota Press.
(2) Cutler, I. (1976). *Chicago: Metropolis of the Mid-Continent* (2nd edition). Dubuque, IA: Kendall/Hunt.
(3) Greene, R. P., and Pick, J. B. *Exploring the Urban Community: A GIS Approach,* Chapter 4.
(4) Pick, J. B., and Butler, E. W. (2000). *Mexico Megacity.* Boulder, CO: Westview.
(5) Piet, P., and Greene, R. P. (2000). "Population growth and the transportation link: Large city comparisons between the humid east and arid west." *Forum of the Association of Arid Lands Studies* 16: 45–52.

Exercise Procedure Flowchart

Part One:

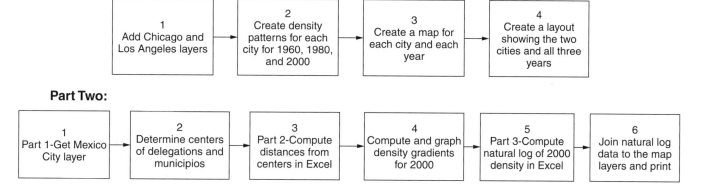

Part Two:

Exercise Procedure

Part 1: Comparing Chicago and Los Angeles Density Patterns over Time

First, you will set up two data frames and add data layers.

1. Start ArcMap with Start Using ArcMap With A New Empty Map selected.

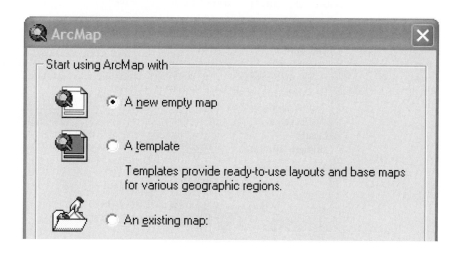

2. Right-click Layers.
3. Click Properties and click the General tab.
4. Change the name Layers to Chicago 1960 and click OK.
5. Click Add Data and navigate to the exercise data folder.

Note: *If you do not see the folder where you placed the exercise data, you will need to make a new connection to the root directory that holds the data.*

6. Add chic_msa.shp from the Chapter 4 folder by selecting it and clicking Add.
7. Select Data Frame under Insert to insert a new data frame.
8. Right-click New Data Frame, click Properties, and change the data frame's name to Los Angeles 1960, and click OK.
9. Click Add Data and navigate to the exercise data folder.
10. Add losa_msa.shp from the Chapter 4 folder by selecting it and clicking Add.

(Creating the Density Patterns by Year for Each Metropolitan Region)

11. Right-click Chicago 1960 and select Activate to activate the Chicago 1960 data frame. These are 2000 census tracts for the 13-county Chicago CMSA (prior to redefinition in 2003). The census tract data tables are consistent for 1960, 1980, and 2000.
12. Change the layer name chic_msa by right-clicking the name, clicking Properties, clicking the General tab, and changing the name to 1960 Density. (Don't exit the Layer Properties dialog box yet.)
13. Click the Symbology tab.
14. Select Quantities, select Graduated Colors, and select POP60_SQMI as the Value Field.
15. Click Classify.
16. Change the classification method to Manual. Set the breaks to be:

    0
    2499.99
    4999.99
    7499.99
    200000.00

17. Click OK.
18. The Label field gives the text that actually appears in the map legend. Now change the cells in the Label field to something more appealing for the legend (see example in screenshot):

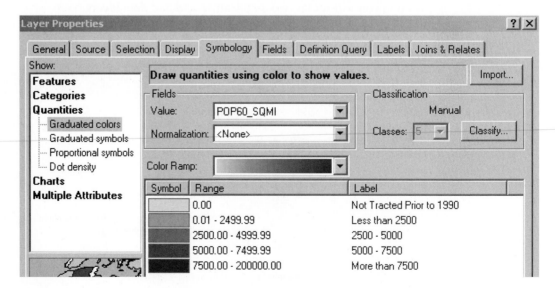

19. Now you want to remove the census tract boundaries. The map will be more effective without the lines. Double-click the symbol for the first category and set Outline Color to No Color by clicking the drop-down arrow. Click OK.
20. Repeat for the remaining classes and click Apply and click OK to see your results.

21. We will copy the 1960 Density layer and paste it into a new Chicago 1980 data frame and then change the classification field from Pop60_sqmi to Pop80_sqmi.

    (a) Click Insert and click Data Frame.
    (b) Change the name of the new data frame to Chicago 1980 by right-clicking the name and clicking Properties and the General tab.
    (c) Right-click the 1960 Density layer in the above data frame and click Copy.
    (d) Right-click the Chicago 1980 data frame and click Paste Layer(s).
    (e) Right-click the Chicago 1980 data frame and click Activate.
    (f) Change the name of the layer from 1960 Density to 1980 Density by right-clicking the name and clicking Properties and the General tab. (Don't exit the Layer Properties dialog box yet.)
    (g) Click Symbology, click the Import tab, choose the default options, and click OK. In the next dialog box, change the field to Pop80_sqmi and click OK.
    (h) Click Apply and click OK.

22. Repeat Step 21 to make a Chicago 2000 data frame and a 2000 Density layer, using the Pop00_sqmi field. Any reference to 1980 should be changed to 2000. At the end of this step your table of contents should like the following for the Chicago data frames and layers:

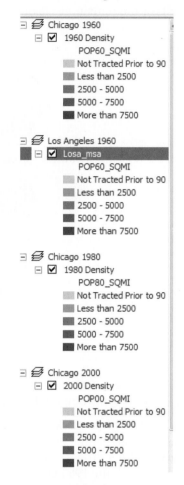

23. Right-click Los Angeles 1960 and select Activate to activate the Los Angeles data frame. These are 2000 census tracts for the 5-county Los Angeles metropolitan area defined as of June 2003. The census tract data tables are consistent for 1960, 1980, and 2000.
24. Change the layer name Losa_msa.shp by right-clicking on the name, clicking Properties, clicking the General tab, and changing the name to 1960 Density. (Don't exit the Layer Properties dialog box yet.)
25. Click Symbology, click the Import tab, choose 1960 Density, and click OK. In the next dialog box, change the field to Pop60_sqmi and click OK.
26. Click Apply and then click OK.
27. We will copy the Los Angeles 1960 Density layer and paste it into a new Los Angeles 1980 data frame and then change the classification field from Pop60_sqmi to Pop80_sqmi.

   (a) Click Insert and select Data Frame.
   (b) Change the name of the new data frame to Los Angeles 1980 by right-clicking the name and selecting Properties and then clicking the General tab.
   (c) Right-click the Los Angeles 1960 Density layer in the above data frame and click Copy.
   (d) Right-click the Los Angeles 1980 data frame and click Paste Layer.
   (e) Right-click the Los Angeles 1980 data frame and click Activate.
   (f) Change the name of the layer from 1960 Density to 1980 Density by right-clicking the name and clicking Properties and the General tab. (Don't exit the Layer Properties dialog box yet.)
   (g) Click Symbology, click the Import tab, choose the default options, and click OK. In the next dialog box, change the field to Pop80_sqmi and click OK.
   (h) Click Apply and click OK.

28. Repeat Step 27 to make the 2000 map for Los Angeles density. Use the Pop00_sqmi field and any reference to 1980 should be changed to 2000.

Creating a Final Map Layout

29. Prior to leaving the Data view you might want to zoom in on the patterns in Chicago and Los Angeles in each of the data frames as they include vast land areas, especially Los Angeles, where there are few or no high-density development patterns on the edge. Make sure to use a standard zoom for each metropolitan area, so that a given area's time series maps will be of a similar scale.

30. You can add a predefined zoom box for Chicago and Los Angeles into their respective data frames (see Ch_zoom.shp and La_zoom.shp in the directory on the CD for the chapter). First activate a data frame and then add the appropriate zoom layer (Ch_zoom for the three Chicago data frames and La_zoom for the three Los Angeles data frames). Once the zoom layer is added to a given data frame, right-click the zoom layer and select Zoom To Layer. Make sure the zoom layer is not turned on (the check box should be cleared). You could then copy the zoom layer to the appropriate data frames and repeat the zoom to layer step (making sure to turn off the zoom layer each time).

31. Once you have all of your data frames prepared, go to the Layout view by clicking its icon at the bottom (icon to right of the globe icon).

32. All six data frames are there. You just need to reposition the six data frames in the following format with associated legend and text.

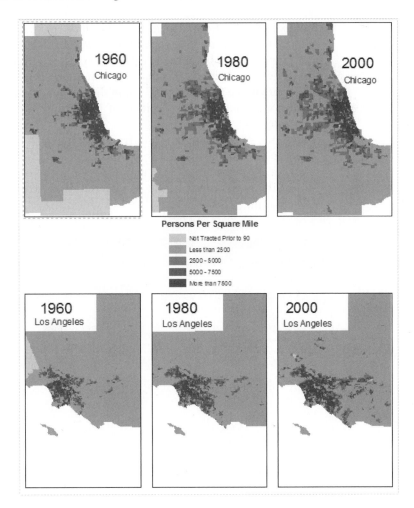

33. To achieve a similar format to the one just shown, you can standardize the *x*, *y* position of the upper left corner of each data frame and assign a standard height and width by right-clicking a data frame, selecting Properties, and clicking the Size And Position tab.

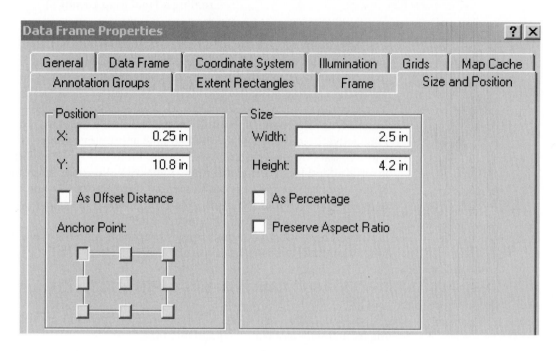

34. In this example, the size included a width of 2.5 in. and a height of 4.2 in., and the *x* position was set to 0.25 in. and the *y* position to 10.8 in. Note that the anchor point that is depressed is the upper left. This is the *x*, *y* position you are specifying on the page. The remaining data frames were all given the same width and height but were assigned an *x* position of 0.25 if they were in the first column, 3.0 if in the second column, and 5.75 if in the third column. Data frames were assigned a *y* position of 10.8 if they were in the top row and 4.92 if in the bottom row.

35. Make one of the Chicago data frames active and click Insert to insert a legend. We only need one legend as they all share the same classification (perhaps make Chicago 1960 active). In the first dialog box, make sure that the only legend item is 1960 density. For instance if Chic_zoom is in the box on the right, select and use the arrow to move it to the left column.

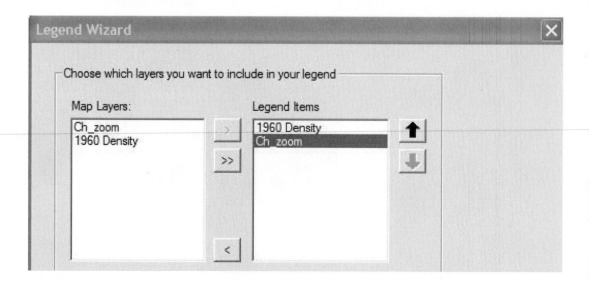

36. Click Next and change the word "Legend" to "Persons Per-Square Mile" and click Next.
37. Click Next in the next several dialog boxes and then click Finish.
38. Double-click on the selected legend. Click the Items tab and in the right column, double-click 1960 Density, click Properties, click the General tab, clear the Show Layer Name and Show Heading check boxes, click Apply, and click OK.
39. Click Insert at the top and insert text for the city names and the dates of the density patterns and print the final layout.
40. Exit ArcMap.

Short Essay

Using the map you created in Part 1, the chapter section on "Analyzing an Urban Issue," and, if available, the background readings cited earlier, write an essay describing (1) how the two metropolitan areas have grown spatially over time from 1960 to 2000 and (2) if the patterns conform to one or more of the classic urban development models (i.e., concentric zone, sector, and multiple nuclei).

Part 2: Computing a Density Gradient for the Mexico City Area and Mapping High-Density Outliers

41. Start ArcMap with a new empty map and add mex_mun03_final.shp.
42. Open the attribute table by right-clicking mex_mun03_final under Layers and then clicking Open Attribute Table.
43. Add two columns by selecting Add Field under Options below the attribute table:
    (a) x_coord as type Float with a precision (length) of 16 and a scale (number of decimal places) of 2.
    (b) y_coord as type Float with a length of 16 with 2 decimals.
44. Calculate: This step calculates the x and y coordinate locations based on the geometric center of each delegation and municipios. You will load some Visual Basic for Applications (VBA) code into the ArcMap Field Calculator. The code has been written for you and it computes a polygon centroid and gives you either the x or y of the centroid.
    (a) Right-click the x_coord field title and then click Calculate Values. Click Yes if a pop-up menu appears on field calculation.
    (b) Click Load and navigate to the Chapter 4 folder and select get_X_Center.
    (c) Click Open.
    (d) Click OK on the Calculator and ArcMap returns the X coordinates.
    (e) Right-click the y_coord field title and then click Calculate Values.
    (f) Note that the previous programming is still there, but it will be replaced automatically by loading of the new code.
    (g) Click Load and navigate to the Chapter 4 folder and select get_Y_Center
    (h) Click Open.
    (i) Click OK on the Calculator and ArcGIS returns the Y coordinates.
    (j) We now need to export this table to a separate table to import into Microsoft Excel for use in the next step.
    (k) Click Export under the Options button below the attribute table.
    (l) Put the file into a folder you can remember and rename it from export_output to Mexden.dbf. Click OK.
    (m) Click No to add the new table to the current map and then close the table.

    (The following are optional steps to compute the 2000 density gradient for the Mexico City megacity in Excel. If you wish to skip these steps, proceed to Step 67 where an intermediate file has been made available to you.)

45. Start Excel and open Mexden.dbf (change file type to All Files to see .dbf files). You might have to adjust column widths to see all columns.
    Note that there are 50 municipios (rows 2 through 51) with the following fields:
    (A) map_index (B) MUN_NAME (C) P00_sqmi (D) x_coord (E) y_coord
46. Add the following fields:
    (F) Distance, (G) den00ln, and (H) pred00

47. Go to the File menu and click Save As. For Save As Type, save it as a Microsoft Excel Workbook. The file will be saved as Mexden.xls.
48. In cell F2, calculate the first muncipio to be the distance to the CBD in miles:

$$= SQRT((d2 - d\$15)\hat{\ }2 + (e2 - e\$15)\hat{\ }2)/1609$$

Where:

> d = x_coord
> e = y_coord
> 15 = row for Cuauhtemoc
> 1609 = conversion factor for meters to miles

This step is a simple geometry formula to calculate the distance from a particular areal unit to the central municipio, Cuauhtemoc.
49. Copy and paste the formula to the other cells down to row 51.
50. Calculate the natural log transformation of population density for 2000 in cell G2:

$$= LN(C2)$$

Note: *C2 is P00_sqmi, which is the population density in 2000.*
51. Copy and paste the formula down to row 51.
52. Still in Excel, go to the Tools menu and select Addins. Place a check mark next to Analysis ToolPak-VBA.
53. Go back to the Tools menu and select Data Analysis and choose Regression.
54. We will compute the 2000 density gradient by filling in the following ranges:

> For Input y-range enter the range for den60ln ($G$2:$G$51)
> For Input x-range enter the range for distance ($F$2:$F$51)
> Check Output Range under Output Options and enter (J1)

Click OK.

55. Excel computes the gradient and puts the output in the location specified (J1).
56. The density gradient (line) can be thought of as the predicted density, or the density predicted given the distance of an areal unit (municipio) to the center of the city.
57. In cell H2, use the regression formula to calculate the 2000 predicted density, multiplying the distance by the slope coefficient and adding it to the Y intercept:

$$=F2*K\$18 + K\$17$$

Where:

> F = distance
> K18 = x variable
> K17 = y intercept

58. Copy the formula down to row 51.
59. Graph the gradient by clicking the Chart Wizard icon. This can also be done by clicking Insert followed by Chart.
60. Choose the XY (Scatter) Plot option and select the first subtype that shows a plot of points.
61. Click Next.
62. Enter the Data Range to be distance, den00ln, and pred00 (in that order).

$$= mexden!\$F\$1:\$H\$51$$

63. Click Next and add the following titles:

> Titles
> Chart Title: Mexico City Density 2000
> X Axis: Distance from CBD (miles)
> Y Axis: natural log of Density 2000

64. Click Finish.

65. Place the graph in the spreadsheet to the far right.

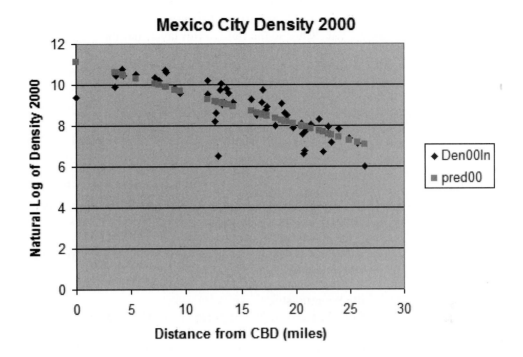

66. We need to put Mun_name, Distance, and Den00ln into a new Excel worksheet:
    (a) Click File, New, and Open New Blank Workbook.
    (b) Toggle back to Mexden.xls using the Window tab.
    (c) Copy the Mun_name column and toggle back to the new workbook.
    (d) Right-click Column A and select Paste.
    (e) Toggle back to Mexden.xls, press the Esc key to unselect Mun_name, and copy Distance.
    (f) Toggle back to the new workbook, right-click Column B, and select Paste Special.
    (g) Choose the Values formats and click OK.
    (h) Toggle back to Mexden.xls, press the Esc Key, and copy Den00ln.
    (i) Toggle back to the new workbook, right-click Column C, and select Paste Special.
    (j) Choose the Values formats and click OK.
    (k) Right-click Column B and click Format Cells.
    (l) Choose number, decimal place = 2, and click OK.
    (m) Right-click Column C and click Format Cells.
    (n) Choose number, decimal place = 2, and click OK.
    (o) Make sure nothing is selected and put the cursor in cell A1.
    (p) On the File menu, click Save As, change File Type to DBF 4, and name it Density00.dbf
    (q) Click OK when informed that Selected File Type Does Not Support Workbooks That Contain Multiple Sheets.
    (r) Click Yes when informed that Density00.dbf May Contain Features That Are Not Compatible With DBF4.
    (s) Close Excel and click No to saves.

    (The following steps require you to return to ArcMap)

67. Start ArcMap and add the Mex_mun03_final.shp and Density00.dbf to the map. If ArcMap is already open, just add mexden00.dbf to the map.

68. Right-click mex_mun03_final.shp, point to Joins And Relates, and click Join.
    (a) Answer the following in the Join Data window:
        i.   What do you want to join to this layer
             1. "Join attributes from a table"
        ii. Choose the field in this layer that the join will be based on
             1. "MUN_NAME"
        iii. Choose the table to join to this layer, or load the table from disk
             1. "Density00"
        iv. Choose the field in the table to base the join on
             1. "Mun_name"
        v.   Click OK (if asked to create an index, click Yes).
69. Click Selection and select By Attributes and enter the following query:

$$\text{Density00.Distance} > 15 \text{ AND Density00.DEN00LN} > 9$$

We chose this by examining the graph and noting a cluster of outliers about 15 miles from downtown that had densities higher than the regression line (about a natural log of 9).

70. Click Apply and then click Close.
71. Add the Federal District Map (Fed_dist03.shp) layer to give some context. Make it a hollow boundary.
72. Right-click mex_mun03_final and click Data and then click Export Data, naming the new file Squatter.shp.
73. Click Yes to Add The Exported Data To The Map As A Layer. Click Selection and Clear Selected Features.
74. Compose a map using the Layout view showing the municipios that are suspected of having squatter settlements due to their unusually high densities at relatively far distances from downtown.

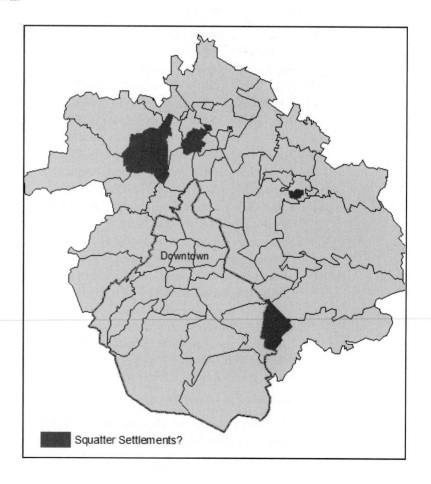

Essay Question on High-Density Outliers

Discuss the location of the high-density outliers relative to the model of Latin American city structure covered in Chapter 4. First, examine how dense these areas were in the year 2000, relative to downtown. These settlements are low-rise, compared to the high-rises in the CBD. Discuss what you think this amount of density means from a living standpoint.

Are these presumed "squatter" settlements located where the Latin American city model predicts they will be?

5

Systems of Cities

The urban systems approach to urban geography considers the interaction of cities within a larger system. Topics typically range from central place theory to the study of cities as nodes with an emphasis on understanding the interaction among the nodes. Marshall (1989) described the importance of the approach as a reminder that a city is not isolated and without relationships to other cities in a system:

> Towns and cities are interdependent: any one town's size, economic character, and prospectus for growth are affected by the nature and strength of its interconnections with other towns. The events that occur in the development of an individual town cannot be understood fully without consideration of trends affecting the entire urban system. (pp. 10–11)

As in other systems research, it is important to emphasize the interdependencies of system components, as a change to one component could have consequences for the others.

Examples of such urban system responses can be drawn from multiple geographic scales. Consider, for instance, political and economic tensions in the early 21st century over global competition in steel imports and exports. Steel producers in U.S. cities like Gary, Indiana, and Pittsburgh, concerned about their domestic clients buying less expensive steel from foreign producers, pressured the Bush administration to impose tariffs on steel imports from places like South Korea. Tariffs were imposed but repealed a year and a half later. Clearly, actions taken by steel producers in cities worldwide have an impact on the economic well-being of their global counterparts. Another regional example of an urban system response was the 1970 slowdown in the steel industry, which resulted in the economic restructuring of places like Pittsburgh and Gary. However, the initial pain was

felt all the way down the urban hierarchy in places like Hibbing, Minnesota, on the Iron Range. Hibbing's livelihood depended on steel producers on the opposite side of the Great Lakes purchasing its iron for steel manufacturing. A local example might be the obvious linkage between a medium city's vibrant new retail strip on its periphery and the declining shopping traffic experienced by retailers in the downtowns of its neighboring cities.

A recent study by Walcott and Wheeler (2001) on the Atlanta metropolitan area demonstrates the urban system approach by establishing the area as a significant telecommunications center as measured by the use of fiber optics. The study finds that Atlanta ranks sixth in the total number of backbone connections with other metropolitan areas and Atlanta's rank in the U.S. urban hierarchy based on telecommunications is considerably higher than that based solely on population. Thus, Atlanta has been reordered within the national hierarchical system of cities, a phenomenon that is reflected in the spatial layout of fiber-optic linkages with other U.S. metropolitan areas (see Figure 5.1). The authors demonstrated that the strong presence of broadband access in both Atlanta's traditional downtown and its developing edge cities attracts new businesses seeking ready access to this backbone to interact with other metropolitan areas in the United States and around the globe. Tellabs' presence in Naperville, Illinois, shown in Figure 5.2, is an example of firms locating in edge cities with good broadband access.

Malecki (2002) further demonstrated a global bias toward world cities in Internet backbone networks, with selected urban areas serving as interconnection points between the backbone networks. Here it is argued that the backbone networks, which ultimately define the outline of the Internet's infrastructure, reinforce the urban system. Nevertheless, Malecki also confirmed that in

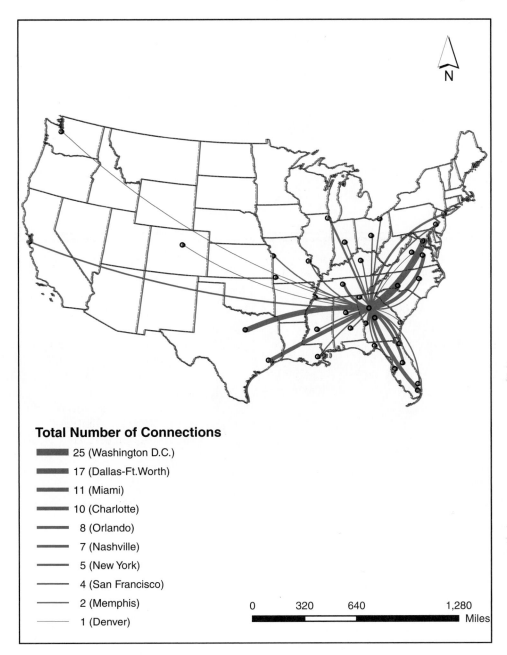

Total Number of Connections

━━━━ 25 (Washington D.C.)
━━━ 17 (Dallas-Ft.Worth)
━━ 11 (Miami)
━━ 10 (Charlotte)
━ 8 (Orlando)
━ 7 (Nashville)
─ 5 (New York)
─ 4 (San Francisco)
─ 2 (Memphis)
─ 1 (Denver)

0   320   640   1,280
Miles

FIGURE 5.1   Atlanta's direct national Internet connections, 2000. Source: Walcott, Susan M. and James O. Wheeler. 2001. "Atlanta in the Telecommunications Age: the Fiber-optic Information Network." *Urban Geography* 22(4):316–339, p. 324, Figure 1, "Atlanta's direction national Internet connections, 2000."

some cases the number of Internet connections did not correspond with the largest cities in population. Both of these studies extend a long tradition in urban systems research of articulating underlying causes for hierarchical distributions of cities. The Internet is perhaps the latest technological influence, and is sure to grow in importance throughout the 21st century.

The chapter begins with a review of central place theory, which is the most classic demonstration of the urban systems approach. *Central place theory* attempts to explain the location, size, and spacing of cities within an urban system. Although an old theory, it continues to provide the theoretical background for current-day applications in location analysis and retail-trade area

FIGURE 5.2   Tellabs in suburban Chicago.

analyses. When many of the rigid assumptions of the classic model are relaxed, it becomes a powerful analytical framework for understanding the complex landscapes of the service and retail economy. Originally, central place theory was applied at a regional scale to understand the relationship between central places and their surrounding hinterlands. Today, the theory is often applied at the scale of an individual metropolitan area to examine consumer transactions occurring at the neighborhood scale all the way to consumers interacting with the large downtown.

The chapter then turns to an evolutionary model of the American urban system. The model emphasizes technological innovations as the principal force for differentiating the urban hierarchy, with each innovation resulting in a new epoch of city building and shifting locational advantages. Prior to the development of the railroad, the large ports in the eastern United States benefited from their position on the Atlantic and Gulf coasts. As the

railroads were built, the port cities continued to grow, but new opportunities opened up in the interior of the United States. When the automobile first came on the scene, urban settlement had already spread across the country, but this new transport technology and its associated highway network allowed for development that was less tied to traditional resource bases and more driven by amenities such as warmer climates.

The chapter concludes with a discussion of global cities, with an emphasis on the concern over the concentration of financial power in a few large cities within a global network of cities. New York, London, and Tokyo are of particular interest, as they were prominent in finance to begin with but became even more specialized in the sector as their manufacturing base receded. Urban scholars have become increasingly interested in examining social and economic dislocation resulting from the concentration of financial resources in a few large cities.

5.1 CENTRAL PLACE THEORY

In central place theory, the role of the city is a provider of services to a surrounding hinterland population. Therefore, the theory is concerned with understanding the spatial arrangement of economic activities and the consuming population of a surrounding market area. We can then characterize the city for purposes of this theory in two ways: (1) as the focus of a hinterland and (2) as the supplier of goods and services to the hinterland. Over time, the theory has been recast in practical terms as relating to goods provided by the city's retail and services sectors to its local market area.

Walter Christaller (1933), a German geographer who worked to explain the size, spacing, and location of cities, is credited as the founder of central place theory. Christaller first tested his theoretical constructs in the 1930s using southern Germany as an example. He developed many theoretical concepts, one of which was *central place,* defined as a place that has developed to provide goods and services to a surrounding hinterland population. This definition would be consistent with most of the larger places in southern Germany in the 1930s, but would exclude towns that did not exist solely to serve a surrounding hinterland population. So for the purposes of his theory, Christaller only considered certain functions of the city and explicitly ignored most manufacturing along with resorts and other such development that did not necessarily fit with the theory. *Centrality* was a concept developed to define central place importance, which was not necessarily correlated with the population size of a place because Christaller noted that a large population did not always signify importance. Rather, Christaller used the number of telephone connections as a surrogate of centrality as most business communication was by way of phones at that time. Perhaps today an important measure of centrality might be the number of Internet connections of a place, especially given that population size does not always predict the number of such connections (Malecki, 2002; Walcott & Wheeler, 2001).

Another important central place concept is *threshold,* which refers to the amount of purchasing power required in the region to support a person engaged in a business activity. Threshold can also be interpreted as the minimum amount of sales necessary for an entrepreneur to break even on a business investment. On the basis of threshold level, the theory distinguishes between two types of goods, *low-order goods* and *high-order goods.* Low-order goods are purchased by a consumer frequently and replenished by an entrepreneur regularly. High-order goods, on the other hand, are purchased by consumers less frequently, and businesses that provide such goods count on a large consumer population from which to draw to offset the lack of frequent patronage by single customers. Thus, because of the hierarchical nature of goods and services, their market sizes are also hierarchical. For instance, because consumers purchase low-order goods and services frequently, entrepreneurs need only a small market size to attain the necessary economies of scale to break even. High-order goods and services, on the other hand, require a larger market size to attain a critical mass of consumers and break even.

The theory also specifies the *range* of a good or outer limit of the market area, which is defined as the maximum distance consumers are willing to travel to purchase the good or service before they decide to do without. The range of a good is said to be the spatial corollary of a threshold because it is the area required for a business to stay viable. Three types of ranges were identified by the theory. The first was referred to as the *inner range* of a good, or the spatial extent of the hinterland necessary for a business to stay viable. If consumers were evenly distributed around an offered good or service, the inner range would be circular and would extend outward from the good until it reached the threshold purchasing power necessary for the business to continue offering the good or service.

The *ideal outer range* of a good is the maximum spatial extent of the trade area defined by the distance consumers are willing to travel to purchase the good or service. The ideal outer range is thus typically larger than the inner range of a good, as it would include a greater area and thus more consumers than the minimum necessary for the entrepreneur to stay in business. In the absence of competition, an ideal outer range extending well beyond the inner range would result in large profits for the business.

The *real outer range* of a good is the actual extent of the market area given that there is competition between central places to attract customers. Berry and Parr (1988) provided actual distances for the real outer ranges of selected goods and illustrated how they varied according to the level of the place (hamlet, village, town, or city) from which they were purchased. Level was measured by the number of central functions (see Figure 5.3):

> Thus, the real range of food stores is just over 5 miles for villages but about 10 miles for towns. Similarly, the real range of clothing stores is just over 10 miles for towns but well over 20 miles for cities. The diagonal line . . . represents an "envelope" curve that traces out variations in the real range (the maximum distances traveled to centers) for different levels of the hierarchy. (Berry & Parr, 1988, p. 26)

Rational economic assumptions about entrepreneurs and consumers dictate that the ideal range will not exist, or at least will not exist for long, as entrepreneurs with full information on the central place landscape and surrounding hinterlands will be quick to offer goods and services to the areas yielding excess profit.

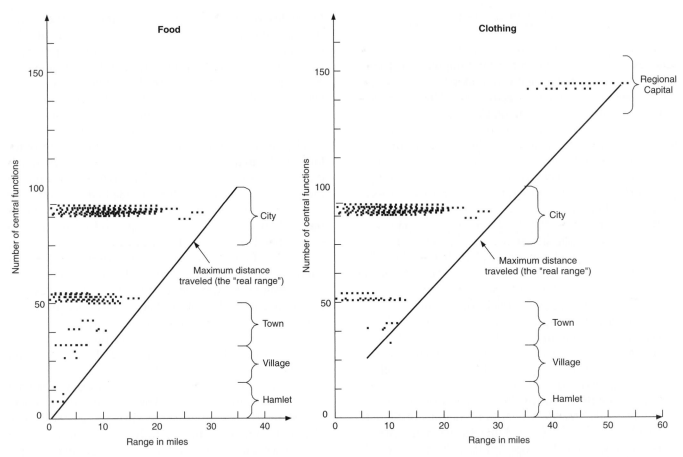

FIGURE 5.3   Examples of range of a good. Source: Berry, Brian J.L. and John B. Parr. 1988. *Market Centers and Retail Location.* Prentice Hall. Fig. 2.3. Page 26. "Real ranges of food and clothing at different levels of the hierarchy."

The outcome of this theory for city systems is that there will always be many more smaller cities in a given landscape than there are large cities because the higher order goods and services offered by large cities require a large hinterland. As a result, there will be a logical hierarchy that emerges (i.e., hamlet, village, town, and city) with substantial distance between larger cities and very little distance between smaller places. At the same time that larger central places offer high-order goods and services, they will also offer low-order goods and services for a smaller area, resulting in a nested structure of market areas.

Many practical applications have been developed from the formal theory. In particular, the urban retail landscape appears to mirror a central place hierarchy. Berry and Parr (1988) noted a strong relationship between regional hierarchies of market centers and that of urban centers, although in the latter case there was more complexity. As seen in Figure 5.4, the typology of business areas within metropolitan areas recognized three types of concentrations: (1) a set of nested hierarchical centers [i.e., central business district (CBD), regional center, community center, neighborhood center, and convenience

center] that was similar to the nesting articulated for market centers from central place theory (i.e., town, village, and hamlet); (2) commercial ribbons composed of shops, restaurants, service stations, and motels; and (3) specialized functional areas of similar businesses held together by close linkages provided by comparative shopping. For Chicago they noted that the size, spacing, and location of these agglomerations were dictated by the same central place principles governing the rural hierarchy of hamlets, villages, towns, and cities.

A decade later, Borchert (1998) demonstrated that Berry's urban classification of shopping centers' business configurations still adequately described contemporary retail landscapes. Borchert used the results of the study to refute recent claims that the classical hierarchy of shopping centers was diminished in light of newer retail configurations:

However, the basic principle formulated by Christaller is applicable as before: goods and services with a high frequency of consumption can be distributed from a denser network of locations than those responding to less frequent

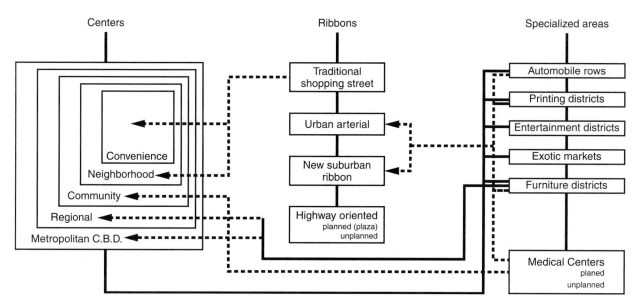

FIGURE 5.4 Typology of business areas within a metropolitan area. Source: Berry, Brian J.L. and John B. Parr. 1988. *Market Centers and Retail Location*. Prentice Hall. Figure 2.16. Page 39 "Typology of business areas within a metropolitan region."

needs. Of course threshold values and the range of goods are changing continuously, but this in no way injures the value of the theory. Actually, Christaller was well aware of the changeability of these variables, which makes the central place theory suitable in dynamic situations. There is no reason why a concept such as threshold should be limited to traditional types of retailing and not extended to innovative forms of retailing. (Borchert, 1998, p. 332)

In addition to explaining the geography of retail, central place theory has been applied to study a wide range of other goods and services. For instance, Blau (1989) studied contemporary cultural patterns in the United States and demonstrated that central place theory helped account for some of the variation in the consumption of both high culture and popular culture from place to place. At the regional scale or at the top of the hierarchy, the theory helps account for why we have a West Coast concentration in the production of films and an East Coast concentration of large cities with a focus on high culture such as opera, dance, and theater. Moreover, Blau (1989) contended that central place theory helps explain why very large places support esoteric and very expensive cultural activities, whereas cultural activities that are in high demand and relatively inexpensive can be found in places of all sizes.

Although many researchers continue to find much explanatory power in central place theory, others find it limiting. Key among the criticisms of central place theory are the economic and landscape assumptions on which the initial theory was based. The economic assumption that consumers are rational is not unique to central place theory but is well developed in neoclassical economics, in which

it has also received much criticism. Christaller's simplistic landscape assumptions were necessary for him to model uniform travel of consumers to central places, yet it is this aspect of the theory that has received much criticism. Dennis, Marshland, and Cockett (2002) summarized some of these concerns and proposed methods for relaxing some of the strict assumptions of Christaller's classical theory for interpreting contemporary retail landscapes.

In the next section, "Analyzing an Urban Issue," we re-create one of the classic central place studies from the 1950s on Snohomish County, which highlights some of the strong empirical evidence in support of the theory. At the same time, the analysis points out deviations from the theory and possible explanations for such statistical outliers. The current-day central place landscape of Snohomish County is presented and compared to the earlier 1950 hierarchy. Methodological issues and expected findings are addressed for a proposed replication of the 1950 study. This urban issue is later pursued in the chapter's geographic information system (GIS) exercise.

5.2 ANALYZING AN URBAN ISSUE

5.2.1 The Central Places Hierarchy in Snohomish County, Washington

Berry and Garrison (1958) selected Snohomish County in the state of Washington to search for evidence of a hierarchy of central places in a system of urban centers. Snohomish County was ideal for this type of study in the 1950s because it had a large number of centers that ranged from very small (hamlets) to medium size (villages and towns) and only had one very large city, Everett, which was excluded from the study. As shown earlier, the hierarchical system of hamlets, villages, towns, and cities is an important aspect of central place theory. That is, the model argues that places will belong to one of these types. Each type of place is characterized by a certain population limit and has a certain level of central place functions that are associated with this given population size (see Table 5.1). For instance, the population threshold of 196 for filling stations could be interpreted to mean that filling stations, at least in the case of Snohomish County in the 1950s, required a market area encompassing 196 persons. By analyzing a variety of urban functions ranging from low-order goods and services (i.e.,

filling stations) to high-order goods and services (i.e., health practitioners), Berry and Garrison were able to classify places into three types.

A GIS project was designed to replicate the county's central place landscape of the 1950s (see Figure 5.5). The GIS database was designed to include all 33 places from the original study as well as the critical attributes of the analysis. For instance, Figure 5.5 highlights the attribute table entry for the central place of Darrington. The type field indicates a value of 2 for Darrington, which would identify it as a village-level central place. The activities field indicates that Darrington had a total of 25 functions identified for it in the 1950s, and the field A1 is the code for filling stations, showing that Darrington had three filling stations at the time. Table 5.1 gives the codes for functions

TABLE 5.1   Population thresholds established by central place function from original Snohomish County analysis

GIS code	Activity	Population threshold	GIS code	Activity	Population threshold
(a) Hamlet-Level Functions					
A1	Filling Stations	196	A4	Restaurants	276
A2	Food Stores	254	A5	Taverns	282
A3	Churches	265	A6	Elementary Schools	322
(b) Village-Level Functions					
B1	Physicians	380	B20	Variety Stores	549
B2	Real Estate Agencies	384	B21	Freight Lines & Storage	567
B3	Appliance Stores	385	B22	Veterinaries	579
B4	Barber Shops	386	B23	Appearel Stores	590
B5	Auto Dealers	398	B24	Lumberyards	598
B6	Insurance Agencies	409	B25	Banks	610
B7	Bulk Oil Distributors	419	B26	Farm Implement Dealers	650
B8	Dentists	426	B27	Electric Repair Shops	693
B9	Motels	430	B28	Florists	729
B10	Hardware Stores	431	B29	High Schools	732
B11	Auto Repair Shops	435	B30	Dry Cleaners	754
B12	Fuel Dealers	453	B31	Local Taxi Services	762
B13	Drug Stores	458	B32	Billiard Halls and Bowling	789
B14	Beauticians	480	B33	Jewelry Stores	827
B15	Auto Parts Dealers	488	B34	Hotels	846
B16	Meeting Halls	525	B35	Shoe Repair Shops	896
B17	Animal Feed Stores	526	B36	Sporting Goods Stores	928
B18	Lawyers	528	B37	Frozen Food Lockers	938
B19	Furniture Stores	546			
(c) Town-Level Functions					
C1	Sheet Metal Works	1076	C6	Photographers	1243
C2	Department Stores	1083	C7	Public Accountants	1300
C3	Optometrists	1140	C8	Laundries and Laundromats	1307
C4	Hospitals and Clinics	1159	C9	Health Practitioners	1424
C5	Undertakers	1214			

Source: Berry, Brian J.L. and William I. Garrison. 1958. "The Functional Bases of the Central-Place Hierarchy." *Economic Geography* 34:145–154. Table 2. "Population thresholds established by central place function."

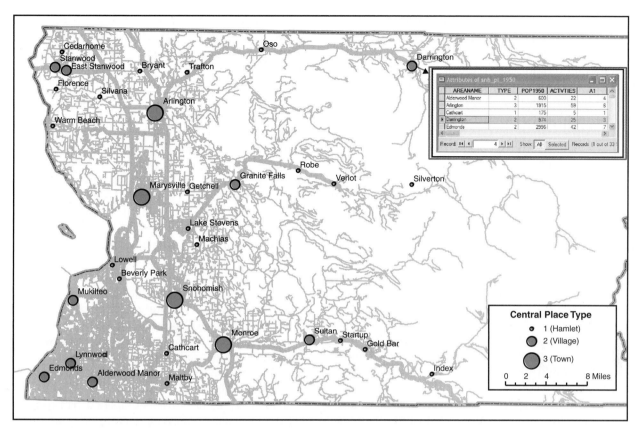

FIGURE 5.5 Snohomish County central place hierarchy in the 1950s, with year 2000 road network. Source: Berry, Brian J.L. and John B. Parr. 1988. *Market Centers and Retail Location.* Prentice Hall. Figure 2.16. Page 39 "Typology of business areas within a metropolitan region."

at the hamlet, village, and town levels. The transportation GIS layer is contemporary and reflects a much more mature network than that of the 1950s; thus it is used only for context and is not incorporated in the analysis.

The GIS design has allowed for the validation of a number of central place principles. First, the visual display of the three levels of the central place hierarchy provides immediate evidence of the principle that there are more small central places in a central place landscape than there are large ones. For instance, with the cartographic technique of graduated symbols it is easy to identify 20 hamlets, 9 villages, and 4 towns in the county. The cartographic display also allows for the immediate visualization of some of the spacing principles of central place theory; for example, it is easy to see that the spacing between hamlets is more equal with respect to other hamlets than it is with the spacing between hamlets and villages or to towns. The distance between the four towns is close to equal. Although not computed for this study, distance measures could be computed directly within the GIS to quantify the variability in these distances.

Each of the three types of central places (hamlets, villages, and towns) identified by Berry and Garrison contained specific groups of central functions and were

characterized by a certain population size. Towns contained all of the low-order central functions (i.e., filling stations, food stores, and churches) of the hamlets and villages plus high-order functions (e.g., photographers, public accountants, and health practitioners) not found in hamlets and villages. Population size was highly correlated with the level of central place functions offered by places, reflecting the threshold concept of central place theory, which states, for example, that a large threshold of population is necessary before a high-order good or service will be offered.

This latter central place principle of thresholds of goods and services was tested within the GIS with several database queries. For instance, Table 5.1 indicates that hardware stores are functions that would be offered by village- and town-level central places but would probably not be offered by hamlets. The following query tests this proposition by querying the database for hamlets that have hardware stores:

$$\text{``TYPE''} = 1 \text{ AND ``B10''} > 0$$

This query returns only one selection; in other words, only 1 out of 20 hamlets (Lake Stevens) had a hardware store. Thus with the exception of Lake Stevens, the threshold concept appears to have been quite valid for this landscape in the 1950s. On closer inspection, Lake Stevens had a population size much more in common with a town-level place, but Berry and Garrison classified it as a hamlet because it was underrepresented by higher order central functions (see Figure 5.6). At least in terms of hardware stores, the GIS query shows that Lake Stevens does seem to fit the character of a town-level place rather than a hamlet-level place.

In the original study, Berry and Garrison discussed both Lake Stevens and Edmonds with respect to being underrepresented with central place functions given their population sizes. To select these two places, on the basis of the above discrepancy, the following query was submitted in the GIS:

"POP1950" > 2000 AND "ACTIVITIES" < 50

One explanation given by Berry and Garrison for the divergence of central place functions in Lake Stevens and Edmonds (along with Beverly Park and Lowell) from their expected function groupings was that they each had experienced rapid population growth at the time (pressure from greater Seattle) and that there is usually a lag in service delivery in such cases. In other words, it was expected that these places would gain more

central place functions as time passed, especially as entrepreneurs recognized their population sizes and began to offer the appropriate goods and services associated with those thresholds.

To illustrate how the central place landscape of Snohomish County has changed since the 1950s, a GIS layer of places in 2000 was created (see Figure 5.7). The creation of the layer also involved some GIS overlay techniques and relational table joins to determine which places are no longer present as well as new central places that have arisen since the 1950s. The first overlay technique identified places that are no longer present in Snohomish County. These places that have been absorbed by other places since the 1950s and are thus not part of the contemporary central place landscape. To determine the places that are new since the 1950s, a relational join was conducted that joined the new 2000 GIS layer with the earlier 1950 GIS layer, which determined that 35 places were added. A total of 23 places had persisted since the 1950s. A *relational join* in a GIS refers to combining two tables into a single table on the basis of a common characteristic shared by both tables.

The GIS activity of updating the central place landscape of Snohomish County from the 1950s to the year 2000 revealed a much more complex urban system. Snohomish County today is much more integrated with the Seattle metropolitan area and many of the older places have coalesced into each other, making it difficult to assign activities to a given center. Similar to the case study of the Chicago area's central place hierarchy by Berry and Parr (1988), consideration might be given to reexamining Snohomish County in terms of its metropolitan context. In a metropolitan context, functions would be grouped into a hierarchy of business centers, highway-oriented ribbons, urban arterial commercial developments, and specialized functional areas. Today, such a study could take advantage of online yellow pages and reduce the amount of data entry time that presumably took place in the 1950s study. On the other hand, many of the earlier central places have grown in size, much greater than town size, and the explosion of categories of economic functions as well as the disappearance of certain functions might make updating the earlier study a challenging project.

5.3 CONTEMPORARY CENTRAL PLACE RESEARCH AND THE GRAVITY MODEL

The gravity model has been applied to the study of central place theory in general and to retail location analysis more specifically. Reilly (1931) was one of the first to apply the gravity concept of physics to the study of retail trade and human interaction among central places. Reilly's formulation is an extension of Ravenstein's

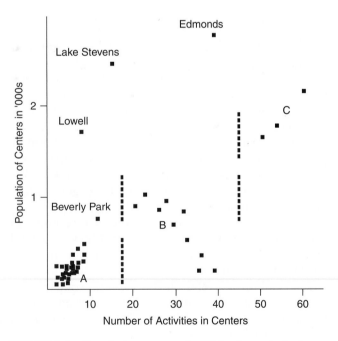

FIGURE 5.6 Population centers (in 1000s of population) versus number of activities in centers. Source: Berry, Brian J.L. and William I. Garrison. 1958. "The Functional Bases of the Central-Place Hierarchy." *Economic Geography* 34:145–154. Figure 2, "Classes of central places in Snohomish County."

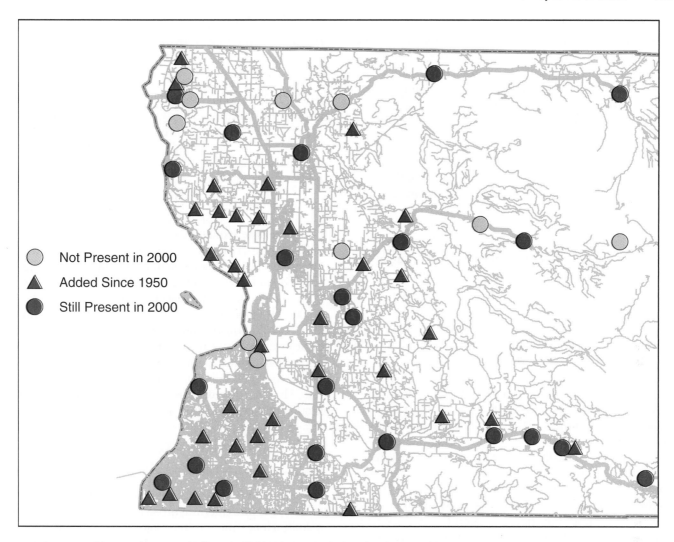

FIGURE 5.7   Changes between 1950 and 2000 in the central place landscape of Snohomish County.

(1885) application of the gravity model to migration studies, which is reviewed in more detail in Chapter 7. The general gravity model holds that any two bodies attract each other with a force proportional to the product of their masses and inversely proportional to the square of the distance between them. Reilly then interpreted this for retail as meaning that two cities would attract consumers from some smaller, intermediate city in direct proportion to their population sizes and in inverse proportion to the square of their distances from the intermediate city. Reilly later went on to develop a technique, now referred to as *Reilly's law of retail gravitation,* which allowed one to determine trade area boundaries around cities based solely on population and distance measures.

Huff (1964) offered an alternative model referred to as the probabilistic model of consumer spatial behavior. Recently, Huff and Black (1997) revisited his earlier

formulation and emphasized that the model is rooted in central place theory:

> The close parallel between the probability model and the central place theory provides a much richer theoretical focus. An attempt to achieve the objective of the central place theory, that is, to derive the optimal size, number, and arrangement of centers to serve a dispersed population, can be made under real-world conditions with the use of the probability model. There are a number of concepts in the central place theory that could be analyzed empirically. The findings from such studies could, in turn, be utilized in a way to enhance the predictive capabilities of the model. (p. 91)

Huff's model results are often depicted as probability contours and interpreted, for instance, as a person from the Lincoln Park neighborhood in Chicago has a probability of 0.9 of going to a nearby coffee house, whereas a person from Little Italy has a probability of 0.2 of going to the same coffee house. The person from Lincoln Park is also assigned a probability of visiting competing destinations, including the one in Little Italy.

Huff first formulated his ideas by establishing an attraction index to be associated with a given retail facility:

$$v(j) = A_j^\gamma D_{ij}^{-\lambda}$$

where A_j is an attraction index associated with a particular retail facility j; D_{ij} is the accessibility of a retail facility j to a consumer located at i; and A^γ and D^λ are empirically derived parameters. The quotient derived by dividing A_j^γ by D_{ij}^λ is regarded as the perceived utility of retail facility j by a consumer located at i (Huff & Black, 1997, p. 84). The contours of the probability that a consumer located at i will choose to shop at retail facility j is determined as follows:

$$P_{ij} = \frac{A_j^\gamma D_{ij}^{-\lambda}}{\sum_{j=1}^{n} A_j^\gamma D_{ij}^{-\lambda}}$$

Today, this probabilistic model of consumer spatial behavior appears in many commercial GIS software packages. If it is not available, it is feasible to employ it directly in a standard GIS system as distance calculations are derived easily with x and y coordinates. To illustrate the latter, consider the application of Huff's probability model for estimating the attendance at three major opera houses in the Chicago, Milwaukee, and Madison metropolitan triangle region (see Figure 5.8). The seating capacity of the three principal opera houses within the region was used as the attraction index: the Lyric in Chicago with 3,563 seats, the Skylight in Milwaukee with 358 seats, and the Oscar Mayer in Madison with 2,200 seats. Census tract centroids (an internal x and y coordinate pair) were used for the origins of those persons attending the opera. In the final computations, each tract was assigned three probability values, one for each opera house its residents were likely to attend (Figure 5.8). As shown on the map, the 50 percent draw for Chicago is much larger than its two competing opera houses.

5.4 CITY SIZE DISTRIBUTIONS

A substantial literature has developed on the topic of city size distributions, which addresses the question of how urban populations of a country are distributed across cities of varying sizes. In Chapter 2, the case of the *primate city* was discussed, which is reflective of a type of city size distribution where one city of a country dominates its urban hierarchy. A very common city size

distribution, however, is the *rank-size distribution*, represented by the following equation:

$$P_r = P_l r^{-q}$$

where P_r is the population of the city ranked r, P_l is the population of the largest city, r is the rank, and q is an exponent. In demonstrating the practical use of this equation, Cadwallader (1996) assumed that $q = 1$ and rewrote the equation as:

$$P_r = \frac{P_l}{r}$$

Using the latter equation, Cadwallader (1996) considered a hypothetical country where there are five cities and the largest of these has a population of 100,000 (see Figure 5.9). Given that the rank-size distribution has an exponent of 1, the second largest city will have a population of 50,000 (100,000 divided by 2), the third city would be 33,333, and so on. As shown in Figure 5.9, city size distributions are plotted on a graph in which rank is on the horizontal axis and population size is on the vertical axis, illustrating a curvilinear form for the rank-size distribution. The rank-size distribution has been shown to fit the United States and a large number of other countries.

5.5 EVOLUTION OF THE AMERICAN URBAN SYSTEM

Borchert (1967) presented an urban historical study on the evolution of the American urban system that revealed technological innovations as the major motive for changing hierarchies over time. He observed that the infrastructures of individual metropolitan areas, accumulated through various historical epochs, were distinguished from each other by different technologies. Borchert defined these epochs by identifying critical dates or transitions on growth curves of selected technologies of transport and industrial energy (see Figure 5.10). These curves represented the following:

1. Rail mileage
2. General cargo tonnage on inland waterways
3. Production of primary energy from:
    (a). Water power
    (b). Coal and oil
    (c). Motor vehicles
    (d). Highway mileage

The critical dates and epochs were then labeled as follows:

1. Sail–Wagon (1790–1830)
2. Steamboat and Iron Horse (1830–1870)
3. Steel Rail (1870–1920)
4. Auto–Air–Amenity (1920–)

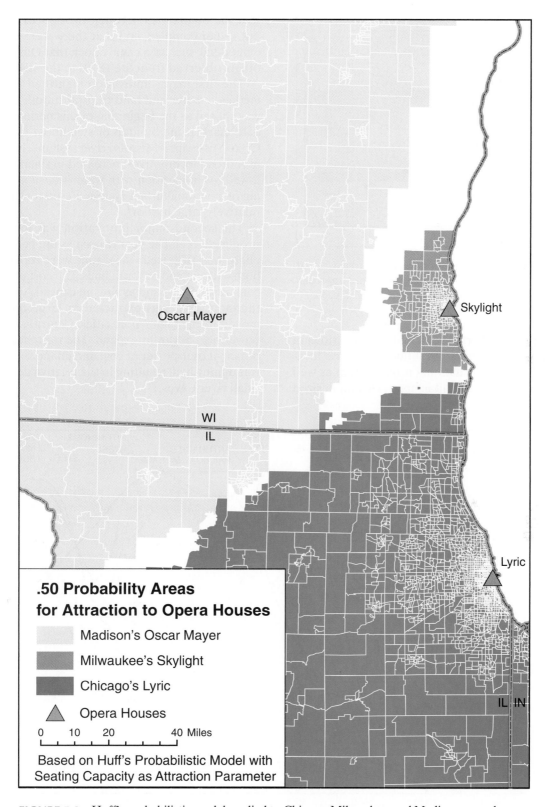

**.50 Probability Areas
for Attraction to Opera Houses**

Madison's Oscar Mayer

Milwaukee's Skylight

Chicago's Lyric

▲ Opera Houses

0   10   20   40 Miles

Based on Huff's Probabilistic Model with
Seating Capacity as Attraction Parameter

FIGURE 5.8   Huff's probabilistic model applied to Chicago, Milwaukee, and Madison opera houses.

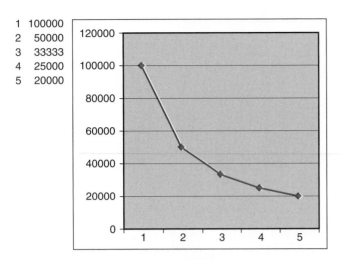

FIGURE 5.9 Rank-size distribution model. Source: Cadwallader, Martin 1996. *Urban geography: an analytical approach.* Upper Saddle River, N.J: Prentice Hall, Inc. Fig. 12.5 p. 305.

According to Borchert (1967), major innovations in transportation and energy were influential in the location of growth within the national urban system as well as within individual metropolitan areas. He emphasized

that the boundaries between the epochs are not sharp but rather complex, and that there were minor epochs embedded, such as a canal epoch that Oliver (1956) and others observed from the 1810s to the 1840s.

The first major innovation identified was the steam engine in water and land transportation. The year 1830 was chosen as the beginning of this period because that date marked the buildup of steamboat tonnage on the Ohio–Mississippi–Missouri river system. The use of steam in railroad locomotives increased in significance and allowed for integration with water-transit systems, resulting in a national urban system biased toward ports with relatively large harbors.

The second major innovation was low-cost, mass-produced steel. The year chosen as the beginning of this phase was 1870 when steel rails were replacing iron on both existing and new rail lines. The cost of transporting coal was reduced and as a result central Appalachia's bituminous coal was extracted in large quantities and moved to the major port cities. Central-station electric power also began to expand. The railroads became the principal transporter of coal and inland waterway traffic was diminished, resulting in the relative decline of many small river towns.

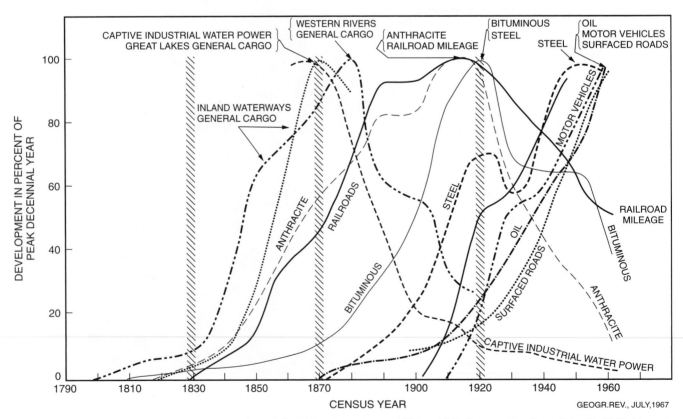

FIGURE 5.10 Borchert epochs in the evolution of the U.S. urban system, 1800 to 1960. Source: Borchert, John R. 1967. "American metropolitan evolution." *Geographical Review* 57:301-332, p. 302, figure "The rise and decline of ten indicators of the technology of transport and industrial energy."

The third major innovation was the introduction of the internal combustion engine in transportation and the beginning year chosen was 1920 when motor vehicle registrations became significant and petroleum production began its steep climb. Borchert argued that this innovation encouraged concentration of urban growth by putting the farmer in an automobile, but more important, it put the farmer on a tractor, which multiplied the land area worked. This sped up urbanization indirectly by increasing the size of the typical family farm. Borchert also stressed the simultaneous trend of the national economy away from primary and secondary employment toward service employment, which advantaged the cities with the highest concentrations of people. Thus, according to Borchert (1967):

> In the auto–air age, even more than in the preceding epoch, growth breeds growth. The second factor is the large and growing amount of leisure time available. As Ullman (1954) pointed out some years ago, this has led to the great importance of amenities as an urban location factor, both for commuting workers and for retired people. (p. 307)

Each of these major innovations led to changes in the physical layout of metropolitan areas in terms of density, location of economic activity, and land use patterns. Each one of these epochs is reviewed for its essential impacts.

5.5.1 Sail–Wagon Epoch, 1790 to 1830

The 1790 Census only recognized 22 urban places (Gibson, 1998) and most of them were port cities on the Atlantic coast or river towns that were linked to them (see Figure 5.11). Some of these early urban places were separate communities at the time of the 1790 Census but have since become part of a larger metropolitan area. For instance, Salem, Marblehead, Gloucester, and Newburyport, all have since become integrated into metropolitan Boston. Borchert (1967) noted that no one city dominated the nation at this point in time, as indicated by the almost equal populations of Boston, Philadelphia, and New York. The smaller inland river towns such as Richmond, Hartford, and Albany were all navigationally linked to the larger coastal cities and their growth depended in large part on this relationship.

5.5.2 Iron Horse Epoch, 1830 to 1870

The 1830 Census recognized 90 urban places and although coastal cities remained prominent, river cities located further into America's interior were beginning to make their mark on the urban settlement system (see Figure 5.12). The largest urban places now extended all the way west to St. Louis on the middle Mississippi. Borchert attributed the growth in this period to the railroads, which were complementary to the waterways as they were built inward from the major ports. Both were based on the steam engine. The development of the railroads accelerated the exploitation of coal, and as a result places like Pittsburgh rose in the urban hierarchy. A very strong river connection among top cities was becoming emphasized; for example, Cincinnati, Louisville, and Pittsburgh were all tied to the Ohio River system (see Figure 5.13). Following the Louisiana Purchase in 1803, New Orleans first appeared on the census list of the largest urban places in 1810, when it was listed in seventh position; by 1830 it had risen to fifth position, just below Boston. An inland waterway location or coastal position was still a very important factor for success in 1830, but when the railroads came, these railroad-linked cities were aided in reaching high positions in the urban hierarchy. New York's population had outdistanced the populations of both Philadelphia and Boston, principally because of a clear advantage with respect to the nation's hinterland by way of rivers, the Erie Canal, and then the railroads. The 1840 Census was the first to record at least 100 urban places of at least 2,500 population.

5.5.3 Steel Rail Epoch, 1870 to 1920

In the 1870 Census, the 100 largest urban places reinforced the patterns established in the iron horse epoch; however, the western limit of the American urban settlement system was considerably more expansive (see Figure 5.14). Added to the list of top urban places were Kansas City, Omaha, San Francisco, and Sacramento. Although added prior to 1870, the largest industrial newcomers to the urban hierarchy since 1830 were Chicago, Milwaukee, Cleveland, and Detroit (see Figure 5.15). According to Borchert, growth resulted primarily from a nationwide system of rail lines and the main industrial metropolitan centers of the Midwest and Northeast retained or advanced their position in the urban hierarchy. Places that dropped down the urban hierarchy, such as St. Louis and Louisville (although still within the top 100 largest places and still growing) fell because they were more central to the innovations of the steam engine and iron horse epochs rather than those of the steel rail epoch.

5.5.4 Auto–Air–Amenity Epoch, 1920–

As Borchert (1967) published his original article in the late 1960s, he deliberately left the date off the end of the auto–air–amenity epoch because America was still in it. The largest 100 urban places in 1920 showed that since 1870 many new places had been added at the expense of others (see Figure 5.16). Borchert noted the declining significance of coal and railroad centers in the

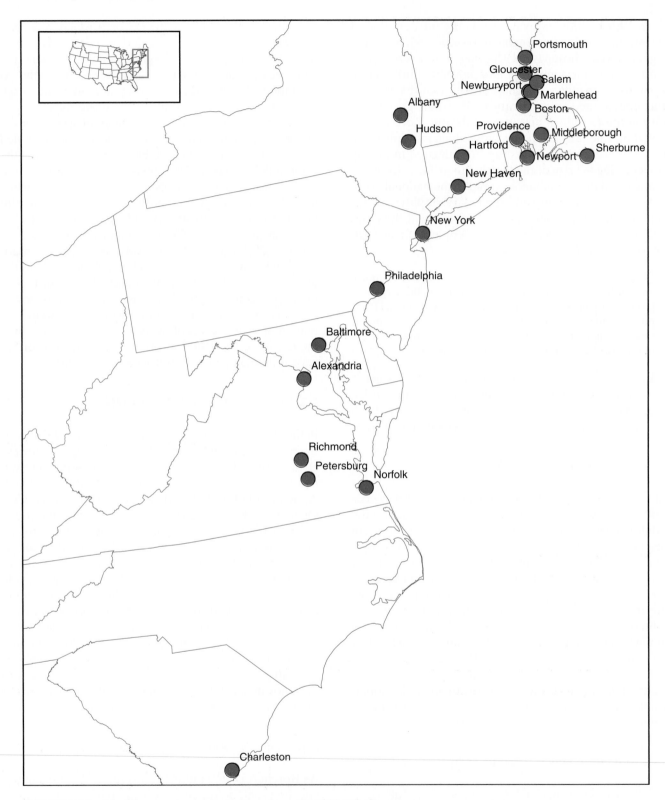

FIGURE 5.11   The 22 recognized urban places from the 1790 U.S. Census.

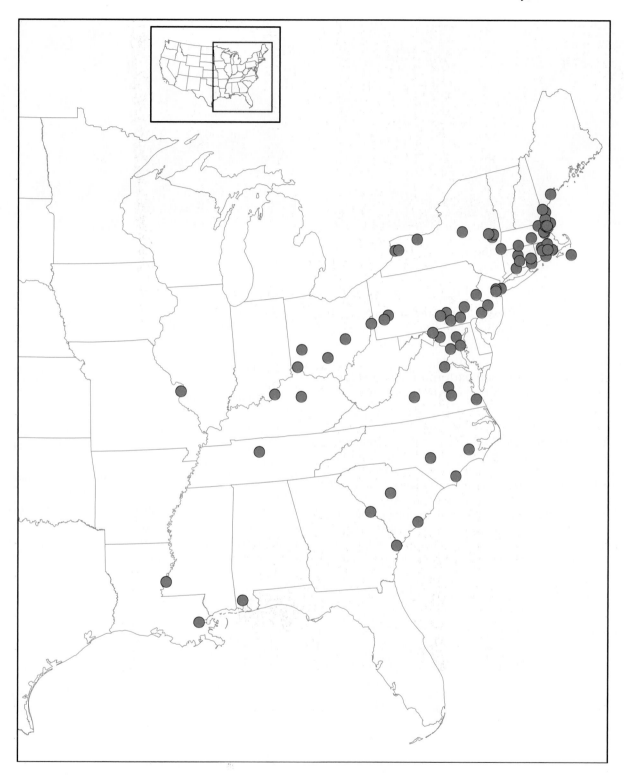

FIGURE 5.12   The 90 recognized urban places from the 1830 U.S. Census.

FIGURE 5.13   Cincinnati riverboat.

FIGURE 5.15   Milwaukee Rail Yards.

Appalachians and across the Midwest, and the replacement by metropolitan areas located in the oil fields from central Kansas to western Texas and the western Gulf Coast, as well as southern Michigan. The shift to the automobile and truck and a supporting highway network resulted in a more dispersed pattern of urban settlement. Borchert suggested that the influence was expressed in two ways:

> One is the entry into the metropolitan ranks of numerous "satellite" cities on the fringes of the historic Manufacturing Belt and within 100 to 150 miles of great metropolitan industrial centers. The other is the effect of suburban dispersal on the definition of a metropolitan area. In the Appalachians, for example, half a dozen "SMSAs" have dropped out because their central cities failed to maintain a population equal to, or larger than, that of other metropolitan centers. In these cases it is typical to find a central city crowded on a valley

floor, blighted by obsolescent buildings, air pollution, narrow streets, and rusty rail lines, and exposed to flood risk. Population and commercial growth have dispersed to the uplands to exploit the resources of open space, a dense rural road net, panoramic views, and relatively clean air. The result is that the metropolitan area has grown and ceased to be "metropolitan" by definition, and the central city has declined and ceased to be "central" in fact. (p. 321)

The 1920 urban hierarchy inaugurated amenities as a major determinant in metropolitan growth, especially in Florida, the states of the Southwest, and southern California, which all had centers high in the urban hierarchy. During the period between 1920 and 1960, there was a net migration of 11.4 million people to California (see Figure 5.17), Arizona, and Florida from other regions (Borchert, 1967, p. 321). Other places that gained during the period were metropolitan areas with universities.

1870 Census

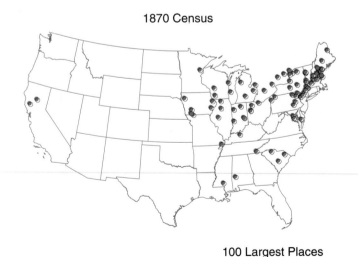

100 Largest Places

FIGURE 5.14   The 100 largest urban places recognized in the 1870 U.S. Census.

1920 Census

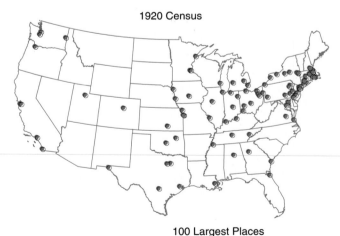

100 Largest Places

FIGURE 5.16   The 100 largest urban places recognized in the 1920 U.S. Census.

FIGURE 5.17 Volleyball players on a Los Angeles beach.

5.5.5 A New Epoch, 1970–

Borchert (1991) would later mark 1970 as the beginning of a new epoch. The critical variables marking this separation were more costly energy together with electronic and jet propulsion technologies. A number of trend reversals were used to choose 1970 as the critical date including (see Figure 5.18) the following:

1. Sharp reversal upwards in the cost curve for energy raw materials
2. Reversal toward greater efficiency of automobile fuel use
3. Growing relative importance of the country's overseas commerce
4. Upsurge in growth of the American overseas cable, radio, phone, and later Internet messages
5. Continuing increase in domestic long-distance calls, with the Internet expanding later
6. Increased per-capita consumption of paper for record keeping, correspondence, magazines, and advertising circulars that has accompanied television, computing, word processing, and desktop publishing

The striking difference between the largest 100 urban places in 1920 and those of 1970 was that the growth pattern driven by amenities and the service economy became more pronounced (see Figure 5.19). In most cases, inertia explained the persistence of many of the largest cities. Those that had claimed good advantage with respect to the railroads in an earlier epoch were now well positioned with respect to the nation's interstate highway system and airports. Los Angeles grew from being the 10th largest city to being the third largest, and Houston and Dallas entered the ranks of the top 10.

Perhaps the most systematic change was the appearance of cities in the top 100 that were once just on the outskirts of the previous epoch's large cities. For instance, Anaheim, Long Beach, Riverside, and Santa Ana, which are all clustered around Los Angeles, were now in the top 100 and all except Long Beach had joined those ranks for the first time in 1970. The trend intensified over next 30 years, with Huntington Beach and Glendale joining the other Los Angeles suburbs by the year 2000 (see Figure 5.20). The phenomenon also appeared in metropolitan areas like Phoenix, with Mesa and Scottsdale now in the top 100 city list, and Dallas-Fort Worth, with Arlington, Plano, Garland, and Irving now included. Meanwhile, the places that fell out of the top 100 between 1970 and 2000 were indicative of the earlier epochs that gave rise to them, including places like Gary, Providence, Syracuse, and Evansville.

5.6 GLOBAL CITIES

Urban system research is becoming increasingly global in scope. This global approach is often referred to as a world cities approach, led by the hypotheses generated by the work of Cohen (1981), Friedmann (1986), and Sassen (2001). The key investigations include the concentration of corporate headquarters offices and subsequent concentration of financial power in a few large cities within a global network of cities. This centralization of top-level management and control functions runs counter to the spatial dispersion of other economic activities such as manufacturing that is typically associated with the process of globalization. Instead, high-level firms concentrate in global cities to command and control these dispersed activities.

According to Sassen (2001), the turnaround occurred in the 1960s with a profound transformation of economic activity in the world economy, including the industrialization of third-world countries and the internationalization of the financial industry into a worldwide network of transactions. New York, London, and Tokyo are the three principal global cities because they were prominent in finance to begin with, but these cities became even more specialized in the sector as their manufacturing base receded quickly. As a result the global cities have emerged as control centers that set the rules for production and marketing worldwide (Knox, 1995). Examining these three cities, Sassen found that they now function in four new ways:

1. As highly concentrated command points in the organization of the world economy
2. As key locations for finance and specialized service firms, which have replaced manufacturing as the leading economic sectors

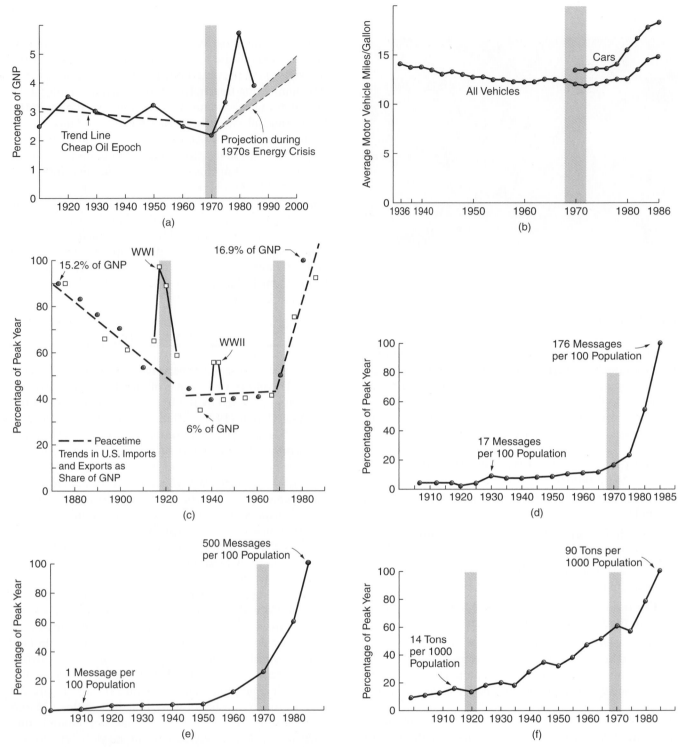

FIGURE 5.18 United States trends having 1970 as a critical year. (a) Long-term trends in the cost of U.S. energy raw materials. Sources: for 1970 and earlier, Borchert 1981, 43; after 1970, *Statistical Abstract of the United States 1988*, Tables 910, 1159, and 1165. (b) Increased efficiency in response to the uncertain outlook for motor fuel cost and supply. Sources: 1970–1986, *Statistical Abstract of the United States 1988*, Table 1003; 1936–1969, *Historical Statistics of the United States*, Series Q 148–162. (c) Peacetime trends and wartime anomalies in total value of U.S. imports and exports expressed in percentage of gross national product. Sources: *Historical Statistics of the United States*, Series U 317–352; *Statistical Abstract of the United States 1988*, Table 1349. (d) Long-trend in overseas messages (cable, radio, telephone) per capita of resident U.S. population. Sources: *Historical Statistics of the United States*, Series R 75–88; *Statistical Abstract of the United States 1988*, Table 881. (e) Long-term trends in U.S. domestic long-distance telephone messages per capita of resident population. Sources: *Historical Statistics of the United States*, Series R 1–12; *Statistical Abstract of the United States 1988*, Table 881. (f) Long-term trend in U.S. per capita of paper, excluding, newsprint, packaging, Industrial, and tissue. Sources: *Historical Statistics of the United States*, Series L 178–191; *Statistical Abstract of the United States 1988*, Table 1129.

182

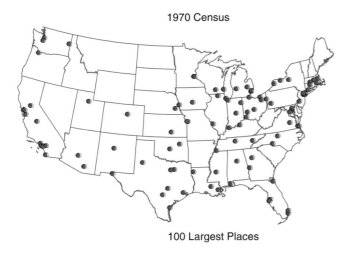

1970 Census

100 Largest Places

FIGURE 5.19   The 100 largest urban places recognized in the 1970 U.S. Census.

3. As sites of production, including the production of innovations, in these leading industries

4. As markets for the products and innovations produced (Sassen, 2001, pp. 3–4)

Sassen (2002) asked the question, why do firms continue to clutch onto perceived benefits of agglomeration economies in light of new telecommunications and computer technologies? Sassen suggested that it is for the old adage of "face-to-face" contact, or what she termed business networks. Sassen further argued that the largest global cities have undergone massive and parallel changes in their economic base, spatial organization, and social structure. It follows from this argument that the new world city system transcends national institutions, politics, and culture and is based more on global economic processes.

Nevertheless, a number of researchers remain skeptical of the world cities approach. Hill and Kim (2000)

contended that fundamental differences in economic base, spatial organization, and social structure persist between major cities in the North Atlantic and East Asian regions. They presented their case by contrasting Tokyo and Seoul with New York and London:

> Seoul, like Tokyo, has not experienced severe manufacturing decline, rapid expansion in producer service employment, extensive foreign immigration or much social and spatial polarisation. Like Tokyo, Seoul is under the sway of a political bureaucratic elite, not a transnational capitalist class. And, as with Tokyo, it would be senseless to claim that Seoul is severing ties of mutual interest with the nation-state; if anything, the capital city is becoming even more integrated with the rest of South Korea. (Hill & Kim, 2000, p. 2186)

Esparza and Krmenec (2000) made a similar claim for the United States, where they noted that at the same time urban system analysts have become increasingly focused on world city hypotheses, large urban centers at the top of the U.S. urban hierarchy still have significant interaction (trade) with other large urban centers in the U.S. urban system, as well as with smaller cities within their regional trade areas. Likewise, McCann (2004) pointed to a polarization in urban studies between research on global cities and cities outside of the global domain. Citing the globalization processes impacting Lexington, Kentucky, McCann suggested that many key aspects of globalization effects on cities are being missed in the dialogue within the global cities literature.

Sassen (2002), also recognized these national and regional interactions, although her emphasis was clearly that global cities become more closely linked to the global economy than they do to their regional or national economies, the consequences of which are an eventual pushing out of firms and people not connected to the internationalized sector. Thus, the world cities or global cities perspective is multifaceted with some attention placed on the consequence of cities not interacting at the top of the hierarchy.

Another key topic within global cities research includes social and economic dislocation resulting from the concentration of financial resources in the few largest cities. Boschken (2003) addressed the concentration of upper middle class population caused by the processes leading to global cities. Boschken devised a classification of a number of U.S. cities along 10 dimensions associated with global cities (see Table 5.2). He found that the urban areas having the greatest number of global-interface functions also tended to have relatively high concentrations of upper middle class residents. Boschken saw the significance of this trend as follows:

> The achievements of UMC [upper middle class]-inspired global-city development have brought the world to the doorstep of those endowed with the educational resources and professional ties that give them entrée to the platform

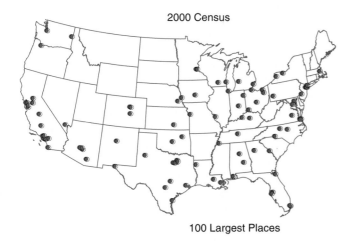

2000 Census

100 Largest Places

FIGURE 5.20   The 100 largest urban places recognized in the 2000 U.S. Census.

TABLE 5.2 Global status for U.S. cities and Upper-Middle-Class factor values

Type of city/ urban area	No. of global dimensions	Upper middle class factor value
Global Cities		
New York	10	−0.66
Washington, DC	10	2.42
San Francisco	10	1.73
Boston	10	1.23
Philadelphia	10	0.06
Chicago	10	−0.77
Los Angeles	8	−0.18
Partial Global Cities		
Atlanta	7	0.81
San Diego	6	0.52
Miami	5	−1.33
Seattle	5	1.06
Houston	5	−0.45
Nonglobal Cities		
San Jose	4	1.55
San Antonio	3	−1.06
St. Louis	2	−0.87
Cleveland	2	−0.88
Indianapolis	0	−0.92
Buffalo	0	−1.18
Milwaukee	0	−1.22

Note: Similar factor values imply similarities in upper middle class characteristics.

Source: Boschken, H.I., 2003. "Global cities, systemic power, and upper-middle-class influence." *Urban Affairs Review* 38:808–830. Table 2 on P. 823. "Global status for U.S. cities and upper-middle-class factor values."

of global connectivity. But globalization is not without trade-offs. As we look outward to a world that, for the most part, seems to be enthusiastically "globalizing American culture and cultural icons" (Friedman, 2000, p. 380), the look inward reveals an unparalleled challenge for American global cities in dealing with nagging problems of dislocation, dissension among cultural variants, and growing competition between values—some global, some provincial. Moreover, in forming urban policy in a global environment, subtle revisions have taken place in social consciousness and in policy makers' striving for social program effectiveness. (Boschken, 2003, p. 824)

Other dislocations have been observed as a result of the global city phenomenon. For instance, Warf and Erickson (1996) noted that the high and rising incomes of the professional class associated with the large firms and banks of these global cities have encouraged neighborhood gentrification by indirectly raising real estate prices within those cities. Displacement of low-income residents by the process of gentrification is well documented (discussed at length in Chapter 6). Chiu and Lui (2004) demonstrated that Hong Kong underwent a process of occupational and income polarization in the 1990s as it was transformed into a global city. The polarization in Hong Kong

also had a gender dimension, as women filled more of the new low-skilled and low-paying jobs associated with the expansion of the service sector (Chiu & Lui, 2004).

5.7 CONTEMPORARY RESEARCH ON URBAN SYSTEMS

The urban system approach continues to be an active realm for research. Consider, for instance, a recent study on urban poverty, a topic that has been frequently approached in the past from an intraurban perspective. Strait (2000) examined rising poverty concentration in American cities as a function of complex employment changes taking place across metropolitan areas and the impact of such changes on poor neighborhoods within cities. Strait (2000) found a significant relationship between metropolitan employment growth and the growth in extreme poverty at the neighborhood level. The latter study follows in the same tradition as the ones reviewed earlier in that it considers city responses to stimuli induced from higher up, at the same level, or from lower down the urban system's hierarchy.

The more classic urban system topics continue to be examined by urban researchers. For instance, Dehghan and Uribe (1999) recently examined the role of central government expenditure in shaping the city size distribution in Mexico. Marshall (1997) pointed out weaknesses of applying the traditional rank-size rule at the scale of a metropolitan region and introduced an alternative he called the constant-Gini model. Berry (1991, p. 31) found that after Borchert (1967) produced his seminal article on the evolution of the urban system, other variations on the theme were soon to follow, such as that by Dunn (1980), who hypothesized five rather than four evolutionary epochs. Conzen (1981) focused on the 19th-century American urban system and provided a more detailed analysis of that period than what Borchert (1967) was able to provide covering five epochs. Focusing on the post–World War II period, Noyelle and Stanback (1983) developed a functional classification of American cities on the basis of the recent economic transformation of the U.S. economy from an emphasis on manufacturing to services, a topic that is addressed in Chapter 9.

Berry (1991) himself attempted to provide an explanation for the timing of the technological innovations of Borchert's (1967) epoch boundaries by applying a long-wave theory (known as Kondratiev waves) of technology successions. Each wave is divided into the four phases of prosperity, recession, depression, and recovery. When economic growth declines, firms become reluctant to invest and unemployment rises. Eventually the trough of the wave is reached and economic activity is stirred up again on the basis of the key technologies identified by Borchert. Short (2004) on the topic of global networks of cities recently introduced the idea of "black holes" in the network, referring to large cities not classified as world cities because of lack of connectivity to the remaining hierarchy. Myer (2003), commenting on the state of research on the global network of cities, contended that the area continues to advance in terms of empirical research, but that the theory of a global network of cities is still very much undeveloped. All of these works demonstrate current interest in the urban system approach and that its tradition, dating back to central place theory, continues to influence the modern-day urban geographer's interpretation of global, regional, and local urban landscapes.

SUMMARY

This chapter has examined urban system approaches to urban geography. It covered basic concepts and principles, and then moved to several examples. It reviewed the earliest and most prominent of these models, central place theory. The "Analyzing an Urban Issue" section used contemporary data and GIS to examine and update a classic study of central place theory in Snohomish County, Washington. City size distributions were discussed with an emphasis on the rank-size distribution, the most frequently found city size distribution. The chapter then discussed an evolutionary model of the American urban system that explained the hierarchy at any given time as a function of the state of transport and energy technology. A global cities approach to urban systems research is rising in popularity as economic transformation and globalization processes result in concentration of financial and high-end professional and executive resources in a select few large cities.

REFERENCES

Berry, B. J. L. (1991) "Long waves in American urban evolution," in *Our Changing Cities.* ed. by J. F. Hart, pp. 31–50. Baltimore, MD: Johns Hopkins University Press.

Berry, B. J. L., and Garrison, W. L. (1958) "The functional bases of the central-place hierarchy." *Economic Geography* 34:145–154.

Berry, B. J. L., and Parr, J. B. (1988) *Market Centers and Retail Location.* Englewood Cliffs, NJ: Prentice Hall.

Blau, J. R. (1989) *The Shape of Culture.* Cambridge, MA: Cambridge University Press.

Borchert, J. G. (1998) "Spatial dynamics of retail structure and the venerable retail hierarchy." *GeoJournal* 45:327–336.

Borchert, J. R. (1967) "American metropolitan evolution." *Geographical Review* 57:301–332.

Borchert, J. R. (1991) "Futures of American cities," in *Our Changing Cities.* ed. by J. F. Hart, pp. 218–250. Baltimore, MD: Johns Hopkins University Press.

Boschken, H. L. (2003) "Global cities, systemic power, and upper-middle-class influence." *Urban Affairs Review* 38:808–830.

Cadwallader, M. (1996) *Urban Geography: An Analytical Approach.* Upper Saddle River, NJ: Prentice Hall.

Chiu, S. W. K., and Lui, T. (2004) "Testing the global city social polarization thesis: Hong Kong since the 1990s." *Urban Studies* 41:1863–1888.

Christaller, W. (1933) *Die zentralen Orte in Suddeutschland.* Jena, Germany: Fischer. English translation by Carlisle W. Baskin *The Central Places in Southern Germany.* Englewood Cliffs, NJ: Prentice Hall, 1966.

Cohen, R. B. (1981) "The new international division of labor, multinational corporations and urban hierarchy," in *Urbanization and Urban Planning in Capitalist Society.* ed. by M. Dear and A. J. Scott, pp. 287–315. London: Methuen.

Conzen, M. P. (1981) "The American urban system in the nineteenth century," in *Geography and the Urban Environment: Progress in Research and Applications* (Vol. IV). ed. by D. T. Herbert and R. J. Johnston, pp. 295–347. New York: Wiley.

Dehghan, F., and Uribe, G. V. (1999) "Analyzing Mexican population concentration: A model with empirical evidence." *Urban Studies* 36:1269–1281.

Dennis, C., Marshland, D., and Cockett, T. (2002) "Central place practice: Shopping centre attractiveness measures, hinterland boundaries and the UK retail hierarchy." *Journal of Retailing and Consumer Services* 9:185–199.

Dunn, E. S., Jr. (1980) *The Development of the U.S. Urban System.* Baltimore: Johns Hopkins University Press.

Esparza, A., and Krmenec, A. (2000) "Large city interaction in the US urban system." *Urban Studies* 37:691–709.

Friedman, T. L. (2000) *The Lexus and the Olive Tree.* New York: Anchor.

Friedmann, J. (1986) "The world city hypothesis." *Development and Change* 17:69–84.

Gibson, C. (1998) "Population of the 100 largest cities and other urban places in the United States: 1790 to 1990." Population Division Working Paper No. 27. Washington, DC: U.S. Bureau of the Census.

Hill, R. H., and Kim, J. W. (2000) "Global cities and developmental states: New York, Tokyo, and Seoul." *Urban Studies* 37:2167–2195.

Huff, D. (1964) "Defining and estimating a trading area." *Journal of Marketing* 28:34–38.

Huff, D., and Black, W. (1997) "The Huff model in retrospect." *Applied Geographic Studies* 1:83–93.

Knox, P. L. (1995) "World cities in a world-system," in *World Cities in a World-System.* ed. by P. L. Knox and P. J. Taylor, pp. 3–20. Cambridge, MA: Cambridge University Press.

Malecki, E. J. (2002) "The economic geography of the Internet's infrastructure." *Economic Geography* 78:399–424.

Marshall, J. U. (1989) *The Structure of Urban Systems.* Toronto: University of Toronto Press.

Marshall, J. U. (1997) "Beyond the rank-size rule: A new descriptive model of city sizes." *Urban Geography* 18:36–55.

McCann, E. J. (2004) "Urban political economy beyond the global city." *Urban Studies* 41:2315–2333.

Myer, D. R. (2003) "The challenges of research on the global network of cities." *Urban Geography* 24:301–313.

Noyelle, T., and Stanback, T. (1983) *The Economic Transformation of American Cities*. Toronto: Rowman & Littlefield.

Oliver, J. W. (1956) *History of American Technology*. New York: Ronald Press.

Ravenstein, E. (1885) "The laws of migration." *Journal of Royal Statistical Society* 48:167–227.

Reilly, W. J. (1931) *The Law of Retail Gravitation*. New York: Knickerbocker Press.

Sassen, S. (2001) *The Global City*. Princeton, NJ: Princeton University Press.

Sassen, S. (2002) "Locating cities on global circuits." *Environment and Urbanization* 14:13–30.

Short, J. R. (2004) "Black holes and loose connections in a global urban network." *The Professional Geographer* 56:295–302.

Strait, J. B. (2000) "An examination of extreme urban poverty: The effects of metropolitan employment and demographic dynamics." *Urban Geography* 21:514–542.

Ullman, E. L. (1954) "Amenities as a factor in regional growth." *Geographical Review* 44:119–132.

Walcott, S. M., and Wheeler, J. O. (2001) "Atlanta in the telecommunications age: The fiber-optic information network." *Urban Geography* 22:316–339.

Warf, B., and Erickson, R. (1996) "Introduction: Globalization and the U.S. urban system." *Urban Geography* 17:1–4.

EXERCISE

Exercise Description

To replicate a 1950s central place study of Snohomish County, Washington. The exercise examines the hierarchical class system of urban centers (i.e., hamlets, villages, and towns) that was identified in the earlier study. Queries identify specific groupings of places within the hierarchical class system. The current-day urban system is mapped and compared to the 1950s central place system.

Course Concepts Presented

Central place theory, central place hierarchy, low-order and high-order goods and services

GIS Concepts Utilized

Spatial queries, graduated symbol maps, and table joins.

Skills Required

ArcGIS Version 9.x

GIS Platform

ArcGIS Version 9.x

Estimated Time Required

1 hour

Exercise CD-ROM Location

ArcGIS_9.x\Chapter_05\

Filenames Required

Snh_pl_1950.shp
Snh_pl_1950b.shp
Snh_pl_2000.shp

Mjrrds.shp
Roads.shp
Snhmsh_cnty.shp
Studyarea.shp
Activity_Types.dbf

Background Information

In this lab, you examine the concepts of central place theory, including the hierarchical class system of urban centers and threshold. You will first re-create the 1950 Snohomish County, Washington central place landscape and then query the landscape to validate rules about threshold population sizes for certain central place functions. You will finish by updating the 2000 central place landscape.

Readings

(1) Berry, B. J. L., and Garrison, W. L. (1958) "The functional bases of the central-place hierarchy." *Economic Geography* 34:145–154.
(2) Greene, R. P., and Pick, J. B. *Exploring the Urban Community: A GIS Approach*, Chapter 5.

Exercise Procedure Flowchart

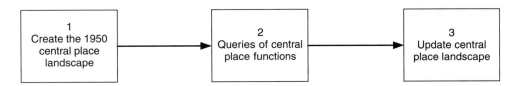

Exercise Procedure

1. Start ArcMap with Start Using ArcMap With A New Empty Map selected. Click OK.

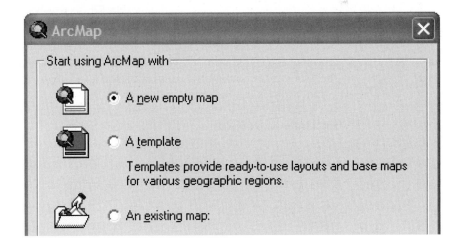

2. Click Add Data and navigate to the exercise data folder.

Note: *If you do not see the folder where you placed the exercise data, you will need to make a new connection to the root directory that holds the data.*

3. Add Snh_pl_1950, Mjrrds, Roads, Snhmsh_cnty, and Studyarea from the Chapter 5 folder. You can add all these files at once by clicking one of the file names, then clicking the others while holding down the Ctrl key on the keyboard and clicking Add.

Map 1: A Hierarchical Class System of Urban Centers

Berry and Garrison (1958) selected Snohomish County, Washington, to search for evidence of a hierarchical class system of urban centers. Snohomish County was ideal for this type of study in the 1950s because it had a large number of centers that ranged from very small (hamlets) to medium size (villages and towns) and only one very large city, Everett, which was excluded from the study. As discussed in Chapter 5, the hierarchical class system of hamlets, villages, and towns is an important aspect of central place theory. That is, the model argues that places will belong to one of these place types. Each type of place is characterized by a certain population limit and has a certain level of central place functions that are associated with a given place size. By analyzing a variety of urban functions ranging from low-order goods and services (i.e., filling stations) to high-order goods and services (i.e., health practitioners), Berry and Garrison were able to classify places into three types. You will now make a map of those three types of places.

4. Rearrange the drawing order of the added map layers into the order shown here. To change the drawing order, click a layer name in the table of contents and drag the layer to a new position.

5. Zoom to the study area by right-clicking the Studyarea layer and click Zoom To Layer. Clear the Studyarea check box as it is only used as a way to zoom in on the portion of Snohomish County that is in the original study area.
6. Click the fill symbol below the snhmsh_cnty layer and in the Symbol Selector dialog box, click the Hollow box option.

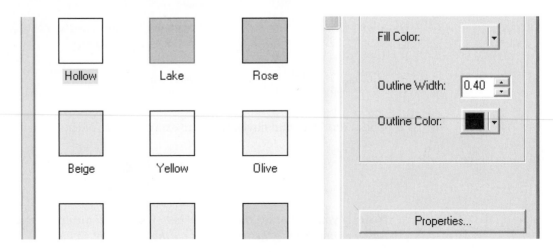

7. In the same Symbol Selector dialog box, click Properties and then click Outline.

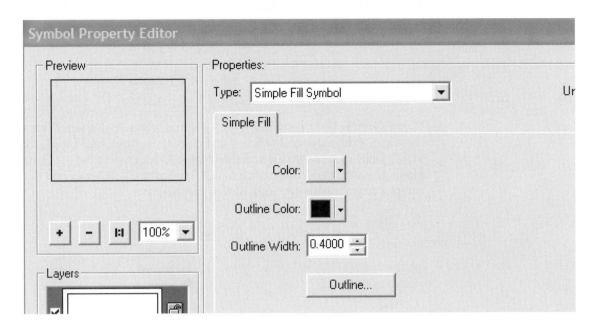

8. In the Symbol Selector dialog box, scroll down and choose the symbol for county boundaries. Click OK on each box as you return to the map.

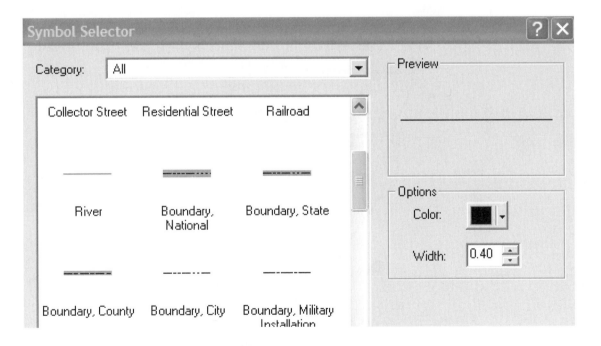

9. Your county boundary should like the following:

10. Click the line symbol below the Mjrrds layer and click the Highway symbol. Next click the Color drop-down box and choose 20% Gray (on the left, two boxes down from white). Click

OK. Click the line symbol below the Roads layer and click the Color drop-down box and select 20% Gray for this as well. Click OK. Your road lines should look like the following:

11. Right-click the layer named snh_pl_1950 and select Properties. Click the Symbology tab of the Layer Properties dialog box. Next click Quantities And Graduated Symbols. Click the Fields Value drop-down list and select Type. Under the Label column, click 1 and change to Hamlet, change 2 to Village, and change 3 to Town. Click Apply and then click OK. Your central place symbols should look like the following:

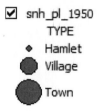

12. Right-click the layer name snh_pl_1950 and select Label Features. Your map should look like the following:

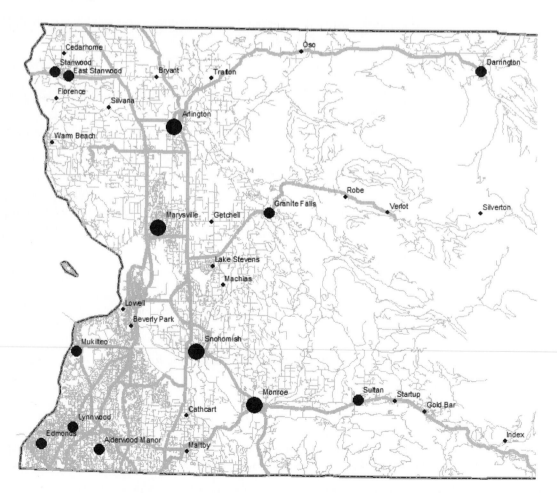

13. Select Layout View from the View menu to compose a map for printing.
14. Click Insert and then click Legend.
15. In the Legend Wizard dialog box, make sure that the only legend item is snh_pl_1950. For instance, for the other layers in the box on the right, select and use the arrows to move them to the left column.

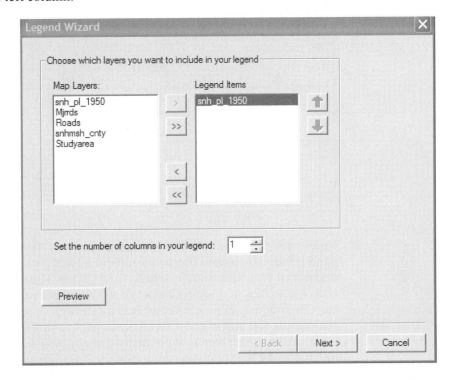

16. Click Next and erase the word "Legend" by backspacing it out.
17. Click Next in the next several dialog boxes and then click Finish.
18. Double-click the selected legend. Click the Items tab and in the right column, double-click snh_pl_1950, click Properties, then click the General tab. Clear the Show Layer Name and Show Heading check boxes, click Apply, and click OK. Click OK as you exit back. Reposition the legend to a suitable spot.
19. Click Insert at the top and insert a title (such as, 1950 Central Place Hierarchy for Snohomish County, Washington). Change the font size to 14 (see the bottom of the screen for the Font Size drop-down box) and reposition the title to a suitable spot.
20. Click File and select Print to print the final map (alternatively you could click File and select Export Map and export to your favorite image format).

Query the Hierarchical Class-System of Urban Centers

(Discussion)

Each of the three types of places (hamlets, villages, and towns) identified by Berry and Garrison (1958) contained specific groups of central functions and were characterized by a certain population size. Towns contained all of the low-order central functions (i.e., filling stations, food stores, and churches) of the hamlets and villages plus high-order functions such as photographers, public accountants, and health practitioners not found in hamlets and villages. Population size was highly correlated with the level of central place functions offered by places, reflecting the threshold concept of central place theory, which states that a large threshold of population is necessary before a high-order good or service will be offered. We will now conduct selected queries of the Snohomish place layer to explore the relationship between population size and central functions offered by a place.

21. Select Data View from the View menu to exit the Layout View mode.
22. Click Add Data and add the Activity_Types.dbf table.

23. Right-click the name Activity_Types and click Open. The table includes fields for a Code, Activity, and Threshold. This is a lookup table as these codes are field titles contained in the snh_pl_1950 layer. The Activity field contains central place functions ranging from low-order goods and services to high-order goods and services. Threshold contains population thresholds for each of the 52 central place activities derived by Berry and Garrison in their study of Snohomish County.

OID	CODE	ACTIVITY	THRESHOLD_
0	A1	Filling Stations	196
1	A2	Food Stores	254
2	A3	Churches	265
3	A4	Restaurants	276
4	A5	Taverns	282
5	A6	Elementary Schools	322
6	B1	Physicians	380

Attributes of Activity_Types

24. Right-click the name snh_pl_1950 and click Open Attribute Table. Move the table and expand it so that you can observe more field names. Note that this table includes field names that match the codes contained in Activity_Types. Thus, we see that Alderwood Manor contains 4 activities that are coded A1. If we look back at our Activity_Type lookup table we see that it equals Filling Stations. It is interesting that the 1950 population of Alderwood Manor was 600 and it had 4 filling stations. The threshold population (see Activity_Type table) is 196, which would imply that we might have only expected to find 3 filling stations in Alderwood Manor given its population of 600. Nevertheless, many would find this to be a minor discrepancy and a good fit between true population and threshold population. The Activity_Type lookup table will be useful for the queries of the snh_pl_1950 map layer in the following steps.

FID	Shape*	AREANAME	TYPE	POP1950	ACTVTIES	A1	A2	A3
0	Point	Alderwood Manor	2	600	22	4	3	
1	Point	Arlington	3	1915	59	6	4	
2	Point	Cathcart	1	175	5	1	0	
3	Point	Darrington	2	974	25	3	2	
4	Point	Edmonds	2	2996	42	7	6	
5	Point	Gold Bar	1	325	9	1	2	
6	Point	Granite Falls	2	600	25	4	3	

Attributes of snh_pl_1950

25. Close the preceding tables, click the Selection tab at the top of the ArcMap screen, and choose Select By Attributes. In the dialog box choose snh_pl_1950 as the Layer. Enter the following query and click Apply and then click Close:

$$\text{"TYPE"} = 1$$

26. Open the attribute table of snh_pl_1950 (right-click the name) and note that 20 of the 33 places were selected, which are all hamlets. Close the table and do a query of the number of hamlets that have hardware stores (we would expect this to be low, as hardware stores require population thresholds more characteristic of villages). First open the Activity_Type table and look up the code for hardware stores (it happens to be B10). Close the table and click the Selection menu at the top again. Enter the following query:

$$\text{"TYPE"} = 1 \text{ AND } \text{"B10"} > 0$$

27. We see that only 1 out of 20 hamlets was selected, Lake Stevens. We will come back to Lake Stevens in a later step, but now repeat Steps 25 and 26 on Type 2 places (villages) and what percentage of them have hardware stores. You should find that there are 9 village-level places and that 7 of them have hardware stores.
28. Are there any other queries you would like to try? How about the presence of health practitioners in towns ("TYPE" = 3 AND "C9" > 0)?
29. In the original study, Berry and Garrison found both Lake Stevens and Edmonds to have been underrepresented with central place functions given their population sizes. The following query should highlight these two places based on this characteristic:

$$\text{"POP1950"} > 2000 \text{ AND "ACTIVITIES"} < 50$$

Update the Central Place Landscape of Snohomish County, Washington for the Year 2000

(Discussion)

One explanation given by Berry and Garrison (1958) for the divergence of central place functions in Lake Stevens and Edmonds (along with Beverly Park and Lowell) from their expected function groupings was that they each had experienced rapid population growth at the time (pressure from greater Seattle) and that there is usually a lag in service delivery in such cases. In other words, it was expected that these places would gain more central place functions as time passed, especially as entrepreneurs would recognize their population sizes and begin to offer the appropriate goods and services associated with those thresholds. We will now update the central place landscape of Snohomish County so that one could conceivably test the latter condition in a concluding challenge step.

30. First, close the Select By Attributes Dialog box and click the Selection menu and click Clear Selected Features.
31. Click Add Data and add snh_pl_2000 and snh_pl_1950b. The snh_pl_2000 are the places for Snohomish County in 2000 and snh_pl_1950b is the 1950 places with a more simplified attribute table than snh_pl_1950.
32. Turn all layers off except for snh_pl_2000, snh_pl_1950b, and snhmsh_cnty. Make sure the drawing order of layers is such that snh_pl_2000 is on top (first click display tab at bottom if you want to adjust drawing order). Next, click the point symbol just below snh_pl_2000 and select the Square 2 symbol. Click OK. Your layer symbols and order should look like the following.

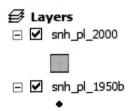

33. Examining the map we see 10 points from snh_pl_1950b that are not covered up by the snh_pl_2000 layer. These represent places that have been absorbed by other places since the 1950s and are thus not part of the contemporary central place landscape of Snohomish County. First click the Selection menu at the top of ArcMap and click Set Selectable Layers. Clear everything except snh_pl_1950b and close. Click the Select tool (see below) and select each of the 10 points by holding the Shift key while you select each one.

34. Once the points are all selected, right-click the name snh_pl_1950b and click Data and then click Export Data, navigate to an appropriate folder, and rename the file snh_old_1950. Click Save and then click OK. Answer Yes to add the exported data to the map as a layer.

35. On the Selection menu, click Clear Selected Features.

36. To determine the number of new central places, right-click the name snh_pl_2000, click Joins And Relates, and select Join. We are joining attributes from a table and we choose AREANAME as the field in this layer that the join will be based on. For entry number two, the table to join to this layer, choose snh_pl_1950b. For entry number three, the field in the table to base the join on, choose AREANAME and click OK (if you get asked to create an index, answer Yes).

37. Right-click the name snh_pl_2000 and click Data and then click Export Data. Name the new file Snh_pl_2000b, and click OK. Answer Yes to add it to the map as a layer.

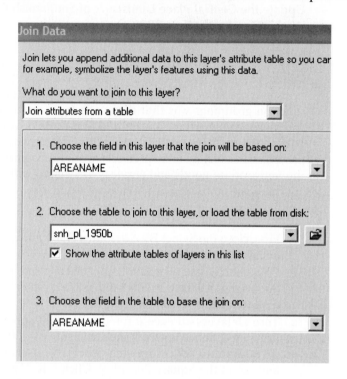

38. On the Selection menu, choose Select By Attributes. Make sure the Layer is set to snh_pl_2000b. Enter the following query:

$$\text{``POP1950''} = 0 \text{ AND ``POPULATION''} > 0$$

39. Click Apply and then click Close. This query selects all of the new places added since the 1950s because we joined the 1950 table to the 2000 table and any place that was added since 1950 will have a zero population for the year 1950 and a population greater than zero for the year 2000. The POPULATION field represents 2000 population.

40. We now want to put these selected points into a new shapefile. Right-click the name snh_pl_2000b, click Data, and then click Export Data. Name the file Snh_new_2000, and click OK. Answer Yes to add it as a layer.

41. To determine the persistent places, on the Selection menu, choose Selection By Attributes. Make sure the Layer is set to snh_pl_2000b. Enter the following query:

$$\text{``POP1950''} > 0 \text{ AND ``POPULATION''} > 0$$

42. Click Apply and then click Close. This query selects all of the places that have persisted from the 1950s to the present.

43. Make a new shapefile with these selected points. Right-click the name snh_pl_2000b, click Data, and then click Export Data. Name the file Snh_pst_2000, and click OK. Answer Yes to add it as a layer.

44. On the Selection menu, select Clear Selected Features.

45. We now have three new shapefiles named Snh_old_1950, Snh_new_2000, and Snh_pst_2000. These layers and the county boundary should be the only layers turned on. Right-click each of these layer names and click Properties. Click the General tab and change the names to more meaningful names that will go into your map legend:

> Change Snh_old_1950 to Not Present in 2000
> Change Snh_new_2000 to Added Since 1950
> Change Snh_pst_2000 to Still Present in 2000

46. On the View menu, select Layout View and make the following map:

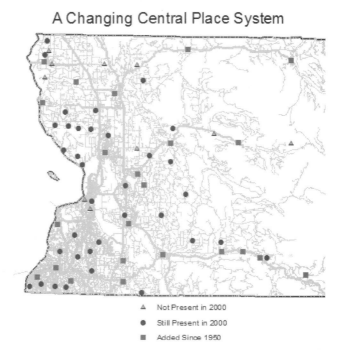

A Changing Central Place System

△ Not Present in 2000

● Still Present in 2000

■ Added Since 1950

A Challenge Step: Update the Central Place Hierarchy of Snohomish County, Washington for the Year 2000

47. Use online yellow pages to determine if Lake Stevens and Edmonds have a number of central functions consistent with their contemporary populations.

48. A more ambitious project would be to use online yellow pages to update the entire central place hierarchy of Snohomish County. Consideration would have to be given to the categories of central place functions, for instance, subtraction of some of the 52 items listed in the Activity_Types table and perhaps addition of some more contemporary functions (i.e., video stores). Also, the number of central places has increased since the 1950s and many of the earlier centers are much larger. Snohomish County today is much more integrated with the Seattle metropolitan area and many of the places coalesce into each other, making it difficult to assign activities to a given center. Consideration might be given to designing the study in terms of a metropolitan context as discussed in Chapter 5. In the latter context, functions are grouped in a hierarchy of business centers, highway-oriented ribbons, urban arterial commercial developments, and specialized functional areas.

Short Essay

Discuss the central place hierarchy that you created for the 1950s, which is represented on the first map you made. Explain the patterns in terms of central place theory. For instance, are there more smaller places in the hierarchy than there are larger places? How did the queries relating central place functions to population size square with the theory? Compare the 1950s central place landscape with the contemporary central place landscape. How would the new population sizes affect the creation of new groupings (i.e., would you have to add more categories to hamlets, villages, and towns)?

6

Neighborhoods

Neighborhoods are the places in cities people consider their community. They are the residential patchwork in the fabric of cities that also provide social, cultural, and recreational outlets for people (see Figure 6.1). The boundaries of neighborhoods can be expressways, roads, or other physical limits but they can also be defined by social and economic networks and therefore have been regarded by some as socially constructed entities (Suttles, 1972). The study of neighborhoods has been undertaken by many urban study disciplines, ranging from urban geography to social psychology. In addition to the question of what defines a neighborhood, research has examined such other issues as the social and economic health of neighborhoods and the changing characteristics of neighborhoods over time. A further issue, neighborhood vitality, has been of interest not only to urban scholars but is also addressed daily by city administrators. The geographic approach to the study of neighborhoods has varied, but in recent decades, much attention has been paid to their definition, stability, economic development, and renewal.

Urban geography has advanced our knowledge of important neighborhood processes such as gentrification. Urban geographers have also extensively studied other processes such as the diffusion of race and ethnicity among and across neighborhoods. Neighborhood associations play an important role in the vitality of neighborhoods, so geographers have had to consider their spatial impacts as well. The concentration of poverty in extremely poor neighborhoods has received much attention, but at the same time urban geographers have investigated such opposite phenomena as gated communities.

The chapter starts with a discussion of the neighborhood concept in general. A number of attempts have been made to delineate the urban neighborhood; however, the definition of a neighborhood is often subjective.

Neighborhoods continue to be an important component of cities as indicated by the continued strength of neighborhood identity. The attachment people show toward their neighborhood is a function of the quality of the neighborhood. Behavioral approaches to the study of neighborhoods have emphasized the role of activism and who is likely to become involved in the community. Many studies have confirmed that the higher their social and economic class, the more likely people are to become active in the neighborhood decision process. Such power concentrated among so few has proved to be detrimental for lower income groups, who have not always had their community interests represented.

A Las Vegas case study later demonstrates that the spirit of community can even be created in the newest of urban landscapes. There are skeptics who argue that community institutions have broken down with the rapid pace of suburbanization. Neotraditional planning is one response to the loss of community, but strong evidence is presented in support of the presence of community, albeit different from traditional conceptions of community, in modern suburbia. A similar concern has been voiced about the rise of the gated community. The gated community is not only on the rise in America, but it is rising rapidly in Chinese cities as well as in the megacities of Latin America. Critics of gated communities fear further diminishment of community and the erosion of public space by the encroachment of private space.

The examination of change in the racial composition of neighborhoods has a long tradition in urban geography and related disciplines. With an early focus of research on White versus Black, racially integrated neighborhoods often used to be viewed as a temporary stage in the transition process from White to Black that would inevitably become entirely Black in some future time period. Early research tried to establish a statistical "tipping point"

197

FIGURE 6.1   Downtown Barcelona neighborhood.

beyond which the proportion of Blacks would rapidly in-crease and become a majority. Recent studies have ques-tioned this deterministic racial succession assumption and argue that the rise of multiethnic neighborhoods re-futes the process. The cause of the continued separation of races and ethnic groups in cities is hotly debated, with some arguing discrimination, whereas others suggest that it is due to own-race preferences by residents. The section "Analyzing an Urban Issue" looks at racial transitions for St. Louis for the 1990 to 2000 period employing a Markov chain analysis in combination with a geographic information system (GIS).

Gentrification continues to be a prominent concept in the neighborhood change literature. The rent-gap explanation of gentrification is reviewed, and recent attempts to measure it are illustrated. The debate continues over what is the best explanation of the process, with economic, cultural, and consumer behavior all playing a part. A number of case studies are reviewed. Urban researchers are addressing more aspects of gentrification, including its varied forms in the developing world.

6.1 CONCEPT OF NEIGHBORHOOD

The concept of neighborhood is largely intuitive; if asked, most people could roughly define the geographic boundaries of their neighborhood. However, the concept itself lacks a succinct definition (Smith, 1985), so it is usually defined by the degree to which the people in the neighborhood have a sense of neighborhood identity. Martin (2003) contended that neighborhoods are a social product created through conflicting perspectives arising out of neighborhood activism. Haeberle (1988) noted that neighborhood definitions are addressed either subjectively by looking at the characteristics of the people or with a place-based approach that tries to demarcate the boundaries with physical features such as traffic arteries. Ahlbrandt (1984) claimed relative success in the physical delineation of 78 distinct neighborhoods in Pittsburgh, a task he completed by holding workshops throughout the city where residents provided input on the boundaries (see Figure 6.2). The very hilly and broken terrain in Pittsburgh, as well as in other similar cities (e.g., Cincinnati and San Francisco), has a good deal to do with forming and preserving neighborhood identity (see Figure 6.3).

Many current city neighborhoods were once separate incorporated places, such as Boston's Charlestown, Roxbury, and Dorchester, with names that are still in common use as neighborhood or area names. Big annexations like New York City's Brooklyn and Boston's Dorchester contain many separate neighborhoods just as they did when they were separate. A sometimes-useful guide to neighborhood extent (and existence in local people's minds) is the local phone book. For instance, how many businesses use a neighborhood's name and how those businesses are distributed geographically can sometimes yield a useful neighborhood boundary. However, Forstall (2004) noted some potential problems with this method:

> For example, the Gold Coast in Chicago, very well recognized by both residents and neighbors, is almost wholly residential (intentionally) and so contains rather few businesses; those that use the name, the last time I checked quite a few years ago, aimed to serve the area but were located well West of what locals consider the Gold Coast.

The importance of neighborhood identity cannot be underestimated. A number of city neighborhoods have at one time or another been widely known across America, for instance, Beacon Hill (Boston), Harlem (New York City), Over-the-Rhine (Cincinnati), Watts (Los Angeles), The Hill (St. Louis), Adams-Morgan (Washington, DC), Little Havana (Miami), and Back of the Yards (Chicago). The Los Angeles City Council perceived the concept of neighborhood identity to be important when they recently changed the name of their South Central community area, made internationally notorious by the Rodney King riots, to South Los Angeles in hopes of changing the image of the area.

On the issue of the differences in the structure and composition of urban neighborhoods, Ahlbrandt (1984) surveyed almost 6,000 residents living throughout Pittsburgh and addressed critical issues about neighborhoods including attachment toward the neighborhood, social support systems, and urban policy implications. On the issue of attachment, the study addressed the relationship between participation in neighborhood activities and the quality of the environment. Neighborhoods that have high crime and lack basic services and other amenities will not attract much community participation. Participation, then, is to some extent a function of the neighborhood's quality of life, which Ahlbrandt depicted in a model of people's feelings about their neighborhood (see Figure 6.4). Among other findings, Ahlbrandt (1984) showed that the neighborhood does not hold the same meaning to all groups in the city and for some it is not even important. However, he found that the elderly, the impoverished, and households with children were more dependent on neighborhoods. In many cases, the impoverished feel despair over the prospects of their neighborhoods even in the best of times, as recently shown by Greenberg (2004), who examined the impact of the economically prosperous years of 1992 to 2000 through a survey and found that neighborhoods were perceived not to have improved during those years, a perception strongest in poor neighborhoods.

Ahlbrandt's (1984) study also addressed the idea that the well-being of people is linked to the neighborhood environment where they live, although he stopped short of suggesting a causative direction for the relationship (see Figure 6.5). The concern that neighborhood well-being could be related to individuals' behavior problems had long been suggested, but resurfaced in the 1980s with William Wilson's (1987) publication of *The Truly Disadvantaged: The Inner City, the Underclass, and Public Policy*. Although Wilson's principal thesis about the rise of an inner-city underclass was linked to theories of economic restructuring, the noted problems of family dissolution and welfare dependency brought on by economic restructuring became the focus of countless studies in which the poor neighborhoods themselves were examined for

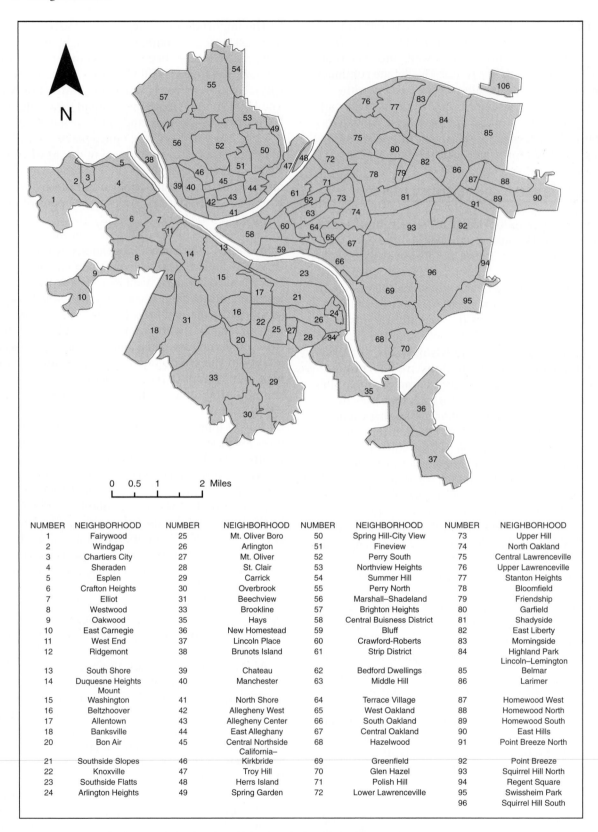

N

0   0.5   1   2 Miles

NUMBER	NEIGHBORHOOD	NUMBER	NEIGHBORHOOD	NUMBER	NEIGHBORHOOD	NUMBER	NEIGHBORHOOD
1	Fairywood	25	Mt. Oliver Boro	50	Spring Hill-City View	73	Upper Hill
2	Windgap	26	Arlington	51	Fineview	74	North Oakland
3	Chartiers City	27	Mt. Oliver	52	Perry South	75	Central Lawrenceville
4	Sheraden	28	St. Clair	53	Northview Heights	76	Upper Lawrenceville
5	Esplen	29	Carrick	54	Summer Hill	77	Stanton Heights
6	Crafton Heights	30	Overbrook	55	Perry North	78	Bloomfield
7	Elliot	31	Beechview	56	Marshall–Shadeland	79	Friendship
8	Westwood	33	Brookline	57	Brighton Heights	80	Garfield
9	Oakwood	35	Hays	58	Central Buisness District	81	Shadyside
10	East Carnegie	36	New Homestead	59	Bluff	82	East Liberty
11	West End	37	Lincoln Place	60	Crawford-Roberts	83	Morningside
12	Ridgemont	38	Brunots Island	61	Strip District	84	Highland Park
13	South Shore	39	Chateau	62	Bedford Dwellings	85	Lincoln–Lemington Belmar
14	Duquesne Heights	40	Manchester	63	Middle Hill	86	Larimer
15	Mount Washington	41	North Shore	64	Terrace Village	87	Homewood West
16	Beltzhoover	42	Allegheny West	65	West Oakland	88	Homewood North
17	Allentown	43	Allegheny Center	66	South Oakland	89	Homewood South
18	Banksville	44	East Alleghany	67	Central Oakland	90	East Hills
20	Bon Air	45	Central Northside California–	68	Hazelwood	91	Point Breeze North
21	Southside Slopes	46	Kirkbride	69	Greenfield	92	Point Breeze
22	Knoxville	47	Troy Hill	70	Glen Hazel	93	Squirrel Hill North
23	Southside Flatts	48	Herrs Island	71	Polish Hill	94	Regent Square
24	Arlington Heights	49	Spring Garden	72	Lower Lawrenceville	95	Swissheim Park
						96	Squirrel Hill South

FIGURE 6.2   Neighborhoods in the city of Pittsburgh, Pennsylvania.

FIGURE 6.3   Squirrel Hill neighborhood, Pittsburgh.

FIGURE 6.5   Housing abandonment in Detroit.

their contribution to the problem behaviors (Jencks & Peterson, 1991). Meanwhile, a whole new literature has emerged that focuses on neighborhood effects (Holloway & Mulherin, 2004; Sampson et. al., 2002).

In a detailed analysis of a Minneapolis neighborhood, Smith (1985) examined whether social status or participation in neighborhood affairs was most important in determining one's sense of identity with a local neighborhood. She found that activism plays a more important role and that the households showing the strongest sense of attachment were those that were both high status and actively involved in neighborhood affairs (Smith, 1985, p. 431). A similar finding was made in Pittsburgh, where it was found that higher income respondents were more likely to participate in voluntary organizations, whereas lower income residents were less likely to interact with their neighbors (Ahlbrandt, 1984, pp. 22–23). Fried (1986) reviewed an extensive literature on neighborhood satisfaction and concluded that the higher the social class position, the greater the degree of neighborhood satisfaction.

Nevertheless, some case studies document local poor residents organizing to create change in their favor. One case, in particular, that has received national attention

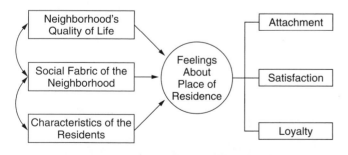

FIGURE 6.4   Allbrandt's model of neighborhood perception: People's feelings about place of residence. Source: Albrandt, Roger S. 1984. "Neighborhoods, People, and Community." New York: Plenum Press. Diagram. "Model of neighborhood perception: people's feelings about place of residence."

involved a Boston neighborhood revitalization effort dubbed the Dudley Street Neighborhood Initiative (DSNI), which was initiated by a trust company in 1984. The trust company sponsored a meeting to unveil their renewal and governing plan, but they were immediately confronted with concerns over the governing structure by some of the residents, one of whom, Che Madyun, a single mother of three, voiced it this way:

> I asked how many of the people up there [running the meeting] lived in the neighborhood. Not one. And then I asked, How can you say the residents are going to be represented? You always have people from downtown or somewhere else telling you what you need in your neighborhood. This is important: planning never happens without people who are going to have to live with the results day to day being involved, from the beginning. (Che Madyun as quoted in Bright, 2003, p. 78)

Che Madyun was eventually elected DSNI board president in 1985 (serving through 1995), and 15 years later, the Dudley Street Neighborhood was considered a success by the planning and urban studies community, due in large part to resident control of the process.

In a study on Sun City West outside Phoenix, McHugh, Gober, and Borough (2002) showed how the wishes of higher status households actively involved in community decision making can sometimes lead to detrimental effects for lower income residents who are not well organized. Sun City West is located in the Dysart School District in northwest Phoenix, composed largely of Latinos of modest economic means (see Figure 6.6). The seniors, who are a minority in the district, were successful at defeating school funding measures and were able to initiate a plan to secede from the district to avoid paying future taxes for the schools (McHugh et al., 2002, p. 639).

A recent case study of the evolution of neighborhood identity in Las Vegas illustrates that the process continues even in the newest of urban landscapes (Gottdiener, Collins, & Dickens, 1999). Las Vegas might be the last

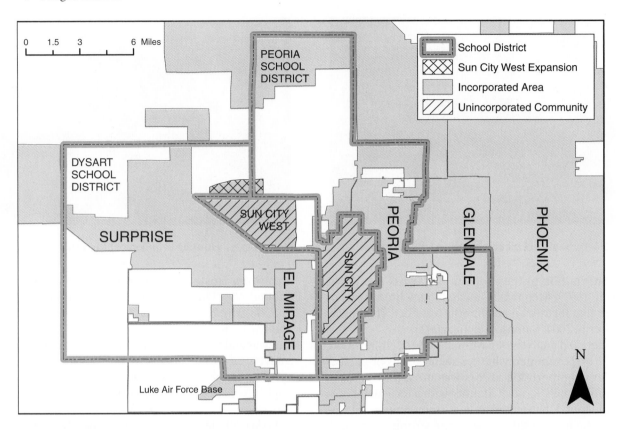

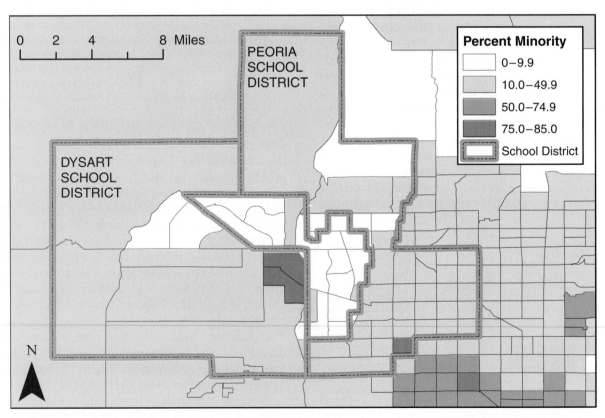

FIGURE 6.6 (A) Boundaries of Dysart and Peoria school districts, Arizona, 2000. (B) Percentage minority population, Dysart and Peoria school districts, 2000. Source: McHugh, Kevin, P. Gober, and D. Borough 2002. The Sun City Wars: Chapter 3. *Urban Geography* 23: 627–648. Fig 3 p. 635 and Fig 5 p. 637.

place one would look for the emergence of community or neighborhood identity, in part because it is often showcased as America's most popular resort town. The metropolitan area of Las Vegas-Paradise now includes 1.4 million people, most of them living in recently built communities, and so provides an excellent test case of how "sense of community" plays out in a growth area.

The first condition for a successful community was residential permanency, which began to occur in volume in Las Vegas in the 1980s. The process of community formation, however, occurred in large part because most of the new housing was built in master-planned communities (Gottdiener et al., 1999, p. 128). Although master-planned communities were introduced early in Las Vegas history, the concept flourished with the 1975 Green Valley Project located south of the Strip in suburban Henderson (see Figure 6.7). The project involved 1,000 acres, divided into 10-acre subdivisions. In the mid-1980s as the region's economy was booming and diversifying, middle-class professionals were drawn to the Green Valley subdivisions. The new residents were attracted by walking trails, imported vegetation, a recreation center, new schools, a shopping center, and many parks (Gottdiener et al., 1999, pp. 133–134). Green Valley was eventually built out, accommodating 30,000 residents. The remaining land was divided into sections to be developed with their own village centers. To many, Green Valley became a successful community with its own unique identity:

> "To its residents the address connotes an upscale, family-oriented community called Green Valley, not the gambling and tourist mecca 15 miles to the northwest" (Gottdiener et al., 1999, p. 138).

Nevertheless, there are skeptics who question whether master-planned communities really live up to the traditional measures of community. Guterson (1992) criticized Green Valley as being too homogenous, bland, and sterile. McHugh et al. (2002) argued that common-interest developments (CIDs), which some master-planned communities can be considered, can have negative consequences for neighboring communities as they attempt to isolate themselves from the more diverse metropolitan region they occupy. For instance, Green Valley is within the city limits of Henderson (formerly the more blue-collar community), and the white-collar residents of Green Valley have squared off with Henderson residents on more than one occasion (Gottdiener et al., 1999, pp. 135–136). A CID combines individual ownership of private housing and shared ownership of common areas.

Summerlin, the second developed master-planned community in the Las Vegas region, fits more the full profile of a CID. Modeled after the Irvine Ranch in southern California, the community is located on the western rim of the Las Vegas Valley and is three times the size of Green Valley. The development of Irvine Ranch in Orange County California helped stimulate rapid growth in the county's largest city, Irvine, in the 1970s and 1980s, but its influence in matters of planning extended well beyond Orange County, as discussed in Chapter 12. The first Summerlin property to develop was the Del Webb retirement community, Sun City (Gottdiener et al., 1999, p. 143). As is typical of CIDs, Summerlin started out fairly homogenous and had a number of covenants, conditions, and restrictions (CC&Rs) that limited the types of alterations that could be made to properties. However, Summerlin, with a deliberate marketing campaign, was able to diversify by attracting families and young

FIGURE 6.7 Henderson, Nevada.

professionals with a village concept emphasizing mixed uses with lots of common space devoted to golf courses:

> With more than 7,700 homes and a population of 14,000 when sales closed in 1998, Sun City Summerlin was the first area in the city of Las Vegas where golf carts were authorized street vehicles. (Gottdiener et al., 1999, p. 144)

The Green Valley and Summerlin case studies by Gottdiener et. al. (1999) illustrate that community functions do appear in new or newly developed urban settlements. In comparing the modern community with the community of the past, it is frequently claimed in the popular media that those in the past were more appealing and functional with respect to providing a "sense of place." These sentiments toward neighborhoods of the past are often seen as nostalgic reactions to earlier community institutions that are claimed by Putnam (2000) to have anchored a community. Similar claims by the "New Urbanists" have been made linking "sense of community" to urban design, and they have helped maintain interest in the neighborhood as a planning unit (Madanipour, 2001). However, Talen (1999) noted that the concept of sense of community as espoused by the new urbanists has not been satisfactorily explained and needs more clarification. Clark et. al. (2002) noted in a comment about gentrification that "the important local amenities are no longer schools, churches, and neighborhood associations, as in the urban mosaic of the Chicago School" (p. 500). Similarly, the most important new community institutions for the master-planned communities of Green Valley and Summerlin might, in fact, be recreational centers and golf courses.

6.2 SOCIAL LIFE IN THE CITY AND URBANISM AS A WAY OF LIFE

Urban sociologists have a long tradition of examining lifestyles in cities, which is taken to mean the "patterned ways of living that distinguish people in the urban arena" (Macionis & Parrillo, 1998, p. 214). There are several key characteristics that help define these lifestyles, including social class, age, gender, race, and ethnicity. The topics of race, ethnicity, and gender are taken up separately in Chapter 8. However, the topic of lifestyles in the city is important enough to devote a section to two of the developers and leading thinkers on the topic, Louis Wirth and Herbert Gans.

Louis Wirth, another University of Chicago sociologist, wrote a classic essay on the topic of lifestyles in the city titled "Urbanism as a Way of Life" (Wirth, 1938). His perspective was pioneering at the time as he suggested that the city shapes lifestyles. The exceptionally high population densities of cities result in a new type of behavior. In particular, residents become rational, self-interested, specialized, reserved, and more tolerant (Macionis & Parrillo, 1998). Relationships in the city were much more functional as opposed to being less structured in rural settings. Merton (1968) extended Wirth's concept of the cosmopolitan lifestyle by introducing the term *local* to encapsulate the opposite type of lifestyle. The life of the localite is centered on an immediate geographic area with social interaction limited to that area, whereas the cosmopolites are more footloose and open to a wider range of social interactions.

Herbert Gans would later challenge Wirth's ideas on lifestyle and the extensions such as that offered by Merton. In particular, Gans (1968) argued that it is not the city that produces the cosmopolite and localite, rather the city is diverse and made up of both and a variety in between. This diversity, according to Gans, was a product of other factors including class, age, and gender. To illustrate his view, Gans developed a typology of urban lifestyles that consisted of four types:

1. *The Cosmopolites*   This was a similar lifestyle to that identified by Wirth to capture the sophisticated urban resident.

2. *The Unmarried or Childless*   This lifestyle often overlaps with the cosmopolite category, which consists of single adults or couples without children living in city apartments. Today some refer to this group as yuppies (young urban professionals) or dinks (dual income, no kids).

3. *The Ethnic Villagers*   These lifestyles are exhibited by some ethnic groups that have not shed their cultural heritages and maintain their homeland lifestyle in ethnic enclaves in the city.

4. *The Deprived and Trapped*   These lifestyles are exhibited by those caught in a cycle of poverty in the city. They are often left with no choice but to live in the most dilapidated housing stock of the city.

Gans recognized that this was not an exhaustive list of urban lifestyles. In fact, he would argue exactly that as a way to dismiss Wirth's idea that the city produces a uniquely urban lifestyle.

6.3 GATED COMMUNITIES

Gated communities are neighborhoods that physically restrict access by the general public with walls or fences and admission only through gates (see Figure 6.8). The number of Americans living in gated communities is estimated at 9 million (Webster, 2001, p. 151). There are many types of gated communities and the residents who

FIGURE 6.8 Gated community in the suburbs. Source: Woodfin Camp & Associates.

live in them do so for a variety of reasons (Blakely & Snyder, 1998). The motives often include protection from crime, feelings of prestige, and securing privacy. Blakely and Snyder (1998) identified three major types of gated communities:

1. *Lifestyle Communities*   Country clubs and retirement developments emphasizing extensive recreational facilities such as golf courses.

2. *Prestige Communities*   Standard urban subdivisions developed with gates to enhance status.

3. *Security Zone Communities*   Residents of the existing communities have added gates to control crime and traffic.

Gated communities have been around for a long time, taking on a variety of forms (Figure 6.8). However, the concept grew in popularity with the advent of master-planned retirement communities in the 1960s and 1970s (Blakely & Snyder, 1998). The concept then spread to resorts, country clubs, and eventually to middle-class suburbia. Once enjoyed only by the upper class, the gated community is now growing rapidly among the middle class; for instance, Frantz (2001) estimated that only 13 percent of Phoenix's gated communities are upper class. The rise of gated communities is attributed

to the rise of disposable income among the middle and upper middle class, and the resulting wish to have their neighborhoods designed to symbolize their status (Blakely & Snyder, 1998).

Blakely and Snyder (1998) measured two aspects of gated community perception by their residents, crime and community spirit (see Figure 6.9). The perception that gates increase security was strong, and most residents of gated communities find their neighbors to be friendly. Luymes (1999) developed a typology of access control and perimeter control that classifies gated communities based on their perceived security and neighborhood status (see Figure 6.10). He found that the typology of control is often established along socioeconomic lines, with the most affluent gated communities having the securest of walls and existing within a larger gated community with 24-hour security (Luymes, 1999, p. 199).

Webster (2001) referred to the occupants of gated communities as the "club realm" and argued that the class is here to stay and will even grow. Therefore, developers and planners should work with the concept. Webster reviewed the leading works on gated communities and noted two viewpoints with respect to this growing trend. The first is an antigate perspective seen in its extreme in Davis's (1990) *City of Quartz*, which decries what he saw as a growing privatization of exclusive residential space in Los Angeles.

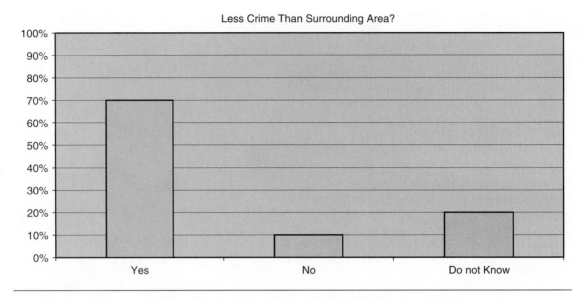

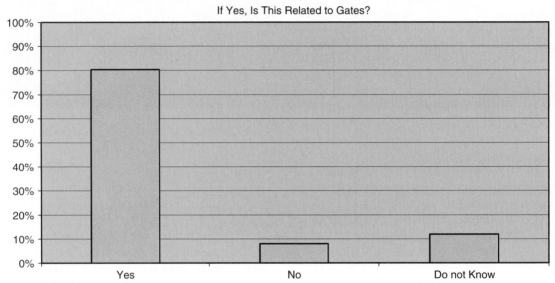

FIGURE 6.9a   Perceptions of the level of crime in gated communities. Source: Fig 6-9 Blakely, Edward J. and M. G. Snyder. 1998. Forting up: Gated communities in the United States. *Journal of Architectural and Planning Research* 15: 61–72. Graphs of Survey of Gated Communities. Figure 1 and 2.

The other extreme is represented by Foldvary's (1994) *Public Goods and Private Communities*, in which he used neoclassical economics to argue that gated communities represent a more efficient mode of urban development. However, the critics of this increasingly popular form of community seem to outnumber those in favor of the trend. Luymes (1999) observed that most critics of gated communities see them as either a symptom or a cause of the loss of civic life at the metropolitan scale.

Cities of the southwestern United States have the largest number of such communities; for instance, the Phoenix metropolitan area had an estimated 641 in 1999 (Frantz, 2001). Luymes (1999) examined real estate advertisements for selected cities throughout the United States for four different years and noted that the Southwest and Florida had the highest concentration of gated communities for the earlier years (see Table 6.1). However, as time passed, advertisements for gated communities spread to the Midwest and the Northeast, particularly to the Chicago, New York City, and Washington, DC metropolitan areas. Although the diffusion of the concept has occurred to other regions of the United States, the highest concentrations continue to be found in Sunbelt cities. Luymes (1999) observed other spatial patterns at the metropolitan scale while conducting field work in some of the cities with high numbers of gated communities. For instance, he found a

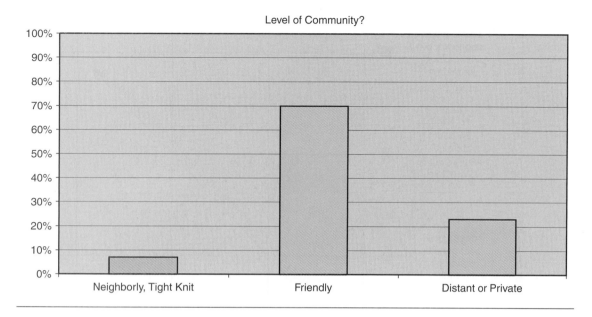

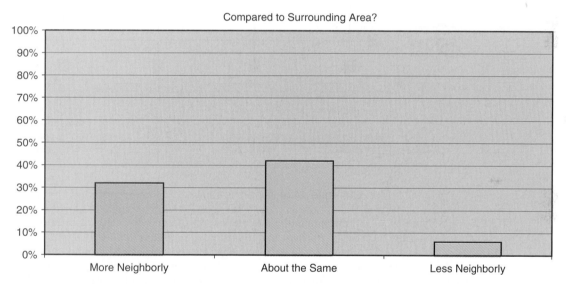

FIGURE 6.9b Level of community feeling in gated communities. Source: Fig 6-9 Blakely, Edward J. and M. G. Snyder. 1998. Forting up: Gated communities in the United States. *Journal of Architectural and Planning Research* 15: 61–72. Graphs of Survey of Gated Communities. Figure 1 and 2.

landscape of continuous walls of new gated communities in the north-central section of San Antonio.

The gated community idea has spread beyond the United States into parts of the developing world. Miao (2003) observed a sharp increase in gated communities in Chinese cities after the 1978 economic reform (see Figure 6.11). Gated communities serve not only the upper and middle class in China but all income levels. The government advocates gated communities because of their perceived role in controlling crime, which in effect contributes to stability (Miao, 2003, p. 49). Gated communities are also increasing in Latin American megacities, including on the outskirts within large edge-city-like projects (Coy &

Pohler, 2002). As in other places, they are being built for their perceived benefit of security; however, according to Coy and Pohler (2002, p. 366), they are forming islands of prosperity in a sea of poverty, increasing the already deep division between the upper class and the rest of the people. Echoing a similar concern of economic polarization in the United States, Diamond (2005) stated:

> If conditions deteriorate too much for poorer people, gates will not keep the rioters out. Rioters eventually burned the palaces of Maya kings and tore down the statues of Easter Island chiefs; they have also already threatened wealthy districts in Los Angeles twice in recent decades (p. A21).

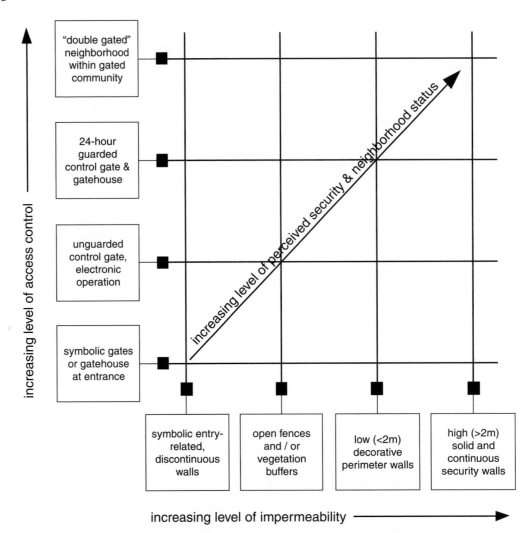

FIGURE 6.10 Typology of gated communities. Source: Luymes, Don. 1999. "The fortification of suburbia: investigating the rise of enclave communities." *Landscape and Urban Planning* 39:187–203. p. 198 Figure 5. "A typology of enclave neighborhoods, organized around the variables of the level of access control (gates), and the level of perimeter impermeability (walls), showing the relationship between these designed control systems, perceived security and status."

6.4 CHANGE IN NEIGHBORHOOD RACIAL COMPOSITION

There is a long tradition in urban geography and sociology of examining the change of White inner-city neighborhoods to Black neighborhoods as Whites moved to newer areas or to the suburbs (Clark, 1993). The transition from White to Black neighborhoods is often described as a *racial succession*, a term that applies to any neighborhood change of racial composition. Colloquially, the process was referred to as "White flight," especially by those directly affected. The term *tipping point* was introduced to measure the critical proportion of Blacks in a neighborhood beyond which it

would rapidly change from mixed to majority Black (Goering, 1978).

"White flight" is still in use today; however, with Black suburbanization increasing in the 1980s and 1990s, some now position the issue as one of class rather than race. Some studies have provided evidence contradicting the assumption that racially mixed neighborhoods will inevitably tip in the direction of more Black over time (Ottensmann, Good, & Gleeson, 1990). Wood and Lee (1991) drew similar conclusions after examining Los Angeles, where they noted an increase in multiethnic neighborhoods with combinations of White, Hispanic, Black, and Asian (see Figure 6.12). Clark (1993) also analyzed the data for Los Angeles; although he found some evidence of the multiethnic trend

TABLE 6.1 Survey of newspaper advertising of gated communities

This table reviews real estate marketing for new single-family communities in major newspapers during selected periods. It indicates the proportion of communities advertised that were gated.

Selected newspapers / Real estate marketing	April 1980			April 1987			April 1993			August 1994			Totals		
	No. comm.	No. gated	% gated	No. comm.	No. gated	% gated	No. comm.	No. gated	% gated	No. comm.	No. gated	% gated	No. comm.	No. gated	% gated
Boston Globe	2	0	0	17	0	0	11	2	18.2	11	3	27.3	41	5	12.2
New York Times	4	1	25	6	1	16.7	10	3	30	7	2	28.6	27	7	25.9
Washington Post	36	0	0	24	0	0	22	2	9.1	26	6	23.1	108	8	7.4
Atlanta Journal-Constitution	8	0	0	18	2	11.1	14	2	14.3	21	1	4.8	61	5	8.1
Miami Herald	n/a			24	5	20.8	10	3	30	36	21	58.3	70	29	41.4
Chicago Tribune	26	2	7.7	30	0	0	5	1	20	16	4	25	77	7	9.1
Denver Post	n/a			n/a			n/a			15	1	6.7	15	1	6.7
Los Angeles Times	45	7	15.5	15	12	80	18	12	66.7	12	8	66.7	90	39	43.3
San Francisco Chronicle	n/a			8	2	25	9	2	22.2	12	2	16.7	29	6	20.7
Houston Post	n/a			n/a			n/a			31	10	32.2	31	10	32.2
San Antonio Exp.-News	n/a			n/a			n/a			22	12	54.5	22	12	54.4
Totals	121	10	8.3	142	22	15.5	99	27	27.3	209	70	33.5	563	129	22.9

No. comm. = number of communities

Source: Luymes, Don. 1999. "The fortification of suburbia: investigating the rise of enclave communities." *Landscape and Urban Planning* 39:187–203. P. 193, Table 1. "A review of real estate marketing for new single-family communities in selected newspapers for selected periods, showing the total number of new communities advertized, and the number and percentage of gated communities."

(a)

(b)

FIGURE 6.11 Gated communities in suburban Shanghai. Source: Miao, Pu. 2003. "Deserted streets in a jammed town: the gated community in Chinese cities and its solution." Journal of Urban Design 8:45–66.

identified by Wood and Lee (1991), he also found an ample number of cases that fit the traditional succession and invasion model of neighborhood change:

The process of neighborhood transition is not creating mixed integrated tracts in very large numbers and, although there are differences between the old automatic patterns of white to minority neighborhood transition, there is still a strong tendency for continuing change to minority predominance once it has been initiated. The change back from largely black to Hispanic tracts is the new dimension at least for the Los Angeles metropolitan area. (Clark, 1993, pp. 170–171)

In a more recent study of immigrant gateway cities, Clark and Blue (2004) found that integration is greater

FIGURE 6.12   Optical shop in Chinatown in Seattle.

at higher education levels and that suburban areas were more integrated than urban cores. Johnston (2003), using 2000 Census data, showed that cores of extreme segregation remain in Los Angeles, despite some reduction in purely White and purely Black areas.

Another stream of literature on racial transitions in neighborhoods addresses the causes for continued separation of races. At one extreme are those that attribute the process to discrimination in a variety of forms, such as Farley et. al. (1997), whereas at the other extreme are arguments for own-race residential preferences led by Clark (2002). Clark saw continuing evidence of very strongly held own-race residential preferences as the most probable cause for racial and ethnic groups showing a continued bias toward separation. Examining results on residential preference by race in four cities, Clark (2002, p. 247) found that on their first choice Blacks opt for a more integrated neighborhood, but he put higher credence on their second choice, which shows that in all of the cities the leaning is toward own-race residential areas (see Figure 6.13). The findings were also consistent for Whites, Hispanics, and Asians, leading Clark (2002) to refer to the preference for own-race residential areas as ethnocentrism. Such a motive stands in stark contrast to motivations attributed to discrimination.

Meanwhile, others contend that it is not choice but discrimination that continues to play a major role in the separation of the races. *Redlining* is a process where lending institutions refuse to loan in a specific area due to perceived mortgage risk, which is the most well-documented form of discrimination that resulted in the separation of races as far back as the 1930s (Harris & Forrester, 2003). Cohen (2003) found evidence of redlining by examining Veteran Administration home loans granted in four New Jersey counties between 1945 and 1960. She found that by 1960 the residents of the racially diverse Hudson County had received only $12 of mortgage insurance per capita compared to the $601 per capita received in Nassau County, home to suburban Levittown and Garden City (Cohen, 2003, p. 205). Looking for more current forms of discriminatory practices, Cashin (2004) examined the movement of upper middle income Blacks to suburban Black enclaves, and she argued that they are still being steered to the enclaves by real estate agents. She pointed to more subtle legal practices than redlining, such as zoning laws and building codes, which perpetuate racial and class segregation.

The next section examines neighborhood change with respect to race for St. Louis for the decade from 1990 to 2000. This urban issue is examined further in the chapter's GIS exercise.

6.5 ANALYZING AN URBAN ISSUE

The GIS exercise at the end of this chapter is designed to measure change in the African American neighborhoods in the St. Louis area between 1990 and 2000 using Markov chain analysis. To apply this methodology, it is first necessary to classify the neighborhoods (census tracts) of the St. Louis area into categories or "states" of African American neighborhoods; for instance, neighborhoods with low, moderate, and high proportions of African Americans. In the exercise, three states of African American census tracts are chosen: high (90–100 percent), medium (10–89 percent), and low (0–9 percent). The selection of these state thresholds was based on the observation that there were many more census tracts in the upper 10 percent and lower 10 percent than in the middle of the range of percentage African American (see Figure 6.14).

Having established the three types of African American neighborhoods, we now return to the question of how neighborhoods move from one state of African American proportion to another. A number of techniques can be used to quantify the changes in characteristics of neighborhoods through time, but a Markov chain has both descriptive and predictive properties that are especially appropriate for the analysis of neighborhood change (Bourne, 1971; Clark, 1965; Greene, 1994; Hunter, 1974). The Markov analysis is then applied to the three

A	Atlanta		
split	1st choice	2nd choice	
1 All White	3	1	
2	2	9	
3 White/Black	54	18	
4	20	66	
5 All Black	21	6	
Total	100	100	

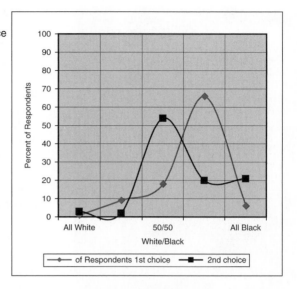

B	Boston		
split	1st choice	2nd choice	
1 All White	5	2	
2	9	20	
3 White/Black	52	22	
4	19	48	
5 All Black	15	8	
Total	100	100	

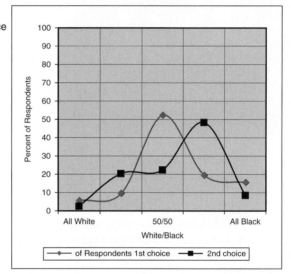

C	Detroit		
split	1st choice	2nd choice	
1 All White	2	3	
2	6	10	
3 White/Black	59	22	
4	10	61	
5 All Black	2	4	
Total	79	100	

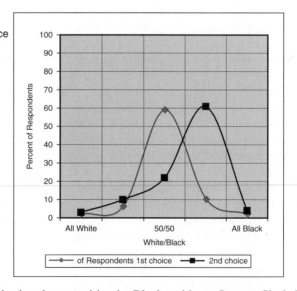

FIGURE 6.13   First and second preferences for neighborhood composition by Black residents. Source: Clark, William A.V. 2002. "Ethnic preferences and ethnic perceptions in multi-ethnic settings." Urban Geography 23:237–256. Figure 4. Page. 247. "Percent of Black preferences for White neighbors in Atlanta (A), Boston (B), Detroit (C), and Los Angeles (D)."

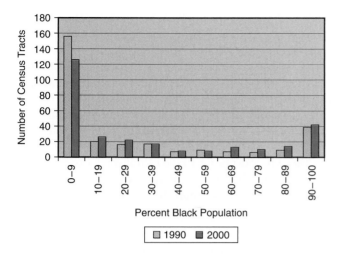

FIGURE 6.14 Percentage Black population in census tracts in St. Louis City and St. Louis County. Source: U.S. Census, 1990 and 2000.

states derived from 1990 and 2000 census-tract distributions of the St. Louis area. Next, an examination is made of every two-state combination (e.g., low-to-low African American) for the 1990 to 2000 transition period.

As a result, a total of 9 two-state combinations are converted into a probability matrix of the form:

2000 State of Percentage African American

		(1) Low	(2) Medium	(3) High	Total
1990 State of Percentage African American	(1) Low	P_{11}	P_{12}	P_{13}	1.0
	(2) Medium	P_{21}	P_{22}	P_{23}	1.0
	(3) High	P_{31}	P_{32}	P_{33}	1.0

where P_{11}, the first cell, represents the probability of a neighborhood starting in 1990 with a low percentage of African Americans and ending the period in 2000 with a similarly low percentage of African Americans. Cell entries in the matrix are calculated by the relative frequency definition of probability:

$$P_{ij} = \frac{f_{ij}}{\sum_{j=1}^{3} f_{ij}}$$

where P_{ij} represents the probability of a neighborhood moving from one African American level to another, f_{ij} is the observed number of transitions from state i to state j. 3 is the number of states in this example.

The results of the counted transitions are entered into the matrix:

$$P =$$

		(1) Low	(2) Medium	(3) High	Total
	(1) Low	124	33	0	157
	(2) Medium	2	84	5	91
	(3) High	0	1	37	38

Using the relative frequency definition of probability we calculate:

2000 State of Percentage African American

		(1) Low	(2) Medium	(3) High	Total
1990 State of Percentage African American	(1) Low	.79	.21	.00	1.0
	(2) Medium	.02	.92	.06	1.0
	(3) High	.00	.03	.97	1.0

We find that a low probability (.06) that a tract with a medium percentage of African Americans will transition to a high percentage. In fact, if a tract is of a medium percentage African American, it is highly likely to remain medium in the next decade (.92).

GIS shows an interesting pattern for the five tracts that made the transition from medium (B) to high (C) (see Figure 6.15). All of these tracts are adjacent to neighborhoods that were high African American in both 1990 and 2000. This would suggest that although medium-percentage African American tracts have a low probability of making the transition to the high-percentage group, they have a much higher probability if they are adjacent to high-percentage African American tracts. In addition, the pattern conforms to patterns observed in earlier studies where African American enclaves formed on one side of town. In the case of St. Louis, this was north of downtown. Thus, the growth of an African American enclave has traditionally moved outward from an established cluster.

The mapped pattern also suggests evidence of an active tipping point; that is, as the Black area expands northwestward, the proportion of White population decreases and the neighborhood tips in the direction of

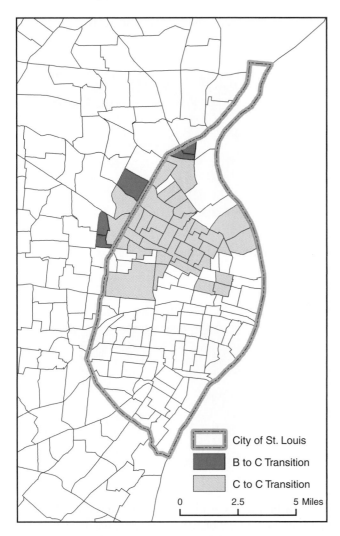

FIGURE 6.15 Racial transitions in census tracts in St. Louis, 1990 to 2000. Note: B represents medium percentage African American. C represents high percentage African American.

Map legend:
- City of St. Louis
- B to C Transition
- C to C Transition

Scale: 0 — 2.5 — 5 Miles

majority African American. The 1990 to 2000 trend also conforms to previous periods where the observed racial integration is short term and the trend is for neighborhoods to be all White or all Black in the long run. Once the wave of transition is complete in one community, adjacent communities will likely experience the same process. Yizhaq, Portnov, and Meron (2004) provided a more elaborate mathematical model of segregation that specifies various contexts and thresholds for tipping points.

There are several limitations to the use of a Markov chain analysis. The first limitation concerns the decision on where to draw the break points for the states. For instance, in a more detailed analysis of the same data, states were broken into 10 states using 10 percent breaks (see Table 6.2). This was useful for analyzing what occurred in the 10 percent to 89 percent change range, where it was

learned that more than half of the census tracts saw an increase in their African American population proportions by 2000.

In addition, the size of the geographic unit used to measure transitions can have an impact on the analysis of integration, racial or other type. As Clark (1965) showed in Los Angeles, the level of integration can be much less at the block level than at the tract level. It is quite likely that there will be variation at the block level; for example, those blocks closest to the established African American core in St. Louis might have a higher African American percentage than the blocks in the same tract further from the core.

6.6 GENTRIFICATION

Gentrification refers to the renovation of deteriorated neighborhoods by upper and middle-income people. It often involves the gradual displacement of the current low-income residents of the area, who are unable to afford the increased rents as the neighborhood climbs in status. The term originated in London, where it was used to describe a new "gentry" typically composed of affluent, childless, two-worker households settling into downtown neighborhoods, where they undertook refurbishing of the older housing stock (Berry, 1999). The process itself became a topic of debate in the early 1980s and mushroomed as a research agenda among social scientists thereafter, partly because it represented a countertrend to the widespread neighborhood decline taking place in cities. Gentrification can be found in cities across the world and has been associated with the processes of globalization (Smith, 1996).

Smith (1979) advanced a controversial theory, the rent gap, to help explain why gentrification occurs in cities. The rent-gap theory is a Marxist interpretation, holding that a gap exists when there is a difference in the capitalized ground rent and potential ground rent generated by a property (see Figure 6.16). The first concept, *capitalized ground rent*, can be defined as the value of a plot of land given its present use, whereas *potential ground rent* is the profit that could be gained by the same plot of land under a "higher and better use." Neighborhood revitalization or gentrification occurs when investors discover a wide enough gap between the two ground rents (see Figure 6.17). Smith (1987) reiterated the simplicity of the theory by suggesting that gentrification is just one means by which the rent gap can be closed. Hammel (1999a), on the other hand, saw the theory as being much broader, with linkages to other urban processes. For example, suburbanization compounds the rent gap of cities, as increased investment on the urban fringe can often imply further disinvestment in the core where gentrification is ripe.

TABLE 6.2 Changes in black population in St. Louis by tract, 1990 to 2000

Year		2000										
		a	b	c	d	e	f	g	h	i	j	Total
Number of tracts												
	a	123	17	5	9	1	1					156
	b	1	6	7	3	1	2					20
	c	1	3	4		3	1	3	1			16
	d			5	5	2	2	2	1			17
1990	e					1	2	1	2	1		7
	f			1				3	2	3		9
	g							4	1	2		7
	h								1	3	2	6
	i								1	5	3	9
	j	1							1		37	39
Transition probability matrix												
	a	0.788	0.109	0.032	0.058	0.006						
	b	0.050	0.300	0.350	0.150	0.050	0.100	0	0	0	0	
	c	0.063	0.188	0.250	0	0.188	0.063	0.188	0.063	0	0	
	d	0	0	0.294	0.294	0.118	0.118	0.118	0.059	0	0	
1990	e	0	0	0	0	0.143	0.286	0.143	0.286	0.143	0	
	f	0	0	0.111	0	0	0	0.333	0.222	0.333	0	
	g	0	0	0	0	0	0	0.571	0.143	0.286	0	
	h	0	0	0	0	0	0	0	0.167	0.500	0.333	
	i	0	0	0	0	0	0	0	0.111	0.556	0.333	
	j	0.026	0	0	0	0	0	0	0.026	0	0.949	

a = 0 to 9.99, b = 10 to 19.99, c = 20 to 29.99, d = 30 to 30.99, e = 40 to 49.99, f = 50 to 59.99, g = 60 to 69.99, h = 70 to 79.99, i = 80 to 89.99, j = 90 to 100.

Source: U.S. Census, 1990 and 2000.

Using 130 years of land-price data for nine redeveloped parcels in Minneapolis, Hammel (1999a) found a rent gap, but he concluded cautious of a one-to-one correspondence between the closing of the gap and the process of gentrification. The study differed from most other rent-gap analyses because it addressed the importance of scale and the geographic unit necessary to examine the rent gap. Promoting Clark's (1988) use of the parcel as a unit of analysis, Hammel (1999a) was able to examine specific locations, whereas most studies examined large districts within cities, which is not regarded as an appropriate resolution for an analysis of land rent. In spite of some of the shortfalls of attempting to test such an abstract concept with parcel detail, Hammel (1999a) was able to establish rent gaps for each of his parcels. For instance, a parcel near Orchestra Hall experienced its largest rent gap during the post-World War II period of suburban expansion of the Twin Cities and it was subsequently redeveloped (see Figure 6.18). The area extending from Orchestra Hall all the way to Loring Park has been an area where gentrification has become intensified (see Figure 6.19). Hammel (1999b) addressed other confusions that had erupted over the issues of geographic scale by suggesting that capitalized rent should be measured at the neighborhood scale, whereas potential rent should be measured at the metropolitan scale. According to Hammel (1999b), a blending of these two scales is critical to an understanding of how the rent gap plays out in cities.

Some urban scholars have wondered why the gentrification process has received so much attention. For instance, Berry (1985) considered some of the early attention to neighborhood renewal trends to be subordinate to larger decentralization trends and coined the often-quoted phrase "islands of renewal in seas of decay." The implication, of course, was that gentrification was

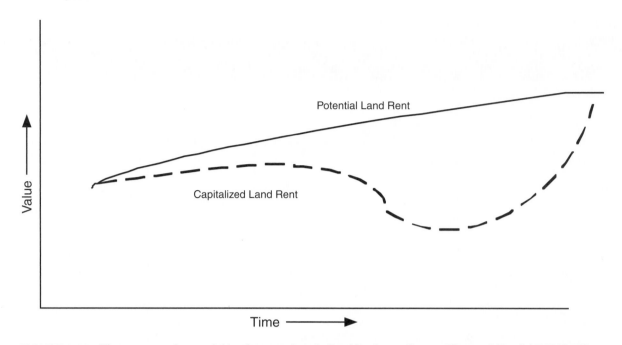

FIGURE 6.16   The rent gap of potential land rent and capitalized land rent. Source: Hammel, Daniel J. 1999. "Gentrification and land rent: a historical view of the rent gap in Minneapolis." Urban Geography 20: 116–145. Figure 1. Page 119. "The rent gap is formed by divergence in potential and capitalized land rents as capital investment in the parcel declines."

small in comparison to the amount of land abandoned by middle-class flight from cities. Bourne (1993) also argued that gentrification's impact was relatively slight, especially as it related to the urban housing stock. In defense of the importance of gentrification as an urban process, Wyly and Hammel (1998) highlighted its impact on the socioeconomic makeup of cities even while recognizing its trivial magnitude in terms of geographic scope. In a later study, the authors suggested that the financial impact should also not be overlooked by showing that gentrified neighborhoods attracted mortgage capital at a rate that

grew 2.3 times the comparable suburban rate between 1992 and 1997 (Wyly & Hammel, 1999).

Other studies have cast doubt on a key component of the debate, the issue of displacement resulting from gentrification. For instance, Engels (1999) showed that social replacement occurred alongside social displacement in a gentrifying neighborhood in the inner Sydney suburb of Glebe between 1960 and 1986. Engels's study suggests that gentrification is a complex process that includes both benefits and negative aspects. This aspect of gentrification has led one leading urban scholar to designate the process a conundrum:

FIGURE 6.17   Neighborhood gentrification in Lincoln Park area of Chicago.

It would be hard to argue that, in the net, gentrification has not benefited the cities and in some cities has even reversed decades of residential decline. Yet it remains a conundrum. For every positive point, there seems to be a counterpoint. Gentrification has brought about neighborhood stability through greater ownership, but removed rental units important to those not in a position to purchase. It has raised inner-city property values and allowed many to notch up their living standard but priced out others. It has attracted national chains such as Starbucks, Crate and Barrel, Gap, Barnes and Noble, and TGI Friday's, which have slowly replaced sometimes shabby local coffee houses, thrift and vintage clothing stores, used book shops, and mom and pop bars. While the national chains offer these neighborhoods products and services appropriate to their new, more upscale residents, such commercial homogenization may be wiping out the very character

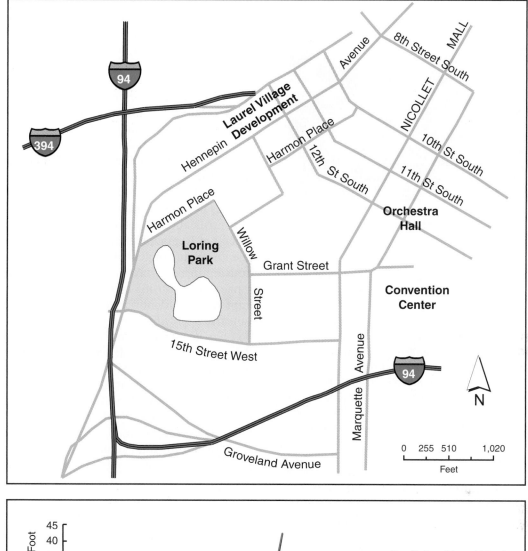

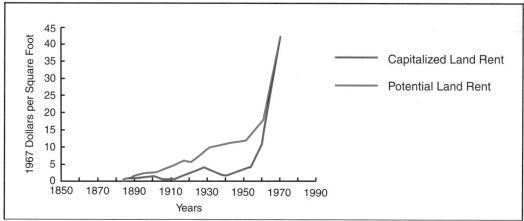

FIGURE 6.18   A century of properties in the Loring Park neighborhood of Minneapolis. Source: Hammel, Daniel J. 1999a. Gentrification and land rent: a historical view of the rent gap in Minneapolis. Urban Geography 20: 116–145. Fig. 3 on p. 127.

that made many inner-city neighborhoods so appealing. All of this is really subjective, though, in that what is viewed as good or bad, positive or negative, will be determined not by hard data on reinvestment or types of goods and services establishments, but largely by people's ideology,

values, and beliefs of what urban neighborhoods should be. (Kasarda, 1999, p. 780)

Recently, authors have been turning to other interpretations of the process. One study argues that gentrification

FIGURE 6.19 Loring Park, Minneapolis.

is indicative of the globalization processes that have been reorienting the city as an "Entertainment Machine":

> Here the stress is on consumption, not production: CNN, MTV, and Hollywood movies are seen as bringing a standard world fare that may encourage more globally homogeneous consumption. It can raise and refocus economic aspirations by redefining consumption desiderata. But much of consumption is driven by local specifics: cafes, art galleries, geographic/architectural layout, and aesthetic image of a city define its unique attractions. (Clark, Lloyd, Wong, & Jain, 2002, p. 494)

In this case, more emphasis is placed on consumption as an interpretation of gentrification with less stress on economic production factors such as finance or jobs. For instance, Ley's (1986, p. 529) conclusion that changes in the economic base of cities have the strongest explanatory power for gentrification is characteristic of the proposed economic interpretations. A consumption interpretation would emphasize that gentrification stems from changing preferences of residents and workers who are choosing where to live less with respect to the location of their jobs and more on where they want to recreate and consume (Clark et al., 2002). In support of this consumption interpretation, Bridge and Dowling (2001) found evidence that the retail spaces of gentrified neighborhoods reflect the consumption tastes of the gentrifiers.

Smith (2002) echoed the concerns that have been raised regarding the definitions of gentrification. Alternative definitions proposed by him would incorporate migration and population processes and would extend the spatial influence of gentrification to include affluent household movement to locations other than those near the downtown, for instance, in aging inner suburbs. Phillips (2004) examined rural gentrification in the British countryside to demonstrate the breadth of gentrification research. Badcock (2001, p. 1560) argued in the same vein by stating that it makes no sense to separate gentrification conceptually from the more comprehensive process of urban revitalization. Smith (2002) offered a number of research questions that would be opened up if the definition of gentrification were made more inclusive (see Table 6.3). Many of these questions focus on the migratory processes and forces influencing gentrification.

A number of studies have presented techniques for delineating or measuring gentrification. Wilson and Mueller (2004) examined newspaper accounts of gentrification to show reporters as key actors in the process. Figueroa (1995) promoted the use of a GIS that incorporates housing market variables as the main indicators of gentrification. Applying the technique in the Canadian city of Regina, Figueroa was able to successfully delineate an area of gentrification by querying areas where housing prices increased from the low to the middle range of the housing market over time. Atkinson (2000) improved the identification of the displacement that results from gentrification by using longitudinal data as opposed to aggregated census data. By utilizing longitudinal data, Atkinson was able to examine individual flows of displacement as the data set contained tracking information of individuals and families over time.

Still other studies have looked at unusual examples of gentrification. Forsyth (1997) examined the gentrification process for the small town of Northampton, Massachusetts, an unusual focus because case studies have largely been drawn from bigger cities. The study assessed

TABLE 6.3 Gentrification research questions

- Where are the places of origin and destination of groups of gentrifiers and displaces lower income households, respectively?
- Do places of origin and destination and the distances moved, or the frequency of nonmigration from the gentrified place, change over time as the process of gentrification unfolds?
- Do these processes involve different inmigrant, outmigrant, intramigrant, or nonmigrant groups?
- Why do gentrifier and displaced households move into and out of gentrified places?
- Are migratory motives and forces differentiated by social groups who move at different phases of the process of gentrification?
- How do gentrifier and displaced households move into, out of, or within the gentrified place, and do the practices change over time as the process unfolds?
- Where (and why) do gentrifiers and nonmigrants settle within gentrified places?
- Do these internal enclaves and settlement patterns of (non)migrants change over time?

Source: Smith, Darren P. 2002. Extending the temporal and spatial limits of gentrification: A research agenda for population geographers. International Journal of Population Geography 8: 385–394. Table 3 on gentrification research questions for population geographers.

FIGURE 6.20   Gentrification in the Zona Rosa of Mexico City.

the success of gentrification by examining both the role of a strong lesbian presence in the community and the community's juxtaposition relative to strong regional markets. Jones and Varley (1999) were among the first to look at gentrification in a developing country. Their study on conservation efforts for the historic center of Puebla, Mexico, drew corollaries to the gentrification processes identified in cities of the developed world (see Figure 6.20). Brown and Wyly (2001) considered Russian immigrants to Brighton Beach in New York City as gentrifiers, which broke the stereotype of gentrifiers as a young urban professional class.

6.7 NEIGHBORHOOD ACTIVISM

Neighborhoods have their own issues and problems to confront. Community and neighborhood organizations are able to tap into the energy and motivation of ordinary citizens to contribute to this problem solving. *Neighborhood activism* refers to active participation by local citizens in their neighborhood and its organizations to address local issues and problems. This section focuses on neighborhood activism as a part of urban geography. It examines the types of problems, steps to solution, and key roles that neighborhood and community organizations can play. Several short examples are given of neighborhood activism in St. Paul and Minneapolis. The section also examines the

pluses and minuses of GIS as a tool to help activist organizations be more productive.

Neighborhood activism is more commonly a collective form of expression, rather than individual. Groups of neighborhood residents organize to express a group's goals and desires, which are stronger than those of individuals. For example, a neighborhood crime-watch organization expresses the active interests of its members in detecting, preventing, and controlling local crime.

Neighborhood activism in the United States complements or contends with other public, nonprofit and private-sector organizations and institutions. In the public sector, parts of neighborhood issues and problems are handled by metropolitan councils of governments, metropolitan planning commissions, city governments and their planning departments, and federal agencies such as the Department of Housing and Urban Development. In the nonprofit sector, economic and civic development organizations promote business expansion, job growth, and cultural enhancements in cities. The private sector provides neighborhood financing and other services.

At the base, these organizations reflect the pluralism of democracy in America, and in many other countries as well. Studies have indicated neighborhood activism in the United States in the latter half of the 20th century was stimulated by the political trend of neoliberalism (Elwood, 2002; Martin, 2003 Fraser, 2004). In particular, neoliberal government policies have encouraged grants, funding, and support for neighborhood organizations. With so many neighborhood entities, community activism both cooperates and contends with the public sector and other organizations. This creates a tension that is often not easily resolved. Sometimes, neighborhood activism must go along with government initiatives and projects, leading to criticism by some of giving in to state priorities or city boosterism (Elwood & Leitner, 2003). At other times, it leads to protests and dissent with government.

One study of neighborhood activism focused on a case study of neighborhood activism in four organizations in the Frogtown neighborhood of St. Paul, Minnesota (Martin, 2003). The study utilized the theory of collective-action frames to better understand the organizations and why place was important to them. Frogtown (Thomas-Dale/District 7 officially) is located just northwest of the downtown, and has over half non-Hispanic minorities, a 35 percent poverty level, and high crime rates.

Two of the four organizations are discussed here, namely, the Thomas-Dale District Seven Planning Council (District Council) and the Thomas-Dale Block Clubs (T-D Block Clubs). The District Council, located in a small office, plays a liaison role between neighborhood constituents and the city's agencies for long-range planning, zoning, and land use (Martin, 2003). The T-D Block

Clubs have the broad objective to foster pride in the neighborhood through block clubs that organize and motivate residents for neighborhood improvements, for instance, a "beautiful block" contest to recognize the best appearance of yards and homes (Martin, 2003). The clubs are funded by foundations, donations, and contracts from the city. The organizations were studied by participant observation and gathering of documents and materials (Martin, 2003).

The conceptual approach utilized was a *collective-action frame*, which is a group of descriptions of dimensions for actions by the collective organizations. In this case, the three frames utilized are motivational, diagnostic, and prognostic (see Table 6.4). The motivational frames indicate how an organization is exhorting actions and describes who and what parts of the community are served (Martin, 2003). The diagnostic frames describe the problems and where the problems have originated (at times, blame). The prognostic frames describe the actions to be taken by the organization to solve the problems. The ways the District Council and T-D Block Clubs fit into these three frames are given in Table 6.4. It is clear that the District Council has a focus on neighborhood physical cleanup and planning, whereas the T-D Block Clubs are directed toward improving the neighborhood appearance and fostering its pride and interactivity. As

TABLE 6.4   Collective-Action frames from two neighborhood organizations in frogtown neighborhood, St. Paul, 1996–1997

Organization	Motivational frames[1]	Diagnostic frames[2]	Prognostic frames[3]	Importance of place-frame

Thomas-Dale District Seven Planning Council	Exhortations
	 Plan for future
	 Clean up neighborhood
	 Support/protect children
	Create community
	
	Descriptions
	 Racial, cultural,
	  economic, diversity
	Historic homes
	 (railroad, working class)
		Lack of green, public space
		Cycle of disinvestment
		Broader processes/decisions
		 affect local conditions
			Plan for future development
			 industrial
			 social
			 economic
			 infrastructure
			 long-range
			  comprehensive plans
			Cleanup days
				All three prior frames focus on the neighborhood place

Thomas-Dale Block Clubs	Exhortations Keep neighborhood and houses clean Individuals are responsible for community	Garbage in streets, yards Poor attitudes, behavior, and lack of responsibility	Foster residential interactions and neighborhood pride Clean up area, plant flowers, and build pocket parks Protest criminal behavior	All three prior frames focus on the neighborhood place Crime may be more widespread in region.
	Win prizes Descriptions: Cultural, religious, racial, diversity Children		Work with police Old property owners accountable for tenants, clean up	

[1]Motivational frames exhort action and describe the community (who, what).
[2]Diagnostic frames describe the problems and assign cause or blame.
[3]Prognostic frames describe the specific actions for solutions.
Source: Martin, Deborah G. 2003. "Enacting neighborhoods. Urban Geography 24:361–385. Figure "Collective-Action Frames for two neighborhood organizations in Forgtown neighborhood, St. Paul, 1996–97."

seen in the right column of Table 6.4 the frames for both cases emphasize the importance of place, predominantly that of the neighborhood.

Neighborhood organizations benefit by the use of GIS (Dangermond, 2002; Elwood, 2002; Elwood & Leitner, 2003; Obermeyer, 1998). A key question, however, is whether their GIS applications should be aligned with government GIS systems or be unique applications driven by neighborhoods' own concerns (Craig & Elwood, 1998; Elwood & Leitner, 2003). Neighborhood and community organizations utilize GIS to generate new knowledge about the community and its conditions, to compare GIS output on a neighborhood with government-generated GIS output, to research the processes and relationships underlying the neighborhood, and to monitor and project neighborhood conditions (Elwood & Leitner, 2003).

In a study of GIS use in dozens of neighborhood and community groups in Minneapolis (Elwood & Leitner, 2003), the groups found support from GIS for strategic, tactical, administrative, and constituent-organizing purposes (Elwood & Leitner, 2003). For example, one neighborhood organization in the Seward area of Minneapolis was able to provide maps and data on housing and financing conditions, so buyers and remodelers could achieve better results. In the long term, this improved the whole housing stock and livability of the neighborhood. The study determined that a key to the organizations' success is that the government continues to provide free or low-cost access to its spatial data. This allows the organizations the potential to come close to or match the research capabilities of the public and private sectors, or to reach a level playing field (Elwood & Leitner, 2003).

Another potential for GIS use is public participation geographic information system (PPGIS). PPGIS is a "variety of approaches to making GIS and other spatial decision-making tools available and accessible to all those with a stake in official decisions" (Obermeyer, 1998, p. 65). With the Internet, today's PPGIS can include access to and interaction with community data and planning directly by the neighborhood constituents. It has the potential to encourage democracy at the grassroots level (Dangemond, 2002; Obermeyer, 1998). At the same time, lack of access to the technology by disadvantaged constituents might reduce the potential of PPGIS.

In summary, neighborhoods in the United States and many other countries have organizations supported by activists and other constituents that focus their attention on neighborhood and community planning, information dispersal, improvements, services, and grassroots democracy. They can coordinate or differ from the public-sector goals and initiatives. GIS is a useful tool for these organizations to foster more information and better decision making, as long as the community can participate and there is latitude to differ from the information and decision advice of public spatial systems.

SUMMARY

The neighborhood has an important function in the city, providing support and identity to its residents. Some have argued that the role of community is threatened by the rapid growth of contemporary suburbia. Nevertheless, evidence is provided in support of the argument that community is alive and well in even the newest of our urban neighborhoods. The interest in measuring the transition of racial composition in city neighborhoods is still strong in urban geography, although more studies are presented today addressing future outcomes and reasons for continued racial separation among neighborhoods. The process of gentrification occurs worldwide and has varying outcomes that involve the effects of displacement and replacement of residents. New areas for urban geography research have been opened up because of a more inclusive definition of the process.

REFERENCES

Ahlbrandt, R. S. (1984) *Neighborhoods, People, and Community*. New York: Plenum.

Atkinson, R. (2000) "Measuring gentrification and displacement in greater London." *Urban Studies* 37:149–165.

Badcock, B. (2001) "Thirty years on: Gentrification and class changeover in Adelaide's inner suburbs, 1966–96." *Urban Studies* 38:1559–1572.

Berry, B. J. L. (1985) "Islands of renewal in seas of decay," in *The New Urban Reality*. ed. by P. Peterson, pp. 69–96. . Washington, DC: The Brookings Institution.

Berry, B. J. L. (1999) "Comment on Elvin K. Wyly and Daniel J. Hammel's 'Islands of decay in seas of renewal: Housing policy and the resurgence of gentrification'—Gentrification resurgent?" *Housing Policy Debate* 10:783–788.

Blakely, E. J., and Snyder, M. G. (1998) "Forting up: Gated communities in the United States." *Journal of Architectural and Planning Research* 15:61–72.

Bourne, L. S. (1971) "Physical adjustment processes and land use succession." *Economic Geography* 47:1–15.

Bourne, L. S. (1993) "The demise of gentrification? A commentary and prospective view." *Urban Geography* 14:95–107.

Bridge, G., and Dowling, R. (2001) "Microgeographies of retailing and gentrification." *Australian Geographer* 32:93–107.

Bright, E. M. (2003) *Reviving America's Forgotten Neighborhoods: An Investigation of Inner City Revitalization Efforts*. New York: Routledge.

Brown, K., and Wyly, E. (2001) "A new gentrification? A case study of the Russification of Brighton Beach, New York." *Geographical Bulletin–Gamma Theta Upsilon* 42:94–105.

Cashin, S. (2004) *The Failures of Integration: How Race and Class Are Undermining the American Dream*. New York: Public Affairs.

Clark, E. (1988) "The rent gap and transformations of the built environment: Case studies in Malmo 1860–1985." *Geografisca Annaler* 70B:241–254.

Clark, T. N., Lloyd, R., Wong, K., and Jain, P. (2002) "Amenities drive urban growth." *Journal of Urban Affairs* 24:493–515.

Clark, W. A. V. (1965) "Markov chain analysis in geography: An application to the movement of rental housing areas." *Annals of the Association of American Geographers* 55:351–359.

Clark, W. A. V. (1993) "Neighborhood transitions in multiethnic/racial contexts." *Journal of Urban Affairs* 15:161–172.

Clark, W. A. V. (2002) "Ethnic preferences and ethnic perceptions in multi-ethnic settings." *Urban Geography* 23:237–256.

Cohen, L. (2003) *A Consumers' Republic.* New York: Vintage Books.

Coy, M., and Pohler, M. (2002) "Gated communities in Latin American megacities: Case studies in Brazil and Argentina." *Environment and Planning B: Planning and Design* 29:355–370.

Craig, W. J., and Elwood, S. S. A. (1998) "How and why community groups use maps and geographic information." *Cartography and Geographic Information Systems* 25(2):95–105.

Dangermond, J. (2002) "Mutualism in strengthening GIS technologies and democratic principles: Perspectives from a GIS software vendor," in *Community Participation and Geographic Information Systems.* ed. by W. J. Craig, T. T. M. Harris, and D. Weiner, pp. 297–308. London: Taylor & Francis.

Davis, M. (1990) *City of Quartz: Excavating the Future of Los Angeles.* London: Verso.

Diamond, J. (2005) "The ends of the world as we know them." *The New York Times* (Op-Ed), January 1: A21

Elwood, S. A. (2001) "GIS and collaborative urban governance: Understanding their implications for community action and power." *Urban Geography* 22(8):737–759.

Elwood, S. (2002). "The impacts of GIS use for neighbourhood revitalization in Minneapolis," in *Community Participation and Geographic Information Systems,* ed. By W.J. Craig, T.M. Harris, and D. Weiner, pp. 77–88. London: Taylor & Francis.

Elwood, S., and Leitner, H. (2003) "GIS and spatial knowledge production for neighborhood revitalization: Negotiating state priorities and neighborhood visions." *Journal of Urban Affairs* 25:139–157.

Engels, B. (1999) "Property ownership, tenure, and displacement: In search of the process of gentrification." *Environment and Planning A* 31:1473–1495.

Farley, R., Feilding, E., and Krylsa, M. (1997) "The residential preferences of Blacks and Whites: a four metropolis study." *Housing Policy Debate* 8:763–800.

Figueroa, R. A. (1995) "A housing-based delineation of gentrification: A small area analysis of Regina, Canada." *Geoforum* 26:225–236.

Foldvary, F. (1994) *Public Goods and Private Communities: The Market Provision of Social Services.* London: Edward Elgar.

Forstall, R. L. (2004) Personal communication.

Forsyth, A. (1997) "NoHo: Upscaling main street on the metropolitan edge." *Urban Geography* 18:622–652.

Frantz, K. (2001) "Gated communities in metro-Phoenix." *Geographische Rundschau* 53:12–18.

Fraser, J. C. (2004) "Beyond gentrification: mobilizing communities and claiming space," *Urban Geography* 25:437–457.

Fried, M. (1986) "The neighborhood in metropolitan life: Its psychological significance," in *Urban Neighborhoods: Research and Policy.* ed. by R. B. Taylor, pp. 331–363. New York: Praeger.

Gans, H. (1968) "Urbanism and suburbanism as ways of life: A re-evaluation of definitions," pp. 34–52 in *People and Plans.* New York: Basic Books.

Goering, J. M. (1978) "Neighborhood tipping and racial transition: A review of social science evidence." *Journal of the American Institute of Planners* 44:68–78.

Gottdiener, M., Collins, C., and Dickens, D. (1999) *Las Vegas: The Social Production of an All-American City.* Oxford, UK: Blackwell.

Greenberg, M. (2004) "Was 1992–2000 the best of times for American urban neighborhoods?" *The Geographical Review* 93:81–96.

Greene, R. P. (1994) "Poverty-area instability: the case of Chicago." *Urban Geography* 15:362–375.

Guterson, D. (1992) "No place like home." *Harper's Magazine* May:55–61.

Haeberle, S. H. (1988) "People or place: Variations in community leaders' subjective definitions of neighborhood." *Urban Affairs Quarterly* 23:616–634.

Hammel, D. J. (1999a) "Gentrification and land rent: A historical view of the rent gap in Minneapolis." *Urban Geography* 20:116–145.

Hammel, D. J. (1999b) "Re-establishing the rent gap: An alternative view of capitalized land rent." *Urban Studies* 36:1283–1293.

Harris, R., and Forrester, D. (2003) "The suburban origins of redlining: A Canadian case study, 1935–54." *Urban Studies* 40:2661–2686.

Holloway, S. R., and Mulherin, S. (2004) "The effects of adolescent neighborhood poverty on adult employment. *Journal of Urban Affairs* 26:427–454.

Hunter, A. (1974) "Community change: A stochastic analysis of Chicago's local communities, 1930–1960." *American Journal of Sociology* 79:923–947.

Jencks, C., and Peterson, P. E. (eds.). (1991) *The Urban Underclass.* Washington, DC: The Brookings Institution.

Johnston, R. (2003) "And did the walls come tumbling down? Ethnic residential segregation in four U.S. metropolitan areas, 1980–2000." *Urban Geography* 24:560–581.

Jones, G. A., and Varley, A. (1999) "The re-conquest of the historic centre: Urban conservation and gentrification in Puebla, Mexico." *Environment and Planning A* 31:1547–1566.

Kasarda, J. D. (1999) "Comment on Elvin K. Wyly and Daniel J. Hammel's 'Islands of decay in seas of renewal: Housing policy and the resurgence of gentrification.'" *Housing Policy Debate* 10:773–781.

Ley, D. (1986) "Alternative explanations for inner-city gentrification: A Canadian assessment." *Annals of the Association of American Geographers* 76:521–535.

Luymes, D. (1999) "The fortification of suburbia: Investigating the rise of enclave communities." *Landscape and Urban Planning* 39:187–203.

Macionis, J., and Parrillo, V. (1998) *Cities and Urban Life.* Upper Saddle River, Englewood Cliffs, NJ: Prentice Hall.

Madanipour, A. (2001) "How relevant is 'planning by neighbourhoods' today?" *Town Planning Review* 72:171–191.

Martin, D. G. (2003) "Enacting neighborhoods." *Urban Geography* 24:361–385.

McHugh, K., Gober, P., and Borough, D. (2002) "The Sun City wars: Chapter 3." *Urban Geography* 23:627–648.

Merton, R. (1968) *Social Theory and Social Structure*. New York: Free Press.

Miao, P. (2003) "Deserted streets in a jammed town: The gated community in Chinese cities and its solution." *Journal of Urban Design* 8:45–66.

Obermeyer, N.J. (1998). "The evolution of public participation GIS." *Cartography and Geographic Information Systems* 25(2):65–66.

Ottensmann, J. R., Good, D. H., and Gleeson, M. E. (1990) "The impact of net migration on neighbourhood racial composition." *Urban Studies* 27:705–717.

Phillips, M. (2004) "Other geographies of gentrification." *Progress in Human Geography* 28:5–30.

Putnam, R. (2000) *Bowling Alone*. New York: Simon & Schuster.

Sampson, R. J., Morenoff, J. D., and Gannon-Rowley, T. (2002) "Assessing neighborhood effects: social processes and new directions in research." *Annual Review of Sociology* 28:443–478.

Smith, D. P. (2002) "Extending the temporal and spatial limits of gentrification: A research agenda for population geographers." *International Journal of Population Geography* 8:385–394.

Smith, N. (1979) "Towards a theory of gentrification." *Journal of the American Planning Association* 45:538–548.

Smith, N. (1987) "Gentrification and the rent gap." *Annals of the Association of American Geographers* 77:462–465.

Smith, N. (1996) *The New Urban Frontier: Gentrification and the Revanchist City*. New York: Routledge.

Smith, R. L. (1985) "Activism and social status as determinants of neighborhood identity." *Professional Geographer* 37:421–432.

Suttles, G. D. (1972) *The Social Construction of Communities*. Chicago: University of Chicago Press.

Talen, E. (1999) "Sense of community and neighbourhood form: An assessment of the social doctrine of new urbanism." *Urban Studies* 36:1361–1379.

Webster, C. (2001) "Gated cities of tomorrow." *Town Planning Review* 72:149–170.

Wilson, D., and Mueller, T. (2004) "Representing 'neighborhood': Growth coalitions, newspaper reporting, and gentrification in St. Louis." *The Professional Geographer* 56:282–294.

Wilson, W. J. (1987) *The Truly Disadvantaged: The Inner City, the Underclass, and Public Policy*. Chicago: University of Chicago Press.

Wirth, L. (1938) "Urbanism as a way of life." *American Journal of Sociology* 44:1–24.

Wood, P., and Lee, B. (1991) "Is neighborhood racial succession inevitable? Forty years of evidence." *Urban Affairs Quarterly* 26:610–620.

Wyly, E. K., and Hammel, D. (1998) "Modeling the context and contingency of gentrification." *Journal of Urban Affairs* 20:303–326.

Wyly, E. K., and Hammel, D. (1999) "Islands of decay in seas of renewal: Housing policy and the resurgence of gentrification." *Housing Policy Debate* 10:711–771.

Yizhaq, H., Portnov, B. A., and Meron, E. (2004) "A mathematical model of segregation patterns in residential neighbourhoods." *Environment and Planning A* 36:149–172.

EXERCISE

Exercise Description

Measuring neighborhood change with Markov chains: Show how 1990 and 2000 census tract information for the St. Louis area can be used to construct a Markov model illustrating African American neighborhood transitions. The transition patterns can be mapped and analyzed.

Course Concepts Presented

Markov chains, population analysis, neighborhood change, race, and ethnicity.

GIS Concepts Utilized

Table editing, calculation, string manipulation, queries, relational database, and spatial query.

Skills Required

ArcGIS Version 9.x

GIS Platform

ArcGIS Version 9.x

Estimated Time Required

2 hours

Exercise CD-ROM Location

ArcGIS_9.x\Chapter_06\

Filenames Required

stl90_00.shp
stlouis_city.shp

Background Information

Markov Chains

This is a frequently asked question in urban geography: How does the ethnic and racial composition of neighborhoods change through time? To answer this question using Markov chains, one needs to first classify neighborhoods based on a characteristic of interest. For instance, to examine the transition question for African American neighborhoods between 1990 and 2000, one would first establish an appropriate percentage threshold for a neighborhood to be considered an African American neighborhood. In this exercise three states of African American neighborhoods are established: high (90–100 percent), medium (10–89 percent), and low (0–9 percent). Next, an examination is made of every two-state combination (e.g., low-to-low African American) for the 1990 to 2000 transition period. As a result, a total of nine two-state combinations are converted into a probability matrix of the form:

2000 State of Percentage African American

		(1) Low	(2) Medium	(3) High	
1990 State of Percent African American	(1) Low	P_{11}	P_{12}	P_{13}	1.0
	(2) Medium	P_{21}	P_{22}	P_{23}	1.0
	(3) High	P_{31}	P_{31}	P_{33}	1.0

where P_{11}, the first cell, represents the probability of a neighborhood starting in 1990 with a low percentage of African Americans and ending the period with a similarly low percentage of African Americans. Cell entries in the matrix are calculated by the relative frequency definition of probability:

$$P_{ij} = \frac{f_{ij}}{\sum_{j=1}^{3} f_{ij}}$$

where P_{ij} represents the probability of a neighborhood moving from one African American level to another, f_{ij} is the observed number of transitions from state i to state j. 3 is the number of states in this exercise.

Readings:

(1) Bourne, L. S. (1971) "Physical adjustment processes and land use succession." *Economic Geography* 47:1–5.

(2) Clark, W. A. V. (1965) "Markov chain analysis in geography: An application to the movement of rental housing areas." *Annals of the Association of American Geographers* 55:351–359.

(3) Clark, W. A. V. (1993) "Neighborhood transitions in multiethnic/racial contexts." *Journal of Urban Affairs* 15:161–172.

(4) Greene, R. P. (1994) "Poverty area instability: The case of Chicago." *Urban Geography* 15:362–375.

(5) Greene, R. P., and Pick, J. B. *Exploring the Urban Community: A GIS Approach*, Chapter 6.

(6) Hunter, A. (1974) "Community change: A stochastic analysis of Chicago's local communities, 1930–1960." *American Journal of Sociology* 79:923–947.

Exercise Procedure Flowchart

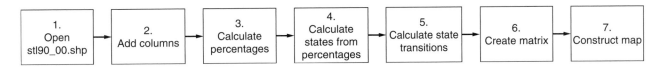

Exercise Procedure

Part 1

1. Start ArcMap with Start Using ArcMap With A New Empty Map selected. Click OK.

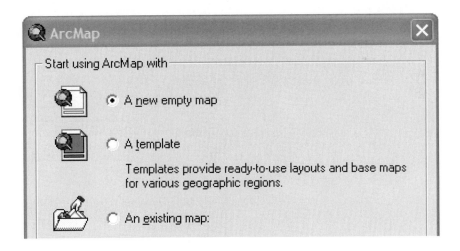

2. Click Add Data and navigate to the exercise data folder.

Note: *If you do not see the folder where you placed the exercise data, you will need to make a new connection to the root directory that holds the data.*

3. Add stl90_00.shp from the Chapter 6 folder.
4. Right-click the name stl90_00 and click Open Attribute Table.

5. Add five columns by clicking Options and selecting Add Field.
   5.1. Add Zblack90 as float with a precision (number of digits) of 12 and a scale (number of decimal places) of 3.

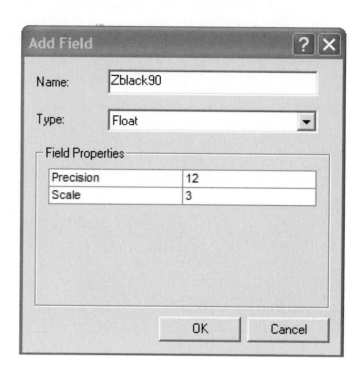

5.2. Add Zblack00 as float with a precision of 12 with 3 decimals.

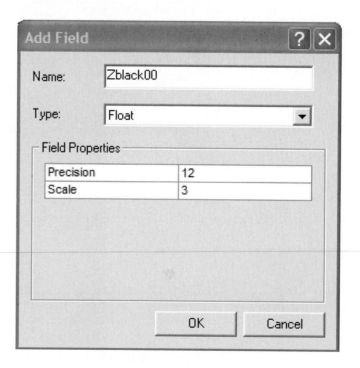

5.3. Add Class90 as text with a length of 1.

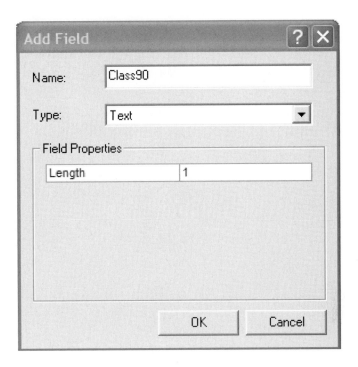

5.4. Add Class00 as text with a length of 1.

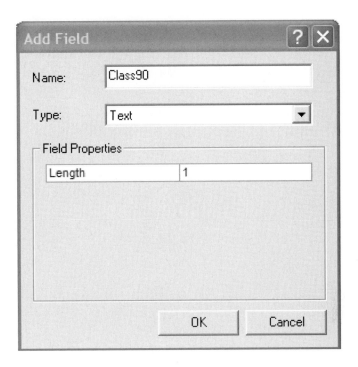

5.5. Add Chg9000 as text with a length of 2.

6. Calculate the proportion of African Americans in 1990 and 2000:
    6.1. Click Options, Select By Attributes, Pop_90 > 0, and then Apply. Click OK and then click Close.
          (This step is necessary because a tract might have zero population and you cannot divide by zero in the next step.)

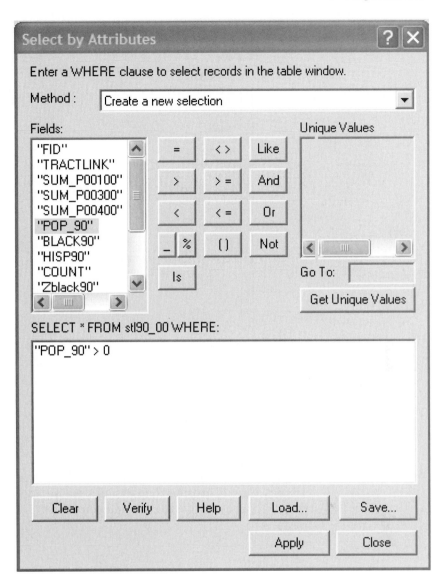

6.2. Scroll over to the right in the table, right-click the Zblack90 field title, and then click Calculate Values. Answer Yes to calculate outside an edit session and enter the following calculation: Black90/Pop_90 * 100.

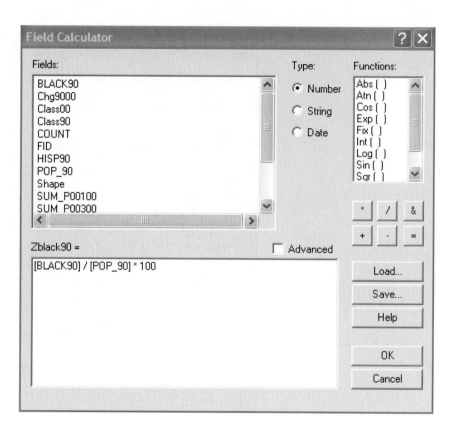

6.3. Click OK.
6.4. Click Options, Select By Attributes, SUM_P00100 > 0, and Apply. Click OK and then click Close. This step is necessary because a tract's 2000 population (SUM_P00100) might have zero population and you cannot divide by zero in the next step.

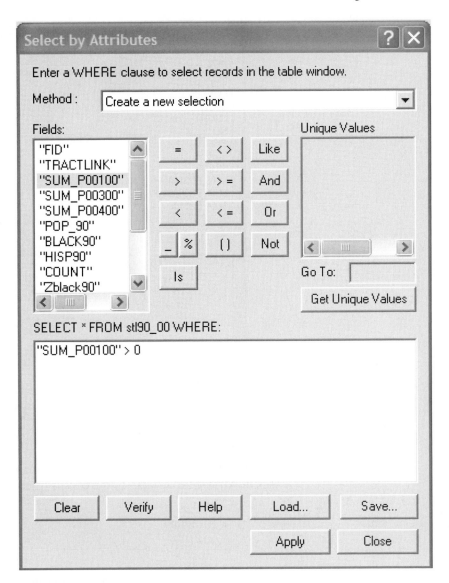

6.5. Scroll over to the right in the table, right-click the Zblack00 field title, and click Calculate Values. Answer Yes to calculate outside an edit session and enter the following calculation: SUM_P00300/SUM_P00100*100. Click OK. The SUM_P00300 field is the 2000 African American population and SUM_P00100 is the total 2000 population.

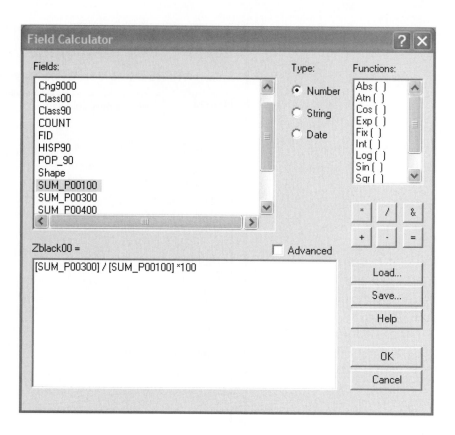

6.6. For Class90 and Class00, we want to assign a letter A, B, or C based on the percentages of African Americans in the census tract.

$$A = 0{-}9.99$$
$$B = 10{-}89.99$$
$$C = 90{-}100$$

1990 A Neighborhood Calculations

6.6.1. Click Options and click Select By Attributes.

6.6.2. Double-click the field ZBlack90, click the operation less than (<), and type 10. Your query should read "Zblack90" < 10

6.6.3. Click Apply and then click Close to generate a new selection.

6.6.4. Click Selected, which appears to the right of Show at the bottom of the attribute table.

6.6.5. Right-click the heading of the Class90 field and click Calculate Values. Type "A" in the calculation space and click OK.

1990 B Neighborhood Calculations

6.6.6. Click Options and click Select By Attributes. The next query should read "Zblack90" > =10 and "Zblack90" < 90

6.6.7. Click Apply and then click Close to generate a new selection.

6.6.8. Click Selected, which appears to the right of Show at the bottom of the attribute table.

6.6.9. Right-click the heading of the Class90 field and click Calculate Values. Type "B" in the calculation space and click OK.

1990 C Neighborhood Calculations

6.6.10. Click Options and click Select By Attributes. The next query should read "Zblack90" > =90

6.6.11. Click Apply and then click Close to generate a new selection.

6.6.12. Click Selected, which appears to the right of Show at the bottom of the attribute table.

6.6.13. Right-click the heading of the Class90 field and click Calculate Values. Type "C" in the calculation space and click OK.

2000 A Neighborhood Calculations

6.6.14. Click Options and click Select By Attributes. The next query should read "Zblack00" < 10

6.6.15. Click Apply and then click Close to generate a new selection.

6.6.16. Click Selected, which appears to the right of Show at the bottom of the attribute table.

6.6.17. Right-click the heading of the Class00 field and click Calculate Values. Type "A" in the calculation space and click OK.

2000 B Neighborhood Calculations

6.6.18. Click Options and click Select By Attributes. The next query should read "Zblack00" > =10 and "Zblack00" < 90

6.6.19. Click Apply and then click Close to generate a new selection.

6.6.20. Click Selected, which appears to the right of Show at the bottom of the attribute table.

6.6.21. Right-click the heading of the Class00 field and click Calculate Values. Type "B" in the calculation space and click OK.

2000 C Neighborhood Calculations

6.6.22. Click Options and click Select By Attributes. The next query should read "Zblack00" > =90

6.6.23. Click Apply and then click Close to generate a new selection.

6.6.24. Click Selected, which appears to the right of Show at the bottom of the attribute table.

6.6.25. Right-click the heading of the Class00 field and click Calculate Values. Type "C" in the calculation space and click OK.

7. Click Show All at the bottom of the attribute table.
8. Click Options and click Clear Selection.
9. Right-click the heading of the Chg9000 field and click Calculate Values. Calculate Chg9000 = Class90 + Class00 (under Type, choose String) and click OK.
10. Determine the probability of a B tract in 1990 making the transition to a C tract in 2000.

   10.1. Under Options, click Select By Attributes and enter the following query:
   Chg9000 = 'BA' (note that 2 get selected)
   Click the Get Unique Values tab to see all values
   Click Apply
   Repeat for next B transition:
   Chg9000 = 'BB' (note that 82 get selected)
   Click Apply
   Repeat for next B transition:
   Chg9000 = 'BC' (note that 5 get selected)
   Click Apply

10.2. Close the table.

We can enter the above three entries into our matrix:

		(1) Low	*(2)* Medium	*(3)* High	
$P =$	*(1)* Low				
	(2) edium	2	82	5	**89 total**
	(3) High				

Using the relative frequency definition of probability we calculate:

2000 State of Percentage African American

		(1) Low	*(2)* Medium	*(3)* High	
1990 State of Percent African American	*(1)* Low	P_{11}	P_{12}	P_{13}	1.0
	(2) Medium	.02	.92	.06	1.0
	(3) High	P_{31}	P_{32}	P_{33}	1.0

We find that the probability of a tract with a medium percentage of African Americans has a low probability of making a transition to a high percentage (.06). In fact, if a tract is of a medium percentage African American, it is highly likely to remain medium in the next decade (.92). Fill in the remainder of the probability matrix using the selection procedure previously above for the B transitions.

11. Mapping the five tracts that made the transition from B to C (BC) reveals an interesting pattern, namely that they are all concentrated around neighborhoods that were highly African American in both 1990 and 2000. This would suggest that although medium-percentage African American tracts have a low probability of making the transition to high-percentage African American tracts, they have a much higher probability if they are adjacent to high-percentage African American tracts. Let's make the map to see this visually.

11.1. Right-click Stl90_00.shp and click as follows:

11.1.1. Properties

11.1.2. Symbology

11.1.3. Categories

11.1.4. Unique Values

11.1.5. Change Value Field to chg9000

11.1.6. Click Add Values tab

11.1.7. Select 'BC' and 'CC' and click OK

11.1.7. Click Apply and then click OK

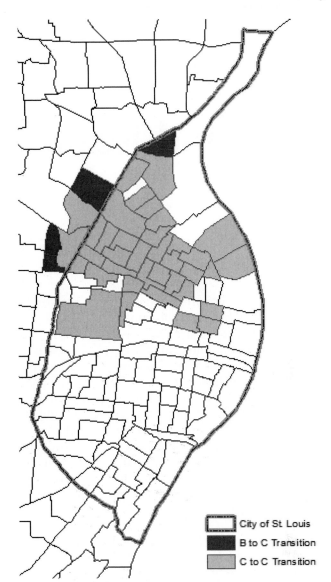

City of St. Louis

B to C Transition

C to C Transition

12. Experiment by mapping other transitions. Map at least two transitions. Note that you can remove a value from the map by again performing steps 11.1.1 and 11.1.2, and then right clicking on the value and selecting "remove value(s)." To replicate this map, add the St. Louis city layer to the map and change its symbol from a solid fill to a hollow fill with a thick outline.

Part 2: Short Essay

Using the maps and probability matrix you created in Part 1, the Chapter 6 section, "Analyzing an Urban Issue," and readings cited in the Background section, write a short essay describing the spatial pattern of African American neighborhood transitions in the 1990s in the city of St. Louis. Try to explain transition patterns and trends. Discuss the strengths and weaknesses of the methodology employed to measure these transitions.

7

Migration and Residential Mobility

Population processes regulate the growth and decline of cities. Within a city, patterns of births, deaths, mobility, and migration contribute to the city's size, shape, immigrant communities, housing supply and demand, and educational needs. In national systems, cities and nonmetropolitan areas are in dynamic flux determined by migration flows. The Sunbelt flow in the United States is a well-known example of migration-driven national flows, with impacts on sending and receiving cities. Between nations, immigration has impacts on sending and receiving countries, and augments the ethnic diversity that is emphasized in the next chapter. This chapter focuses on migration and residential mobility, including effects on city processes, such as transport, education, and labor supply. Migration is on the rise in the world, as trade barriers fall and larger market systems lead to bigger and more mobile labor pools (United Nations, 2002). This has also prompted some nations to discourage and restrict migration from other counties. Other countries such as China have restricted internal migration.

The chapter first considers the basic concepts and factors of migration—the push and pull factors that lead individuals, families, and groups to pull up stakes and take the risks of moving. It next turns to residential mobility, which consists of shorter distance moves that often depend on a combination of housing, family life cycle, and jobs. Building on the Chapter 4 discussion of the structure of cities (e.g., that the CBD or other zones attract workers), this chapter examines the bias of groups of residential moves toward poles of attraction in cities, a concept termed *directional bias*.

The "Analyzing an Urban Issue" section addresses the problem of how accessible a location is, based on the method of population potential. The practical problem is how accessible four proposed sites in Los Angeles are for siting a center for education about the environment for non-English-speaking residents. You will base your decision on the site's accessibility for the eligible population living in the city areas surrounding each site.

National migration flows in the late 20th century for the United States and Australia are examined and compared. They are driven by economic and geographic forces that differ for each nation and have important impacts on cities.

International immigration is on the rise in the world, which has 175 million lifetime international migrants, mostly living in cities (United Nations, 2002). A lifetime international migrant refers to the total of migrants who have migrated from their country of origin to another country at any point during their lifetime. At the same time, 40 percent of nations have set policies to lower immigration (United Nations, 2002). Those countries represent a range of sizes and development levels. The complex forces of international migration include economics, politics, cultural divisions, conflict, and warfare. In the United States, the largest recent groups of immigrants are Hispanics and Asians, streams that have impacted cities and national life as well. A case looks at the large stream of immigrants from the Dominican Republic to the United States, especially to New York and New Jersey, and their assimilation and adjustments.

Finally, Chinese rural-to-urban migration is studied. China has a distinctive system that controls migration through residence registration requirements, but the country is experiencing greater movement through its change to a socialist market economy. The resultant "floating population" (i.e., rural-to-urban migrants lacking residence papers) is not supported by the government and often hard-pressed to make ends meet. The chapter should alert you to how migration and residential mobility alter the size and spatial arrangement of city populations, and provide you with a basis to understand upcoming chapters on ethnic changes and the urban fringe.

7.1 CONCEPTS AND FACTORS OF MIGRATION

Migration in its fundamental aspect is the process of human movement from one community, state, or nation to another. It constitutes one of the three standard components of population change. The other change components are births and deaths. For cities, annexation can be added. For a geographic unit with fixed boundaries, that is, without annexation, the population change is calculated as births minus deaths plus entering migrants minus departing migrants. Migration differs from births and deaths in not being a biological event, although all are social events. It has complex causes and can be more variable than the other components of population change.

A useful classification of migration divides it into four categories, by whether it is recurrent versus nonrecurrent and local versus long distance (Jones, 1990, adapted from Duncan, 1959). *Recurrent movement* is movement that repeats on a regular basis, such as commuting, daily crossing of an international border to work, or seasonal movements of farm labor. It is commonly based on the desire to reach a place of work that might be distant from the worker's residence. Thus, local recurrent movement constitutes *commuting*, whereas long-distance movement is *seasonal work*.

In Table 7.1, *nonrecurrent movement* represents a one-time move, for example, a family moving from Mexico to the United States. This is divided into local moves, referred to as *residential mobility,* and long-distance moves, referred to as *migration* (see Figure 7.1). Local is commonly taken to mean within the metropolitan area boundary, although it can be defined as movement within a county, province, or other unit.

For measurement of local movement, the size of the local geographic unit is critical. For a large unit such as a big county, more movement will be classified as local and less as long distance than with a small county. Thus, the local versus long-distance designations are to some extent arbitrary and depend on the unit size.

The U.S. Census considers residential mobility as a change of place of residence, either within a one-year period or five-year period (U.S. Census, 2003). It considers migration a change of a person's residence from five years earlier that crosses a jurisdictional boundary, so for the census, mobility is a form of migration. The census classifies migration into migration between counties

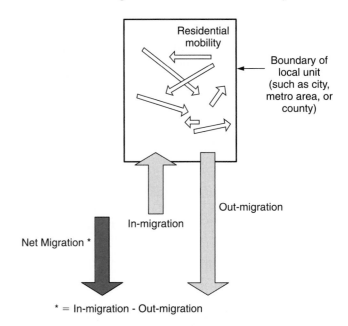

FIGURE 7.1 Migration and residential mobility.

(intercounty), between states (interstate), or between countries (migration to or from abroad; U.S. Census, 2003).

In this textbook, we refer to movement within metropolitan areas or other urban units such as urban areas and cities as *mobility*. Movement between metropolitan areas or between large units such as states and nations is referred to as *migration*.

For a geographical unit, as seen in Figure 7.1, a stream of migrants arrives (immigration) and another stream leaves (out-migration). The difference of in-migration minus out-migration is called *net migration,* which can be either positive or negative depending on whether immigration exceeds out-migration or not. Owing to unavailability of specific migration data, most planning and analysis studies are done in terms of net migration. However, many times, in-migration and out-migration are more meaningful. Consider the difference between the stream of migrants from Mexico to the United States and that from the United States to Mexico. Migrants in the two streams are different in average age, occupation, income, nationality, and motives for migration. For certain purposes, such as producing population estimates and projections, it might be more meaningful to focus on net migration. For other purposes, such as assessing the impacts and adjustments of Mexicans in the United States, the gross number of immigrants in a single directional stream is better.

7.1.1 Factors in Migrating

This section examines economic, social, and geographic factors that encourage or discourage residential mobility and migration. It also considers forced migration.

TABLE 7.1 Categories of movement

	Non-recurrent	Recurrent
Local	Residential mobility	Commuting
Long distance	Migration	Seasonal work

Source: Jones, Huw. 1990. Population Geography. 2nd Ed. Paul Chapman Publishing Ltd. P. 179. Table 8.1. "A classification of spatial movements of population."

Economic Factors

The economic situations at the migrant's origin and destination are crucial to decisions about moving. Factors at the origin, often called *push factors,* include economic downturns, unemployment, and low wages, whereas those at the destination are *pull factors,* including prosperity, presence of jobs, and especially presence of jobs that fit the immigrant's skills. An example of a combination of push and pull is the hourly income differential between the United States and Mexico, which is approximately eightfold. Such a huge economic differential might make up for migration obstacles, and in fact has led to huge immigrant flows. Low-wage workers in Mexico are shown in Figure 7.2.

Social and Demographic Factors

These include age, education, cultural background, language, social networks, and proclivity to chronic moving. For age, numerous studies in many countries and regions have shown that younger adults, especially those in their 20s, are most likely to move. Many move related to their college and graduate education, or to initial job opportunities. Because they often have young families, there are associated high moving rates for young children. From ages from the 30s into the 80s, there is a steady drop in migration rates due to life-cycle changes. For example, in the United States during 1999–2000 (see

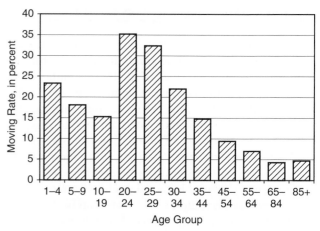

Moving Rate is the ratio of persons changing residence to the total number of persons, expressed in percent.

FIGURE 7.3 Moving rate by age, United States, 2000–2001. Source Schachter, 2001. Data from U.S. Census Current Population Survey of 2000.

Figure 7.3), 35 percent of 20- to 24-year-olds moved, whereas only 5 percent of people 85 or older did so (Schachter, 2001). The youngest children moved at a 23 percent rate, nearly as high as the rate for young parents. Older people, most of them with community and family ties, established careers, or retirement arrangements in their locale, are much more rooted.

FIGURE 7.2 Low-wage workers in Mexico City.

With higher levels of education, people tend to have greater job market opportunities and geographic reach, leading to higher moving rates and longer distances. By contrast, uneducated and often unskilled persons have limited geographic market opportunities. Similarities in culture and language at origin and destination are positive moving factors, as they help in assimilation and acculturation at the destination. A migrant's social network also is a positive factor. The potential inclusion into a group of new friends and family is a pull; the network potentially offers identity, support, advice, and even a head start on employment and social mobility. At the same time, it is often hard for the prospective migrant to obtain data on such networks. Immigrant families, such as the one seen in Figure 7.4, usually go through many years of settlement and adjustment at their destination and are helped by networks. Some movers, known as *chronic movers,* might move dozens of times during their lives. A chronic mover has an elevated proclivity to move on the basis of his or her prior experience. This behavioral trait is more common for shorter distance residential mobility.

Geographical Factors

Locations of origin and destination, topography, transportation networks, climate, proximity to oceans, and other geographical features can impact moving. For instance, Duluth, Minnesota, receives less migration than some otherwise comparable cities because of a frigid winter climate and remote location. Directional bias, which is the tendency for residential moves throughout an area to point toward metropolitan poles, is discussed later.

Rural-to-urban migration refers to migration from rural to urban areas. Such migration streams can continue for decades, and are more prevalent in developing economies where a large share of the population is still rural. As rural agriculture becomes more mechanized and efficient, traditional agricultural workers have fewer jobs available and their livelihood suffers. For many, the economic attraction of cities pulls them from impoverished rural settings to seek work in the city. For instance, in the United States in most of the 19th century, there was continuing flow from rural areas to cities, massive immigration of Europeans mostly from their rural zones, and the high rates of natural increase mentioned in Chapter 2. Many large American cities expanded as a result. In Mexico from the 1920s to today, there has been steady rural-to-urban migration, constituting a major factor in the growth of Mexico City, Guadalajara, the northern border cities, and others. Many of today's Mexican farmers, such as the one appearing in Figure 7.5, face continuing pressures to migrate to urban areas. A recent study of rural-to-urban migration in West Africa found that once the rural migrants arrive they adapt quite well to their new urban setting (Beauchemin and Bocquier, 2004).

FIGURE 7.4   Immigrant family from India, in their home. Source: © James Marshall/CORBIS.

FIGURE 7.5   Mexican farmer in Chiapas, Mexico. Source: Getty Images Inc.–Image Bank.

7.1.2 Forced Migration

Forced migration is migration through coercion at the point of origin. The coercion might stem from a wide variety of factors, such as extreme economic privation, environmental catastrophe, ethnic and racial conflict, persecution by despotic rulers, war, or decisions by immigration authorities to force certain residents to be deported or repatriated (see Figure 7.6). To be forced, the extreme economic privation goes beyond severe poverty to also involve risk to health and existence. Normal factors of age, education, employment, and marital status, among others, might condition a migrant's decision to submit to coercion factors. The scale of the forced migration might vary. For instance, the causes can apply to an individual or to a whole segment of a population such as a tribe or ethnic group. Fundamental to forced migration is the belief by individuals that flight is mandatory (Wood, 1994). Forced migrants might enter another nation as legal or illegal immigrants, or

under the special categories of refugee or asylum seeker Wood, 1994).

The number of lifetime forced migrants worldwide in the mid-1990s was estimated at 43 million (U.S. Committee on Refugees, 1993). "Lifetime" means that they were forced to migrate at some point during their life. It is probably roughly at the same level today. Origin nations sending out the most refugees are the poorest ones, usually subject to intense military conflict, and mostly located in the Middle East, Africa, or southeast Asia (Wood, 1994; U.S. Committee for Refugees, 2003). In 2002, Afghanistan was the largest nation as a source of current refugees at 3.5 million (U.S. Committee for Refugees, 2003). Forced migrations can involve mass exodus from cities or mass arrivals. An example was the 1970s forced mass immigration of more than 100,000 refugees from Vietnam to the United States, with a large share eventually ending up in Orange County, California. After the Vietnam War ended in 1975, many Vietnamese immigrants came to the United States through Camp Pendleton, just south of Orange County, and later mostly were sent to sponsors in many parts of the country (Allen & Turner, 2002). However, eventually a large number of naturalized families returnmigrated to Orange County, forming the large Vietnamese community centered in Westminster and Garden Grove known as Little Saigon, which is pictured in Figure 7.7. The location of Vietnamese population in southern California is shown in Figure 7.8. Today, the migrants and their descendents form the largest single Vietnamese immigrant community in the United States at 135,000 and have a significant impact on Orange County's social services, employment, economy, and culture (Allen & Turner, 2002). This community represents one eighth of the country's 1.12 million Vietnamese residents.

Many of today's proven migration factors were recognized more than a century ago by the British statistician E. G. Ravenstein. He identified the migration principles of rural-to-urban migration, economic push and pull, outmigration versus in-migration, return migration, patterns of short-distance moves, and many others. Consider these excerpts from three of the "laws" of Ravenstein (1885), along with comments on their relevance today.

1. "Each main current of migration produces a compensating countercurrent."

   This refers to the well-known tendency for return migration, or the proclivity of a migrant from a certain point of origin to return to that origin.

2. "Migrants proceeding long distances generally go by preference to one of the great centers of commerce and industry."

   This refers to the tendency for immigrants to enter large urban areas. For instance, in the past two decades U.S. immigrants have tended to enter Los Angeles, Houston, Chicago, Washington, DC, and other large urban areas.

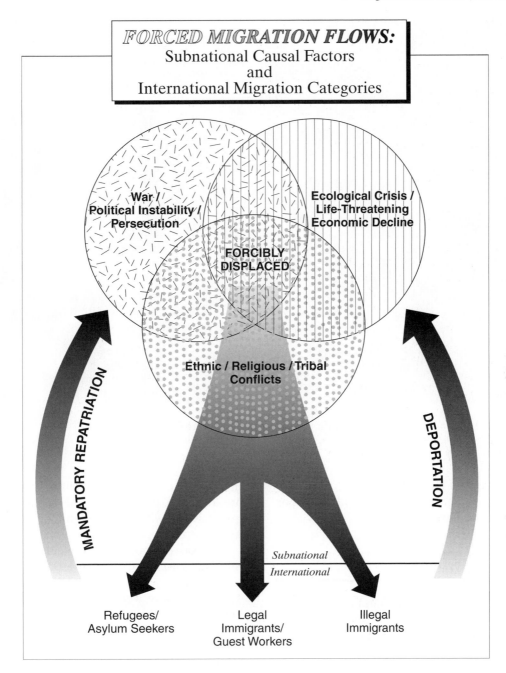

FIGURE 7.6   Model of forced migration.

3. "The great body of our migrations only proceed a short distance."

The preponderance of movement is residential mobility, involving short-distance moves often of less than 10 miles. For instance, in the U.S. Census of 2000, 43 percent of Americans changed residence in the same geographic area over the prior five years, but only 3 percent immigrated from abroad.

What differs today from Ravenstein's time is the collection by governments and international agencies of large amounts of migration data from such varied sources as population registries, censuses, special surveys, and administrative sources, including drivers' licenses and health care systems. The Current Population Survey (CPS), a national survey conducted annually by the U.S. Census, has been valuable in documenting multiyear trends in U.S. migration including frequency, social determinants, and regional flows. Likewise in 2002, the U.N. Population Division surveyed international migration for 239 countries or areas to yield an unprecedented

FIGURE 7.7   Mall food court in Little Saigon, California. Source: PhotoEdit.

range of worldwide data (United Nations, 2002). Hundreds of other nations' surveys, censuses, and reports on migration are also available annually.

7.2 MEASUREMENT AND MODELING OF MIGRATION

This section concentrates on the measurement of net migration and the gravity model. Both of these are useful for estimating migrants coming into or out of urban areas. *Net migration* is calculated by the equation

$$M_n = (P_{t+n} - P_t) - (B_n - D_n)$$

where M_n represents the migrants between time t and $t + n$,
  P is the population at a particular time, and
  B and D are births and deaths during period n.

The *vital statistics method* utilizes this equation to calculate net migration. The method is fairly easy to apply in the United States for counties or larger regions, as county population totals are available from the Census Bureau and county births and deaths are collected by the National Center for Health Statistics

in cooperation with states. If sufficient data by age group are available, they can also be applied for a single age group, which will yield the number of migrants for that age group.

An example would be to calculate the number of net migrants into Big City from 1995 to 2000. The population of Big City in 1995 was 1,250,000 and in 2000 was 1,550,000. There were yearly averages of 25,000 births and 18,000 deaths during the five years. Net migration is calculated as follows:

$$M_{1990-2000} = (1,550,000 - 1,250,000) - 5*(25,000 - 18,000)$$
$$= 265,000$$

Another fundamental method for modeling migration is the *gravity model* (Jones, 1990; Zipf, 1949). The equation for this model is:

$$M_{ij} = k(P_i \times P_j)/D_{ij}$$

where M_{ij} represents the number of migrants flowing between point i and j,
  P is the population at point i or j,
  D_{ij} is the distance between points i and j,
  and k is a constant.

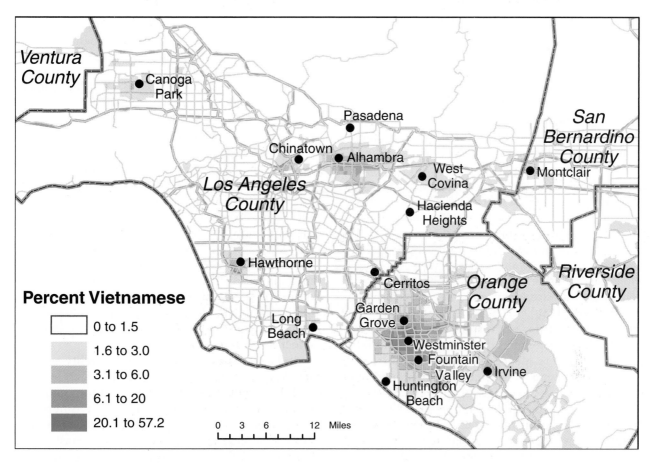

FIGURE 7.8   Location of Vietnamese population in southern California, 2000.

P_i is the population at origin, which represents a pool of possible movers. In other words, it represents the people at origin subject to "push" factors. The P_j at destination represents its population attraction ("pull"). The distance between points i and j serves as a proxy for the magnitude of intervening factors that reduce the likelihood that migration will take place. The longer the distance, the harder to break the personal bonds at origin, establish new ones and a social network at destination, and obtain complete information about the destination and its opportunities, among other things. Sometimes in the model, the population (P) variables are replaced by economic (E) variables.

The gravity model is useful in gauging migration flows into or out of urban areas. It helps in conceptualizing that a longer distance of move reduces migration and that both origin and destination population or economic features influence moves. For instance, international migration from Chile to the United States is at a very low level, as reflected in the resident population of only 68,849 persons of Chilean origin in the United States in 2000, one of the smallest populations from Latin America (U.S. Census, 2001). The gravity model predicts a small amount of

migration because the population of Chile at 15.8 million in 2003 is relatively small (P_j in the numerator) and the distance between the United States and Chile (roughly 5,200 miles) in the denominator is large.

The gravity model was developed further by Stouffer (1940) and other later modelers, who substituted the number of intervening opportunities for distance. Intervening opportunities, such as competing employment centers, government restrictions, startup costs at the destination, and transportation costs, interfere with successfully moving between the origin and destination. In the example of migration from Chile to the United States, the intervening opportunities include Chilean restrictions on out-migration, U.S. restrictions on immigration, moving costs, capital costs of housing in the United States, and better opportunities for employment in Chile or closer to it. Other kinds of models to measure migration change, such as regression, forecasting, and use of administrative databases, are beyond the scope of this book, but they are useful to students who go further in urban geography (Jones, 1990; Siegel & Swanson, 2004). The two basic models are data friendly and fairly easy to apply.

7.3 BEHAVIORAL PROCESSES FOR RESIDENTIAL MOBILITY

Residential mobility is movement to a new residence within a city, metropolitan area, county, or province. It is different from migration in that the mover does not move across the boundary of the geographic unit of origin. These moves tend to be short and more frequent than for migration. Residential mobility has significant social and behavioral impacts on people, although, unlike migration, it often leaves intact the social network and familial ties, and frequently leaves intact a job. Residential mobility is driven by *adjustment causes*, that is, response to temporary economic and market fluctuations, and *induced causes*, that is, responses to permanent or semipermanent aspects of jobs, the economy, or the life cycle of workers (Clark, 1986). These factors in residential mobility are summarized in Figure 7.9, modified from Clark (1986).

Adjustment factors are responses to temporary changes in housing, neighborhoods, and accessibility. Consider your own residence and whether you are satisfied with your present housing unit, neighborhood, and transportation access to your usual destinations. If you are dissatisfied with them, you are more likely to move. For housing adjustment, the subfactors include its housing space, design, financing, and maintenance. For instance, a family with new children might outgrow its home, necessitating a move. Housing, seen in Figure 7.10 for a neighborhood in northern Seattle, can be a factor in family mobility. Housing prices in your neighborhood might rise, implying it is a good time financially to move to another part of the city. As discussed in Chapter 6, neighborhood attractiveness or unattractiveness to its residents depends on the neighborhood's position in its cycle, physical appeal, public services such as schools and parks, income, crime levels, and social and ethnic characteristics. Accessibility factors include physical access to work, friends, family, schools, shopping, and leisure activities.

Government policy influences the adjustment factors. For instance, the federal government impacts mortgage markets through the mortgage institutions it sponsors such as the Federal Housing Authority, Fannie Mae, Freddie Mac, and veterans mortgage benefits. Those markets play a role in determining housing cost. Another example is the interstate highway program that was initiated in the 1950s by the Eisenhower administration. The presence of these major highways in metropolitan areas has affected residential commuting patterns and the factor of accessibility. State and local government policies also affect the factors.

Induced reasons for residential mobility are, broadly speaking, long-term, unavoidable ones. They include jobs and the family life cycle. Changes in jobs and employment that are involuntary stem from business and job market factors that are often beyond the control of the individual, such as downsizing, layoffs, transfers of people and facilities, and mergers. Any of these might necessitate residential mobility to be closer to a transferred job or to change jobs.

The family life cycle approach refers to the impact of an individual's life course on residential change. A general model of this is seen in Figure 7.11. For example, let us consider an individual who is making residential moves throughout her life cycle. Growing up in the original parental household, this individual moved several times with her family. At college, she changed residence to live in a college dormitory. After college she started a sequence of jobs and changed apartments several times to be near her work. After getting married, she and her husband moved to a small home. With the birth of her second child, the family outgrew their first home, and upgraded by selling it and buying a larger one. Later on, the family did well economically, enabling them again to upgrade to a larger home in a nicer neighborhood. Unfortunately, the parents later divorced, and the husband moved to a condominium across the city. The wife remarried and her new husband moved into her home. Late in life, that couple, now empty nesters, decided they had too much space, downsizing to an apartment. Finally they moved into an assisted-living facility. In all, this family life cycle consisted of a dozen residential moves, which were strongly induced at particular time points.

The most important reasons for moving are housing adjustment, neighborhood adjustment, and life-cycle

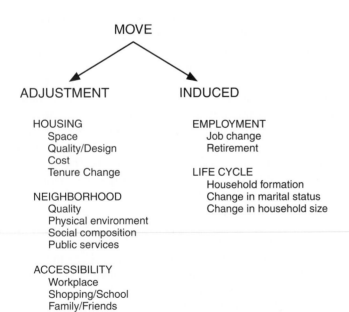

FIGURE 7.9   Factors in residential mobility. Source: Clark, William A.V. 1986. *Human Migration*. Beverly Hills: SAGE Publications. P. 44. Figure 2.8. "Reasons for household relocation."

FIGURE 7.10 Housing in north suburban Seattle: migration factor.

change, each accounting for 15 to 45 percent of reasons for residential change (Clark, 1986). All this leads to substantial residential mobility in the United States. The number of individuals one-year old and older moving in the United States during the 1990s remained steady at 41 million to 44 million yearly (Schachter, 2001), and the percentage who moved, including migrants, was consistently between 16 and 17 percent (U.S. Census, 2001). Compared with other countries, this ratio of one in six persons moving each year is quite high. It compares to Australia's 18 percent residential mobility rate in 2000–2001 (Hugo, 2003), but is double the rates of Britain and Japan. For any of these nations, the residential causes can be ascribed to the reasons given earlier. Britain's small land area might constrain mobility.

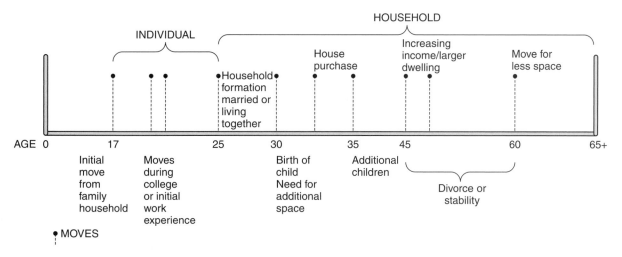

FIGURE 7.11 General model of family life cycle for residential mobility. Source: Clark, William A.V. 1986. *Human Migration*. Beverly Hills: SAGE Publications. P. 39. Figure 2.6. "A Life Cycle Perspective on Mobility."

The reasons the particular nations vary in residential mobility are complex, and depend on many geographical, cultural, and demographic factors. For instance, Japan has one of the world's oldest population age structures, with 19 percent of the Japanese population over age 65 versus 12 percent for the United States and 13 percent for Australia (Population Reference Bureau, 2004). Earlier the chapter mentioned that old age tends to reduce migration rates. Japan's rooted family traditions are another factor. For Mexico, culture is also important. Mexico's residential mobility rate of 5 percent every five years (measured between 1995 and 2000 and summarized by López Villar, 2003) is much lower than that for the advanced nations

mentioned, even though its age structure is much younger, with only 5 percent over age 65. More than offsetting this, however, are Mexico's strong cultural traditions of family and residential stability.

7.4 DIRECTIONAL BIAS IN MOVING

Residential mobility consists of many short-distance moves. Figure 7.12 (top) shows a hypothetical set of directional residential moves for a section of Portland, Oregon. If all the moves are lined up and altered to have the same direction (Figure 7.12, bottom), it is clear that most of the moves are for shorter distances, with only a few

Part A. This shows the residential moves for city section.

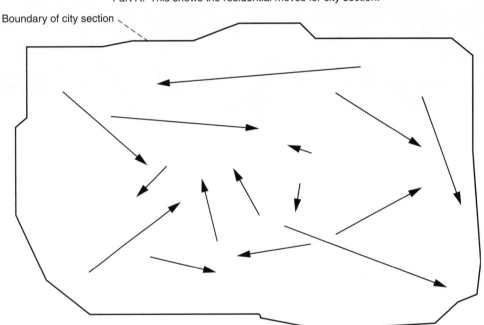

Part B. This shows the moves shifted to a common point of origin.

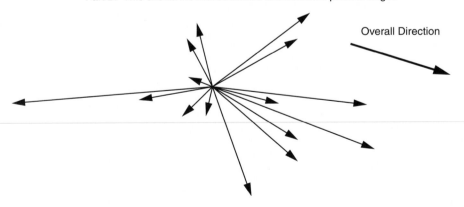

FIGURE 7.12 Hypothetical residential moves in a section of Portland, Oregon. Source: Clark, William A.V. 1986. *Human Migration*. Beverly Hills: SAGE Publications. P. 30. Figure 2.1a. "A sample of household moves in the San Fernando Valley of Los Angeles." Figure 2.1b. "Samples of moves from Figure 2.1a plotted with common (hypothesized) origin."

for long distances. The curve for the distribution of residential moves resembles an exponential distribution (Clark, 1986). It is beyond the scope of this book to discuss the formula for this curve and why it is known to apply (see Clark, 1986). The important point is that there are many more short-distance residential moves. The short distance stems from people and families trying to maintain existing neighborhood and social network ties as much as possible.

Another way to rearrange this map of moves is to line all the moves up at the same point of origin, while retaining their original directions (Figure 7.12, bottom). Now it is evident that the moves as a group have an overall average direction, which is shown by the thicker arrow. In this illustration, the overall direction points toward the central city, which is the urban area's largest accessible employment center.

Directional bias is the overall average direction of a group of residential moves. It was originally studied in the late 1960s, starting with seminal papers including that of Adams (1969). Studies have shown that directional bias in cities tends to be toward the CBD or toward suburbs that have employment opportunities.

Today, directional bias also needs to take into account a wider range of models, including the large developing world cities. For instance, as discussed in Chapters 2 and 3, Mexico City expanded in the past three decades through urban rings that have expanded outward while the city center has depopulated. The rings contain new employment centers, such as the booming Santa Fe corporate center. The city's directional bias has not been studied, but it is likely some of its direction is toward these expanding rings, rather than either the CBD or the poverty-stricken outer periphery. More directional bias studies would be helpful in examining many developing world cities.

Research on directional bias ideally requires an inventory to be kept of the origins and destinations of all residential moves for a particular area and time period. Neither the U.S. Census nor state or local governments are likely to be able to gather this information, because respondents would object. The Internal Revenue Service might have some of the information but is prohibited from utilizing it for studies. Because of lack of data, the technique has not been widely used by city and regional planners. With better information sources, it would become a useful planning tool, because it sheds light with fine detail on the directional mobility trends within cities.

7.5 ANALYZING AN URBAN ISSUE: ACCESSIBILITY INDEX IN LOS ANGELES

We have seen that an important aspect of residential mobility is accessibility, or how accessible a residential location is to jobs, schools, friends, and relatives. A closely related concept is how accessible offices, businesses, schools, government offices, and so on are to the people they serve. Rather than a residence, here the central point is a business, social, or educational site.

To analyze accessibility, the following basic information is needed:

- Spatial location of the base site where the residence, business, or other unit is located.
- Spatial locations of all the access points.
- Measurement of the population associated with each access point.

Accessibility to such a base point can be analyzed by a method known as *population potential* (Geertman & van Eck, 1995; Handy & Niemeier, 1997; Liu & Zhu, 2004). The formula for population potential is

$$v_j = \sum_{i=1}^{r} (P_i/d_{ij}) \qquad (1)$$

where population potential v_j at any base site j is found by first taking the population, P_j, of every subarea i in the study area and dividing by the distance d_{ij} from its centroid to the base site. Then these ratios are summed up for all r subareas, $i = 1, 2, \ldots, r$. The *centroid* is the average of the x,y coordinate points for a zone, based on some characteristic.

Thus, if there is a large population with access to a site, but it is nearly all far away, this formula will considerably diminish the impact of the population on the site.

A simple use of population potential would be to decide where to locate a pizza parlor in a small city. Population potential could be calculated for 10 different prospective sites, chosen on other criteria such as transport and zoning. The one with the highest population potential would have top priority to be chosen.

The urban issue for this chapter applies population potential to determine the best among four potential existing building sites in Los Angeles in which to locate a center for environmental education for community clients largely from an ethnic minority population. Because the center serves a disadvantaged population, the population potential is calculated on the basis of minority population only. This chapter's exercise explains further how the four population potentials are calculated.

For the exercise, making siting decisions on the basis of real-world data and robust methods is better than just siting such centers on the basis of a "seat of the pants" approach. Geographic information systems (GIS) adds to the strength of the method, as it is possible to visualize the four sites and add in other layers, such as transport, terrain, high schools, environmental reserves, or protected zones.

More advanced accessibility analyses can add further refinements to the calculation of access. Among the other factors (Handy & Niemeier, 1997; Liu & Zhu, 2004) that can be taken in to account for calculating access are the following:

- Characteristics of the transportation system
- Timing of the activity (time of day or day of the week)
- Characteristics of the individual or group seeking the access
- Types of activities occurring at the access points
- The rate of decline of the travel attractiveness with distance.

An example of a more advanced population potential problem would be the following. Fast-food restaurants are being located at 20 test sites to serve the Phoenix metropolitan area. A more advanced population potential will be calculated for each site. The population potential for a fast-food location in this market is known to decline by the square of the distance from a given site. Again, a population potential formula is used to estimate the population potential of each of the 20 locations. The formula is based on the number of customers at the location. The formula for this example of population potential is:

$$v_j = \sum_{i=1}^{r} (C_i * t_{ij}/d_{ij}^2) \qquad (2)$$

where the population potential v_j at any site j is found by first taking the number of customers C_i of every fast-food area i in the study area; multiplying it by the traffic ease-of-flow value t_{ij} and dividing by the square of the distance d_{ij} from the study site to the base site.

As a population potential calculation gets more complicated, as seen in this example, the problems of data collection become more difficult. In a real-world planning situation, it is often better to stick with a simpler model because it is easier to understand. GIS is a great tool to use for estimating accessibility and population potential, because the user can look at the spatial locations of the base site, access points, and transportation systems before, during, and after the analysis. The exercise keeps it fairly straightforward, utilizing the simple formula (1). After building and running the model, in the essay questions, you evaluate the strengths and weaknesses of the model, and whether it makes sense to expand it.

7.6 MIGRATION PATTERNS AND TRENDS IN THE UNITED STATES AND AUSTRALIA

This part of the chapter explores recent trends in migration in the United States and Australia, especially during the 1990s. Both nations have had traditions of elevated internal migration. At the same time, their patterns are different, depending on contrasting economic, social, and geographical factors. The focus here is on the national patterns and how they affect cities and urban areas.

7.6.1 U.S. Internal Migration

The United States experienced high rates of interstate migration over the past quarter century. As seen in Table 7.2, about a fifth of Americans migrated to a different state every five years from 1975 to 1995, and another 4 to 5 percent of migrants were international migrants (Schachter, 2001). The overall rate of any type of movement was consistently about 45 percent.

Regionally, on a net basis, migrants left the Northeast and Midwest and moved to the South and West, until the 1990s, when the West's net migration lowered to about break even (see Table 7.3). The table also shows large-scale net movement of population from the other three regions to the South. Those migration streams are large, as seen by the 1.6 million persons who migrated both

TABLE 7.2 Moving rates in the United States, five-year periods from 1975 to 1995

Migration time period	Population 5 Yrs. and older (in 1000s)	No. of movers (in 1000s)	Moving rate (percent)				Percent of movers with previous residence in			
			Total	Inter-county	Inter-state	From abroad	Same country	Different county, same state	Different state	Abroad
1975–1980	210,323	97,629	46.4	19.5	9.7	1.9	54.0	42.0	20.9	4.0
1980–1985	216,108	90,126	41.7	17.8	8.7	1.8	53.1	42.6	20.8	4.3
1985–1990	230,446	107,649	46.7	19.1	9.4	2.2	54.5	40.8	20.1	4.7
1990–1995	241,805	106,616	44.1	17.0	8.2	2.2	56.7	38.5	18.5	4.9
Source: U.S. Census, 2000.										

TABLE 7.3 United States migration flows, in-migration, out-migration, and net internal migration by region, 1990–1995 (numbers in 1000s)

| From residence in 1990 | To residence in 1995 | | | | |
	Northeast	Midwest	South	West	Total out-migration
Total in-migration	1,162	2,191	4,682	2,269	
Northeast	—	387	1,586	505	2,478
Midwest	281	—	1,613	749	2,643
South	613	1,025	—	1,015	2,653
West	268	779	1,483	—	2,530
Net migration	–1,316	–452	2,029	–261	

Source: U.S. Census, 2000

from the Northeast to the South and from the Midwest to the South, and 1.5 million from the West to the South from 1990 to 1995 (Table 7.3). This is consistent with the long-term trend of migration to the South and West discussed in Chapter 1. A consequence has been the movement of the nation's centroid of population (i.e., weighted mean center) from Northeast to South. This is seen in Figure 7.13, which shows the center moving from south central Indiana in 1900 toward the southwest, reaching south central Illinois in 1950 and presently in south central Missouri.

U.S. Census data, in a different series emphasizing migration over a one-year period, demonstrated that singles and highly educated people are more likely to be

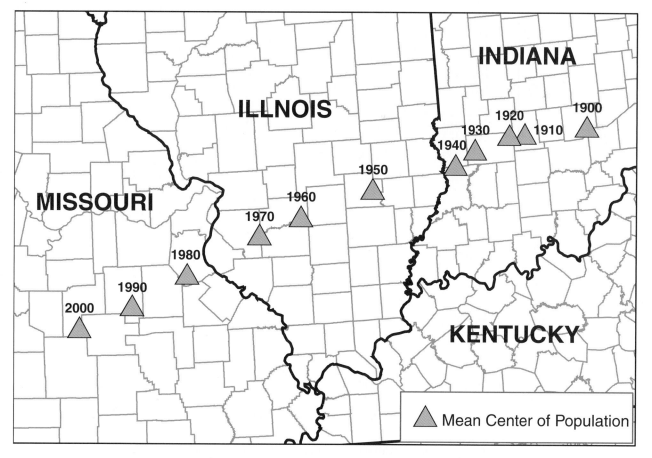

FIGURE 7.13   Shift in U.S. population centroid in the 20th century. Source: Census Special Report, 2002.

interstate migrants (U.S. Census, 2001). For instance, from 1999 to 2000, 4.2 percent of singles migrated, versus 2.6 percent of married people (U.S. Census, 2001). The same age range differences are evident for interstate migrants as for residential mobility, for example, 3.7 percent of children under 10 migrated to a different state in 1999–2000, compared to 6.2 percent of those in their 20s, whereas only 1 percent of those over 65 migrated. Note that sometimes the Census measures migration over five-year periods and at other times over one-year periods, with higher migration rates for the five-year durations.

Why Americans migrate also depends on their level of education. In the prior study, 2 percent of non–high-school graduates moved to a different state, compared to 3.5 percent of people with graduate degrees (U.S. Census, 2001). Figure 7.14 shows the reasons for moving for intracounty movers and intercounty migrants in the year 1999, cross-classified by education. Less-than-high-school-educated

migrants tended to do so three quarters of the time for housing and familial reasons, whereas only 18 percent of the migrants reported work-related reasons. By comparison, for migrants with graduate and professional education, almost half of the migration was work-related. As discussed earlier, residential movers do so mostly for life-cycle reasons, as seen by the near 90 percent proportion of housing- and family-related reasons for them, regardless of education (U.S. Census, 2001). Work only accounted for 5 or 6 percent of intracounty moves. Economic reasons are more important for long-distance migrants, because it is more likely for long-distance moves that migration is accompanied by a change in job, employer, or type of work.

As seen in Table 7.4, net migration for U.S. metropolitan areas in 1999–2000 was practically even at a gain of 137,000 persons, whereas nonmetropolitan areas lost that amount. Within metropolitan areas, a net of 3.3 million people moved from the central cities to suburbs

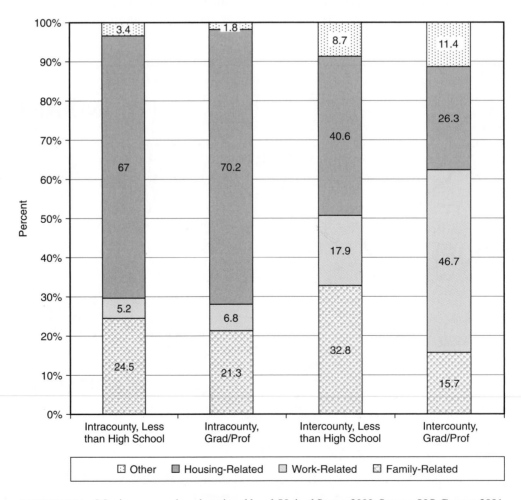

FIGURE 7.14   Moving reasons by educational level, United States, 2000. Source: U.S. Census, 2001.

TABLE 7.4   United States migration flows, 1999–2000

| Geographic area | Internal migration | | | Movers from abroad (1) |
	In-migrants	Out-migrants	Net internal migration	
Type of Residence				
Total	3,951	3,951		1,745
Metropolitan areas	2,044	1,907	137	1,639
Central cities	3,670	6,928	−3,258	845
Outside central cities	7,376	3,981	3,395	794
Nonmetropolitan areas	1,907	2,044	−137	106
Region				
Total	3,106	3,106		1,745
Northeast	363	615	−252	292
Midwest	722	640	82	238
South	1,258	1,031	227	612
West	763	820	−57	604

These estimates are from the Current Population Survey for March 1999 to March 2000.

(1) These numbers from the Current Population Survey include both temporary and permanent movers to the Untied States, among the civilian population.

Source: U.S. Census Bureau, Current Population Survey, March 2000, and Population Estimates Program.

(U.S. Census, 2001). This reflects a pattern during the 1990s of continuing loss of population from the central cities to suburbs (U.S. Census, 2001). Chapter 10, which concerns edge cities and the urban fringe, accounts in more depth for this shift away from the centers of metropolitan areas.

The table's columns on in-migrants and out-migrants reveal a high volume of migration movement. For instance, underlying the 137,000 net migration to metro areas are inflows and outflows of about 2 million migrants each. This underscores the point made earlier in the chapter that sometimes it is more important to consider net impact (i.e., net migration), whereas other times concentrating on the impacts and ramifications of individual in- and out-migration streams.

To conclude, U.S. interstate migration from 1999 to 2000 remained steady at a high level and reflected the Sunbelt migration discussed in Chapter 1. Single, more educated, or young adults with accompanying young children were more likely to migrate. Metropolitan and nonmetropolitan migration were nearly even, but within the metropolitan areas, migrants tended to settle in the suburbs. Migrants were economically motivated, especially for the highly educated. U.S. domestic migration is discussed further in Chapters 8 and 10.

7.6.2 Australian National Migration in the 1990s

Australian migration is presented as a comparison and contrast to the United States. Like the United States, Australia is an advanced nation economically, but differs in its topography, climate, world location (i.e., as a separate continent in the Southern hemisphere), and system of cities. It is intended as an alternative example to the United States. The further contrast of internal migration and its effects on cities in a developing nation—China—is discussed in the last section. Australia is a large nation, nearly the physical size of the United States without Alaska, but its population is much smaller, at 19.0 million people in 2001. Australians are a very mobile population, like U.S. residents. As seen in Figure 7.15, the major metropolitan areas of Australia are located along the eastern and southeastern coastal areas. A total of 58.5 percent of the population is located in the southeastern states of Victoria and New South Wales, which include the nation's two largest cities, Melbourne and Sydney. This unusual distribution is the result of the bulk of the continent's interior being mostly inhospitable in climate and very dry.

Its internal migration system shows similarities and differences from the American system. Australia has slightly higher residential mobility than the United States, with 18 percent of its population changing residence in the year 2000–2001 (Hugo, 2003). In addition to its mobility, 27 percent of the Australian population in 2001 was foreign born, which is more than double the 11 percent for the United States (Australian Bureau of Statistics, 2001; U.S. Census, 2003). The foreign born come from a wide variety of nations, with the largest groups from Great Britain and Ireland combined at 6 percent of total population, New Zealand at 2 percent, and Italy at 1 percent (Australian Bureau of Statistics, 2003). Over the past several decades, Australia has received

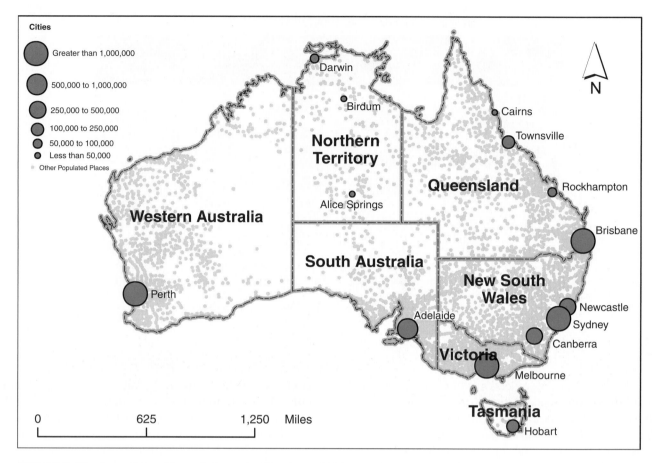

FIGURE 7.15   Australia and its major metropolitan areas, 2001.

more immigrants from large Asian nations, such as China and India, through push factors of weaker economies, perceived lower quality of lifestyle, and often fewer opportunities and freedoms in sending nations, with pull factors from Australia opposite to the push ones. The impact of foreign immigration is further seen in an additional 19 percent of Australian population that is native born but has at least one parent born overseas. Over the past several decades, the foreign-born immigrants have tended to settle in the biggest cities, Sydney and Melbourne, pictured in Figure 7.16 (Hugo, 2003). Many later move on to smaller cities.

Australia's geographical distribution of population has not changed much over the last 90 years. As seen in Figure 7.17, since 1911 the centroid location for the national population has remained relatively fixed in the center of the state of New South Wales. The location of this point only 300 miles from both Sydney and Melbourne demonstrates the continuation of the original settlement of Australia along its east and southeast coasts. Since 1911, the slight 65-mile shift in the centroid northward and slightly west reflects the more rapid growth of the northern state of Queensland, and to a lesser extent,

the state of Western Australia. Queensland has grown because of its warmer climate and economic expansion, especially in tourism and related businesses.

From 1996 to 2001, Australia's growth was 1.3 million persons, relative to its 1996 population of 17.9 million. Fifty-six percent was from natural increase and 44 percent from foreign immigration (Australian Bureau of Statistics, 2003). Two thirds of the 576,000 foreign immigrants settled in New South Wales and Victoria, the states of Sydney and Melbourne, and predominantly within those cities (Hugo, 2003). Vietnamese immigrants to Melbourne are shown in Figure 7.18.

The interstate migration flows from the 1996 to the 2001 censuses are seen in Figure 7.19 (Hugo, 2003, based on the Australian Bureau of Statistics, 2003). It shows all 42 interstate migration flows during this five-year period. The width of an arrow is proportional to the size of flow. What is evident immediately is the location of major migration flows along the eastern flank of country, not surprising as Australia's population is also concentrated in the east. There are large flows out of the populous state of New South Wales and to a lesser extent South Australia and the island state of Tasmania and into Queensland.

FIGURE 7.16 Melbourne. Source: Getty Images, Inc.

The flow from New South Wales to Queensland is represented by the big northward arrow. Another important but lesser destination is the Perth area in Western Australia. Queensland, the northeastern state with the large city of Brisbane and growing tourism and development along its "Gold Coast," resembles in some ways Florida and the Southeast in the United States. Thus, the latest migration might parallel to some extent the U.S. movement away from the older, established cities into newer regions. One difference from the United States is that many foreign immigrants into the United States have flowed directly into the Sunbelt's "port of entry" cities such as Los Angeles, San Francisco, Houston, and Miami, whereas in Australia, the foreign immigrants have moved initially into the two most dominant, established cities, even as natives have tended to migrate to Queensland. Although the regional shifts appear large on the map, it must be remembered that they have only slightly altered the national population centroid over the last 90 years.

Another distribution that has not been greatly affected by the migration changes is the distribution of metropolitan and nonmetropolitan settlement (see Table 7.5). Over the past 35 years, the main shift has been a sharp increase in the proportion of Australian population in urban areas of 20,000 to 499,000, up from 12 percent to

19 percent, and reduction by 2 percent in its very large cities and and by 3 percent for rural population. Although the 20th-century trend was to reduce the rural proportion, in the 1970 Australia experienced a decade of turnaround, with migration away from the urban areas. In the 1990s, rural-to-urban migration returned to about zero. Future shifts might depend on the future environmental and amenity aspects of the largest cities, as well as on government investment in Australia's rural areas to make them more accessible. Because Australia is so large and lightly populated, it has not built up the necessary infrastructure to give modern transport access to much of its territory, especially the inhospitable interior regions (Hugo, 2003).

What about the impact of Australia's migration patterns on its cities? The country's foreign immigrant population has settled in the large ethnic enclaves of Sydney and especially Melbourne. There are cultural and political ramifications of recent ethnic minority settlement for a nation that before World War II experienced limited and mostly European immigration. For instance, there is a question of use of foreign languages in schools and the workplace versus English, and the question of government support for poor minority immigrants. The political issue of foreign immigration has become part of

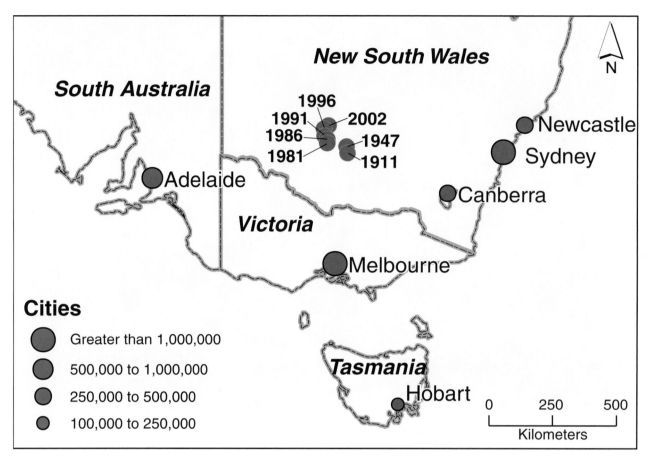

FIGURE 7.17   Shift in Australia's population centroid, 1911–2002. Source: Hugo, Graeme. "Recent Trends in Internal Migration and Population Redistribution in Australia." Paper presented at Meeting of Population Association of America, Minneapolis, Minnesota, May 2003. "Figure 1. "Shifts in the Australian population centroid, 1911-2002."

TABLE 7.5   Distribution of Australia's population by settlement size, 1966–2000

Settlement size	Number of urban centers			Percentage of population		
	1966	1996	2001	1966	1996	2001
500,000 and over	5	5	5	56.0	53.1	54.0
100,000–499,999	4	8	9	5.4	9.2	10.8
20,000–99,999	22	50	43	6.8	9.8	8.8
2,000–19,999	250	366	364	12.4	11.3	10.4
1,000–1,999	178	312	285	2.2	2.5	2.1
Total urban	459	741	706	82.9	86.0	86.3
Total rural				16.9	14.0	13.7
Total population				100.0	100.0	100.0
Total number				11,599	17,892	18,972

Totals include migratory population. Total number is in thousands.

Source: Hugo, Graeme. "Recent Trends in Internal Migration and Population Redistribution in Australia." Paper presented at Meeting of Population Association of America, Minneapolis, Minnesota, May 2003. Table 5. Distribution of population by settlement size, 1966, 1996, and 2001."

FIGURE 7.18   Vietnamese immigrants to Sydney. Source: Black Star.

the national debate, as it has in the United States and many European countries.

Queensland, which is growing especially quickly, needs to provide housing, services, government offices, and infrastructure. Politically, Queensland and Western Australia have increased in importance, as together they expanded from 21 percent of the national population in 1947 to 29 percent in 2001 (Australian Bureau of Statistics, 2003). This is analogous to the growing American political importance of Florida and the Southwest.

Some cities in the 1990s experienced little net migration change. For instance, Adelaide, the metropolis of South Australia, had a negligible net migration change during the period from 1996—to 2001 (see Table 7.6). However,

TABLE 7.6   Internal migration 1996–2001: Australian metropolitan statistical divisions

Statistical division	In-migration	Out-migration	Net migration
Sydney	175,732	233,685	−57,953
Melbourne	169,710	157,343	12,367
Brisbane	186,035	132,914	53,121
Adelaide	74,450	77,245	−2,795
Perth	100,290	89,695	10,595
Hobart	19,540	19,907	−367
Canberra	43,808	36,102	7,706

Modified from Hugo, 2003, based on unpublished data from 2001 Australian Census.

Source: Hugo, Graeme. "Recent Trends in Internal Migration and Population Redistribution in Australia." Paper presented at Meeting of Population Association of America, Minneapolis, Minnesota, May 2003. Table 14. Metropolitan statistical divisions: internal migration 1996-2001."

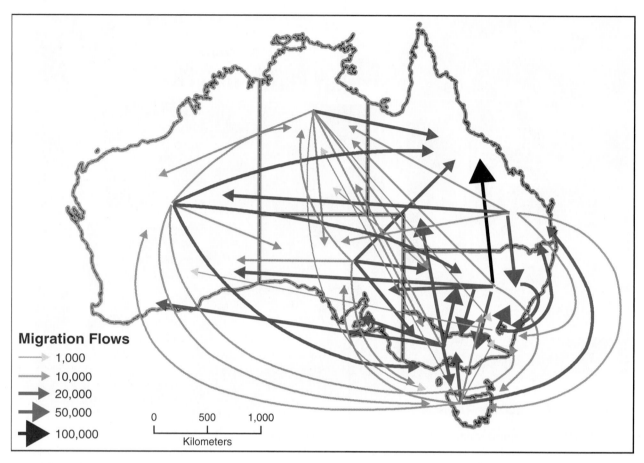

Migration Flows
- 1,000
- 10,000
- 20,000
- 50,000
- 100,000

0   500   1,000
Kilometers

FIGURE 7.19   Australian interstate migration flows, 1996–2001. Source: Hugo, Graeme. "Recent Trends in Internal Migration and Population Redistribution in Australia." Paper presented at Meeting of Population Association of America, Minneapolis, Minnesota, May 2003. Figure 8. "Adelaide: percentage change in population between the 1976 and 1981 censuses."

as seen in Figure 7.20, Adelaide grew mostly in its core areas and extreme south, with the middle ring declining. This reflects gentrification of the city with a return to the center (Hugo, 2003). Australia's other cities have responded to growth and change with specific patterns of intracity mobility and redistribution, which the Australian Census of 2001 has data for, but this is beyond the scope of this chapter.

Hugo (2003) identified the following as the major factors affecting Australian migration: infrastructure, especially transport and communications; economic structural change; differences in housing and living costs; environmental issues; accessibility and remoteness of interior lands; and the migration patterns of Australia's baby boomers, who in 2001 were in the age range of 25 to 44 (Australian Bureau of Statistics, 2003). These resemble general factors covered earlier in the chapter for migration and mobility. They also resemble somewhat the factors noted for the United States, although America has

greater transport accessibility, a more developed interior infrastructure, and its boomers are about 10 years older.

7.7 INTERNATIONAL IMMIGRATION

Immigration, the migration of people from one nation to another one, has occurred in historical periods. For instance, Greeks migrated to Persia to support Greece's attempt to dominate it several centuries B.C., while Romans migrated to other regions in the expanding empire. More recently, Italians migrated in large numbers during the late 19th century and early 20th century to Argentina. At the formation of Pakistan in 1947, numerous Muslim residents of postpartition India migrated to Pakistan and many Hindu residents migrated from Pakistan to India.

International migration is important to urban geography because it influences the ethnic composition, population size, and cultural features of cities. Immigrants

and later by remittance payments sent home by migrants. This section gives an overview of the current world situation in immigration, followed by the U.S. situation, and a discussion of immigration impacts. The next section details a case on immigration from the Dominican Republic to the United States.

The world has seen a large increase in international immigration, so that in 2000 there were 175 million international migrant stock, represented 2.8 percent of the world's population (United Nations, 2002). *Migrant stock* is population that has ever migrated from another country to their present country. The migrants have contributed workers, cultural customs and traditions, and childbearing to receiving nations, whereas the sending countries have sometimes lost valuable workforce and productivity. At the same time, immigration has sometimes put burdens of welfare, support, and conflict on receiving nations.

As seen in Table 7.7, the United States accounts for the largest destination of international migrant stock, with about 35 million in 2000, followed by the Russian Federation at 13 million, Germany at 7.3 million, and the Ukraine at 6.9 million. Saudi Arabia's migrant stock of 5.3 million, in eighth place, is remarkable given its small population of 24 million. The presence of approximately 20 million migrants in Russia and the Ukraine is partly explained by the breakup of the Soviet Union in the early 1990s, during which about 30 million new "immigrants" were created overnight from what had been internal migrants. Overall, according to the United Nations (2002), about 104 million migrant stock are in developed nations, versus 71 million in less developed ones.

The highest proportions of international immigrants by percentage are in the Middle East (see Table 7.8), especially in the United Arab Emirates, Kuwait, Jordan, Israel, Oman, and Saudi Arabia. In these countries, immigrants are brought in to make up for shortages of local workforce. Israel's recent migrants consist mostly of refugees from Russia, but earlier ones came from many countries in Europe, North Africa, the Middle East, and some from the United States. Russian immigrants to Israel are seen in Figure 7.21. As these Middle Eastern nations modernize, key questions involve the length of stay and adjustment of these immigrant populations, whether they achieve local citizenship, and the extent of return migration.

Another important factor for international migration is the remittances that migrants send back to their families in the homeland. This applies mainly to migration from developing to developed countries, for which the wage scales are much higher at the destination. Mexicans in the United States, for instance, currently are sending back about $9 billion in remittances annually. For poor

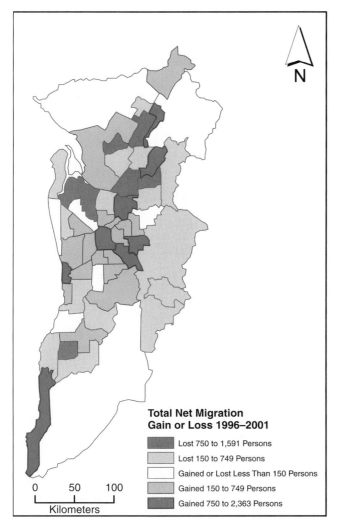

FIGURE 7.20 Migration changes in Adelaide, Australia, 1996–2001. Source: Hugo, Graeme. "Recent Trends in Internal Migration and Population Redistribution in Australia." Paper presented at Meeting of Population Association of America, Minneapolis, Minnesota, May 2003. Figure 8. "Adelaide: percentage change in population between the 1976 and 1981 censuses."

from other countries tend to settle in cities because of job markets and social networks that might exist there, among other things. They contribute to forming neighborhoods, social and cultural organizations, and enterprises that shape the city, and can lead to social cooperation as well as conflicts. The topic is built on further in Chapter 8 concerning ethnicity, Chapter 9 on industrial location including border cities, and aspects of Chapter 12 on planning.

Besides the impacts of immigration on the receiving countries and their urban areas, the sending countries can be impacted through labor losses, family separations,

TABLE 7.7 Countries with the largest international migrant stock by number and percent of population, 2000

Nation	Migrant stock (in thousands)	Country population in 2002	Percent ratio of migrant stock to 2002 population
United States	34,988	287,400	12.2
Russia	13,259	143,500	9.2
Germany	7,349	82,400	8.9
Ukraine	6,947	48,200	14.4
France	6,277	59,500	10.5
India	6,271	1,049,000	0.6
Canada	5,826	31,300	18.6
Saudi Arabia	5,255	24,000	21.9
Australia	4,705	19,700	23.9
Pakistan	4,243	143,500	3.0
United Kingdom	4,029	60,200	6.7
Kazakhstan	3,028	14,800	20.5
Ivory Coast	2,336	16,800	13.9
Iran	2,321	65,600	3.5
Israel	2,256	6,600	34.2
Poland	2,088	38,600	5.4
Jordan	1,945	5,300	36.7
United Arab Emirates	1,922	3,500	54.9
Switzerland	1,801	7,300	24.7
Italy	1,634	58,100	2.8
Total for top 20 nations	118,480	2,165,300	
Total worldwide	174,781		

Note: International migrant stock refers to residents of a given country who migrated from another country.

Source: United Nations. (2002) *International Migration Report.* Publication ST/ESA/SER.A/220. New York: Author.

TABLE 7.8 Major source nations for international immigration by percent, 2000

Nation	Migrant stock as percentage of total population
United Arab Emirates	73.8
Kuwait	57.9
Jordan	39.6
Israel	37.4
Singapore	33.6
Oman	26.9
Estonia	26.2
Saudi Arabia	25.8
Latvia	25.3
Switzerland	25.1
Australia	24.6
New Zealand	22.5
Gabon	20.3
Canada	18.9
Kazakhstan	18.7
Lebanon	18.1
Ivory Coast	14.6
Gambia	14.2
Ukraine	14.0
Belarus	12.6
Average for top 20 nations	27.5
Average worldwide	2.6

Source: United Nations. (2002) *International Migration Report.* Publication ST/ESA/SER.A/220. New York: Author.

nations sending migrants, such as El Salvador, Nicaragua, Somalia, and Yemen, seemingly small remittances of several thousand dollars appear very large in the origin nation, often exceeding the mean annual income.

Of the world's 175 million international migrant stock, about 7 percent are refugees (United Nations, 2002). They are mostly located in Africa and Asia, followed by the developed nations (United Nations, 2002).

We estimate that 75 percent of the 175 million migrants are located in urban areas, commonly in larger cities. This implies both major benefits and major burdens. For instance, the large city might benefit by low-skilled workforce and augmented multiculturalism. Another benefit for developed nations is the ability to counter the prospect of population decline that is affecting most of them. This stems from the large decline in fertility that has occurred worldwide in developed nations as well as in many developing ones (United Nations, 2002). If deaths exceed births, as is now the case in many developed nations, to maintain an even population level the balance must be made up by net migration. Because the developed nations today are sending out relatively few emigrants, there must be at least some in-migration to avoid population decline.

In the United States, based on the past period, census projections assume a long-term net international

FIGURE 7.21   Russian immigrant family arriving in Israel. Source: Contact Press Images Inc.

migration of 900,000 persons annually from now until 2050, whereas the European Union needs an inflow of 1 million international migrants per year to counter population decline (Kent & Mather, 2002). However, it is not clear now whether European policies will allow in so many immigrants, leading to an estimate that Kent and Mather (2002) considered more realistic of 340,000 immigrants annually.

In spite of workforce and other economic needs, some countries are resisting large-scale migratory influxes. Consequently, today 44 percent of developed countries have established government policies to discourage immigration, whereas 39 percent of developing nations have such policies (United Nations, 2002). The policies were not often approved amicably, but involved political tensions and often difficult compromises. Only 4 percent of nations have policies aimed at increasing immigration (United Nations, 2002). In the post-9/11 world, security concerns could further add to restrictive policies on immigration.

7.7.1 Immigration in the United States

The United States has international immigration policies that date back to before the founding of the republic (Frey, Abresch, & Yeasting, 2001). The topic of the forced immigration of slaves to the United States is considered in the next chapter on ethnicity. As seen in Table 7.9, the United States at its beginning granted widespread citizenship to free White people while also importing many nonfree Blacks. Until after World War I, the doors were largely open to foreign immigrants. The initial groups immigrating came from northern Europe—Britain, Ireland, Scandinavia, Germany, and the Netherlands. Late in the 19th century they were replaced mostly by southern and eastern Europeans—Italians, Poles, Russians, Jews from many countries, and some Chinese and Japanese. Figure 7.22 pictures European immigrants arriving in the United States. In 1882, the Chinese Exclusion Act was passed, restricting Chinese immigrants on the basis of ethnic origin, the first such act in the nation's history. From the beginning of the United States to 1922, the great majority of immigrants from Europe to the United States moved from rural areas in their origin nations to settle in cities in the United States. Beginning in 1917, Blacks in the South, mainly from rural areas, replicated this European pattern by migrating to northern cities. This contrasted with immigration before the American Revolution, which was mainly to rural areas.

During the 20th century, there have been fluctuating policies of openness and closed doors, ending up with the elimination of the quota system in 1965, which specified that nations would bring in the same number of

TABLE 7.9   Major milestones in the history of U.S. immigration since 1790

1790. Start of naturalization. Immediate citizenship granted to free White persons who had resided in the United States for two or more years.

1882. Chinese Exclusion Act. This stopped the immigration of Chinese. This was the first instance of the United States restricting immigration based on race.

1921. Quota system started. Its principal purpose was to limit migration from Eastern Europe. Annual immigration was limited to 3 percent or less of the 1910 population of the country of origin.

1965. Quota system eliminated. Instead, every country allowed to send the same number of immigrants annually, based on the number who were in the United States at the time of the 1910 Census. Exception: Several nations from Western Hemisphere were permitted to have unlimited immigration.

1986. Immigration Reform and Control Act (IRCA). The Act gave amnesty to millions of illegal immigrants. It included strong steps to combat illegal immigration.

1996. Illegal Immigration Reform and Immigrant Responsibility Act. Reinforced government agencies that reduce illegal immigration. Removed public-benefits rights of immigrants who are legal.

Source: Immigration and Naturalization Service, 2002.

immigrants each year. This started a wave of immigration that grew in the last third of the century to reach nearly a million immigrants entering per year in the 1990s. There were so many Mexican immigrants arriving and living undocumented in the country that in 1986 Congress passed the Immigration Reform and Control Act (IRCA). It provided amnesty to millions of undocumented immigrants in the United States, but at the same time toughened illegal immigration controls. After amnesty took effect, the illegal immigrant controls have had limited success, and the immigration streams from Mexico in particular, have fallen slightly, if at all.

In recent years, almost a million legal immigrants have been admitted to the United States yearly. As seen in Table 7.10, they were admitted mostly for purposes of family reunification, although employment and refugees or asylum seekers were also important. In addition, there are huge numbers of temporary visitors, students, and workers to the country—about 14 million each year. Not shown here are the estimated 750,000 illegal immigrants entering the country each year.

The result of recent decades of immigration has been the presence of a huge foreign-born population in the United States, totaling 35 million in the year 2000 (United Nations, 2002). This foreign-born population is largely

FIGURE 7.22   European immigrants arriving in the United States at the turn of 20th century. Source: Library of Congress.

TABLE 7.10 U.S. immigrants and temporary visitors, 1995–2000

Category	Number (in 1000s)	Percent
Family reunification	3,121	68
Employment	535	12
Refugees and asylees	516	11
Domestic diversity	292	7
Other	108	2
Total immigrants	4,572	100
Temporary visitors (business, travel)	130,643	91
Students and their families	2,757	2
Temporary workers, trainees, and families	2,223	2
Other	7,171	5
Temporary visitors and students - Total	142,794	100

Source: Immigration and Naturalization Service, 2002.

TABLE 7.12 Large proportional diasporas in the United States, 1997

Country	Country's foreign born population in U.S. (in 1000s)	Population of the origin country	Ratio
Jamaica	379	2314	0.164
Belize	27	190	0.142
El Salvador	607	5,119	0.119
Haiti	440	5,054	0.087
Dominican Republic	632	7293	0.087
Mexico	7,017	81,250	0.086

Source: Frey, William, Bill Abresch, and Jonathan Yeasting. 2001. American by the Numbers, New York, The New Press, Table P. 58. Unnumbered Table. "Measuring the Diasporas, 1997."

located in metropolitan areas. For instance during the 1990s, 4.9 million immigrants arrived in ten major metropolitan areas in the United States, (see Table 7.11). The largest of them, the seven "port of entry" metropolitan areas of New York, Los Angeles, San Francisco, Miami, Chicago, Washington, DC, and Houston together accounted for 4.4 million immigrants. As a consequence, those cities have developed large ethnic communities, needed to bolster their service and support systems, struggled to provide education, expanded ethnic events, and experienced more frequent political changes. Another important aspect has been the formation of diasporas (see Table 7.12). A *diaspora* (scattering) refers to that part

of the population of a country of origin that has migrated overseas. Mexico's largest diaspora today is located in the United States, and is estimated at 7 million. This and other diasporas retain connections to the home country, for instance, in communications and visits to the homeland, sending of remittances, and sometimes even voting rights from the country of origin.

7.8 TRANSNATIONAL IDENTITIES

Immigration brings about mixed identities. *Transnational identities* refer to identification of immigrants with their countries of both origin and destination. A broader concept is that of *transnationalism,* which refers to the in-betweenness and ambivalence of national identity, for people, organizations, and places. An example is an immigrant family that arrives in America from Brazil. The family learns and participates in the new customs and traditions of the United States, but maintains some Brazilian cultural tradition. Within the family, different members might have different extents of transnational identity. If, after arrival, a new child is born in America, he or she will be raised with a stronger U.S. identity than Brazilian, whereas the non–English-speaking grandmother who came with the family has a transnational identity that is much more Brazilian than American. Today, because the United States has the highest level of foreign-born population in 75 years, reaching 11 percent in 2000, the issue of transnationalism is a prevalent one.

Over the generations, transnational identities change (Portes & Rumbaut, 1996). Assimilation processes often encourage adoption of the traditions of the new culture and destination language. However, in other cases, the transnational identity persists and full assimilation does not occur. Assimilation is controversial—it might or might not be advantageous. Some urban political geographers

TABLE 7.11 Number of immigrants arriving in ten major U.S. metropolitan areas, 1990–1999

Metropolitan area	Number of immigrants arriving 1990–1999
New York	1,408,543
Los Angeles	1,257,925
San Francisco	494,189
Miami	420,488
Chicago	363,662
Washington	267,175
Houston	214,262
Dallas-Fort Worth	173,500
San Diego	159,691
Boston	137,634
Total[a]	4,897,069

[a] This total represents more than 60 percent of all arriving.

Source: Frey, William H., Bill Abresch, and Jonathan Yeasting. 2001. America by the Numbers. New York, The New Press. Table P. 51. Unnumbered Table. "A history of immigration."

emphasize the benefits of remaining unassimilated, pointing to social, cultural, and economic reasons.

The economic context of transnational identities is important. In the broad context, the identities of many immigrants are economically driven. They might retain certain ties with their homeland, or have stronger involvements at the destination, due to economic advantages. Globalization trends discussed in Chapter 5 have encouraged the economic aspects, due to the increased flows of capital and people worldwide. Because of the historical factors of immigration and culture, transnational identities are different among ethnic groups and can differ by gender (Mitchell, 2003). In border communities between nations, transnational identity is often more prominent and more persistent (Arreola, 2002; McConville, 1983). Assimilation toward only one national identity might be precluded along international borders, where residents are enmeshed every day in both cultures.

Two brief examples illustrate the basic concepts, as well as differences, in transnational identities, and the next section on Dominicans in the United States explores this topic in greater depth. The first example is immigrants from Hong Kong to Vancouver. In the early 1980s, Canada set up programs to attract wealthy immigrants. At the same time, many Hong Kong residents sought opportunities to establish foreign citizenship due to the upcoming Chinese assumption of sovereignty in 1997. After arrival, Hong Kong immigrants had many problems adjusting to Vancouver, including issues and conflicts related to "landscape design, house demolitions, house size, architectural style, tree removal, and downzoning" (Mitchell, 2003, p. 76). Another issue for some Hong Kong immigrant families was that the wife established residence in Canada and the husband commuted back and forth, tending to economic interests at place of origin. The woman maintained the home and family life, while the husband had a transnational economic-based life (Mitchell, 2003). This example demonstrates the influence of globalization and economic and gender aspects.

A second example is transnational identity in the border city of Laredo, Texas. This city of 177,000 people in 2000 is a rapidly growing major trucking port of entry for the United States. It has a sister city in Mexico, Nuevo Laredo, with which it was joined until 1848. Laredo has among the highest proportions of Hispanic population (94 percent) among U.S. cities. There are not problems of discrimination and resentment because of the dominant majority. In fact, Anglo residents of Laredo tend to assume some Mexican traits (Arreola, 2002). Residents of Laredo identify culturally with both the United States and Mexico (Arreola, 2002). There are many families that have extended across both Laredos for generations. Not only is there a high extent of ability to speak Spanish (about 95 percent), but also, economically, residents of Laredo, Texas, are heavily dependent on Mexico, which provides shoppers and trucking business, aspects that are long term. In sum, the transnational identities for residents of Laredo are strong and persistent over time, and stem from a mixture of history, language, border proximity, and familial and economic ties (Arreola, 2002).

The following case study concerns immigration of Dominicans during the 1980s and 1990s, mainly to New York City and New Jersey, and the adjustments entailed in the new country. The impact of immigration on U.S. cities is discussed further in the next chapter on ethnicity and poverty.

7.9 CASE OF IMMIGRATION IMPACT ON A CITY: DOMINICANS IN NEW YORK CITY

Immigrants from the Dominican Republic constitute one of the fastest growing groups in the United States. In 1990, there were 347,858 U.S. residents born in the Dominican Republic; this number leaped to 672,000 in 2000 (U.S. Census, 2002). Similarly, in 1990, there were 520,151 persons of Dominican ancestry, growing to 764,945 of Dominican origin in 2000 (U.S. Census, 2002). Among Hispanic groups in the United States, Dominicans in 2000 were third, as measured by country of foreign birth, behind Mexicans and Cubans. The size of this in-migration has made the Dominican migration to the United States a major diaspora group. Dominicans in the United States were equivalent to almost 8.7 percent of the remaining population of the Dominican Republic homeland (Frey et al., 2001; U.S. Census, 2001; CIA World Factbook, 2003), a percentage nearly identical to the corresponding values for Haiti and Mexico, although less than the top three diaspora nations.

The Dominican Republic as origin country provided "push" factors leading partly to this large exodus. It is a developing country of 8.7 million people that shares an island with Haiti (CIA Factbook, 2003). It has a mixed agricultural economy involving coffee, sugar, and tobacco, combined with tourism. The economic situation in the Dominican Republic has fluctuated for several decades, suffering at times in the 1970s and 1980s. The push stemmed from those privations. Although in the 1990s, the Dominican economy had a very fast growth rate, it was dampened by huge economic losses from Hurricane Georges in 1998 (CIA Factbook, 2003). The United States and its vast labor markets formed a pull factor. As more and more Dominicans have settled in the United States, there is additional pull from the presence of this large social network.

Dominican migrants to the United States have tended to be from urban areas with low education by U.S. standards, although some have been better educated, and even well off. However, once they arrive in the United States, the educational levels of their offspring often drop to a relatively low level. Most Dominicans have entered the United States illegally, commonly by having false documents or overstaying a tourist visa (Grasmuck & Pessar, 1996). In some cases, such overstayers planned to return when events improved in the origin nation, but often this did not happen and a temporary visit turned into a long stay and eventually naturalization.

Dominicans have predominantly settled in New York City or northern New Jersey. As pictured in Figure 7.23 and mapped in Figure 7.24 for New York, they have settled mostly in Manhattan's Washington Heights, nearby areas just to the east in the Bronx, in Queens just to the south of La Guardia Airport, and along the Brooklyn–Queens boundary mostly just to the southwest in Brooklyn. They have added their own flavor and cast to these areas, especially sections of Queens, where there are Dominican restaurants, cultural events, magazines, and other media (Grasmuck & Pessar, 1996).

However, in adjusting to their new country, many of the first-generation Dominican immigrants have suffered from low-income jobs that were mostly menial, job discrimination, and family disorganization. This low status stems from their relative lack of skills for the competitive

U.S. job market. The immigrants have gravitated toward low-paying jobs, mostly in retailing and services, including work in hotels and restaurants and, mostly for women, in the garment industry (Grasmuck & Pessar, 1996). Their income has averaged about one third of the U.S. average, with an extraordinary poverty rate of 30 percent, about three times the U.S. average. This poverty stems from lack of relevant skills for the U.S. workplace; for instance, about 60 percent of Dominicans over 25 years old had not finished high school in 1990 (Grasmuck & Pessar, 1996).

The situation has led to a high divorce rate and family disorganization. For example, in 1990, 40 percent of Dominican households had a single female parent. Family breakdown tends to perpetuate poverty, often into the second generation. At the same time, however, some women who have gained employment in this process have experienced a feeling of liberation, albeit at a low occupational level (Grasmuck & Pessar, 1996).

Another adverse factor is the high proportion of Dominicans lacking English- speaking ability. As seen in Table 7.13, the proportion of substandard or no English is more than 50 percent for recent immigrants and even for those immigrating 30 years ago it was more than one third. This serves as a major impediment to the advancement of Dominicans in their new country. At the same time, a

FIGURE 7.23 Census enumeration in a Dominican American area of New York City. Source: The Image Works.

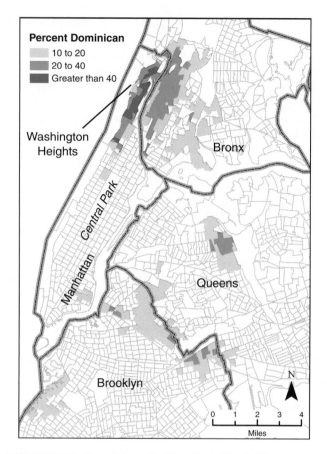

FIGURE 7.24   Dominicans in New York City, 2000.

Other aspects of the adjustment process for Dominicans include overcoming racial discrimination, establishing their ethnic identity, whether to be self-employed, and how to continue social and economic ties with the homeland (Grosmuck & Pessar, 1996). One of the differences of Dominicans versus some other entering Latino groups is that many have an African background. In census self-attribution, only about a quarter identified themselves as White, with 50 percent mixed race, and 25 percent Black. As a result, many Dominicans have experienced discrimination against Blacks in addition to anti-Latino discrimination. The dual-race factor has influenced their lack of success in New York City, including in the housing and employment markets.

A related issue is that of ethnic identity. As with any immigrant, does the Dominican resident identify himself or herself with the homeland, with Latino culture in the United States, or with Black identity? A person's working through this over time might lead to personal satisfaction, but sometimes also to neighborhood and work tensions and conflicts.

As with some other immigrant groups, Dominicans have shown a tendency to self-employment, and especially to start small ethnic-based concessions and businesses (Grasmuck & Pessar, 1996). The problem is the riskiness of succeeding in a competitive metropolitan market, and again there is a high rate of failure. Nevertheless, this might offer them more hope than being bogged down in low-level service jobs.

Dominicans have maintained close ties with their homeland, stimulated by the large diaspora factor. For instance, a large proportion of immigrants send remittances to family members in the homeland. Even though these immigrants are mostly poor by U.S. standards, the money they send has much more relative value back in their home nation. Estimates indicate that as much as 25 percent of Santo Domingo residents are receiving remittances (Grasmuck & Pessar, 1996). Moreover, the

positive sign for immigrants is the low but increasing proportion of college graduates in this group, which signifies that more educated people are leaving the homeland, although the proportion is still quite low. As with other immigrant groups, the second generation has a nearly 100 percent rate of English speaking. However, so far, the arrival of ever increasing immigrant streams has maintained a low overall average of English speaking performance.

TABLE 7.13   English ability among adults of Dominican origin, by timing of entry to United States and status

Year of entry	Percent college graduates	English ability "Not well"	English ability "Not at all!"
Born in U.S.	27.1	7.8	2.0
1965–1969	4.5	27.8	8.8
1970–1974	4.2	26.8	14.4
1975–1979	6.5	25.6	25.2
1980–1984	5.1	32.5	23.5
1985–1990	11.3	29.2	34.2

Source: Grasmuck, Sherri and Patricia Pessar. 1996. "Dominicans in the United States: First' and Second-Generation Settlement, 1960–1990," in Pedraza, Silvia and Ruben G. Runbaut (eds.), Origins and Destinies: Immigration, Race, and Ethnicity in America. Wadsworth Publishing Company, P. 284, Table 2. "College graduates and English-Speaking ability by year of entry for adult Dominican ancestry population over the age of 25."

small proportion of immigrants who have had financial success in the United States tend to invest significantly back in the homeland. Another sign of the close ties to the homeland are low rates of naturalization of Dominicans versus other entering Latino groups (Grasmuck & Pessar, 1996).

In summary, this case example points to a complex process of immigrant adjustment to a new country. It has not been an easy road for Dominicans, nor for most immigrants in America, yet many among the second generation have gained more education and skills and been able to move up from their poor beginnings. The ties with the homeland help in maintaining identity, and immigrants provide a source of revenue to send back to the mother country. There has been a major urban impact especially on social, cultural, neighborhood, and service sectors in New York City.

7.10 RURAL–URBAN MIGRATION AND CHINA'S CITIES

This last section examines internal, rural-to-urban migration in the developing world, which was seen earlier to characterize nations especially during their development phase. This case involves the world's most populous nation, China, which had a population of 1.300 billion in

2004 (Population Reference Bureau, 2004). Rural-to-urban migration has created an urban component of 70 to 110 million referred to as *floating population* (Goodkind & West, 2002). This population consists of rural-to-urban migrants who have arrived in China's urban areas, but have not gained a permanent residency permit; hence they have restricted or no papers and are referred to as floating. It constitutes about a quarter of the nation's urban population, and is in the population range of Germany or Mexico. Figure 7.25 shows Chinese rural migrants working in a city.

This case demonstrates that migration patterns can vary for nations at different development levels across world regions, and with different types of government regimes. The case forms a major contrast to the open, free, and unregulated internal migration in the highly mobile United States and Australia. It also provides an example of rural-to-urban migration, which is prevalent in the developing world today, but also was a major cause of growth in U.S. and European cities over the last two centuries. China's floating population constitutes a major national social problem that will lead to more discontent, resentment, and conflict, unless it can be better planned and provided for.

This situation has arisen historically based on a residential registration system established in 1951, known as

FIGURE 7.25   Rural migrant metal workers in a Chinese city. Source: Peter Arnold, Inc.

hukou, that has been weakened but not ended in the past two decades by China's thrust toward a market economy. Under the Communists, China underwent a series of periods of supporting and withdrawing support for urban development. From 1952 to 1965, China emphasized urban industrialization and consequently most of the cities grew, some explosively. With the advent of the Cultural Revolution in 1966, China's leaders emphasized returning large numbers of mostly educated urban residents, who supposedly had taken on Western ways, to the rural areas to learn to be true Communists. The cities languished without much support. From 1978 to today, China has been emphasizing new economic reforms. It has opened up a market economy, increased agricultural efficiencies, and adjusted trade policies sufficiently to gain admittance to the World Trade Organization. However, better agricultural efficiencies have caused many farm workers to depart and seek low-skilled jobs in cities. At the same time, the loosening of central control has enabled the large cities to become somewhat more autonomous and exert greater control over their residents.

All these changes have led to rapidly expanding Chinese urbanization (see Table 7.14). From only 144 million urban residents in 1970, there were an estimated 389 million by 1999 (Song & Zhang, 2002). In fact, the urban growth over the 29 years approaches the size of the entire U.S. population. Likewise, there has been growth in the number of million-plus cities, from 15 over 1 million in 1980 to 37 in 1999 (Song & Zhang, 2002).

The case demonstrates connections between migration, urbanization, and industrialization that relate generally to nations undergoing economic development. As cities grow and industrialize, their job markets demand labor that exceeds what is available locally or what can be provided through natural increase. Migrants are attracted to the expanding job markets and settle there, which draws population from rural to urban areas, causing the country to become more urbanized. The introduction to these complex processes here will be elaborated on in Chapter 9 on industrial location.

The *hukou* registration system was started for cities in 1951 and became a permanent system for both cities and rural areas in 1958 (Chan & Zhang, 2003). It has two parts: (1) a resident is officially registered into a permanent residence location, which he or she is limited to except for special reasons, and (2) the resident is assigned an agricultural or nonagricultural designation. Until the loosening of this system in the early 1990s, there was strict enforcement of registration, and persons not living at their registered location were returned back to their registered place. Since 1992, however, the system has loosened somewhat by including exception categories that enable certain rural migrants to gain more or less long-term residence in cities (Chan & Zhang, 2003). For instance, one of the exceptions to traditional *hukou* is the Blue Stamp *hukou,* which commenced in 1992, and determines eligibility for city residence based on the migrant's contributions to the urban economy. At the root, the exceptions are due to the development of a market economy, which requires mobile labor. In spite of many exceptions, however, the *hukou* system remains in effect today, and constrains rural-to-urban migration.

Another reason for the *hukou* system stems from the Chinese government's long-term favoritism of the industrial over the agricultural sector. This has contributed to higher wages and more attractive jobs in industry, and an artificially wide differential between industry and agriculture. This is sometimes termed the "dual economy" or "dual society" (Chan & Zhang, 2003). As seen in Table 7.15, urban industry not only receives a lot of state support, but was traditionally state- or local-government owned, although some of it is slowly being privatized. In the dual society, urban households are under government control, which gives them more stability, but also restricts residence, whereas rural households are self-reliant and not subject to much government control. Local collectives provide the employment and support to rural workers. This dualism has been somewhat weakened, although not eliminated, by the market economy. Government control of urban areas and industries is not as

TABLE 7.14 Urban population of China, 1952–1999

Year	Total population (in millions)	Urban population (in millions)	Average yearly growth rate in urban population	Urbanization (%)
1952	574.8	71.6		12.5
1962	672.9	116.6	4.88	17.3
1970	829.9	144.2	2.66	17.4
1980	987.0	191.4	2.83	19.4
1990	1143.3	301.9	4.56	26.4
1999	1259.1	388.9	2.81	30.9

Source: Song, S., and Zhang, H. K. (2002) "Urbanization and city size distribution in China." *Urban Studies* 39(12):2317-2327.

TABLE 7.15 Dual economy and dual society in China (up to early 1980s)

Dual economy	
Urban industry	**Agriculture**
Priority sector	Nonpriority sector
State owned	Nonstate sector
State supported and controlled	Self-reliant
Monopoly driven	Provides cheap resources to state sector
Dual society	
"Non-agricultural" households	**"Agricultural" households**
Subject to political stability and control	Self-reliant, with less central control
Employment and welfare provided by state	Employment and welfare based on local collectives
Restricted entry	Tied to farming and land

Source: Chan, Kam Win, and Li Zhang. 2003. "The Hukou System and Rural-Urban Migration in China: processes and Changes." Working Paper. Seattle: Department of Geography, University of Washington. P. 14. Table 3. "Dual economy and dual society in the pre-reform era."

tight as it used to be, and rural areas have some other options available. On the latter point, China's government has made attempts over many years to provide jobs for displaced farm workers through rural industrialization. However, those programs have largely failed, and the migrants continue to go to the middle-sized and large cities.

What have been the characteristics and experiences of rural-to-urban migrants? A recent study (Liang, Chen, and Gu, 2002) has indicated that the motivation of rural migrants to go to cities is mostly business or factory work (37 percent), job transfer or job assignment (15 percent), and marriage (22 percent). Clearly, economic pull reasons dominate, and marriage is second. Comparing the rural migrants to those who stay (see Table 7.16), the movers are younger, more likely to be single, and better educated, although gender is not a distinguishing trait. This is consistent with the chapter's migration principle

TABLE 7.16 Characteristics of Chinese internal migrants versus nonmigrants

	Migrants	Non-migrants
Male (percent)	51	52
Age (average)	28	33
Single (percent)	36	25
Illiterate (percent)	9	19
Elementary or higher education (percent)	29	41
Agricultural worker (percent)	71	89

Source: Liang, Zai, Yiu Por Chen, and Yasmin Gu. 2002. "Rural Industrialization and Internal Migration in China." *Urban Studies* 39(12):2175–2187. P. 2182. Table 2. "Descriptive characteristics of Chinese internal migrants versus non-migrants."

that younger, unmarried persons are generally more inclined to move. Also, because they aspire to city jobs in industry, education and skills are a big plus. A geographical aspect of this migration is that the destinations are predominantly on China's prosperous east coast, especially the largest cities in the southeast. There, new free economic zones continue to be developed on a large scale, with labor demand that cannot come close to being satisfied by the local workforce alone.

The few available studies of the assimilation and acculturation of rural migrants would indicate that although the pay is good, there is discrimination and resentment, and that the jobs are often physically demanding (Goodkind & West, 2002). These migrants are often not provided with normal urban services, even including access to schools (Chan & Zhang, 2003; Goodkind & West, 2002).

SUMMARY

This chapter has introduced migration, which constitutes one of the major growth and change factors for cities. After discussing the basic concepts and principles of migration and residential mobility, the chapter presented the background of the lab exercise on accessibility for proposed environmental centers in Los Angeles, using the population potential index. The U.S. and Australian populations were seen to be highly mobile. International immigration, on the rise today, is caused by a combination of push and pull factors. The overall trend is for movement from sending nations in the developing world toward the developed world. The case example of immigration from the Dominican Republic to the United States profiles first-generation immigrants who were "pulled" by perceived economic opportunity, and have been marginalized in some serious respects, while retaining a close bond with their homeland through a large diaspora. The final case concerns rural-to-urban migration within China. The resultant floating population is likely to become larger and more problematic unless the government can provide it with more resources and legitimacy.

REFERENCES

Adams, J. (1969) "Directional bias in intra-urban migration." *Economic Geography* 45:302–323.

Allen, J. P., and Turner, E. (2002) *Changing Faces, Changing Places: Mapping Southern California.* Northridge, CA: The Center for Geographical Studies, California State University Northridge.

Arreola, D. D. (2002) *Tejano South Texas.* Austin, TX: University of Texas Press.

Australian Bureau of Statistics. (2003) *Selected Social and Housing Characteristics.* Canberra: Australian Bureau of Statistics, 2001 Census.

Beauchemin, C., and Bocquier, P. (2004) "Migration and urbanization in Francophone West Africa: An overview of the recent empirical evidence." *Urban Studies* 41:2245–2272.

Chan, K. W., and Zhang, L. (2003) *The Hukou System and Rural-Urban Migration in China: Processes and Changes.* Working Paper. Seattle, WA: University of Washington, Department of Geography.

CIA. (2003) *CIA World Factbook 2003.* Washington, DC: Central Intelligence Agency. Available 2004 at *www.cia.gov.*

Clark, W. A. V. (1986) *Human Migration.* Beverly Hills, CA: Sage.

Duncan, O. D. (1959) "Human ecology and population studies," in *The Study of Population: An Inventory and Appraisal ...* ed. by P. Hauser and O. D. Duncan, pp. 678–716. Chicago TL: University of Chicago Press.

Frey, W. H., Abresch, B., and Yeasting, J. (2001) *America by the Numbers: A Field Guide to the U.S. Population.* New York: The New Press.

Geertman, S. C. M., and van Eck, J. R. R. (1995) "GIS and models of accessibility potential: An application in planning." *International Journal of Geographical Information Systems* 9:67–80.

Goodkind, D., and West, L. A. (2002) "China's floating population: Definitions, data, and recent findings." *Urban Studies* 39(12):2237–2250.

Grasmuck, S., and Pessar, P. (1996) "Dominicans in the United States: First- and second-generation settlement, 1960–1990," in *Origins and Destinies: Immigration, Race, and Ethnicity in America.* ed. by S. Pedraza and R. G. Rumbaut, pp. 280–292. Belmont, CA: Wadsworth.

Handy, S. L., and Niemeier, D. A. (1997) "Measuring accessibility: An exploration of issues and alternatives." *Environment and Planning A* 29:1175–1194.

Hugo, G. (2003) "Recent trends in internal migration and population redistribution in Australia." Paper Presented at the Annual Meeting of the Population Association of America. Minneapolis, MN, April 1–3.

Immigration and Naturalization Service. (2002) *2000 Statistical Yearbook of the Immigration and Naturalization Service.* Washington, DC: Immigration and Naturalization Service.

Jones, H. (1990) *Population Geography* (2nd edition). London: Paul Chapman.

Kent, M. M., and Mather, M. (2002) "What drives U.S. population growth?" *Population Bulletin* 57(4):40.

Liu, S., and Zhu, X. (2004) "Accessibility analyst: An integrated GIS tool for accessibility in urban transportation planning." *Environment and Planning B: Planning and Design* 31:105–124.

López Villar, D. A. (2003) *Migración en México: Datos de 1990 al 2000.* (Migration in Mexico: Data from 1990 to 2000). Aguascalientes, Mexico: Instituto Nacional de Estadísticas, Geografía, e Informática. Available in 2004 at *www.inegi.gob.mx.*

McConville, J. L. (1983) "Border culture overview," in *Borderlands Sourcebook: A Guide to the Literature on Northern Mexico and the American Southwest.* ed. by S. Ellwyn, pp. 245–247. Norman: University of Oklahoma Press.

Mitchell, K. (2003) "Cultural geographies of transnationality," in *Handbook of Cultural Geography ...* ed. by A. Kay, D. Mona, S. Pile, and T. Nigel, pp. 74–87. London: Sage.

Poortes, A., and Rumbaut, R.G. (1996) *Immigrant America* (2nd edition). Berkeley, CA: University of California Press.

Population Reference Bureau. (2004) *World Population Data Sheet.* Washington, DC: Author.

Ravenstein, E. (1885) "The laws of migration." *Journal of Royal Statistical Society* 48:167–227.

Schachter, J. (2001) "Geographical Mobility: Population Characteristics," Current Population Reports Publication P-20-538. Washington, DC: U.S. Census Bureau. Available in 2004 at *www.census.gov.*

Siegel, J. S., and Swanson, D. A. (eds.). (2004) *The Methods and Materials of Demography* (2nd edition). St. Louis, MO: Elsevier Science and Technology.

Song, S., and Zhang, H. K. (2002) "Urbanization and city size distribution in China." *Urban Studies* 39(12):2317–2327.

Stouffer, S. (1940) "Intervening opportunities: A theory relating mobility and distance." *American Sociological Review* 5:845–867.

United Nations. (2002) *International Migration Report.* Publication ST/ESA/SER.A/220. New York: Author.

U.S. Census. (2001) *The Hispanic Population.* Census 2000 Brief. Washington, DC: Author.

U.S. Census. (2003) *Selected Appendixes: 2000.* 2000 Census of Population and Housing. Census Publication PHC-2-A. Washington, DC U.S. Census Bureau. Available in 2004 at *www.census.gov.*

U.S. Committee for Refugees. 1993. *1993 World Refugee Survey.* Washington, DC, United States Committee on Refugees. Available in 2004 at *www.refugeesusa.org.*

Wood, W. B. (1994) "Forced migration: Local conflicts and international dilemmas." *Annals of the Association of American Geographers* 84(4):607–634.

Zipf, G. (1949) *Human Behavior and the Principle of Least Effort.* New York: Hafner.

EXERCISE

Exercise Description

The goal is to evaluate the accessibility of four sites proposed for an environmental educational center to the non-English-speaking population of Los Angeles County.

Several local conservation groups have challenged the idea that environmental education is equally accessible to the population of Los Angeles County. One concern that has been raised by these groups is that a large share of the population of Los Angeles County is made up of recent immigrants who are underserved by environmental education centers where conservation concepts are communicated in English.

In this project, we assume the following hypothetical arrangement. As part of its marketing strategy, called *Making Environmental Education Accessible with Smart Locations,* Los Angeles County has appointed you to help evaluate the location of four proposed sites to house a new graphic-intensive environmental education center for the non-English-speaking population of Los Angeles County. The County has identified four available office buildings that would serve as excellent high-tech education centers, but only one can be selected as the center. The four candidate locations are in Torrance, Santa Monica, Pasadena, and Burbank. Los Angeles County wants you to consider the four sites and determine the one that is most accessible to the non-English-speaking population of the county.

Course Concepts Presented

Population potential, gravity model, and accessibility.

GIS Concepts Utilized

Table editing, calculation, string manipulation, queries, relational database, and spatial query.

Skills Required

ArcGIS Version 9.x

GIS Platform

ArcGIS Version 9.x

Estimated Time Required

1 hour

Exercise CD-ROM Location

ArcGIS_9.x\Chapter_07\

Filenames Required

edsites.shp
lacounty.shp
La_tracts.shp

Background Information

Population Potential

We employ a technique referred to as a population potential (Warntz, 1964) or more generally known as an accessibility index (Plane & Rogerson, 1994):

$$v_j = \sum_{i=1}^{r} (p_i/d_{ij})$$

where v_j, at any site j is found by first taking the population, P_i, of every subarea i in the study area and dividing by the distance d_{ij} from its centroid to the site. Then these ratios are summed up for all r subareas, $i = 1, 2, \ldots, r$.

The larger the accessibility value for a location, the more accessible that location will be to a population distribution.

Readings

(1)   Coffey, W. J. (1981) *Geography: Towards a General Spatial Systems Approach.* New York: Methuen.

(2)   Greene, R. P., and Pick, J. B. *Exploring the Urban Community: A GIS Approach,* Chapter 7.

(3)   Hansen, W. (1959) "How accessibility shapes land use." *Journal of the American Institute of Planners* 25:72–77.

(4)   Plane, D. A., and Rogerson, P. (1994) *The geographical analysis of population: With applications to planning and business.* New York: Wiley.

(5)   Warntz, W. (1964) "A new map of the surface of population potentials for the U.S., 1960." *Geographical Review* 54:170–184.

Exercise Procedure Flowchart

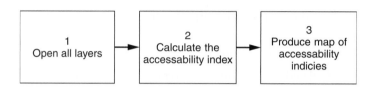

Exercise Procedure

Part 1

1.   Start ArcMap with Start Using ArcMap With A New Empty Map selected. Click OK.

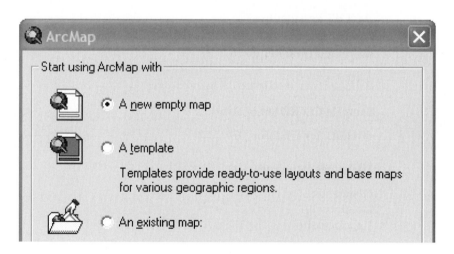

2.   Click Add Data and navigate to the exercise data folder.

   Note: *If you do not see the folder where you placed the exercise data, you will need to make a new connection to the root directory that holds the data.*

3.   Add edsites.shp and La_tracts.shp from the Chapter 7 folder. (Do not add lacounty.shp; you will add that later.) You can add the two layers at once by clicking on one of the file names,

clicking on the other one while holding down the Ctrl key on the keyboard, and then clicking Add.

4. Right-click La_tracts and click Zoom To Layer. For drawing order, edsites should be on the top and La_tracts on the bottom. If it is not you can click a layer name in the table of contents and drag the layer to a new position.

5. Open the attribute table for La_tracts by right-clicking its name and then clicking Open Attribute Table.

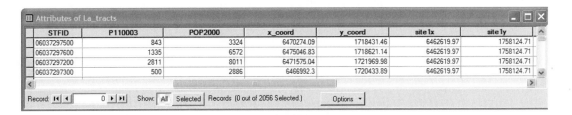

6. Note that the table includes the following fields:

    P110003: 2000 Population for which language spoken at home is other than English.
    POP2000: 2000 Population.
    X_coord: The x coordinate for an internal point within a census tract.
    Y_coord: The y coordinate for an internal point within a census tract.

7. Scroll over and note the next 8 fields:

    Site1X: The x coordinate for the point representing the Torrance site (site 1).
    Site1Y: The y coordinate for the point representing the Torrance site (site 1).
    and so on.

site1y	site2x	site2y	site3x	site3y	site4x	site4y	access1	access2	access3	access4
1758124.71	6413726.63	1829210	6511656.15	1882349.7	6459350.89	1883166.97	0	0	0	0
1758124.71	6413726.63	1829210	6511656.15	1882349.7	6459350.89	1883166.97	0	0	0	0
1758124.71	6413726.63	1829210	6511656.15	1882349.7	6459350.89	1883166.97	0	0	0	0
1758124.71	6413726.63	1829210	6511656.15	1882349.7	6459350.89	1883166.97	0	0	0	0

Record: 0   Show: All Selected   Records (0 out of 2056 Selected.)   Options ▾

Part 2: Calculating the Accessibility Index

The following formula is used for calculating the accessibility index for each site:

1. $A_1 = (noneng_1/dist_1$ to $Site_1)+(noneng_2/dist_2$ to $Site_1)+ \ldots (noneng_{2056}/dist_{2056}$ to $Site_1)$
2. $A_2 = (noneng_1/dist_1$ to $Site_2)+(noneng_2/dist_2$ to $Site_2)+ \ldots (noneng_{2056}/dist_{2056}$ to $Site_2)$
3. $A_3 = (noneng_1/dist_1$ to $Site_3)+(noneng_2/dist_2$ to $Site_3)+ \ldots (noneng_{2056}/dist_{2056}$ to $Site_3)$
4. $A_4 = (noneng_1/dist_1$ to $Site_4)+(noneng_2/dist_2$ to $Site_4)+ \ldots (noneng_{2056}/dist_{2056}$ to $Site_4)$

where A_1 is the accessibility index for the first education site, $noneng_1$ is the non-English-speaking population for census tract 1, $dist_1$ is the distance between the non-English-speaking population of census tract 1 and the location of the first education site, $noneng_2$ refers to census tract 2, and so on ..., and $noneng_{2056}$ refers to census tract 2056 (there is a total of 2,056 census tracts in Los Angeles County). Similar calculations are used for the remaining three education sites.

1. Scroll over and note the final 4 fields of the table:

    Access1: Field to compute accessibility for the site at Torrance
    Access2: Field to compute accessibility for the site at Santa Monica

Access3: Field to compute accessibility for the site at Pasadena
Access4: Field to compute accessibility for the site at Burbank

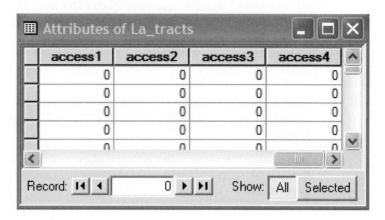

2. To calculate Access1, for Torrance, right-click the field name and click Calculate Values.
3. Click Load and open the site1 formula. Click OK.

   *The site1 formula is a script for the term P_i/d_{ij} for site 1 in Torrance. This way you do not have to type the four formulas. Take a look at one of the formulas to gain an understanding of it. Note that the distance is being divided by 5,280, which is a conversion factor of feet to miles.

4. Repeat steps 2 and 3 for Access2, Access3, and Access4, loading their respective formulas (site2, site3, and site4).
5. To calculate the accessibility of each site, right-click the field names (access1-4) and click Statistics.

   Record the sum for each:

Site 1	Torrance	357,902
Site 2	Santa Monica	339,914
Site 3	Pasadena	416,231
Site 4	Burbank	447,272

   Clearly the proposed Burbank Educational Center is the most accessible to the population that speaks a language other than English at home. Close the table.

6. Add a field to edsites called Access and calculate it to equal the above values.

   6.1. Right-click edsites and select Open Attribute Table.
   6.2. Click Options, and the click Add Field.
   6.3. For name type, select Access.
   6.4. For type, select Long Integer.
   6.5. For precision, select 8.

7. Leave the table open and right-click on the gray menu bar at the top of the screen and turn on the Editor toolbar. Alternatively, you can select Toolbars from the View menu.

   7.1. Click the Editor tab on the Editor toolbar and click Start Editing.
   7.2. Make sure edsites is the target.
   7.3. Click the topmost cell in the field column for Access. This corresponds to site 1, Torrance. Type in the value 357902.
   7.4. Repeat this process to enter the remaining values for Access.
   7.5. Click the Editor tab on the Editor toolbar, then click Stop Editing. Answer Yes to Save Edits.
   7.6. Close the table.

8. Right-click the edsites layer name and select Properties. Click the Symbology tab of the Layer Properties dialog box. Next click Quantities and then click Graduated Symbols. Click

the Fields Value drop-down list and select Access. Click Apply and then click OK. Your map should look like the following:

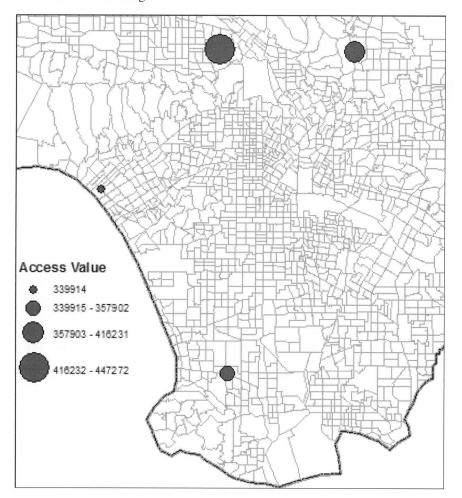

9. Select Layout View from the View menu to compose a map for printing.
10. Click Insert and then click Legend.
11. In the Legend Wizard dialog box, make sure that the only legend item is Edsites. For instance, select La_tracts in the right column and use the single arrow symbol to move it to the left column.
12. Click Next, then erase the word Legend by backspacing, and replace it with Access Value.
13. Click Next in the next several dialog boxes and then click Finish.
14. Now double-click on the selected legend. Click the Items tab and in the right column, double-click Edsites, click Properties, and click General. Clear the Show Layer Name and Show Heading, click Apply, and click OK. Click OK as you exit back. Reposition the legend to a suitable spot.
15. Click Insert at the top and insert a title for your map. Reposition the title if necessary.
16. To achieve the same map as just shown, you could make the LA tract symbol hollow with a gray outline and add the lacounty layer, also making it hollow with a county outline symbol.

Part 3: Essay

In your essay, review the chapter section and the readings on population potentials and discuss the strengths and weaknesses of this method for evaluating the four potential government-sponsored environmental education sites in Los Angeles County. Suggest ways to enhance the method. An example of something to consider is whether the method recognizes transportation, and, if not, how transport can be reckoned with.

8

Race, Ethnicity, Gender, and Poverty

This chapter concerns the geography of race, ethnicity, gender, and poverty. These fundamental aspects of cities go back millennia. For instance, Roman cities had racial and ethnic groups and some residents lived in poverty. The chapter focuses on these topics in the context of the United States; there is some discussion of the issues in Mexico, Central America, and Europe. The chapter considers the geographic background of ethnic enclaves and identity, followed by a section on the concepts, processes, and measurement of race, ethnicity, and segregation. It discusses the U.S. experiences with immigration, ethnic distribution, and segregation. A growing knowledge about gender in cities is examined, including studies leading to better understanding of gender and sexuality in the city and the feminist view of geographic information systems (GIS). Gender differences lead to varied perceptions, experiences, and priorities in cities. One section defines poverty and explains how to measure it. Then, poverty is considered for the United States and its cities, including poverty in the ghetto. The last section considers a different type of poverty—that of the *colonias,* that is, largely unplanned communities on the peripheries of U.S. border cities.

Race refers to a large portion of humanity unified by common physical and genetic traits. Often, race involves some type of common history or culture. Examples of races are African American, White, or Asian. *Ethnicity* refers to smaller groups of people classed according to common cultural, national, tribal, religious, or linguistic background. Both race and ethnicity are culturally and socially determined phenomena, although race is more likely to involve common physical characteristics. Examples of ethnic categories are Russian, Cuban, and Hispanic. Jews are an example of an ethnic group determined by religion. Ethnic groups are not sharply divided, and often a person can belong to more than one ethnic group. For instance, a person can be both Argentinian and Hispanic,

or both Polish and Jewish. Hispanic and Latino have different meanings to some and the same meaning to others. We have chosen in this book to regard them as the same and use them interchangeably. The U.S. Census has standardized on Hispanic and does not utilize Latino in its census attributes (U.S. Census, 2003).

A person has both race and ethnicity; for instance, a White Scandinavian, or a Black French person. These concepts are influential in the geography of cities for many reasons. For example, people of racial or ethnic backgrounds often group together in the same neighborhoods or areas, they have shared traditions and cultural backgrounds, and they sometimes act as a voting group in urban politics.

Ethnic or racial distributions in cities are not static but change dynamically as new immigrants arrive and existing groups move to other neighborhoods or depart the city. The multiple ethnic identities of cities give them much of their interest and excitement but at the same time can lead to tensions, conflict, separation, or segregation.

Intermarriage across ethnic or racial lines also influences racial and ethnic change. Some cities that historically had numerous distinctive ethnic neighborhoods today are more homogenous as a result of intermarriage. An example is a comparison of the racial and ethnic mix of the city of Chicago at the start of the 20th century versus the city today. In 1900, Chicago had many distinctive European immigrant communities, among them Poles, Germans, Hungarians, Italians, and Irish. Today, the city still has substantial ethnic groups, but the composition of ethnicities has changed, with a larger proportion of Hispanic ethnic groups and Blacks, and some assimilation of earlier European immigrant population, so they identify more as Americans rather than Poles or other groups. An Italian American neighborhood in Chicago is seen in Figure 8.1.

FIGURE 8.1 Italian American neighborhood in Chicago.

The second section of the chapter focuses on poverty, mostly in the United States but also in Mexico and Europe. *Poverty* refers to the situation in which individuals, households, or families have below-average living conditions, including access to food, clothing, shelter, and money. It is usually measured in terms of the money value, often that of a minimum basket of basic items for living. There is no single worldwide standard for measuring poverty. Rather, many nations have established their own definitions. Those affected by it cannot gain access to amenities that tend to make life more comfortable, not to mention luxuries that many of us can enjoy from time to time. At the same time, people in poverty often work productively. In fact, people in poverty exhibit quite a bit of diversity and specifically do not fit a widespread image of unproductive and antisocial behavior (Jargowsky, 1996). For instance, many students are poor while attending college and some even would be placed under the official poverty line. In the United States and other countries, certain racial and ethnic groups are overrepresented in poverty. For instance, African Americans in the United States and indigenous peoples in Mexico are more likely than average to be impoverished. This chapter examines the history of how poverty became associated with certain ethnic groups.

8.1 ETHNIC ENCLAVES, SOCIAL, AND COMMUNITY IDENTITY

Many cities are known for their ethnic districts, neighborhoods, and sections. Who would think of Boston without its Italian and Irish neighborhoods, and Miami without Cuban sections? Not only do people of the same ethnic identity in a city often tend to reside in the same neighborhoods, but their ethnicity is manifest in the shopping, public events, social life, festivities, restaurants, art, and even music of the area. Figure 8.2 shows a market in such a neighborhood, Greek Town in Chicago. Geographers, sociologists, and urban historians have long been studying ethnic identity and its relationship to city geography and spatial processes. The range of ethnic and racial identities contributes to the larger concept of "urbanism as a way of life" (Wirth, 1938). This concept refers to the total atmosphere of living in the complexities of a city, of which ethnic differences are a part. Along with

FIGURE 8.2 Greek Town in Chicago.

other factors, ethnicity fosters in cities a particular urban feeling, that is, urbanism, that distinguishes them from rural areas (Frey & Zimmer, 2001).

In New York City, the Russian community of "Little Odessa" is centered in Brighton Beach in southern Brooklyn next to Lower New York Bay adjoining the Atlantic Ocean (see Figure 8.3). There are other smaller Russian immigrant zones in central Brooklyn and in the Washington Heights area of upper Manhattan. The Russian community in New York City has developed from the stream of Russian immigrants arriving in the United States since the 1970s. About a quarter of all Russians arriving in the United States settled in New York City and 70 percent of those in Brighton Beach (Miyares, 1998). Brighton Beach, in the 1860s a beach resort, later changed to an entertainment area, and in the 1960s became primarily residential with local businesses (Miyares, 1998). The Russians have transformed it further. They have purchased most of the businesses, and Russian language and signs are commonplace. Russian immigrants in Brighton Beach are seen in Figure 8.4. Many of the immigrants were highly educated, including professionals,

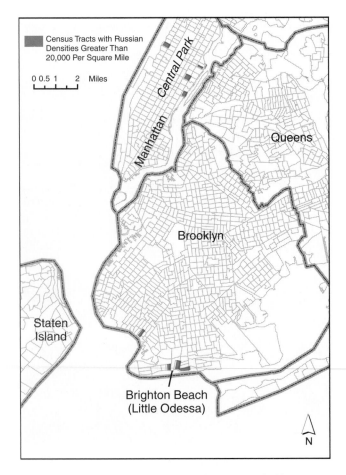

FIGURE 8.3   Russian communities in New York City.

and as a result the process of assimilation has involved less social-service aid than with other immigrant groups (Miyares, 1998). Many of the Russian immigrants have shifted their professional training and management experience from the Soviet Union into entrepreneurial small businesses, which have done well, including retail outlets, professional offices, restaurants, and insurance firms. There is an entrepreneurial spirit, which some have even called a "hustling mentality" (Gold, 1988). The success of Little Odessa and its immigrants has been notable and has contributed to the contemporary economy and ethnic mix of New York.

The Chinese community in Los Angeles represents a longer established ethnic and immigrant group. In 2000, there were 329,000 ethnic Chinese in Los Angeles County, as well as 85,000 in the surrounding metropolitan counties of Orange, Riverside, San Bernardino, and Ventura (U.S. Census, 2003). Chinese had first arrived in the 19th century, and the original Chinatown had moved around to several locations near downtown Los Angeles (Allen & Turner, 2002). Today, a large concentration of Chinese live in the western San Gabriel Valley, extending from Monterey Park northward (see Figure 8.5). This area has tended to attract affluent Chinese immigrants. It serves as the Chinese financial and business center in southern California and some would say in the nation (Zhou, 1998). A second large concentration is located to the southwest of the first one in Hacienda Heights, Rowland Heights, and Walnut. It has extensive Chinese businesses, commerce, and residences. Other areas with Chinese settlement in the Los Angeles area include Cerritos and the Palos Verdes Peninsula, the northern San Fernando Valley, and small clusters around the major universities of University of California at Los Angeles, University of Southern California, Cal State Fullerton, and University of California at Irvine.

In recent decades, the Chinese immigrants to the Los Angeles metro area have tended to be well-educated, professionals, and usually middle class or wealthy. They are often from Taiwan or Hong Kong (Zhou, 1998). This contrasts with Chinese immigrants to New York City, who have tended to be working class and more likely to have come from Mainland China. Los Angeles's attraction to its prosperous group is better financial support to certain of the migrants, as well as economic ties, because there is a connection between the ethnic Chinese economy of greater Los Angeles and the dynamic economy of China (Zhou, 1998). Many Chinese immigrant families have been successful in business and excelled educationally. In greater Los Angeles, this large and successful ethnic group has demonstrated that the factors for economic success are present for an arriving ethnic group, if the circumstances are right. Many other cities in the United States have had significant Chinese immigration and domestic in-migration, leading

FIGURE 8.4   Immigrants in Little Odessa in Brighton Beach, New York. Source: Getty Images, Inc–Liaison.

to development of communities, such as that shown for Seattle's Chinatown in Figure 8.6.

The original ethnic identities have been preserved in many ethnic neighborhoods and zones of cities. Often, use of the native language by immigrants has been reduced greatly after the first generation (Portes & Rumbaut, 2001), but many aspects of culture have been maintained over several generations. In other nations, cities have received foreign and domestic ethnic groups, which have settled into distinct parts of cities and influenced their identity and culture. The city of Montreal started out French and is heavily influenced by the French heritage of the majority of its citizens. Mexico City has communities such as Milpa Alta and Chalco in the southeast and Huixquilucan in the west with high concentrations of indigenous people, from a mixture of specific Indian groups (Pick & Butler, 1997).

8.2 CONCEPTS AND PROCESSES OF RACE, ETHNICITY, AND SEGREGATION

Race, ethnicity, and segregation are measured and analyzed through simple ratios, as well as by more complex techniques. This section introduces you to measuring these phenomena. The techniques include simple percentages, charting of ethnic and racial distributions, the index of dissimilarity for segregation, and the centrographic method to measure spatial location and extent. The centrographic method is utilized later in the chapter's GIS exercise to evaluate the degree to which African Americans are more segregated than Hispanics in St. Louis.

To measure a single racial or ethnic category, the *percentage of a group in a base population* is a common measure. For instance, the percentage Asian for all persons in Los Angeles is calculated as follows:

$$\text{Percentage Asian} = 100 \times \frac{total\_number\_of\_Asians}{total\_population}$$

A real example is the Chicago–Gary–Kenosha consolidated metropolitan statistical area (CMSA) in 2000, which had 385,410 Asians out of a total population of 9,157,540. This example uses the older concept of CMSA, not the combined statistical area (CSA) discussed in Chapter 3. Percentage Asian is as follows:

$$100 \times \frac{385,410}{9,157,540} = 4.2 \text{ percent}$$

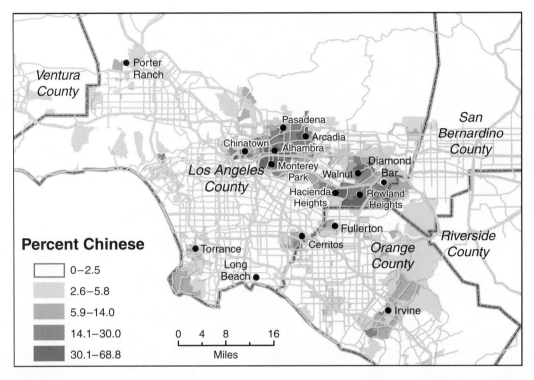

FIGURE 8.5 Percentage of Chinese in the the Los Angeles urban area, 2000. Source: U.S. Census, 2003.

The same approach can be used to calculate the percentage of Puerto Ricans among all Latinos in the Chicago CMSA in 2000. That is estimated as follows:

Percentage Puerto Ricans among Latinos

$$= 100 \times \frac{total\_number\_of\_Puerto\_Ricans}{total\_number\_of\_Latinos}$$

$$= 100 \times \frac{152,229}{1,421,543} = 10.7 \text{ percent.}$$

The relative size distributions of different ethnic groups can be shown by simple *proportionate distribution*. For instance in the New York–Northern New Jersey–Long Island CMSA in 2000, out of 21,104,292 total residents, there were 11,863,430 non-Hispanic Whites, 3,849,990 Hispanics, 3,406,201 non-Hispanic Blacks, 1,423,211 non-Hispanic Asians, 429,324 of two or more races other than Hispanics or Blacks, and 132,136 other races (U.S. Census, 2002). The non-Hispanic categories are present because any census respondent can declare himself or herself as Hispanic or not. Using the non-Hispanic categories means that a respondent will not be counted twice. The distribution can be pictured in a pie chart in percentages (see Figure 8.7).

Racial segregation represents substantial overconcentration of a particular racial group, relative to another racial group, in particular zones of a city and with corresponding substantial underconcentration in other zones. Just where the cutoff becomes substantial is up to the researcher or analyst to decide. This measure is a statistical one based on numbers of people. It does not measure psychosocial aspects of segregation or the intent by parties to create or maintain segregation. An example of the data to estimate segregation is the simplified data set on percentage Black versus the percentage White shown in Table 8.1.

Racial segregation can be calculated by the *index of dissimilarity*, which has the formula

$$Index\_of\_Dissimilarity = 0.5 \sum |m_i - n_i|$$

where m_i is the ratio of the number of minority persons in the ith geographic zone to all minority persons, and n_i is the ratio of the number of nonminority persons in the ith geographic zone over all nonminority persons.

On the basis of the table, we calculate

Index of Dissimilarity

$$= 0.5 \times (|0.8 - 0.2| + |0.1 - 0.4| + |0.5 - 0.25| + |0.5 - 0.15|)$$

$$= 0.75$$

When you formulate this, the most important decisions are which group to select as the minority group and

FIGURE 8.6   Seattle's Chinatown.

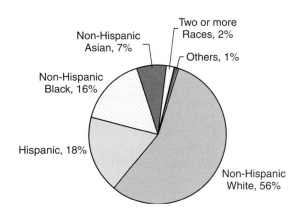

FIGURE 8.7   Racial distribution in the New York–New Jersey–Long Island consolidated metro region 2000. Source: U.S. Census, 2003.

which comparison category to select as the nonminority group. Values for the index of dissimilarity range from 1 (maximum dissimilarity) to 0 (maximum similarity). Values for the index are given later in the chapter for racial segregation in the largest U.S. metropolitan areas in 1990.

The index of dissimilarity is useful in measuring segregation within a single geographic area community, but it does not summarize segregation for a spatial group of communities. We are sometimes interested in asking how spatially concentrated or dispersed racial or poverty attributes are across an area. This topic of

spatial segregation is covered in the next section, "Analyzing an Urban Issue," which corresponds to the GIS exercise at the end of the chapter.

8.3 ANALYZING AN URBAN ISSUE

The *centrographic method* overcomes the problem that the index of dissimilarity does not measure the spatial distribution of population. On the other hand, the centrographic method measures segregation by taking into account the reference community, as well as the surrounding area; it measures the extent of concentration or dispersal around a core point. This method can answer questions about which racial groups in a city are more or less dispersed.

The *mean center* locates the central core for the racial, ethnic, or poverty attribute. It is specified by an *x-y* point location. The *standard radius* measures the amount of dispersion of the attribute across the area. This dispersion is most commonly the standard deviation (Wong, 2003). To apply the method, we need to calculate the weighted mean center and the standard radius. The standard radius is the mean distance from the center for one standard deviation, or for about two thirds of the population. The

TABLE 8.1   Example: segregation data and calculations

Zone of example city	Number of blacks	Number of whites	Total population in the zone	Proportion of blacks	Proportion of whites
A	80	20	100	0.8	0.2
B	10	40	100	0.1	0.4
C	5	25	100	0.05	0.25
D	5	15	100	0.05	0.15

radius forms a circle around the mean center that shows how concentrated or dispersed a characteristic is spatially. The weights are proportional to the concentrations of the attributes at particular x,y coordinate locations (Wong, 2003).

For example, Japanese households in a U.S. city might have a mean center located 12 miles north and 4 miles west of the CBD (x-y point location). The weights represent the average number of individuals per household for the small areas being measured. The standard radius is calculated to be 5 miles. This means that about two thirds of the Japanese individuals are located within 5 miles of the mean center. A large standard radius shows wide dispersion, whereas a small standard radius implies concentration in one portion of the area.

The formula for x and y coordinates of the weighted mean center is the following:

$$x_w = \frac{\sum xw}{\sum w}, \; y_w = \frac{\sum yw}{\sum w}$$

where x and y are geographic coordinates and w is the concentration weight for a particular racial, ethnic, or poverty attribute, and x_w and y_w are the x,y coordinates for the weighted mean center.

The formula for standard radius is as follows:

$$r = \sqrt{\left(\frac{\sum x^2}{n} - x_w^2\right) + \left(\frac{\sum y^2}{n} - y_w^2\right)}$$

where x and y are geographic coordinates and n is the total number of points.

If standard deviational circles are calculated for two ethnic groups, the size of the circles represents roughly how concentrated the groups are. If the circle for the first group is larger than the circle for the second group the first group is more dispersed throughout the area.

The exercise at the end of this chapter provides practice with the centrographic method. Specifically, you will compute the mean center and standard radius for African Americans and Hispanics in the city of St. Louis and neighboring St. Louis County for the year 2000. The results for St. Louis are shown in Figure 8.8. As part of the exercise, you will discuss the relative sizes of the circles and the amount of nonoverlap or overlap of the two groups.

For all these methods, it is important to consider the scale of the phenomenon in relation to what is being measured, that is, the size of the unit of observation. For instance, segregation can be measured by index of dissimilarity for a city on the basis of its census tracts, or for a ZIP code based on its block groups. Each is measuring

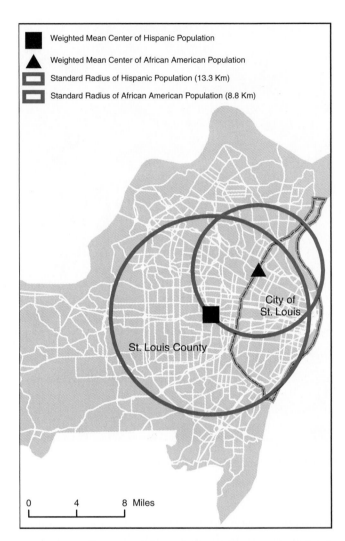

FIGURE 8.8   Centrographic results for ethnic groups in St. Louis.

a somewhat different social phenomenon. The more microlevel segregation of block groups might have different causations compared to segregation of ZIP code areas.

More advanced centrographic techniques compute *standard deviational ellipses* instead of standard deviational circles (Wong, 2004; Gong, 2002). To do this, for each ellipse, the method provides calculations of the weighted mean center, the directional orientation of the ellipse, and the dispersions for the long and narrow axes of the ellipse. As with circles, the extent of integration or segregation can be approximated by the size of the two or more ellipses, each representing an ethnic, racial, or poverty group. In addition, the ellipse offers the added value of indicating the spatial direction of greatest dispersion. In sum, a variety of centrographic techniques can be utilized to measure ethnic, racial, and poverty levels and distributions.

8.4 CITIES, IMMIGRANTS, AND ETHNICITIES

Many growing cities are built up partly through immigration over time, whereas others grow through natural increase. Periods of high immigration can occur at different time points. For instance, more southern European immigrants arrived in New York City in large numbers at the turn of the 20th century; German cities in the 1970s and 1980s received large numbers of Turkish immigrants; Israel's cities received Russian immigrants in the 1980s; and America's recent port-of-entry cities such as Los Angeles, Houston, Chicago, and Washington, DC have received large volumes of Hispanic and Latino immigrants since 1980.

As seen earlier, migrants arriving in cities bring along cultural, language, occupational, artistic, and business traits, skills, and experiences that impact the way of life of the city. At the same time, immigrants tend to become more assimilated and acculturated over generations. For instance, it is well known that most immigrant children arriving in U.S. cities will rapidly learn English and adapt to American customs and culture, while retaining some native cultural customs, whereas their grandparents will often continue to depend on their native languages and to maintain customs from their homelands (Portes & Rumbaut, 2001). Different immigrant groups vary in their speeds of adaptation.

Many scholars have studied the extent and speed with which the "melting pot" in the United States absorbs different ethnic groups (Glazer & Moynihan, 1970; McKee, 2000a; Schlesinger, 1992). They do not agree on the extent or rapidity of assimilation and acculturation, or even on whether the melting pot is a plus or minus. However, they largely agree that assimilation depends on the time of arrival, institutions and processes present for assimilation, support from the native population, closeness or distance of the original country, culture, language, and other factors (McKee, 2000b). In the United States, for instance, the presence of a widespread public school system that stresses native language and culture has been a consistent stimulus to assimilation and acculturation (Portes & Rumbaut, 2001). On the other hand, resentment against arriving immigrant groups by dominant groups has sometimes slowed the process (McKee, 2000a). Examples in the United States were resentment of Chinese arriving in the late 19th century and resentment of immigrant groups sometimes associated with wartime enemies, such as Japanese Americans during World War II. The geographical aspect of this process is examined later in the chapter. This has been true recently with the new Hispanic and Asian immigrants of the 1990s, who often arrived in Los Angeles and large Texas cities, but now are dispersing to medium-sized cities stretching from the midSouth to the Pacific Northwest (Frey, 2002a). Thai

FIGURE 8.9 Thai restaurant in Seattle.

immigrants have settled in Seattle, as seen from the Thai restaurant shown in Figure 8.9. There is also a progressive movement of immigrant groups and their descendents over time within metropolitan areas. The example of residential mobility in Chicago was looked at in Chapter 7.

8.5 ETHNIC IMPACTS OF WAVES OF IMMIGRATION TO THE UNITED STATES

In a broad perspective, nations and their cities are built up and changed through waves of immigration, that is, surges from a particular origin. In the past century in the United States, a wave of southern and eastern Europe immigrants arrived, particularly in the first decade of the 20th century. This immigration surge began as the result of wars and economic privation in Europe in the 19th century, combined with the draw of a relatively prosperous and peaceful America. As seen in Figure 8.10, in 1900, out of America's population of 76 million, there were 10.3 million foreign-born population, or nearly 14 percent. Most of the immigrants settled in cities, especially in large cities such as New York, Philadelphia, Boston, and Chicago (U.S. Census, 2001). After 1900, the foreign-born

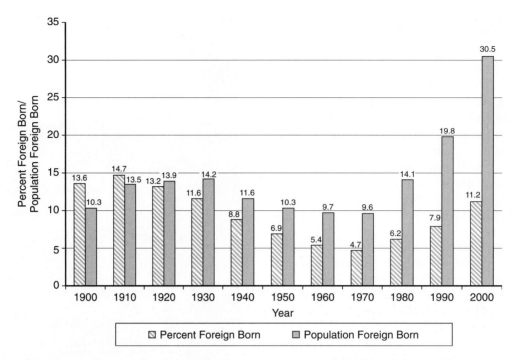

FIGURE 8.10   Foreign-born population, percentage and total, in United States, 1900–2000.

proportion grew somewhat to 14.7 percent in 1910, and then declined steadily to only 4.7 percent in 1970, as the earlier wave of immigrants died off. The reasons for this 60-year slide are varied. The U.S. experience with World War I led the nation to become more isolationist in the late 1920s and 1930s. During World War II, restrictions remained tight, although there were some refugees admitted, and the war itself cut off many traditional sources of immigrants. The 1950s through 1970s were the decades of the Cold War and relatively restrictive immigration policies, although millions of immigrants continued to arrive (U.S. Census, 2001).

Starting in the 1980s, immigration into the United States grew again. During the 1980s, the U.S. population age distribution became steadily older, opening up jobs for younger workers, many of which could be filled by young, unskilled immigrants. The reasons are several. First, since the 1980s, the U.S. population has aged, so demand for unskilled jobs has grown. Immigrants tend to be young and fill in this gap. Second, international trade and other forms of global economic exchange expanded, and a global workforce developed. Third, wars, refugees, economic recessions, and privations in other nations created "push" factors to send immigrants into the United States. By 2000, the percentage of foreign-born population in the United States was at 11.2 percent, the highest level since 1930, reflecting an all-time record number of 30.5 million foreign born residents.

As immigration numbers have fluctuated, so too have the origins of immigrants. As seen in Figure 8.11a, the

wave of mid-19th-century immigrants to the United States came predominantly from northern and western Europe and from Canada. However, in the late 19th and early 20th centuries, the immigrant proportions shifted until half the immigrants were from southern and eastern Europe (Figure 8.11b). By the mid-20th century (1931–1960), the volume of immigration had greatly subsided, but the mix shifted again, to a greatly expanded proportion from Latin America, mostly from Mexico (Figure 8.11c). In the 1960s and 1970s, Latin American immigrants accounted for 39 percent of inflow, and Asian immigrants 24 percent (Figure 8.11d). In the 1980s and 1990s, half of immigrants were from Latin America, a third from Asia, and only 13 percent from Europe, as seen in Figure 8.11e. This late-20th-century flow was primarily economically driven.

This progression of sending nations and regions has impacted American cities in varying ways over the 180 years shown. As the immigrant waves have come and been supplanted, the ethnic mix and geography of settlement patterns within particular cities and metropolitan areas have changed. Understanding for a given city the changing mixture of immigrants and the spatial context of their arrival, dispersion, and assimilation is part of the study of urban geography.

The geographical arrival patterns are different today from a century ago. Today, the largest arrival metro areas for overseas immigrants are the Los Angeles consolidated metro area (CMA), followed by the New York CMA, San Francisco–Oakland–San Jose, Miami–Fort Lauderdale, and Chicago (see Table 8.2). The first three account for

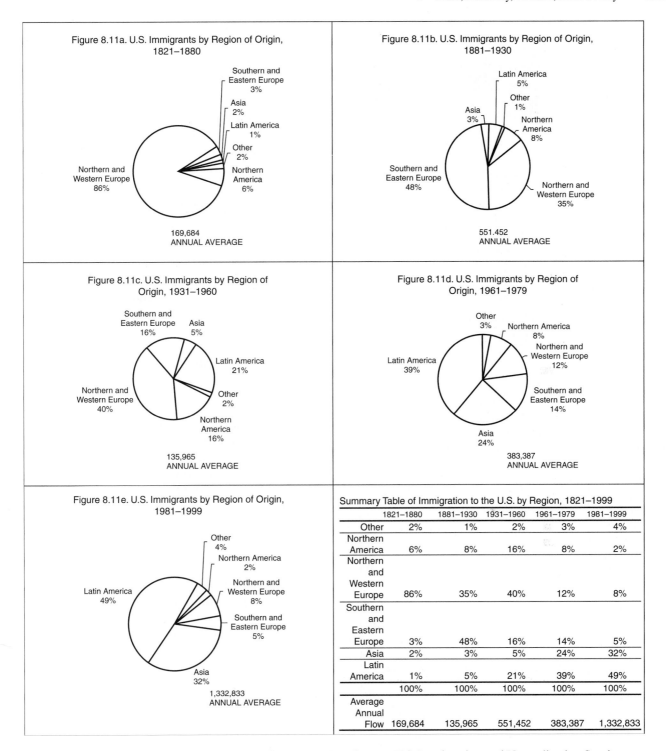

FIGURE 8.11   U.S. immigration by region of origin, 1821–1999. Source: U.S. Immigration and Naturalization Service.

11 million of the nation's foreign born, or 37 percent. All five of these cities are known for ethnic diversity and multicultures. A century ago in 1910, eastern and midwestern cities accounted for a large share of the foreign-born population. At that time, most immigrants settled in New York, Chicago, Philadelphia, Boston, and the younger city of Cleveland, all of which were known for their ethnic and racial diversity. These were "older" and more traditional cities, except for Chicago and Cleveland. They attracted immigrants because of their prosperous economies and jobs, as well as the networks of foreigners of the same origin already present.

TABLE 8.2   Metropolitan areas with the largest number of foreign-born population, 1910 and 2000

1910		2000	
City	Number of foreign born	Metropolitan area	Number of foreign born
New York	1,927,703	Los Angeles-Riverside-Orange Country CMSA	4,708,000
Chicago	781,217	New York-NJ-CT-PA CMSA	4,690,000
Philadelphia	382,578	San Francisco-Oakland-San Jose CMSA	2,007,000
Boston	240,722	Miami-Fort Lauderdale CMSA	1,647,000
Cleveland	195,703	Chicago-Gary-Kenosha CMSA	1,070,000

CMSA = metropolitan statistical area.
Source: Spain, 1999; U.S. Census, 2002.

This is typified by contrasting Chicago and Los Angeles over the past 130 years in their levels of foreign-born residents (see Figure 8.12). For a century, between 1870 and 1970, Chicago attracted more immigrants, resulting in a higher foreign-born proportion than Los Angeles.

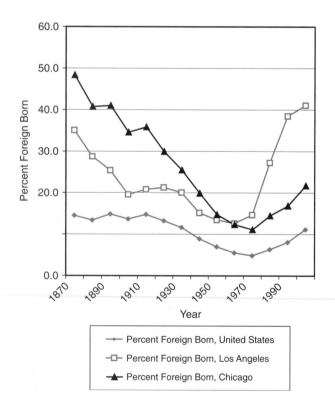

FIGURE 8.12   Percentage foreign born in Los Angeles, Chicago, and United States, 1870–2000.

Since then, Los Angeles has experienced immigration leading to twice the percentage foreign born. Because Chicago was much larger during most of this period until after World War II, there were a much larger number of immigrants coming through Chicago.

Cities in other nations have also been influenced by waves of immigration. Buenos Aires and other large cities in Argentina received huge immigration inflows from Italy and southern Europe in the late 19th century and early 20th century. The influence is seen today in many elements of Argentine culture, arts, customs, and architecture. Australia and its cities received substantial immigration from Asia in the late 20th century. As seen in Chapter 7, waves of immigration are dynamic and influenced by push–pull and political factors that send immigrants from certain nations, attract them through economic opportunities, and regulate their flow through policies stemming from the countervailing political influences such as isolationism, globalization, national security, organized labor resistance, and workforce needs.

8.6 IMMIGRATION AND MIGRATION OF AFRICAN AMERICANS AND HISPANICS TO THE UNITED STATES

The immigration and internal migration surges have affected all major ethnic groups in the United States and their geographical distribution in cities. This section gives particular attention to the U.S. history of geographic change for African Americans and Hispanics or Latinos. We emphasize the influences of these immigration waves on U.S. cities.

8.6.1 Blacks

Blacks were first forcefully brought to the United States in 1619 as part of the slave trade. That trade continued over the next two centuries, and slaves resided mostly in the rural areas of the South, where their labor would be most useful to agriculture. After the abolition of slavery at the end of the Civil War, most Blacks remained in the South but others began to migrate to the North. Following the Civil War, there was segregation, although not at the levels we think of today. In southern cities, Black tended to live nearby their White employers, whereas in northern cities, Blacks were sometimes accepted nearly as the equal of Whites (Massey & Denton, 1993). In the late 19th century, African Americans started a long flow of migration from the rural South to the urban northern cities, and later also to growing southern cities. On arriving in the North, however, many migrants found that their dreams were not realized. One reason was the resistance of receiving cities to accepting larger proportions of Blacks. This resistance was also a factor leading to the start of the urban Black ghetto in the early 20th century.

A *ghetto* is defined as a neighborhood or group of neighborhoods nearly entirely inhabited by a single racial or ethnic group, and within which a large percentage of the city's members of that group live, especially due to social, legal, or economic pressure. It is important to emphasize that the Black urban ghetto was not present in the 19th century, a time when Blacks and Whites lived together in the North with a moderate amount of harmony (Massey & Denton, 1993). The major underlying cause of the increasing dominance of ghettos was the industrialization of the North, which required population to support new transformative industries. This implied the need for low-paid manufacturing workers, which offered opportunities to many arriving African Americans. It was convenient to build housing for these workers near the manufacturing plants, resulting in high concentrations of Blacks and, eventually, ghettos.

After World War I, the United States became increasingly isolationist and, partly as a result, the flow of European immigrants was sharply reduced (Massey & Denton, 1993). This opened up the opportunity to fill the gap through a much higher migration of southern Blacks to the North. Between 1910 and 1902, 525,000 of them migrated to northern cities, and from 1920 to 1930, 877,000 southern African Americans entered this flow. The arrival of massive numbers of Blacks led to resistance from some Whites that added to the tendency to form ghettos, increased housing barriers for Blacks, and added to tensions that in certain cases boiled over into race riots (Massey & Denton, 1993). The color line became stronger and segregation increased. Segregation of African Americans in northern cities grew from an average index of dissimilarity of 46 in 1860 to 59 in 1910 to 89 in 1940 (Massey & Denton, 1993).

In the era from 1940 to 1970, tensions increased. Especially during the 1950s, there was White flight from northern central cities to the suburbs. *White flight* represents movement of White population from the central city to the suburbs, often due to the perceived need to get away from economic, educational, racial, and other types of problems and conflicts. Meanwhile, African American migration from the rural South continued, and most of the large Black ghettos continued to grow in size. Even though major urban riots involving mainly Black protest occurred between 1963 and 1968, the civil rights movement of the 1960s brought about legal and regulatory improvements. Some discriminatory practices such as "redlining" were abolished. *Redlining* refers to financial institutions withholding home loans or insurance from residents in neighborhoods regarded as poor financial risks, often for cost-related reasons.

The Fair Housing Act of 1968 made discrimination in the sale or rental of housing illegal. From the 1960s to today, some improvements have taken place as the suburbs have opened up more to middle-class Blacks,

especially in certain suburban sectors. Some additional discriminatory real estate practices have been curtailed. However, urban ghettos have persisted as the result of continuing economic and occupational barriers.

In 2000, African Americans numbered 34.6 million, constituting 12.3 percent of the country's population (U.S. Census, 2003). Fifty-five percent of Blacks lived in the central cities of metropolitan areas, versus only 22 percent for non-Hispanic Whites (U.S. Census, 2003). Other characteristics are given for African Americans and non-Hispanic Whites in Table 8.3. The percentage of families that are married couples reflects family stability. A lower percentage implies that there is more divorce, separation, and widowhood. Studies of Black families have indicated a higher proportion of single-parent families headed by the mother, compared to non-Hispanic Whites. This can add to the economic burden on Black families because there is only one breadwinner. It is clear that, although much progress has been made in status, including an expanded Black middle class, overall a gap remains socioeconomically for the cross-section of Blacks. An urban, lower socioeconomic neighborhood in Chicago that is mostly African American is pictured in Figure 8.13.

8.6.2 Hispanics and Latinos

The Hispanics and Latinos in the United States originated from many different Latin American countries. However, of Hispanics and Latinos indicating a specific country of origin in the 2000 Census, 71 percent were of Mexican origin, followed by Puerto Rican (11.6 percent), Central American (5.8 percent), South American (4.6 percent), Cuban (4.2 percent), and Dominican (2.7 percent). This section focuses on the surges of Hispanic immigration and domestic migration after arrival. Besides African

TABLE 8.3 Selected characteristics in the United States. African Americans versus non-Hispanic Whites in year 2000

Characteristic	African Americans (%)	Non-Hispanic Whites (%)
High school education, persons 25 years and older	77	88
Bachelors degree	15	28
Managerial/professional occupation	17	32
Family annual income over $75,000	23	33
Poverty	26	8
Married couple families	47	82
Families with woman family head without a spouse	45	13
Source: U.S. Census, 1999.		

FIGURE 8.13   Inner-city neighborhood, Chicago. Source: Corbis/Bettmann.

Americans and Hispanics, other ethnic groups (e.g., Greeks, Vietnamese, Russians, and Armenians) arriving in America have had highly interesting and unique immigration histories, for which the student should refer to the chapter references (see many histories in McKee, 2000a; Spain, 1999).

The first Mexican Americans came along with the U.S. annexations of Texas in 1845 and of Arizona, New Mexico, California, western Colorado, Utah, and Nevada in the Treaty of Guadalupe Hidalgo in 1848 and the Gadsden Purchase of 1853, although the population was very sparse in these purchased territories. All this territory had previously been a part of the nation of Mexico since its independence in 1821 and before that the Spanish colony of Mexico. During the late 19th century, Mexican American population expanded slowly in these annexed territories.

In the 20th century, a variety of U.S. policies on immigration and economic push and pull factors have had varying effects on the volume of Mexican immigration. From 1900 to 1930, there was migration, largely legal, to take low-paying jobs in the United States. From the

1940s to today, a number of U.S. government programs and laws have affected immigration. For example, the Bracero Program, which lasted from 1942 to 1964, involved legal and temporary migration of experienced Mexican agricultural workers or *braceros,* to the United States to fill labor shortages. It was spurred originally by the farm-labor shortages in the United States stemming from the domestic workforce demands of World War II. Some braceros ran into resistance from racist and extremist groups in the United States. More than 3 million braceros entered the United States under this program. It was finally ended in 1964 because of a perceived excess of agricultural laborers and the increasing use of the mechanical harvester for cotton (Border Agricultural Workers Project, 2004).

In the last two decades of the 20th century, there occurred a vast and unprecedented form of migration of Mexicans to the United States. These immigrants were a mixture of legal, naturalized, and undocumented. It is estimated that more than 10 million Mexicans arrived in the United States during this time. The motivation was to fill jobs, mostly unskilled, in manufacturing, low-level services, and agriculture. Before immigrating to the United States, Mexicans often migrated first to Mexican border cities, such as Tijuana, seen in Figure 8.14. Mexican American populations grew particularly in the border states of California, Arizona, New Mexico, and Texas. As a result, the cities today with the largest Mexican American populations are Los Angeles, San Diego, San Antonio, Houston, Chicago, El Paso, Dallas, and Phoenix. All these cities, except Chicago, are in border states or fairly close to the international border with Mexico.

Los Angeles is a special case. By 2000, Los Angeles CMSA, with a total population of 16.4 million, had a Hispanic population of 6.6 million, of which 5.3 million were estimated to be Mexican (United States Census, 2002; Allen & Turner, 2002). It is remarkable that metropolitan Los Angeles' Hispanic population grew by 1.8 million during the 1990s, building up the urban area as the nation's largest Latino center. This growth was due not only to migration but also to high fertility. The enhanced fertility stems from cultural patterns in Mexico of higher fertility than in the United States. There are recent signs, however, that overall in the United States, the proportion of Mexican immigrants settling in Los Angeles is lowering compared with the remainder of the country (Allen & Turner, 2002).

One offshoot of the successive immigration surges is a cultural mixing that is quite evident in the U.S.–Mexico border region, a mixing that is sometimes referred to as "MexAmerica" (Garreau, 1981; Langley, 1987). This implies a blending of Mexican and American cultures to constitute a new area of North America. The mixing can be seen in cultural traditions, cuisine, mannerisms, and work habits common to both sides of the border that give

FIGURE 8.14   Pedestrians in Tijuana, Mexico.

greater affinity within MexAmerica than with the rest of the two countries. How far this melding will go and its ultimate form are yet unknown and might depend on future intermingling, cross-border cooperation, and new forms of economic exchanges.

Residential distributions within cities are also changing as a result of immigration. In the 19th century, for Mexican Americans in U.S. border cities, the residential location of population was around central plazas, as was true in Mexico. As border cities developed more mixed character (i.e., Mexican Americans mixing with Anglo Americans), the Mexican American areas near the plazas became known as *barrios,* which translates in English as neighborhood. These neighborhoods developed a distinctive Mexican demographic composition and cultural aspect. As the border cities grew rapidly, Mexican Americans residing near plazas of adjacent towns became absorbed by the expanding city. As a result, some barrios today are located away from the center of the bigger city (Arreola, 2000).

The barrios over time became segregated, as had the ghettos for Blacks. The reason was that many Whites moved out because of higher Mexican American percentages. Whites perceived that they would be better off economically and educationally with reduced problems

away from the barrio. In a succession process, the barrio's housing stock deteriorated and became cheaper, leading to a replacement influx of more Mexican Americans (Arreola, 2000). In metropolitan Los Angeles, a "mega-barrio" has developed that in 2000 was nearly 200 square miles in area (Arreola, 2000). Part of this heavily Latino zone is east of downtown Los Angeles (see Figure 1.4). Its present-day subareas of Boyle Heights, Lincoln Heights, and City Terrace demonstrate this succession process; they were formerly Jewish areas, but as the former residents moved out to the already heavily Jewish west side areas, more Latinos moved in, transforming these areas into barrios (Arreola, 2000).

Another example of a largely Hispanic city is San Antonio, Texas. In 2000, the metropolitan area was 51 percent Hispanic or Latino. In San Antonio, other minority groups are relatively small; in fact all the other non-White racial and ethnic groups combined, including Blacks, together comprise only 20 percent of residents. It is a city demographically dominated by Hispanic population, giving a strong cultural aspect to the city. Miami is an analogous case, but with a different cast. The metropolitan area is 57 percent Hispanic or Latino, of which 29 percent are Cuban, with 28 percent other Hispanic or Latino (U.S. Census, 2002). Cubans number 651,000. The immigration of Cubans into Miami came as a refugee stream following the Cuban Revolution in 1959. Subsequently, Cuban migrants were attracted to Miami from other major American metro areas (Gober, 2000). On average, the Cuban community in Miami has done very well, with the second generation increasingly moving into the middle class and city leadership (Portes & Stepick, 1993). Miami has many Cuban neighborhoods that reflect cultural traditions, such as Little Havana. It is an example of transplanted culture, with acquisition of some American characteristics and retention of others that are Cuban. Part of the reason for maintenance of the traditions is that many Cuban migrants, at least in the early years, intended to return to Cuba after the Castro regime changed.

In summary, Hispanic and Latino immigration has grown as a factor in the United States ever since the abrupt annexation and purchase of a large part of Mexico more than 150 years ago. Waves of immigration have occurred that have depended on economic factors, labor shortages, and in the case of Cuban immigrants, forced refugee streams. The Hispanic and Latino population has until now settled predominantly in major cities in the border states, Chicago, and Miami. However, that pattern today is changing with settlement across a broad variety of medium-sized cities.

8.7 RACIAL SEGREGATION

Racial segregation was defined earlier in the chapter as a geographic overconcentration of a particular racial group. It is often accompanied by the intent by parties to create or maintain segregation. This can include exclusion by one party of another from jobs, education, housing, social privileges, and economic assets.

There are also noncoercive factors that influence persons of the same race or ethnicity to live in close proximity to other similar persons. These factors include similarities in culture and language; social networks; and support for settlement, acculturation, and assimilation. Most of these factors have to do with immigrant or migrant groups, which can adjust and find greater support in segregated areas. An example of cultural factors in segregation would be Italian communities that are present in many U.S. cities; people of Italian background might value the support networks, common cultural and language features, and greater social networking opportunities, and thus prefer to live in somewhat segregated Italian neighborhoods.

Racial segregation was present in the United States at its origins; slavery existed here from 1619 until 1865. After the Civil War, legal segregation of Blacks continued for many decades, until changes resulting from the civil rights movement of the 1960s.

Other ethnic groups have also experienced varying extents of segregation. In recent decades, Latinos and Asians have been segregated residentially, although not to the same degree as Blacks. Earlier in the country's history, Jews experienced residential segregation; were segregated socially by exclusion from certain organizations such as social clubs; and were restricted from some universities, colleges, and to some extent, high-level professions. In the 19th century, American Indians were segregated onto reservations and many remain so today in a lesser form. Segregation has occurred in many other nations also. In extreme forms, it was present for Jews in 1930s Germany, and for Blacks, Indians, and mixed-race groups in South Africa. In milder forms, it is present, for example, in Mexico for indigenous people and in Turkey for Kurds.

Residential segregation can be analyzed geographically. Determining its distribution of intensity across cities and within cities can help us understand how segregation occurred, what its impacts are on communities, and what policies can help resolve it.

A widely used measure of racial segregation in U.S. cities is the index of dissimilarity. Several studies done in the 1990s introduced this measure for studying the urban geography of U.S. cities (Darden, 2001; Massey & Denton, 1993). An example compares each major ethnic group with the non-Hispanic White population. Using this index, all of the major metropolitan areas in the United States in 1990 (Darden, 2001) show African Americans to be the most segregated (index value 68), followed by Hispanics (46) and Asians (41). The index values for these three groups are shown in Table 8.4 for the 15 largest metropolitan areas. As a reminder, a value of 0 means no residential segregation and a value of 100 implies complete segregation (i.e., separation). Black residential segregation varies from extreme segregation in Detroit and Chicago, to values of less than 50 in Anaheim–Santa Ana and Riverside–San Bernardino, California. Generally metro areas in the East and Midwest have the highest Black segregation, whereas western and southwestern areas are lower. However, Los Angeles is high at 73. Atlanta, Houston, and Washington are somewhat lower at 66 to 68, followed by Dallas and Minneapolis. The Hispanic segregation index varies considerably, ranging from a high of 66 percent in New York to a low of 23 percent in St. Louis (which has very few Hispanics).

For both Hispanics and Blacks, segregation is higher in cities with a larger percentage in the minority group (the correlations between the segregation index and percentage in a minority group are 0.498 and 0.622, respectively). An example is the New York Primary Metropolitan Statistical Area (PMSA), which had relatively high percentages of around 22 to 23 percent for both Blacks and Hispanics, and concomitant high segregation indexes of 86 and 63, respectively. PMSA was a concept used in the 1990 U.S. Census, but not continued for 2000. There were distinctive areas in 1990 in New York for most of the populations of these groups. Asian segregation varies less

TABLE 8.4 Racial residential segregation indices in the 15 largest U.S. metropolitan areas, 1990

	Black segregation index	Black percentage of population	Asian segregation index	Asian percentage of population	Hispanic segregation index	Hispanic percentage of population
Los Angeles-Long Beach, CA	73	10.5	46	10.2	61	37.8
New York, NY	82	23.2	48	6.2	66	22.1
Chicago, IL	86	21.7	43	3.6	63	12.1
Philadelphia, PA-NJ	77	18.8	43	2.1	63	3.6
Detroit, MI	88	21.4	43	1.3	40	1.9
Washington, DC	66	26.2	32	5.0	41	5.7
Houston, TX	67	18.1	46	3.7	49	21.4
Boston, MA	78	6.0	37	3.0	36	5.0
Atlanta, GA	68	25.8	40	1.8	34	2.0
Riverside-San Bernardino, CA	44	6.5	33	3.6	36	26.5
Dallas, TX	63	15.8	41	2.5	50	14.4
San Diego, CA	58	6.0	48	7.4	45	20.4
Minneapolis-St. Paul, MN-WI	62	3.6	41	2.6	35	1.5
St. Louis, MO-IL	77	17.2	38	0.9	23	1.1
Anaheim-Santa Ana, CA	37	1.6	33	10.0	50	23.4
AVERAGE	68	14.8	41	4.3	46	13.3

Note: The Segregation Index is 100 times the index of dissimilarity for minority groups, compared to the nonminority (White) population, as defined in Chapter 8. The metropolitan areas are not consolidated ones.

Source: Darden, 2001, based on data from 1990 U.S. Census.

among cities and there is little if any relationship with Asian presence.

Similar types of segregation comparisons can be made between cities and suburbs, between different parts of the central city, and among cities from different nations, although differences in international definitions require such comparisons be made carefully.

Regarding African American segregation, studies have pointed to housing as the key intermediate "gateway" device that determines the extent of segregation. In a huge study of Black segregation that focused on the largest U.S. metropolitan areas in 1990, Massey and Denton (1993) found that White citizens were willing to accept into their neighborhoods only a limited proportion of Blacks, and they identified the housing barrier as the critical one (Massey & Denton, 1993).

This has persisted in spite of the passage in 1968 of the U.S. Fair Housing Act and subsequently in 1988 of the U.S. Fair Housing Amendments Act, which strengthened the enforcement mechanisms of the original act. The 1968 Act made it illegal for White real estate agents to prevent Blacks from living in a particular neighborhood of the city in any way. Although Whites accepted this law, they have not put into practice its provisions (Darden, 2001). For example, the way around the law for some White real estate agents is a technique called *racial steering*. In this method, agents steer away Black homebuyers who are seeking a home in a predominantly White neighborhood; likewise, the agents steer White homebuyers from Black areas. Racial steering is difficult to monitor, so it persists even though it is illegal. It is difficult even

to study, because it requires a large sample to identify the Black–White differences. This can be done by a method called *paired testing* (Darden, 2001; U.S. HUD, 2002). Here, the researchers role-play being African American and White buyers. Each is set up with fictitious demographic and socioeconomic characteristics (e.g., family size, income, and age) that on average are similar between races. Each buyer approaches several actual real estate agents (even the same ones) about buying a home in a segregated neighborhood they do or do not racially match. The extent of success for each is measured and compared. These studies have consistently demonstrated that many real estate agents engage in such discriminatory steering (Darden, 2001). A recent large-scale paired testing study (U.S. HUD, 2002) indicated that discrimination against Blacks in the renting of housing was present for 22 percent of test cases, and for 25 percent of cases for Hispanics. The good news was that the percentage for Blacks had declined from 26 percent in a comparable study in 1989. However, these are still high amounts, pointing to a great challenge ahead.

Through steering and other real estate techniques, African Americans have experienced lowered opportunities to move out of the central-city ghetto into White areas. The result reinforces and increases segregated Black communities, mainly in the central city, and segregated White ones, mainly outside. Some interest groups are concerned about these barriers and are working to attack and lower them.

Geography is crucial to understanding these and many other forms of segregation because residential segregation involves a form of physical separation, and hence a geographic perspective. GIS is a good tool for governments, affected parties, and interest groups to use to see how and where segregation has occurred and how resources can be best deployed to mitigate it.

8.8 EMERGING ETHNIC HOT SPOTS IN THE SUBURBS

As a result of changes in the 1990s, America is undergoing a further shift in its metropolitan–nonmetropolitan balance that is altering the urban experience and lifestyles for many people. This section considers how cities or metropolitan areas have been affected by the immigration shifts already discussed, how the balance of suburbs versus central city is changing, and how ethnic minorities have become more prevalent in suburbs as a consequence. Finally, it looks at a new trend of departure of White population from suburbs to what Frey (2002a;2002b) called *exurbs,* that is, small cities with natural and cultural attractions, removed from the big cities, in states such as South Carolina and Oregon.

Necessary background for understanding these changes is the hypothesis put forward in the mid-1990s by demographer William Frey of "new demographic Balkanization" (Frey, 1995). This phrase refers to classification of cities and states in the United States based on very distinctive characteristics, that is, separate areas, or Balkans, deriving from the very distinctive countries in the Balkan region (Frey, 1995). This concept implies unevenness in ethnic and racial diversity across the states and cities. The particular categories of metro areas (or states) that Frey (1995) put forth are the following:

1. *High-immigration metro areas.* Metro areas with high immigration from overseas during the past five years (1985–1990). This group is typified by Los Angeles, New York City, San Francisco, and Miami. Frey (1995) termed them the new "ports-of-entry" into the country.

2. *High internal migration metro areas.* Metro areas that grew more by a high level of domestic migration from other parts of the United States rather than by immigration from abroad. Examples of these metro areas are Atlanta, Tampa-St. Petersburg, Seattle, and Phoenix.

3. *High out-migration metro areas.* Metro areas that experienced high flows of out-migration, without compensating inflows of immigrants. They tend to be cities suffering from loss of manufacturing industry or cities affected by the late 1980s oil bust. Examples are Detroit, Pittsburgh, Cleveland, and New Orleans.

4. *Other metro areas.* Metro areas that cannot be classified in the other three categories. They depend on other patterns. Instances are Columbus, Ohio; Minneapolis-St. Paul; and Indianapolis.

The four metro area types are distinctive. The ports of entry tend to be more diverse ethnically and more dynamic than the others. This has major implications for the future. For instance, if it continues, those cities might foster greater creativity and experience higher productivity (see Florida, 2002). At the same time, they have been susceptible to White flight of lower socioeconomic Whites. "Immigrants are displacing internal migrants at the lower rungs of the socioeconomic spectrum" (Frey, 1995, p. 746). Population is leaving the old manufacturing belt (i.e., part of category 3) and is tending to migrate to ports of entry. This could be partly due to some ports of entry being located in more favorable climates.

The more detailed components of internal and foreign migration from 1985 to 1990 are shown for certain metro areas from Categories 1 and 2 in Figure 8.15, modified from Frey (1995). The ports of entry have economies with jobs that are drawing the immigration of the new minorities, that is, Latinos and Asians, and these metro areas are strongly affected by the immigration. For internal migration areas, such as Las Vegas, Phoenix, Atlanta, and Tampa-St. Petersburg, the new minorities are less important; rather White internal migration is the driver, followed by arrivals of domestic new minorities, and Black migrants for Atlanta and Phoenix. Chicago and New York appear different, with relatively more White immigration from abroad combined with White and Black internal out-migration driving its migration changes.

The theory of demographic balkanization has been influential in understanding the shifts from the much larger volumes of immigration in the 1980s and 1990s. It sets the stage for understanding the central-city, suburban, and exurban fluxes. In the 1990s, the suburbs became much more ethnically and racially diverse. In particular, the overall proportion of ethnic minority population in suburbs expanded from 19 to 27 percent (Frey, 2001). The metro areas in category 1 (i.e., ports of entry) had high minority percentages in their suburbs, including Los Angeles, Houston, New York, Chicago, and Washington, DC (Frey, 2001). In terms of absolute population growth in the suburbs, minorities were responsible for most of this growth, especially in the ports of entry and southern metro areas. Further, it was surprising that for the 102 biggest metro areas, 47 percent of ethnic minorities lived in the suburbs (Frey, 2001).

The implication is that the old concept of suburb is outdated. Ethnic groups are becoming very prevalent in suburbs. A ramification is that certain Whites are being displaced from suburbs; rather than returning to the central city, in the 1990s they migrated to what Frey (2001) called *exurbs.* He defined an exurb as a small to medium-sized city located in one of 13 "national-suburban states" (Frey, 2001), such as South Carolina, Oregon, and Washington. "These white losses [from the suburbs] reflect mostly young people, married couples, parents, and new

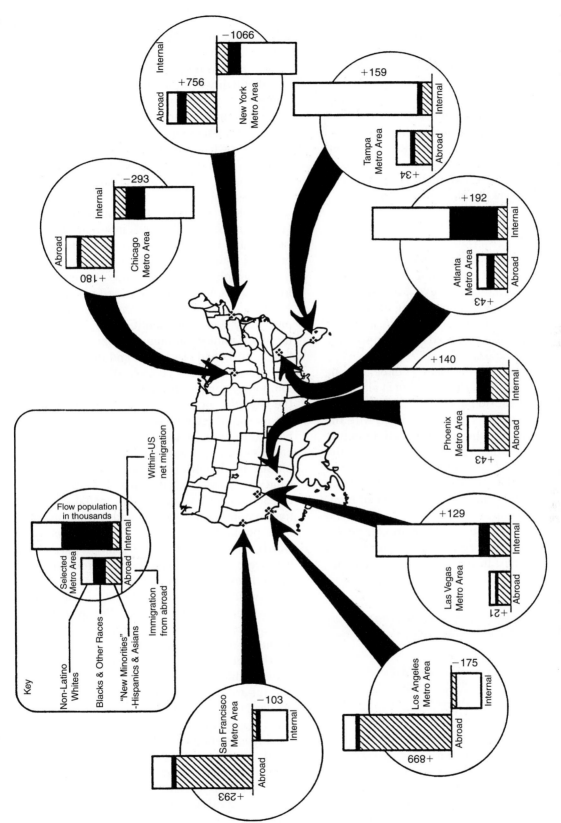

FIGURE 8.15 Foreign and internal components of migration, selected port of entry metro areas, 1985–90. Source: Frey, William H. 1995. "Immigration and Internal Migration 'Flight' from US Metropolitan Areas Toward a New Demographic Balkanisation." *Urban Studies* 32(4–5). P. 741. Figure 3. "Migration components for selected high immigration metros and high internal migration metros, 1985–90."

retirees heading to the 13 new 'national-suburban' states—where they seem to be seeking new lives rooted in more old-fashioned values" (Frey, 2002b, p. 43).

The nation is slowly changing on the basis of a changed suburbia, with a higher proportion of minority groups and the outward spread beyond suburbs that Frey (2001) termed *exurbia*. This section illustrates that the geographic aspects of ethnic change never seem to stop changing. What had been standard ethnic migration patterns for a half-century have now changed, altering the typology of metro areas, readjusting the balance of immigration versus domestic migration, creating more ethnic presence in the suburbs, and leading to White flight to exurbia, dispersed small cities beyond the big metropolitan areas.

8.9 GENDER

This section discusses the emerging understanding of gender and sexuality in the city, and a feminist view of GIS (Hanson, 1992). Some feminists have been skeptical about the value of GIS for understanding gender and other social phenomena of concern to them, although some examples of GIS have modified and enhanced the study of gender (Kwan, 2002a).

Gender is a term that describes aspects of being male or female that are socially determined (Pain et al., 2001), whereas *sex* refers to male–female biological differences. Traditionally, masculine traits were associated with the workplace, competition, public activities, and monetary rewards, whereas feminine traits were related to the home and its private and emotionally based activities. As seen in Table 8.5 (Pain et al., 2001, modified from McDowell & Sharp, 1999), the traditional view of feminine traits versus masculine ones implies less power and reduced social and spatial mobility. The greater power traditionally given to men versus women is sometimes referred to as *patriarchy,* in particular, a political and socioeconomic system in which men have more power and control than women. Patriarchy is often regarded as having domination and subordination, determined through the intersection of race, age, class, sexuality, and gender, not just gender alone.

Feminism is the theory of the political, economic, and social equality of women. Although centuries old, it was energized by the women's liberation movement of the 1960s. Feminist geographers are particularly concerned with gender issues and have developed a considerable extent of knowledge, especially over the past two decades.

Sexuality in this context refers to the social aspects of being sexual. The interest of urban geographers has been on how sexuality influences cultural, economic, and social geographies and how they in turn influence sexuality (Pain et al., 2001). The focus in urban geography studies of sexuality has been on the social aspects of sexual orientation (Pain et al., 2001).

Geographic studies of gender have included the following areas, among others: the effects of gender inequality; the explanations of gender inequality related to patriarchy; gendered identities; gender and public space; the space–time continuum of women's and men's lives at home, work, and elsewhere; and the relationship of people's bodies to space (Pain et al., 2001). The roots of contemporary gender differentiation in urban space were in the late 18th and 19th century in Europe, at the time of the industrial revolution and growing waged labor, often backed by trade unions. These trends led women to become more restricted in urban space. Many studies emphasize the importance of space, for instance, studies on what public spaces in cities in which women feel unsafe or threatened, and women's daily spatial movements, relative to home, work, power, and fear. Sometimes studies have pointed to women's *spatial entrapment,* in other words, that they are not welcome in parts of the urban landscape and so remain trapped in narrower segments. This relates also to women's commute times from home to work, which studies have indicated are shorter on the average than men's, implying smaller space ranges, as well as the necessities of keeping track of children, and fulfilling home roles (Mitchell, 2000). Another feminist theme in urban geography studies has been women's and men's bodies (Mitchell, 2000). For instance, the body is an important aspect in American shopping centers, including the display of pictures of bodies, and many venues that involve the body (Longhurst, 1998). Women are often more socially identified with their bodies than men (Mitchell, 2000). Another perspective is that the location of a person (i.e., his or her body) can be tracked throughout the day or week, as it relates to social and urban phenomena.

The methods that urban feminist geographers use include analyzing distance and measures of separation, studying localities by gender and sexuality, examining

TABLE 8.5   Gender: masculine and feminine traits

Masculine traits	Feminine traits
Public	Private
Outside	Inside
Work	Leisure
Rational	Emotional
Earning	Spending
Production	Consumption
Empowered	Disempowered
Freedom	Constraint

Source: Pain, R., Barke, M., Fuller, D., Gough, J., MacFarlane, R., and Mowl, G. (2001) "Geographies of gender and sexuality," in *Introducing Social Geographies*, pp. 120–140. London: Arnold Publishers, P. 122, Box 6.3; modified from McDowell and Sharp, 1999

microgeographies of the body, conducting interviews, and recently applying GIS methods, although it has received limited acceptance so far in this subfield.

Several specific studies of gender that are considered are women and public places, gender aspects of the high-technology industry, and gender and sexuality in the urban workplace. Women are known to have more fear of public places than men (McDowell, 1997; Day, 2001; McDowell, 2001). This fear stems from real data on experiences and incidents that have occurred, as well as other types of communications about fear, including media warnings and advice and protectiveness from men (Day, 2001). The specific types of women's public-place fears are of sexual assault, harassment, crime, sex discrimination, and hate crimes (Day, 2001). Women's reactions to these real or perceived threats include avoidance of particular public spaces at specific times (McDowell, 1995; McDowell, 1997), requests for escort and protection by men (Day, 2001), changing modes of transit, adoption of more masculine behavior such as learning physical self-defense, activism, protests, and other political activity. One outcome of growing fear in England since the early 1990s is the rising proportion of women who are driving to and from work, even though they have traditionally taken public transportation (McDowell, 2001).

In a study of women's fear of urban parks termed the Woods Project, Burgess (1998) sought to determine why women feared woodlands on the outskirts of British cities. One-gender-only group discussions were conducted with respondents at two urban woodlands, Bestwood Country Park on the urban fringe of Nottingham, England, and Bencroft-Wormly Woods on the northern fringe of London. The women and men were both woodlands visitors and nonvisitors from different ages and socioeconomic classes. From a physical point of view, the parks' enclosures of trees and plants, reduced visibility, dark areas, and narrow paths could increase fear. On the social side, women's encounters with strangers and threatening or violent behavior have grown. Even flashing was a reported problem, leading to varied reactions including anxiety and outrage (Burgess, 1998). Furthermore, the mass media coverage of events in the parks was very influential on respondents' perceptions and levels of fear. This was true even if knowledge of a particular incident was limited. This study also suggested how visitors to the parks, particularly women and ethnic minorities, could cope with their fear, including "voluntary self-exclusion and not visiting the greenspace at all to taking a dog, children, a friend, male partner, or son, so that one is not going alone" (Burgess, 1998, p. 127).

A study by Massey (1998) examined the space–time patterns of mostly male scientists and engineers in the high-tech industry of Cambridge, England. These highly skilled and educated workers put in very long hours—to the extent of being workaholics—and were absorbed in gaining the latest knowledge and responding to very tight deadlines in highly competitive markets. Their work environments were well ordered, abstract, and logical—in other words, masculine—in contrast to their emotional and leisure-oriented homes. The study, based on in-depth interviews, showed that the space–time of the workplace "transgressed" into the home sphere (Massey, 1998). For instance, respondents reported that such workers typically brought their work home, even after long hours in the workplace. They were often distracted by work even when engaged in home-centered social activities and caring for children. They were mostly unmindful of and nonparticipative in household work (Massey, 1998, p. 167). Massey referred to this as a "spatial split between mind and body" (p. 167) A small minority of the workers were "resisters" in that they did not bring work home, but it often increased their hours spent in the workplace. This study raises difficult issues that include the value of such highly skilled cutting-edge work to the worker and society versus its social and family costs. Another issue is how compartmentalized people's lives should be spatially and temporally. The few high-tech workers among the housewives of the male scientists and engineers highlighted these differences. Unlike their husbands, they seemed to merge rather than partition the space–times of work and home.

A study of Merchant Banks in the city of London examined the dominant masculine work environment of mostly young, middle-class, competitive, largely male bankers and the extent to which the fewer women bankers could adjust to it (McDowell, 1997). Women reacted in a number of ways, including changing their body image to conform more to the male environment, but in that case sometimes being harassed for conforming. The study emphasizes the gender issues of body-, place-, and gender-driven cultures.

An early theme among some feminist geographers was a divide between their scholarship and GIS. Some criticized GIS as being too logical and coming from a male-dominated environment, based on technology originating from government and the military (Kwan, 2002b). The problem was that these scholars felt that their conceptual, qualitative, and interpretive studies could not be expressed in terms of GIS (Kwan, 2002c). Some feminist studies, on the other hand, have recently incorporated GIS in different ways. For instance, a study showed a mixed result from GIS adoption by women activist groups: In some cases, the GIS knowledge empowered them to achieve their goals, but in other cases it hobbled their efforts (McLafferty, 2002; Timander & McLafferty, 1997). Kwan (2002b) pointed out that the key to making GIS relevant is to shift its use toward interpretation, visual images, and narrative, so feminist geography scholarship can be enhanced. She demonstrated this in her own studies of the daily life paths of women in Columbus, Ohio (Kwan, 2002b), and Portland, Oregon (Kwan, 2002c). As seen in Figure 8.16, she utilized special 3-D displays and algorithms tailored to variables in the urban environment confronting women to display space–time paths during the day (Kwan, 2002a). The key was gathering unconventional travel diary data and tailoring new algorithms to analyze it. Hopefully, a growing number of gender studies will be able to incorporate such appropriate and novel uses of GIS.

Studies of sexuality in urban geography started in the 1980s by examining gay and lesbian places and networking (McNee, 1984, 1985). This has continued in a large line of research that has concerned the topics of "gay community development, territoriality, neighborhood change, gentrification, social movements, urban politics, and the cultural politics of urban space as they pertain to sexuality" (Brown & Knopp, 2003, p. 317). Like studies of gender, much of the study of sexuality has been conceptual and qualitative, although a few studies have been empirical (Brown & Knopp, 2003). Other types of urban geography studies of sexuality include examining residential architecture as conforming to a heterosexual, nuclear family (Valentine, 1993a), studying the heterosexuality dominance of layouts of home spaces (Valentine, 1993a, 1993b), and examining sexuality in cities in a broader

sense of including all types—heterosexual and homosexual (Nast & Pile, 1998).

An example of an early study of gay neighborhoods in San Francisco (Castells, 1983) shows the large areas by the mid-1980s in downtown San Francisco having gay settlement, as well as the years that settlement originally occurred (see Figure 8.17). The city might have developed its large gay communities originally through its role as a seaport, which brought a population with more men than women and many transitory residents (Mitchell, 2000). Over time, the city also developed a reputation of having bohemian sections that might have attracted gays and lesbians to immigrate. The center of the gay (and to a lesser extent, lesbian) area of San Francisco is Castro Street, shown in the middle of Figure 8.17. These groups often live in gentrifying areas and contribute to gentrification (Brown & Knopp, 2003; Mitchell, 2000). For San Francisco, this led to rising housing values. As a consequence, low-income community members could be forced to leave. A further aspect of gay and lesbian settlement is the political cohesion that might develop on the basis of spatial proximity. This could in turn lead to changes in the city's politics and affect areas of the city much broader than the immediate enclave (Mitchell, 2000).

In summary, gender is a key aspect of the social life of cities that has been receiving increasing attention from urban geographers in the past two decades. The

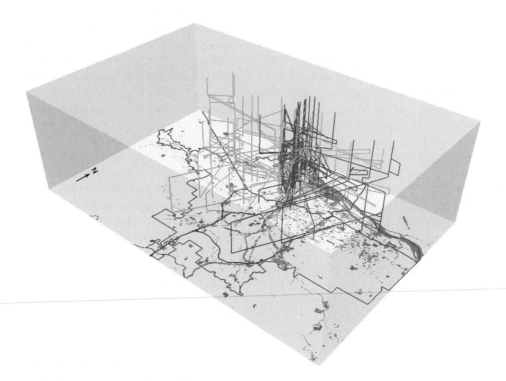

FIGURE 8.16   Space–time paths of a sample of African American women in Portland, Oregon. Source: Kwan, M.-P. (2002b) "Introduction: Feminist geography and GIS." *Gender, Place, and Culture* 9(3):261–262.

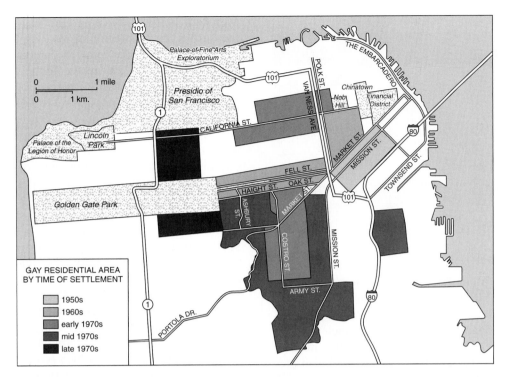

FIGURE 8.17 Gay neighborhoods in San Francisco. Source: Mitchell, D. (2000) *Cultural Geography: A Critical Introduction*. Oxford, UK: Blackwell.

knowledge of many aspects of gender and sexuality has grown. Although early scholars could not find appropriate uses for GIS, recently new data and techniques point to the possibilities for its use in gender studies.

8.10 POVERTY AND ITS MEASUREMENT

The chapter turns for its remainder to look at poverty, its role in cities, and its geography. Poverty is defined as the status of a person, household, or family that lacks a usual or socially acceptable amount of financial assets or material possessions. It is an important topic in urban geography because it is present in virtually every city. It represents a lifestyle that usually restricts opportunities and implies costs to government. Poverty is influenced by changes in the economy and its labor markets. For example, changes in technology might reduce or eliminate certain types of jobs, such as typesetter, pressuring some members of occupational groups under the poverty line. Recessions tend to amplify the numbers and percentage of a country's population that is in poverty. Poverty can enter into politics as governments and politicians try to address or combat it. The definition of poverty varies in its definition depending on the particular historical time and culture. It also is defined differently across nations.

The U.S. definition of poverty was established at the beginning of the War on Poverty in 1963, although the specific statistical cutoffs have been raised regularly in response to inflation. It is based on the concept of a minimal "basket" of basic living items for a family. It has been updated by applying inflation indexes since 1965 and is still in use today. Under the definition, a family is classified as poor if its income falls under an appropriate poverty threshold. The 48 thresholds are arrived at by estimating the average cost for a meal plan that is nutritionally balanced, and, for a family, multiplying it by three to determine the threshold (Lichter & Crowley, 2002; U.S. Census, 2004). There are some adjustments made for family size and age (U.S. Census, 2004); the threshold is lower for smaller families with fewer children. A family's income is measured against the thresholds. Family income is defined for poverty purposes as consisting of money income before taxes, totaled for all family members living in the housing unit, but excluding capital gains.

Since the mid-1960s, the U.S. poverty definition has been adjusted each year for inflation based on the Consumer Price Index (CPI). As seen in Table 8.6, for 2000, the poverty threshold for a family of four with two children under 18 was an annual income of $17,463.

In 2000, 31.1 million Americans were in poverty by this definition, or 11.3 percent of the population of 275.9 million.

There are certain weaknesses in the poverty definition, as a panel of the National Academy of Sciences

TABLE 8.6 Poverty income thresholds in 2000

Family size and no. of children	Income threshold for poverty
One person over 65 years	$8,259
One person under 65 years	$8,959
Two people with head of household under 65 years	$11,531
Four people with two children	$17,463
Four people with no children	$17,761
Eight people, including four children	$30,188
Nine people, including two children	$37,813
Source: U.S. Census, 2001.	

(NAS) pointed out in 1995. The panel suggested taking into account only after-tax income, counting as income noncash benefits such as food stamps, deducting from income certain fixed work-related expenses such as transport and child care, and deducting out-of-pocket medical expenses (U.S. Census, 2002). The NAS panel also proposed adopting higher poverty thresholds for metropolitan areas and for specified regions to reflect differences in cost of living. However, the panel's suggestions have not been adopted, mainly because of political interest groups with stakes in the old measures.

The U.S. poverty measure classifies each family into one of just two categories: poor or nonpoor. It provides no way to recognize a family that is just above or just below the threshold. This can be overcome by using a depth of poverty measure. *Depth of poverty* is defined as the ratio of a family's income to its poverty threshold. For example, Family X's income is $16,500. The family's poverty threshold is $17,500. The depth of poverty is calculated as follows:

Depth of poverty

$$= \frac{income\_for\_Family\_X}{poverty\_threshold\_for\_Family\_X} = \frac{\$16,500}{\$17,500} = 0.943$$

The U.S. definition is primarily an *absolute poverty standard,* in that it is based on a fixed bundle of economic necessities and its dollar amounts are absolute. However, it is not strictly absolute because it is adjusted over time based on national averages. Some nations measure poverty using a relative standard. A *relative poverty standard* determines poverty by the percentage of population underneath the average income or average consumption of a nation, or some multiple of average income or average consumption. An example is Mexico, which does not have an absolute poverty standard. For Mexico, one relative poverty standard is computed as the proportion of

Mexicans below the average income. Another relative measure is the proportion of working Mexicans with no income or with salaries below the national minimum wage, which we define as low income. In the year 2000, this latter relative measure was 20.7 percent, or showed a fifth of the Mexican working population in relative poverty.

The spatial distribution of 2000 poverty in the Mexico City metropolitan area, according to this definition, was shown in Figure 1.7. It is clear that there is a huge contrast in poverty in different parts of the metropolitan area, ranging from 6 percent to over 50 percent. The zones with the least poverty are in the sectors that extend from the central city to the northwest and east, as well as in a few scattered municipios to the north and south. The areas with the highest poverty levels are semi-rural municipios in the far northeast and southeast. These municipios represent a mixture of rural poverty and that of recent immigrants, who in Mexican cities settle on the periphery. Mexican poverty often involves makeshift housing, as seen in Figure 8.18. Many Mexican workers in poverty are not part of the salaried labor force, but are in the informal labor force, which refers to very small enterprises that do not register with the government, such as the street vendor shown in Figure 8.19.

Relative poverty measures are useful in comparing among nations or between cities of difference nations. Data on a fixed bundle of economic assets are rarely available across a large sample of different countries; also, these data would be of limited applicability unless living costs were alike in all the countries. The relative poverty for 19 rich nations (see Table 8.7) is calculated on the basis of the percentage of household population having under 40 percent of median adjusted household income (Smeeding, Rainwater, & Burtless, 2001).

In this table, the United States is seen to have a comparatively high poverty rate for several reasons. Many European nations have comprehensive national social welfare programs that exceed anything provided in the United States. Also, it is more difficult for governments to control poverty in a large, sprawling nation such as the United States. There are many different regions that require varied approaches to reducing poverty, so mitigation programs are more complex. Antipoverty programs are challenging to deploy across vast reaches (Smeeding et al., 2001). This also applies to the large nations of Australia and Canada.

8.11 POVERTY IN THE UNITED STATES

Poverty was officially recognized as a key national problem during the 1960s. As seen in Figure 8.20, the poverty level in 1959 for all Americans was 22 percent, for Whites 19 percent, and for Blacks 55 percent, levels much higher than today. By 1965, the overall national rate had

FIGURE 8.18   Mexican urban poverty: Makeshift housing.

FIGURE 8.19   Street vendor, Mexico City.

TABLE 8.7   Relative poverty in 19 rich nations in the mid-1990s

Nation	Year	Poverty rate*
United States	1997	10.7
Italy	1995	8.9
Australia	1994	7.0
Japan	1992	6.9
Canada	1994	6.6
United Kingdom	1995	5.7
Israel	1992	5.2
Spain	1990	5.1
Netherlands	1994	4.7
Sweden	1995	4.6
Germany	1994	4.2
Switzerland	1992	4.0
Denmark	1992	3.6
France	1994	3.2
Norway	1995	3.0
Austria	1992	2.8
Finland	1995	2.1
Belgium	1992	1.9
Luxembourg	1994	1.3
Average		4.8

* The percentage of individuals in households that are less than 40 percent of median adjusted household income.

Source: Smeeding, Timothy M., Lee Rainwater, and Gary Burtless. 2001. U.S. Poverty in a Cross-national Context, in Danzinger, Sheldon H. and Robert H. Haveman (eds), Understanding Poverty, Russell Sage Foundation, P. 173, Table 5.2. "Poverty rtes in nineteen rich countries, by age group, in the 1990s."

since 1970 have shown ups and downs in this measure, generally tied to the economic cycle. For instance, in the mid- to late-1990s, the poverty rate dropped across all race categories as the economy boomed. The national poverty rate fell to a low of 11.3 percent in 2000, close to the all-time low of 11.1 percent in 1972, in the immediate aftermath of the War on Poverty. However, economic recession led to a small poverty uptick to 11.7 percent for the United States in 2001 (U.S. Census, 2002).

For cities in 2001, poverty was twice as high in the central cities (18.4 percent) versus the suburbs (8.3 percent). The difference between central city and suburbs was greatest in the metropolitan areas of the Northeast and Midwest, and less in the Sunbelt.

Examples of poverty rates for individual metropolitan areas were Los Angeles–Riverside–Orange County, 15.6 percent; Miami, 18 percent; Chicago–Gary–Kenosha, 10.5 percent; city of Chicago, 19.6 percent; San Antonio, 15.1 percent; and New York–Northern New Jersey–Connecticut, 16.0 percent. During the mostly prosperous 1990s, poverty change in the U.S. metro areas was neutral; half of them increased and half of them decreased in poverty rate. Relative to a booming economy, central cities were not doing well in fighting poverty. The situation had been even worse in the 1980s, when the poverty rate rose 75 percent for central cities.

Why did central-city poverty rise? The surge of immigration to the United States mentioned earlier starting in the 1980s brought many millions of new arrivals to U.S. metropolitan areas, mostly from Latin America and Asia. Because most were less skilled and educated than the

fallen to 17.3 percent. The War on Poverty, launched by the Johnson administration in 1965, produced many gains. Although the nation made a lot of progress, in particular in the 1960s on reducing Black poverty, the three decades

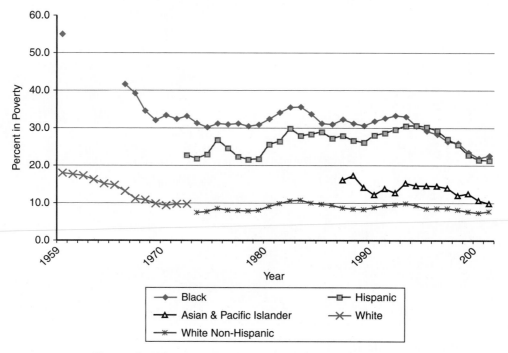

FIGURE 8.20   Changes in U.S. poverty levels, by ethnic group, 1959–2001.

native population, they have contributed to greater poverty. Another factor is the persistence of the Black ghetto, usually in central cities. The family structure in the ghetto, with a large component of single mothers and unmarried childbearing, contributes to higher poverty (Lichter & Crowley, 2002). A countervailing factor reducing poverty has been the presence of many noncash government benefits.

Although our focus is on urban geography, it is important that the poverty rate in 2000 was higher in nonmetropolitan areas versus metropolitan territory (Lichter & Crowley, 2002). Why is this the case? Away from larger cities, poverty tends to be located in small pockets, characterized by residents with very low education and job skills. Nonmetropolitan poverty is further associated with isolation, very poor schools, unsafe drinking water and other environmental conditions, weak public infrastructure, and racial discrimination. The effect of such poverty is seen in the list of states with the highest statewide poverty rates in 2000, being primarily states with large rural populations, in particular, Louisiana (20.3 percent), West Virginia (19.3 percent), and Mississippi (18.2 percent). Some metropolitan states had among the lowest rates (New Jersey at 8.3 percent and Connecticut at 7.9 percent). New Jersey has large suburban areas that are prosperous, and Connecticut has high personal income stemming from its affluent New York suburbs in the southwest; prosperous Hartford, which is an insurance center; and the state's emphasis on high-skilled manufacturing. Even though metropolitan poverty areas have social, health, and public safety problems, their residents often have the advantage versus nonmetropolitan residents in access to larger labor markets.

Another important perspective on poverty is the distinction between chronic poverty and episodic poverty. *Episodic poverty* is the result of a short-term or midterm condition, such as a divorce, temporary job loss from an economic recession, or family losses stemming from a natural disaster. In those conditions, a person can take short-term steps to exit poverty by correcting the condition. In *chronic poverty,* the person is often trapped in poverty by low education, lack of skills, handicaps, and persistent discrimination. People in that category are often referred to as the underclass, a topic that is taken up at the end of the chapter. For the chronically poor, poverty must be viewed as a long-term condition from which it is very difficult to exit.

8.12 EXTREME POVERTY: THE GHETTO AND THE UNDERCLASS

Returning to the United States, at the bottom of the poverty scale there are areas of *extreme poverty.* These are commonly zones that have 40 percent or more of individuals in poverty, contrasted with the recent overall national rates

of 11 to 12 percent. The areas of extreme poverty in cities tend to be in or near ghetto areas (Greene, 1991; Jargowsky, 1996; Jargowsky, 1997), but extreme poverty is not exclusive to the ghetto. The *underclass* refers to those people living in chronically extreme poverty and facing social barriers to extract themselves from it.

The causation of the ghetto has been explained as deindustrialization as well as cycles of entrapment related to lack of education and joblessness. Wilson (1987, 1996) explained a major cause of the ghetto as deindustrialization and accompanying middle-class flight to the suburbs. His studies were based on the city of Chicago. Here, economic trends caused the central city to lose manufacturing jobs and gain service and white-collar jobs (Peck and Theodore, 2001). Another example of this is Pittsburgh, which, starting in the 1920s and 1930s, replaced its former huge steel manufacturing core largely with service jobs. A primary reason was lack of level land near water transportation. In both cases, this meant that impoverished ghetto residents had a harder time finding jobs their usual low skills are suited for. A concurrent trend is that middle-class households, both Whites and African Americans, are leaving the central city to seek improved living circumstances in the suburbs. The people left behind in the poor neighborhoods, according to Wilson (1987), lose their middle-class role models. All these trends together tend to trap the ghetto resident in the ghetto, without the money or educational ladder to escape. An area of Chicago's largely African American ghetto is seen in Figure 8.22.

Wilson's theory has been tested by other scholars (Greene, 1997; Massey & Denton, 1993; Strait, 2000, among others). Although as a whole it has held up fairly well under intense scrutiny, many disagree with parts of it. One criticism is that it does not take into account regional differences. For example, some poverty areas might be located in economically dying areas of a multicounty region, and not be associated with a metro area. Another is that the theory does not recognize the importance of the trend to outsourcing that has been affecting American industry for more than a decade (Strait, 2000). Outsourcing might take away low-skilled jobs from ghetto residents.

The underclass is seen as increasingly trapped in the ghetto, without an obvious way out. "The deck is stacked against black socioeconomic progress, political empowerment, and full participation in the mainstream of American life" (Massey & Denton, 1993, 148). The trapping takes place because the social aspects of segregation are so strong that they pressure young ghetto residents against violating the ghetto culture by receiving too much education or by obtaining too high level a job. Cultural evidence of this "inside" behavior includes speech patterns and use of distinctive language. The cycle perpetuating the ghetto consists of strong ghetto

Poverty in Central American Cities

Urban poverty is on the increase in Central American metropolitan areas and cities. The rapid urbanization of Central America is contributing to this process. The entire region is anticipated to rise in urbanization (i.e., the percentage of population in urban areas) by 20 percent from 1970 to 2010 (World Bank, 2002). Some nations are expected to increase even more rapidly over the 40 years, particularly El Salvador (projected to rise from 39 to 70 percent), Honduras (29 to 61 percent), and Costa Rica (39 to 64 percent). A study by the World Bank (2002) examined the problem of poverty in three rapidly growing cities in the region: Panama City, Panama; Tegucigalpa, Honduras; and San Salvador, El Salvador. Because of the rural-to-urban immigration and the natural population increase, the primate or capital cities of Central American nations are becoming outsized. *Outsized* refers to a metropolitan area's growth beyond the capacities of its natural environment to support it. In some respects, they are moving along the track of Mexico City, which has suffered from too much growth. Because the countries are smaller than Mexico, there are fewer competing cities, which adds to their growth potential. The three study cities are growing rapidly and the number of poor households is expanding at twice the rate of well-off households (World Bank, 2002). The poor in these cities tend not to belong to the formal labor market but rather are unemployed or work in the informal labor force, that is, in small concessions such as street vendors, car cleaners, and so on. Poverty in Central America is shown in Figure 8.21. Only a third of the poor belong to and are able to benefit from the social security systems.

Another set of problems relates to lack of infrastructure and city services. The cities usually lack potable water, sewers, and solid waste disposal service (Tarmann, 2002). The lack of drinking water harms public health. The government agencies providing the services often

FIGURE 8.21   Poverty in downtown El Salvador, 2000. Source: Yuri Cortez/Agence France Presse/Getty Images.

have faulty information systems so they do not know who or where their poor customers are. As a result, a poor household might have to wait five years to be connected to water and other utilities (World Bank, 2002) and, meanwhile, be forced to scavenge to transport water, firewood, and other essentials.

Corruption might play a role, too. Land developers try to sell false land titles to these poor people. The land that is available is often inadequate and on the periphery,

sometimes in steep locations like ravines or hillsides. The poor can suffer financial losses as a result of such "pirate" land developers (Tarmann, 2002). As in Mexico City and other Latin America cities, the poor often construct their own homes initially out of cheap materials and continue to improve them over a long period of time, even a generation, making use of better materials and construction. Sometimes, this is a way out of poverty if their investment increases in value.

culture, separation, isolation, and poverty (Massey & Denton, 1993).

Other scholars have new dimensions of study on extreme poverty. Jargowsky (1996) used extensive data to examine the characteristics of people who live in extreme poverty throughout the country. He found, for instance, that although extreme-poverty neighborhoods were 50 percent African American, they were also 22 percent White, 14 percent Mexican, 6 percent Puerto Rican, and 8 percent other races. The industry distribution of workers in high-poverty areas did not differ from city neighborhoods

outside these areas (Jargowsky, 1996), although unemployment was high. Many ghetto residents held jobs, attended school, and were not criminals or drug addicts. The one characteristic, however, that was present throughout the extreme-poverty areas was the mother-headed households, with the father absent (Jargowsky, 1996).

Greene (1991) looked at the relationship between extreme poverty and population and poverty growth. As seen in Figure 8.23, most extreme-poverty neighborhoods in New York City had declining population and increasing poverty (labeled B in the map). The main concentration

FIGURE 8.22   Ghetto neighborhood in New York City. Source: AP Wide World Photos.

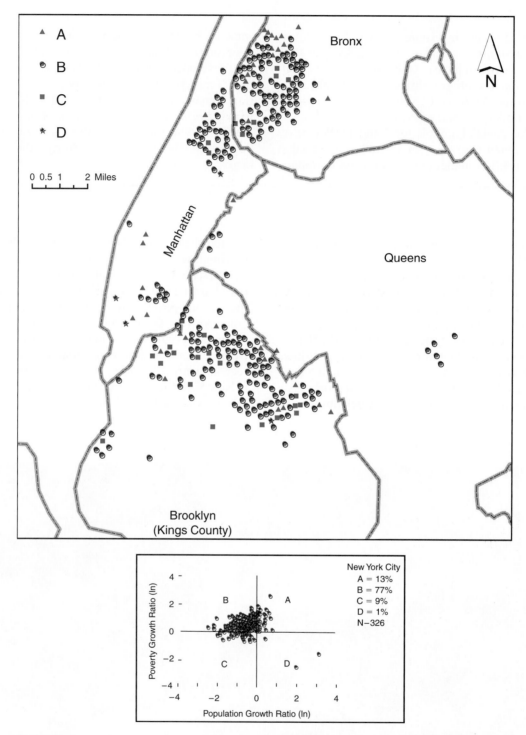

FIGURE 8.23   Extreme-poverty neighborhoods in New York City. Source: Greene, Richard P. 1991. Figure 4 on P. 536. "Poverty Area Diffusion: The Depopulation Hypothesis Examined." *Urban Geography* 12:526–541.

of these areas was in the ghettos of the Bronx and Brooklyn. These areas corresponded to Wilson's theory of greater poverty amidst White flight. It is interesting that small frequencies of other types of extreme-poverty groups (A, C, and D in Figure 8.23) were scattered near the edges of the ghetto areas. In repeating this analysis for Los Angeles, there was less support for the dominance of group B, as Wilson had thought. Wilson had studied Chicago, and might have missed other patterns in newer, Sunbelt cities.

The underclass continues into the 21st century. It is a social group with great suffering and often-reduced motivation in trying to break out of the cycle.

8.13 ANOTHER TYPE OF POVERTY AREA: COLONIAS ALONG THE MEXICAN BORDER

Colonias are unincorporated places on the U.S. side of the Mexico–United States border that are poor, lack basic utilities, and have serious health problems. It is estimated that about 400,000 Texans live in colonias (Texas Secretary of State, 2002) and there are tens of thousand of additional colonia residents in New Mexico and Arizona. The majority of colonia residents have U.S. citizenship, yet their primary language is Spanish, which restricts their educational advance. An estimated 15 percent are not legal residents of the United States (Texas Natural Resource Conversation Commission, 1999). There are many first-generation families who have invested their life savings in homes and businesses in colonias. The housing structures are often crude and built by the owners in stages with a patchwork of materials that include bricks, plywood, and cinder blocks (Pepin, 1998; Texas Secretary of State, 2002). Housing in a Texas colonia is seen in Figure 8.24. The word *colonia* in Mexico has a different meaning—it refers to a government-defined neighborhood in a city, which could be at any income level. In Mexico, the concept more analogous to the meaning of a U.S. colonia is a squatter settlement, located in cities mostly at the periphery. Squatters are immigrants who arrive generally with limited means, claim land, and initially construct makeshift housing, often claiming or "squatting" on the land (Ward, 1998a).

The level of income is very low, and in one study of Texas colonias, the median family income was only $9,000 per year (KLRU and PBS Online, 2002). About 43 percent of Texas colonia residents are classified as living in poverty (Ward, 1998b). The Texas counties with the highest colonia prevalence are among the poorest in the United States.

The first colonias appeared in the 1950s in Texas. Developers saw opportunities to purchase lots outside city limits, often along floodplains. For example, the distribution of colonias on the outskirts of El Paso, Texas, is seen in Figure 8.25. The lots could be legally subdivided, and there were no regulations requiring developers to put in standard utilities. Developers for many years benefited by the use of *contracts for deed,* a technique of selling land where the buyer puts only a small amount down and has low monthly payments. However, the buyer does not receive the title until the final payment has been made. This practice was restrained by the 1995 Colonias Fair Land Sales Act in Texas, but similar difficulties remain (Texas Secretary of State, 2002). Also, along with the curtailment, financial institutions are less likely to lend to colonia residents, as the risk of a foreclosure is higher.

Their inadequate utilities and infrastructure are the most unfortunate aspects of colonias. In 1990, it was estimated that colonias in Hidalgo, Cameron, and El Paso counties had utilities and services provided to the home in the following percentages: potable water connections

FIGURE 8.24   Housing in Texas colonia near Mexican border. Source: Pepin, Madeleine. 1998. "Texas Colonias: An Environmental Justice Case Study. Available 2003 at *http://itc.ollusa.edu/ faculty/pepim/philosophy/cur/colonias.htm.*

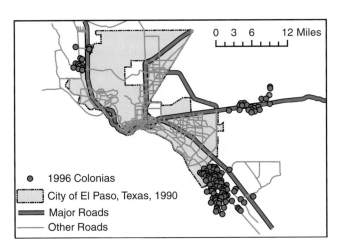

FIGURE 8.25   Colonias in El Paso, 1996. Source: State of Texas, Office of the the Attorney General. Border colonia geography http://204.64.55.40/scripts/csrimap.dll.

at 75 to 95 percent; heating at 50 to 60 percent; sewage at 0 to 5 percent; and trash at 0 percent (KLRU and PBS Online, 2002; Pepin, 1998). This meant that households with deficits had to go to outside providers to obtain these services. For colonia households, the 1990 Census indicated that 50 percent had incomplete plumbing and 40 percent lacked kitchen facilities. Also, a 1990 General Accounting Office report identified that 59.7 percent of colonias had access to a water system, but only 0.4 percent had access to a sewer system (Chapa & Eaton, 1997). This means that sewage and wastewater are released to contaminate the environment. It is not surprising that there is poor health in the colonias, including high rates of tuberculosis, hepatitis A, salmonellosis, dysentery, cholera, and giardiasis (EPA, 2004; Federal Reserve Bank of Dallas, 2004). The poor drainage and threat of flooding in the region add to the public health problems.

Given all these problems, why have colonias developed? The driving force has been the housing needs of Mexican immigrants and Mexican American residents. They lack assets to purchase housing, and so seek something affordable, even at this low level. The colonias appear to have a more traditional family pattern than Texas as a whole (Pepin, 1998). The problems for this form of poverty are concentrated in environmental damages and degraded health conditions. The colonias were not the result of large-scale economic-based immigration and resistance to it, as with the ghetto, but rather they stem from needs for very low-cost housing and settlement. Colonias form a contrast that broadens the notion of U.S. poverty and points to a common thread of very low income.

SUMMARY

This chapter covered the significant urban geography topics of race, ethnicity, gender, and poverty. The cities in every society have these elements. Cities are built up partly through migration, which often is from other nations and frequently brings in new racial and ethnic groups. The ethnic mix contributes to the variety and appeal of city life and urbanism, but also provides the basis for potential conflicts. Cities have geographical patterns of ethnicity that often change over time, as central-city communities expand outward to suburbs, and as successive redistributions and migration changes occur. The chapter looked at the immigration, settlement, and further dispersion of the major ethnic groups of African Americans and Latinos and Hispanics.

Gender is an important aspect of urban geography that has received increased attention in recent years. Gender includes the topics of gender inequality, gender and public space, the work and home patterns of women

and men in the city, gendered identities, and the related topic of sexuality. Some feminist researchers have criticized GIS and recently some have offered promising ways to use it for feminist research.

Poverty can be measured in a variety of ways, one of which is based on the ability to afford basic baskets of goods. In the United States over the past 40 years, poverty dropped following the War on Poverty in the 1960s, but in recent decades appears to have stabilized. Compared with its peer nations, the United States has high rates of relative poverty. The extreme poverty that is present in the ghetto has led to the development of an underclass that is in many respects self-perpetuating. The poverty in colonias along the Mexican border is distinctive in its deep and chronic environmental and health problems. Overall, the poverty problem is a long-term policy issue for governments and international organizations.

REFERENCES

Allen, J. P. and Turner, E. (2002) *Changing Faces, Changing Places: Mapping Southern California.* Northridge, CA: Center of Geographical Studies, Department of Geography, California State University Northridge.

Arreola, D. D. (2000) "Mexican Americans," in *Ethnicity in Contemporary America* (2nd edition). ed. by J. O. McKee, pp. 111–138. Lanham, MD: Rowman & Littlefield.

Border Agricultural Workers Project. (2004). Los Braceros, 1942-1964. El Paso, TX: Border Agricultural Workers Project. Available in 2004 at *www.farmworkers.org*.

Brown, M., and Knopp, L. (2003) "Queer cultural geographies— We're here, we're queer, we're over there, too," in *Handbook of Cultural Geography*. ed. by K. Anderson, M. Domosh, S. Pile, and N. Thrift, pp. 313–324. London: Sage.

Burgess, J. (1998) "But is it worth taking the risk?" in *New Frontiers of Space, Bodies, and Gender*. ed. by R. Alnley, pp. 115–128. London: Routledge.

Castells, M. (1983) *The City and the Grassroots*. Berkeley, CA: University of California Press.

Chapa, J. and Eaton, D. (1997) *Colonia Housing and Infrastructure: Current Characteristics and Future needs*. Austin, TX: University of Texas, Lyndon B. Johnson School of Public Affairs.

Clark, W. A. V. (1993) "Neighborhood transitions in multiethnic/racial contexts." *Journal of Urban Affairs* 12(2):161–172.

Darden, J. T. (2001) "Race relations in the city," in *Handbook of Urban Studies*. ed. by P. Ronan, pp. 177–193. London: Sage.

Day, K. (2001) "Constructing masculinity and women's fear in public space in Irvine, California." *Gender, Place, and Culture* 8(2):109–127.

EPA. (2004) 'Colonias Facts.' Washington, DC: Environmental Protection Agency. Available 2004 at *www.epa.gov*.

Federal Reserve Bank of Dallas. (2004) *Texas Colonias*. Dallas, TX: Federal Reserve Bank of Dallas. Available 2004 at *www.dallasfed.org*.

Florida, R. (2002) *The Rise of the Creative Class.* New York: Basic Books.

Frey, W. H. (1995) "Immigration and internal migration flight from U.S. metropolitan areas: Toward a new Balkanization." *Urban Studies* 32(4):733–757.

Frey, W. H. (2001) "Melting pot suburbs: A census 2000 study of suburban diversity," *Paper in Brookings Institution Census 2000 Series*, p. 17. Washington, DC: The Brookings Institution.

Frey, W. H. (2002a) "Escaping the city—And the suburbs." *American Demographics* 24(6):21–23.

Frey, W. H. (2002b) "The new suburbanization." *The American Enterprise* April/May:43.

Frey, W. H., and Zimmer, Z. (2001) "Defining the city," in *Handbook of Urban Studies.* ed. by P. Ronan, pp. 14–35. London: Sage.

Glazer, N., and Moynihan, D. P. (1970) *Beyond the Melting Pot* (2nd edition). Cambridge, MA: MIT Press.

Garreau, J. (1981) *The Nine Nations of North America.* Boston, MA: Houghton Mifflin.

Gober, P. (2000) "Immigration and North American Cities." *Urban Geography* 21(1):83–90.

Gold, S. J. (1988) "Refugees and small business: the case of Soviet Jews and Vietnamese." *Ethnic and Racial Studies* 11:411–438.

Gong, J. (2002) "Clarifying the standard deviational ellipse." *Geographical Analysis* 34(2):155–167.

Greene, R. P. (1991) "Poverty area diffusion: The depopulation hypothesis examined." *Urban Geography* 12:526–541.

Greene, R. P. (1997) "Chicago's new immigrants, indigenous poor and the changing geography of metropolitan employment." *Annals of the American Academy of Political and Social Science* 551:178–190.

Hanson, S. (1992) "Geography and feminism: Worlds in collision?" *Annals of the Association of American Geographers* 82(4):569–584.

Jargowsky, P. (1996) "Beyond the street corner: The hidden diversity of high-poverty neighborhoods." *Urban Geography* 17(7):579–603.

Jargowsky, P. (1997) *Poverty and Place: Ghettos, Barrios and the American City.* New York: Russell Sage Foundation.

KLRU and PBS Online. (2002) *The Forgotten Americans, Focus: Las Colonias.* Austin, TX: KLRU/TV and PBS Online. Posted at *http://www.pbs.org/klru.*

Kwan, M.-P. (2002a) "Feminist visualization: Re-envisioning GIS as a method in feminist geographic research." *Annals of the American Association of Geographers* 92(4):645–661.

Kwan, M.-P. (2002b) "Introduction: Feminist geography and GIS." *Gender, Place, and Culture* 9(3):261–262.

Kwan, M.-P. (2002c) "Is GIS for women? Reflections on the critical discourse in the 1990s." *Gender, Place, and Culture* 9(3):271–279.

Langley, L.D. (1987) *MexAmerica.* New York, NY: Crown Publishers.

Lichter, D.T. & Crowley, M.L. (2002) "Poverty in America: Beyond welfare reform." *Population Bulletin* 57(2):1–36. Washington, DC: Population Reference Bureau. Available at *www.prb.org.*

Longhurst, R. (1998) "(Re)presenting shopping centres and bodies," in *New Frontiers of Space, Bodies, and Gender.* ed. by R. Alnley, pp. 20–34. London: Routledge.

Massey, D. (1998) "Blurring the binaries? High tech in Cambridge," in *New Frontiers of Space, Bodies, and Gender.* ed. by R. Alnley, pp. 157–175. London: Routledge.

Massey, D. S., and Denton, N. A. (1993) *American Apartheid.* Cambridge, MA: Harvard University Press.

McDowell, L. (1995) "Body work: Heterosexual gender performances in city workplaces," in *Mapping Desire.* ed. by D. Bell and G. Valentine, pp. 75–95. London: Routledge.

McDowell, L. (1997) *Capital Culture: Gender at Work in the City.* Oxford, UK: Blackwell.

McDowell, L. (2001) "Women, men, cities," in *Handbook of Urban Studies.* ed. by R. Paddison, pp. 206–219. London: Sage.

McDowell, L., and Sharp, J. (1999) *A Feminist Glossary of Human Geography.* London: Arnold.

McKee, J. O. (ed.). (2000a) *Ethnicity in Contemporary America* (2nd edition). Lanham, MD: Rowman & Littlefield.

McKee, J. O. (2000b) "Introduction," in *Ethnicity in Contemporary America* (2nd edition). ed. by J. O. McKee. Lanham, MD: Rowman & Littlefield.

McLafferty, S. (2002) "Mapping women's worlds: Knowledge, power, and the bounds of GIS." *Gender, Place, and Culture* 9(3):263–269.

McLafferty, S., and Preston, V. (1997) "Gender, race, and the determinants of commuting: New York in 1990." *Urban Geography* 18(3):192–212.

McNee, B. (1984) "If you are squeamish . . . ." *East Lakes Geographer* 19:16–27.

McNee, B. (1985) "It takes one to know one." *Transition* 14:2–15.

Mitchell, D. (2000) *Cultural Geography: A Critical Introduction.* Oxford, UK: Blackwell.

Miyares, I. M. (1998) " 'Little Odessa' – Brighton Beach, Brooklyn: An examination of the former Soviet refugee economy in New York City." *Urban Geography* 19(6):518–530.

Nast, H., and Pile, S. (eds.). (1998) *Places Through the Body.* London: Routledge.

Pain, R., Barke, M., Fuller, D., Gough, J., MacFarlane, R., and Mowl, G. (2001) "Geographies of gender and sexuality," in *Introducing Social Geographies*, pp. 120–140. London: Arnold.

Pepin, M. (1998) "Texas colonias: An environmental justice case study." Posted at *http://itc.ollusa.edu/faculty/pepim/philosophy/cur/colonias.htm.*

Peck, J. and Theodore, N. (2001) "Contingent Chicago: Restructuring the spaces of temporary labor." *International Journal of Urban and Regional Research* 25.3:471–496.

Pick, J. B., and Butler, E. W. (1997) *Mexico Megacity.* Boulder, CO: Westview.

Portes, A. and Rumbaut, R. (2001) *Legacies: The Story of the Immigrant Second Generation.* Berkeley, CA: University of California Press.

Portes, A. and Stepick, A. (1993). *City of the Edge: The Transformation of Miami.* Berkeley, CA: University of California Press.

Smeeding, T. M., Rainwater, L., and Burtless, G. (2001) "U.S. poverty in a cross-national context," in *Understanding*

Poverty. ed. by S. H. Danziger and R. H. Haveman, pp. 162–189. New York: Russell Sage Foundation.

Spain, D. (1999) *America's Diversity: On the Edge of Two Centuries.* Washington, DC: Population Reference Bureau.

Strait, J. B. (2000) "An examination of extreme urban poverty: The effect of metropolitan employment and demographic dynamics." *Urban Geography* 21(6): 514–542.

Tarmann, A. (2002) *Easing Urban Poverty Key to Economic Growth in Central America.* Washington, DC: Population Reference Bureau. Available at *http://www.prb.org.*

Texas Natural Resource Conservation Commission. (1999) *Colonias: Serving the Unserved.* Austin, TX: Author. Available at *http://www.tnrcc.state.tx.us.*

Texas Office of the Attorney General. (2002) Border Colonia Geography Online. Available at *http://maps.oag.state.tx.us/colgeog/colonias.htm.*

Texas Secretary of State. (2002) *Colonias.* Austin, TX: State of Texas, Office of the Secretary of State. Available at *www.sos.state.tx.us/border/colonias* (10/02).

Timander, L. M., and McLafferty, S. (1998) "Breast cancer in West Islip, NY: A spatial clustering analysis with covariates." *Social Science and Medicine* 46(12):1623–1635.

U.S. Census. (2001) "Profile of the foreign-born population in the United States." *Current Population Reports.* Special Studies, pp. P23–P206. Washington, DC: U.S. Census.

U.S. Census. (2002) *Supplementary Survey Summary Tables: Profile of Selected Social Characteristics 2000.* AT-20. Washington, DC: Author.

U.S. Census. (2003) *Census* 2000. Washington, DC: U.S. Census. Available at *www.census.gov.*

U.S. Census. (2004). *Poverty.* Washington, DC: U.S. Census. Available at *www.census.gov.*

U.S. HUD. (2002) *Discrimination in Metropolitan Housing Markets 1989–2000.* HUD Report No. 02-138. Washington, DC: U.S. Department of Housing and Urban Development.

Valentine, G. (1993a) "(Hetero)sexing space: Lesbian perceptions and experiences of everyday spaces." *Society and Space* 11:395–413.

Valentine, G. (1993b) "Negotiating and managing multiple sexual identities: Lesbian time-space strategies." *Transactions of the Institute of British Geographers* 18(2):237–481.

Ward, P. (1998a). *Mexico City* 2nd edition. New York, NY: John Wiley and Sons.

Ward, P. (1998b). *Colonias in Texas and Mexico: Cross Border Perspectives and Policies.* Austin, TX: University of Texas Press.

Wilson, W. J. (1987) *The Truly Disadvantages: The Inner City, the Underclass and Public Policy.* Chicago, IL: University of Chicago Press.

Wilson, W. J. (1996) *When Work Disappears: The World of the New Urban Poor.* New York: Knopf.

Wirth, L. (1938) "Urbanism as a way of life." *American Journal of Sociology* 44:1–24.

Wong, D. W. S. (2003) "Implementing spatial segregation measures in GIS." *Computers, Environment, and Urban Systems* 27:53–70.

Wong, D. W. S. (2004) "Comparing traditional and spatial segregation measures: A spatial scale perspective." *Urban Geography* 25:66-82.

World Bank. (2002) *Urban Services Delivery and the Poor: The Case of Three Central American Cities.* Report No. 22590. Washington, DC: Author.

Zhou, Y. 1998. "How do places matter? A comparative study of Chinese ethnic economies in Los Angeles and New York City." *Urban Geography* 19(6):531–553.

EXERCISE

Exercise Description

In this project we employ a set of statistical procedures known as centrographic methods. Specifically, we will compute a year 2000 mean center and standard radius (also known as standard distance) for two groups, African American and Hispanic, in the city of St. Louis and neighboring St. Louis County. The standard radius is a spatial measure of dispersion where a large value indicates wide dispersion and a small value indicates concentration. By computing a standard radius for each group, we can determine which group is more spatially concentrated.

Course Concepts Presented

Centrographic methods, segregation, race, and ethnicity.

GIS Concepts Utilized

Spatial statistics.

Skills Required

ArcGIS Version 9.x

GIS Platform

ArcGIS Version 9.x

Estimated Time Required

1 hour

Exercise CD-ROM Location

ArcGIS_9.x\Chapter_08

Filenames Required

stl00.shp
stlouis_city.shp

Background Information

Centrographic Methods

A frequently asked question in urban geography is this: How segregated are groups within cities? To answer this question, nonspatial measures such as the well-known index of dissimilarity have been applied. Although these measures provide a useful statistic for understanding the amount a group would have to be redistributed to be fully integrated, they lack the cartographic element associated with centrographic methods. Centrographic methods can summarize large quantities of spatial data and can be simultaneously represented on a map. The specific centrographic statistic used in this exercise is known as a standard radius (also standard distance). Its value summarizes the dispersion of a group's distribution around its mean center and is directly analogous to a standard deviation in univariate statistics. The standard radius is derived in three steps:

1. To report the findings in common measurements, latitude and longitude coordinates for the centers of each census tract are projected to Universal Transverse Mercator (UTM) coordinates. The east–west UTM coordinates are presented as Eastings (x) and the north–south coordinates as Northings (y). The Easting distances are presented in meters from an east–west reference axis and the Northings in meters from the Equator.
2. A weighted mean center is calculated by multiplying the Eastings and Northings by the group population that is located at those Eastings and Northings. Because we do not know the precise coordinates of each group member, we assign the group population to the Easting and Northing of the center of the census tract in which they are located. The equation for the weighted mean center is:

$$x_w = \frac{\sum xw}{\sum w}, \; y_w = \frac{\sum yw}{\sum w}$$

The weighted mean center of a geographic distribution is analogous to the concept of a center of gravity where the weighted mean center is the point on which the distribution would balance. Perhaps the concept is most popularly demonstrated by the U.S. Census Bureau, which reports the new population center of the United States at each decennial census. In the first U.S. Census of 1790 the U.S. population center was located just outside of Baltimore, Maryland. In the 2000 Census, the U.S. population center was located almost 100 miles southwest of St. Louis, Missouri.

3. A standard radius of the group distribution, defined as the square root of the mean of the sum of the squared distances of the observations from the weighted mean center is calculated:

$$r = \sqrt{\left(\frac{\sum x^2}{n} - x_w^2\right) + \left(\frac{\sum y^2}{n} - y_w^2\right)}$$

where x and y are geographic coordinates, n is the total number of points and x_w and y_w are the x, y coordinates for the weighted mean center.

The standard radius describes dispersion in terms of a circle about the mean center. It is a measure of the spread of a group around its mean center.

Readings

(1) Bachi, R. (1963) "Standard distance measures and related methods for spatial analysis." *Papers, Regional Science Association* 10:83–132.

(2) Greene, R. P. (1991) "Poverty concentration measures and the urban underclass." *Economic Geography* 67:240–252.

(3) Greene, R. P., and Pick, J. B. *Exploring the Urban Community: A GIS Approach*, Chapter 8.

(4) Lefever, D. W. (1926) "Measuring geographic concentration by means of the standard deviational ellipse." *American Journal of Sociology* 32:88–94.

(5) Massey, D. S., and Eggers, M. L. (1990) "The ecology of inequality: Minorities and the concentration of poverty, 1970–1980." *American Journal of Sociology* 95:1153–1188.

(6) White, M. J. (1984) "The measurement of spatial segregation." *American Journal of Sociology* 88:1008–1018.

Exercise Procedure Flowchart

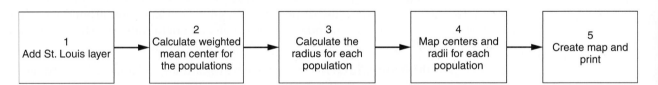

Exercise Procedure

1. Start ArcMap with Start Using ArcMap With A New Empty Map selected.

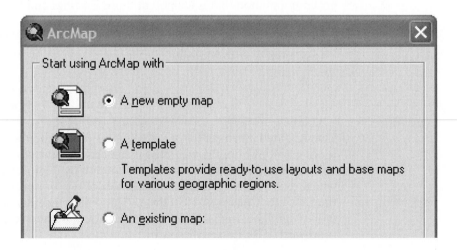

2. Click Add Data and navigate to the exercise data folder.

   Note: *If you do not see the folder where you placed the exercise data, you will need to make a new connection to the root directory that holds the data.*

3. Add Stl00.shp and Stlouis_city from the Chapter 8 folder. You can add the two layers at once by clicking one of the file names, clicking on the other one while holding down the Ctrl key on the keyboard, and clicking Add.

4. Right-click the layer name stl00 and click Open Attribute Table.

   Note that the table has year 2000 information for the total population, African American population, and Hispanic population. We next compute the mean centers for the African American and Hispanic populations.

5. Close the table and click the ArcToolBox button at the top (red button).

6. Under Spatial Statistics Tools and Measuring Geographic Distributions, double click Mean Center.

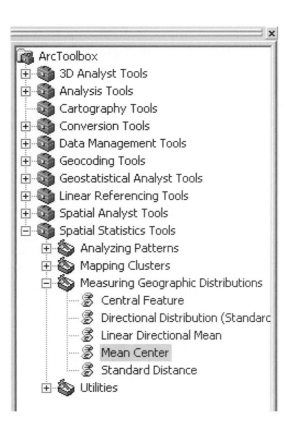

7. Fill in the Mean Center window by specifying an input file (stl00.shp), output file (afamcntr.shp), and weight field (black00). The path names might be different on your machine and the output file is a new file to be created.

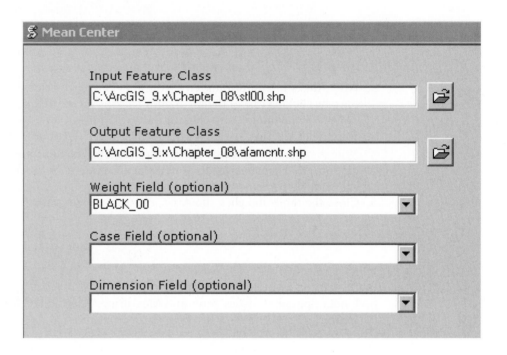

8. Click OK and it will display a processing box. Click Close when completed.

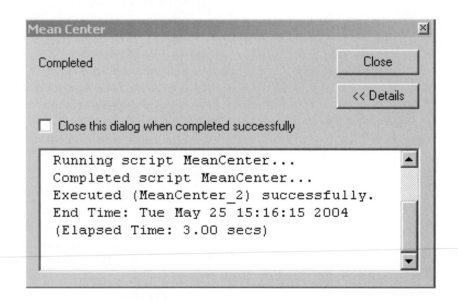

9. Repeat the calculation of a mean center for the Hispanic population.

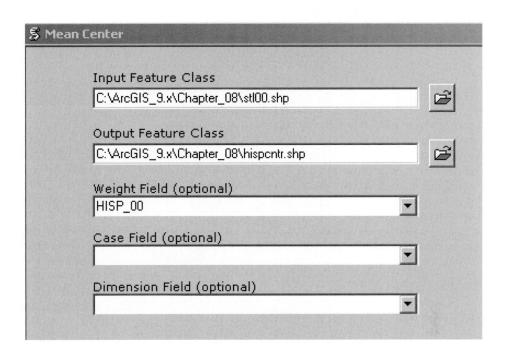

10. Now compute a standard radius (standard distance) for each population. Under Spatial Statistics Tools and Measuring Geographic Distributions, double-click Standard Distance.

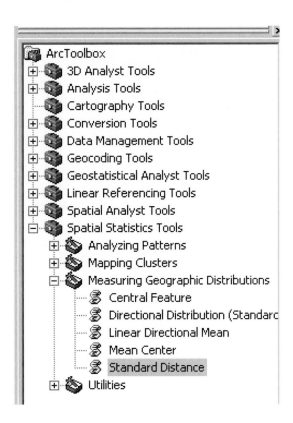

11. Fill in the Standard Distance window by specifying an input file (stl00.shp), output file (afamstd1.shp), and variable (Black00).

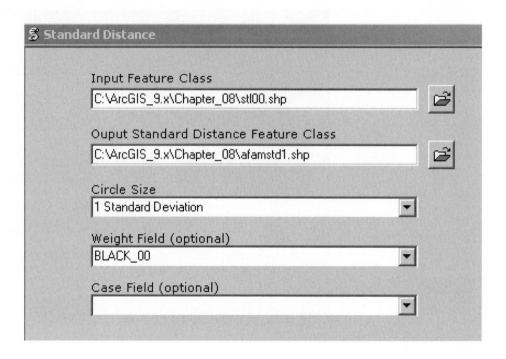

12. Choose 1 Standard Deviation as the circle size and click OK. This will display a processing box. Click Close when completed.

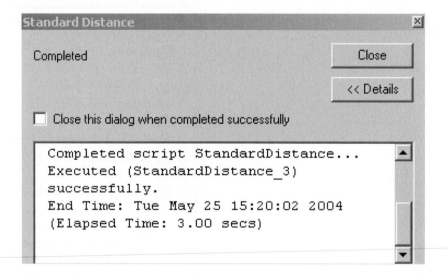

13. Repeat the calculation of a standard distance for the Hispanic population.

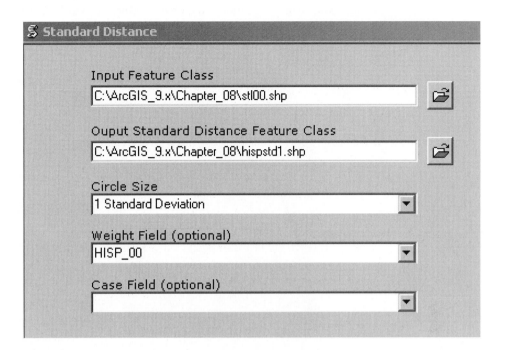

14. If you right-click and open the attributes of the standard distance layers, you can observe the coordinates for each group's mean center and the standard radius (StdDist).
    (African American Population)

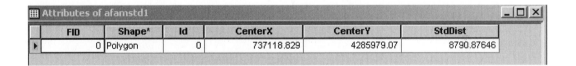

FID	Shape*	Id	CenterX	CenterY	StdDist
0	Polygon	0	737118.829	4285979.07	8790.87646

(Hispanic Population)

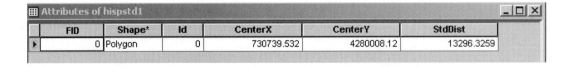

FID	Shape*	Id	CenterX	CenterY	StdDist
0	Polygon	0	730739.532	4280008.12	13296.3259

15. You can make the polygon symbols for the standard radius layers hollow, so that you can see both layers together. Click the symbol box below afamstd1 and in the Symbol Selector dialog box, click the Hollow box, change Outline Width to 2, change Outline Color, and click OK. Repeat the last step for hispstd1.

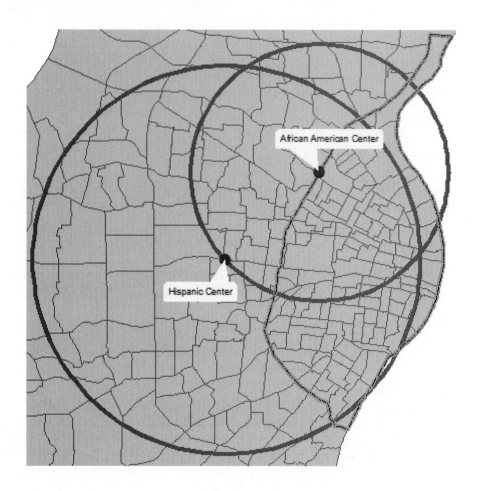

16. Select Layout View on the View menu to compose a map for printing.

Essay

In your essay, review the Chapter 8 "Analyzing an Urban Issue" section. What is the distance between the two mean centers in St. Louis and what does this tell us about the general location of these two groups in St. Louis? Which group is more concentrated? Why might this difference be important for urban planners? Do you see any political or social implications of the centrographic results? What are some of the strengths and weaknesses of centrographic methods?

9

Industrial Location and Cities

Cities are places of both production and consumption. During the Industrial Revolution, cities became the main locus of manufacturing activity and grew rapidly as a result. Although the modern-day city still accommodates manufacturing (See Figure 9.1), cities have become more diversified in the range of industrial activity they perform. Nevertheless, some cities continue to specialize in manufacturing (e.g., Davenport, Iowa), whereas others have become prominent centers of producer services (e.g., New York City), and still others have become more specialized in consumer services (e.g., Las Vegas). This chapter focuses on the role of industry in cities. Cities are important for industry not only because they facilitate the production process by supplying a rich and diverse labor force, but also because they drive a consumption process triggered by a dense demand for products and services.

The chapter examines the location of industry both across cities (interurban) and within cities (intraurban). It presents these location concerns in their historical context as geographers have introduced a number of paradigms over time for analyzing industries in cities. The chapter reviews the recent resurgence of activity among urban scholars on the topic of industry and cities, most notably on the location of manufacturing activity, services, and consumption. However, it begins with the formal theorizing of manufacturing location that can be traced back to Alfred Weber (1929), a pioneer in this endeavor. The urban geography of services is examined with mention of its roots in central place theory, a topic already discussed in Chapter 5. It reviews the rapid growth of the service industry and the subsequent explosion in subcategories within this industry. It considers the full extent of industry, which is a broad term that includes all economic activities, such as manufacturing, retail, wholesale, services, finance, and government.

The chapter focuses on urban industrial topics mostly in the context of the United States with some discussion of industrial issues in Mexico, particularly the *maquiladora* industry in the twin cities along the U.S.–Mexico border. Concepts of globalization and comparative advantage must be considered in explanations for differences in the industrial composition of cities that are adjacent to each other but are nonetheless divided by national borders. An example of consumption exchanges between Mexico and U.S. cities is also presented where each city depends on the other for local business.

The chapter includes a contemporary analysis of interurban patterns of industry that considers the transformation of cities from emphasis on production to increasing shares of service and white-collar employment. It reviews recent studies on cities that have attracted a disproportionate share of high-tech industry and other studies that look at the factors that continue to add to their growth. It reviews the geography of retail services and a new consumer-city hypothesis, which emphasizes the importance of consumption over production for predicting future urban growth and development. The final section considers the geography of producer services.

9.1 INDUSTRY CLASSIFICATION AND LOCATION CONCEPTS AND PROCESSES

This section presents basic definitions and concepts for industrial location within and across cities. These definitions serve as foundations for understanding the follow-up sections on processes and case studies. Urban economies are complex and are made up of activities that are not always easy to classify and/or describe.

FIGURE 9.1   Los Angeles factory.

To help study the relationships between cities and industry we adopt a classification of industry types used by economic geographers:

1. *Primary Activities*   Agriculture, mining, fishing.
2. *Secondary Activities*   Manufacturing and construction.
3. *Tertiary Activities*   Retailing and services.
4. *Quarternary Activities*   Information services and data processing.

Primary economic activities include those activities where workers are directly engaged with raw products such as agriculture, fishing, and mining. Secondary economic activities include manufacturing industries or those industries where workers are involved in making or processing products. The tertiary economic sector consists of activities where workers are involved in providing services, both to producers and to consumers. Some have modified this classification of economic activities to include a quaternary sector made up of professionals engaged in the production and processing of information or knowledge-based activities. With growing diversity in activity types in the economy it is important to recognize more specialized categories of economic activity (see Figure 9.2). Note, for instance, the declining proportion of U.S. workers engaged in manufacturing from 1959 to 1994 and the rising proportion engaged in services over the same time period, especially in finance, insurance, and real estate.

To reflect ongoing economic transformations in the global economy, the U.S. Census Bureau replaced its long-established Standard Industrial Classification (SIC) with the North American Industry Classification system (NAICS) starting with its 1997 Economic Census. The SIC system was first developed in the 1930s when many large U.S. cities had a strong manufacturing base. The NAICS recognizes industry categories that are more commonplace in today's economy, including an explosion of service

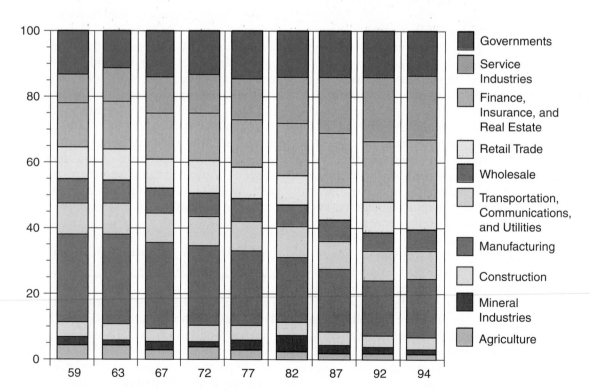

FIGURE 9.2   Gross domestic product: Percentage of total by sector, 1959–1994. Source: The Economic Census–Two Moments of Truth: 1954 and 1997.

TABLE 9.1 New NAICS industries: selected examples

• Semiconductor machinery manufacturing	• Telecommunication resellers
• Fiber-optic cable manufacturing	• Credit care issuing
• Reproduction of computer software	• Temporary help supply
• Manufacture of compact discs, except software	• Telemarketing bureaus
	• Hazardous waste collection
• Convenience stores	• HMO medical centers
• Gas stations with convenience food	• Continuing care retirement communities
• Warehouse clubs	• Casino hotels
• Food/health supplement stores	• Casinos
• Pet supply stores	• Other gambling industries
• Pet care stores	• Bed and breakfast inns
• Cable networks	• Limited service restaurants
• Satellite communications	• Automotive oil change and lubrication shops
• Paging	• Diet and weight-reducing centers
• Cellular and other wireless communications	

Source: U.S. Bureau of the Census, 2003.

categories and new terms to capture activities in the information economy (see Table 9.1). The NAICS recognized 361 industries not previously identified separately by the SIC system. Not only does the NAICS include current and appropriate concepts and definitions for the classification of industry but it also makes the U.S. data more comparable to those of its neighboring countries of Canada and Mexico. Thus the industrial compositions of border cities such as El Paso, Texas, and its Mexican counterpart Ciudad Juarez can be more easily compared.

9.2 MANUFACTURING LOCATION THEORY AND AGGLOMERATION

We begin with manufacturing because many cities, especially older cities, grew as a result of their reliance on this industry type. This section reviews two important theories with respect to manufacturing location. The first is known as least-cost theory and the second is referred to as economic-base theory.

9.2.1 Least-Cost Theory

Alfred Weber (1929), a German economist, was the first to develop a general theory of manufacturing location. He was also the younger brother of the well known sociologist, Max Weber (Stutz and Warf, 2005). Weber explained the location of industry in terms of firms seeking a minimum cost location. The two principal costs isolated in Weber's model were labor costs and transport costs. Between these, Weber regarded transport costs as the most important factor because they increase with distance,

although at a decreasing rate (see Figure 9.3). The initial loading and unloading costs at the origin and destination are referred to as fixed costs, and the distance charges are often set by zone, resulting in a stepping-stone cost structure. In addition, the distance that a good is to be shipped often dictates the mode of transportation. The general rule is that water is best for long hauls, rail for intermediate hauls, and truck for short hauls.

These simple transportation cost principles allowed Weber to simulate a variety of location outcomes for manufacturing activity based on a number of different contexts. The following two examples will help illuminate Weber's approach. In the first, Weber considered the influence of transport costs on the location of a plant that exists in the context of a single market such as a city and a single input or material resource (see Figure 9.4). A firm wishes to build a plant to manufacture a product that requires one raw material input from one source region (SR) and to sell its output in one marketplace (MP). In this situation, there are two possible least-cost locations—SR and MP. Any site in between incurs additional terminal costs (loading and unloading), as well as two short hauls, which costs more per unit of distance than one long haul. TTC on the graphs represents the total transportation cost of shipments. A second example shows, however, that it is likely that shipping the finished product from the source region is more costly than shipping raw resources. In that case, the plant will typically locate at the marketplace or city to avoid the high cost of shipping finished products (see Figure 9.5). Thus, in the second graph, the transport costs are lower at the market place as indicated by the line intersecting the vertical cost bar. Weber's original

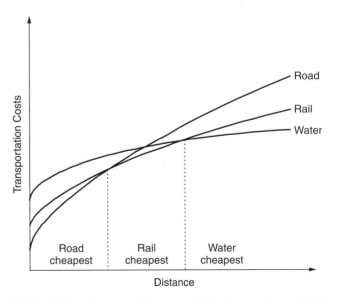

FIGURE 9.3 Transportation costs and distance. Source: Cadwallader, Martin 1996. *Urban Geography: an analytical approach.* Upper Saddle River, N.J.: Prentice Hall, Inc. Fig. 7.2 p. 156.

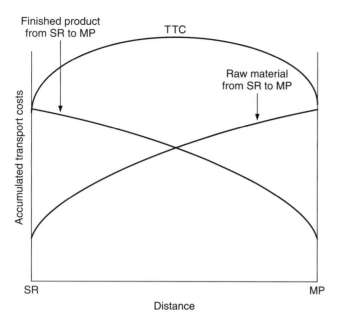

FIGURE 9.4   Transport costs versus distance: Single market and single raw input or raw material resource. Source: Yeates, Maurice 1990. The North American City. Harper and Row Publishers, New York (Fourth Edition). Fig. 3.9 on p. 82.

work considers many more scenarios, each isolating an aspect of the manufacturing location decision process.

One weakness of Weber's model for use in today's U.S. manufacturing landscape is the emphasis on transportation costs, which have been shown to be of declining significance (Webber, 1984, p. 83). For instance, since the time of Weber's work, manufacturing in the United States has

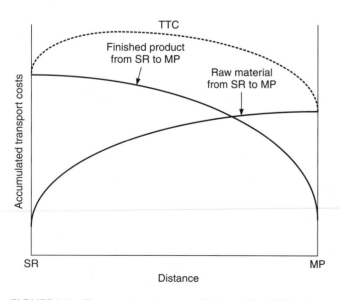

FIGURE 9.5   Transport costs versus distance: Cost differences for shipping finished goods. Source: Yeates, Maurice 1990. The North American City. Harper and Row Publishers, New York (Fourth Edition). Fig. 3.10 on p. 83.

shifted away from heavy manufacturing toward light manufacturing, which requires less bulk on the input side so there is less bulk when transporting the final finished goods as well. Figure 9.6 shows light manufacturing in Mexico City. In addition, many newer methods exist for shipping goods, including growing containerization as previously shown in Chapter 2 (see Figure 2.1). Nevertheless, industrial geographers consider Weber's approach to be a good starting point for understanding the location of industries.

9.2.2 Economic-Base Theory

The economic-base approach rose to prominence in the late 1930s after the publication of *Principles of Urban Real Estate* by Weimer and Hoyt (1939). The approach distinguished between two types of employment, *primary* being industrial employment and *ancillary* representing service employment. Economic-base theory held that the foundation of a city and its ability to grow depended on manufacturing employment because it produced sales outside the community, referred to as exports. As a consequence of a growing industrial base, a city could expect growth in ancillary employment because revenue returning from export sales could support more services. Cleveland, Ohio, is a good example, where great transport access to Lake Erie by way of the Cuyahoga River facilitated growth and development in both its manufacturing base and service base (see Figure 9.7). Employment related to the production of goods and services produced for export outside the community was referred to as basic, whereas employment related to sales within the local community was referred to as nonbasic. Economists at the time attributed urban growth to the basic portion of total employment, whereas ancillary services only played a supporting role.

In the 1940s and 1950s, urban practitioners adopted many of the ideas from economic-base theory for economic development. It followed from economic-base theory that urban growth would proceed fastest if a city concentrated on increasing its basic employment; in other words, one did not need to be too concerned about increasing nonbasic employment, as that would follow naturally. A basic/nonbasic ratio was developed to determine the proportion of service-related jobs that could be expected as a result of growing the manufacturing sector:

$$BA/NBA$$

where BA is basic activity employment and NBA is nonbasic activity employment. If a city determined that its ratio was higher than 1:1, for instance, 1:4, that would mean that for every person involved in basic activity there would be four persons engaged in a nonbasic activity. Once derived, the ratio could be assumed to be static and it was not uncommon to use it as a predictor or what has become known as a multiplier effect. For instance, a 1:4 ratio might be used

FIGURE 9.6   Light manufacturing in Mexico City.

FIGURE 9.7   Cleveland.

to suggest that an additional increase of 100 basic jobs would result in 400 new nonbasic jobs for a total of 500 new jobs.

The economic-base approach is not as simple as it might first seem. Perhaps the most problematic aspect of the approach is determining how activities get classified as basic or nonbasic. Consider, for instance, the services provided by a medical doctor. At first it might seem appropriate to classify them as nonbasic but perhaps on later review it is learned that some doctors in the community are attracting patients beyond the city limits. That raises an additional problem of where to draw the boundary to differentiate sales generated inside versus outside the community. These are all methodological issues that need to be addressed prior to applying the economic-base approach. Nevertheless, the approach is still widely used and many variations have been developed, especially in the area of calculating multiplier effects (Hinojosa & Pigozzi, 1988; Mulligan & Kim, 1991). Today, the economic-base approach is widely used and applied among local economic development practitioners.

9.2.3 Agglomeration Economies

Agglomeration economies are another important topic in considering the location of manufacturing activity. Economic geographers refer to *agglomeration economies* as those savings in production costs that occur when firms locate near one another (see Figure 9.8). Weber (1929) also considered agglomeration as a factor of location. Webber (1984, p. 78) defined three specific agglomeration economies. The first is the cost saving due to scale, such as the rise of businesses providing services to other firms. A second is information, as communication is enhanced by firms locating near each other. The third is social fixed capital or societal infrastructure, which would include railroads, highways, schools, shopping centers, and banks. All of them reduce production costs including associated services.

Agglomeration continues to be explored by contemporary urban scholars. Cervero (2001) considered another agglomeration issue, that of the shape of a metropolis. He contended that the shape of a city carries significant costs and benefits for agglomeration economies. He empirically examined the relationships between labor productivity and certain indicators of urban form. Kenworthy and Laube (1999) found that gross regional product per capita was generally higher in cities that were less dependent on automobiles. Similarly, Cervero contended that large cities that are compact, have good accessibility, and are endowed with efficient transport technology are among the most productive of all urban settlements. Agglomeration economies were cited as a cause of the strong relationship between employment densities and productivity levels.

FIGURE 9.8   Oracle headquarters in Silicon Valley, with other Tech company facilities in background. Source: Corbis/Bettmann.

Agglomeration economies continue to be a topic of much research. Rigby and Essletzbichler (2002) recently examined the impact of agglomeration economies on industry productivity across U.S. metropolitan areas by examining plant-level data as opposed to aggregate data, which previous studies have relied on. They showed that many establishment and city-specific factors influence labor productivity across cities. Mota and Castro (2004)

studied the injection mould industry cluster at Marinha Grande in Portugal and found firms there profited immensely from the emergence of relationships between nearby firms.

Rantisi (2002) showed that New York City's Garment District, or Fashion Center in today's parlance, benefits significantly from agglomeration economies (see Figure 9.9). The Fashion Center is bounded by 40th Street to the north,

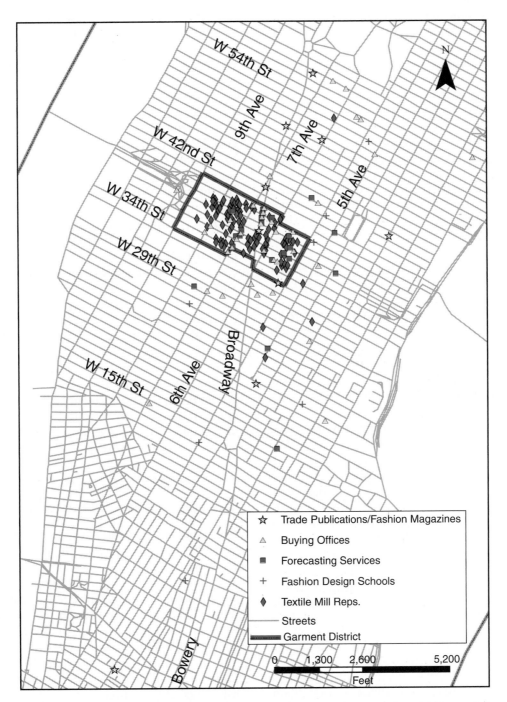

FIGURE 9.9 Agglomeration in New York's Garment District. Source: Rantisi, Norma M. 2002. The competitive foundations of localized learning and innovation: the case of women's garment production in New York City. *Economic Geography* 78:441–462. Fig. 4 p. 450.

34th Street to the south, Fifth Avenue to the east, and Ninth Avenue to the west. Seventy-five percent of Manhattan's apparel industry is located within this four-by-six block area. Two thirds of the businesses are fashion-related, the principal ones being apparel manufacturers and contractors. However, they also include retailers, textile suppliers, and fashion-design schools, all of which depend on each other for services. Hum (2003) saw similar patterns among Asian and Latino immigrant-owned firms across the river in Brooklyn's Sunset Park neighborhood. Scott (2002a) also noted that the garment industry's geographical configuration nearly always appears in the form of a dense agglomeration of firms, even in the contemporary age of globalization (see Figure 9.10). Scott showed that Los Angeles has a similar concentration of interacting businesses within its Fashion District:

> This area represents the functional and spatial pivot of the industry and is the source of a multi faceted stream of positive externalities or agglomeration economies, ranging from the services offered by the many specialized sub-contract shops that lie within a stones throw, to the dense local labour market of garment makers that has developed in the vicinity. (Scott, 2002a, p. 1293)

Scott contended that the Los Angeles Fashion District will soon be on par with those in New York, Paris, London, and Milan (see Figure 9.11).

Even as clothing production gets outsourced from high-cost regions to low-cost regions, agglomeration economies reemerge in the receiving region as demonstrated recently by eastern Europe (Smith, 2003). Many buyers in western Europe have outsourced clothing production to eastern Europe where agglomeration is occurring in regional clusters (see Figure 9.12).

It may seem odd that with advancements in information and communication technologies that there would still be a continued trend of agglomeration of economic activity. In particular, the need of locating close to other firms for shared communication advantages through face to face contact would appear obsolete given the efficiency of such transactions that can be accommodated nearly instantaneously through the Internet. Liu, Dicken, and Yeung (2004) examined this very issue in a case study of Nokia's manufacturing cluster in Beijing and found on the contrary that agglomeration tendencies are being facilitated by the new information and telecommunications technologies.

9.2.4 Intrametropolitan Industrial-Location Theories

A number of theories have been advanced to explain the location of manufacturing within cities. Among these is the incubator theory, which explains the evolution of firms concentrating close to the center and later relocating to the outer edge of the city (Leone & Struyk, 1975). The analogy of the central city as an incubator was proposed because this theory holds that agglomeration economies associated with the high-density core are more favorable to new firms starting out. It was then argued that, once the firm had established itself, it would become more self-sustaining and then be able to relocate to the fringe where other costs including land were substantially less.

Although a critic of the incubator theory, Scott (1982) offered a product-cycle alternative by arguing that in the early stages of a firm, the firm depended on the core of the city for skilled labor and other agglomeration economies. However, with time, the firm would become less dependent on the city and eventually set up branch plant operations in small towns away from the center, to benefit from lower costs and better transportation access. Scott and others later challenged the incubator theory, feeling that it was limiting because it only allowed for births of firms in the city core and could not explain the births occurring outside the core. A study of manufacturing firm relocations in the Chicago region by Stumpf (1996) suggested that the motivations for firm movement from the core to the edge were public policy, infrastructure issues, labor force, and transportation access (see Figure 9.13). The use of trend surface analysis allowed Stumpf to conclude that the dominant movement trend in the Chicago region was biased toward the western suburbs, particularly in the north as illustrated in the map by the expanding width of the trend band toward that direction.

To demonstrate industrial location from an urban geography perspective, the next section, "Analyzing an Urban Issue," examines the distribution of employment in nine industrial sectors within Los Angeles County. This urban issue is later pursued in the chapter's geographic information system (GIS) laboratory exercise.

9.3 ANALYZING AN URBAN ISSUE: INTRAURBAN INDUSTRIAL LOCATION

The dispersion of economic activity has been especially evident in the Los Angeles area (see Figure 9.14). Nevertheless, as Giuliano and Small (1991) noted, there are distinct subcenters within the Los Angeles region even though it has been characterized as the most dispersed settlement pattern among large urban areas in the United States. Others have noted the spatial logic to the apparent dispersion of these activities, especially as one disaggregates the economic activity by sector. For instance, Scott (2002b) showed that manufacturing establishments cluster together in specialized industrial districts, with the region's craft industries concentrating closer to the central city, whereas the high-technology industrial districts take a more peripheral location.

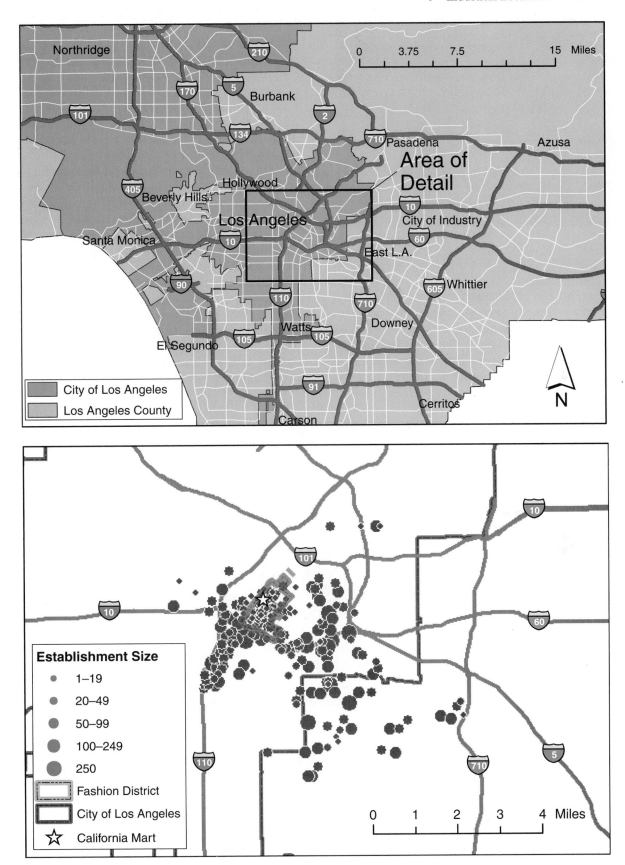

FIGURE 9.10   Agglomeration in Los Angeles's Fashion District. Source: Scott, Allen J., 2002. Competitive dynamics of Southern California's clothing industry: the widening global connection and its local ramifications. *Urban Studies* 39:1287–1306. Fig. 3 on p. 1295.

FIGURE 9.11 Los Angeles's Fashion District.

The conventional explanation for these patterns has been one of agglomeration economies and lower wages near the center and abundant and inexpensive land on the urban periphery (Blackley & Greytak, 1986). A similar finding has been made for the Dallas area where high-technology firms are shown to be avoiding the central business district (CBD) and are concentrating toward suburban Richardson north of downtown, probably to maximize access to a technically skilled labor force (Waddell & Shukla, 1993).

Economic activities outside manufacturing, for instance, professional services or retail, also show clustering, but the amount of clustering and reasons for it are quite different. In this section, we examine the clustering of industrial activity of various types for Los Angeles County. Urban geographers have become increasingly interested in the dispersion of employment away from the central core of cities to outlying employment subcenters, often in the suburbs. The shift of employment to the suburbs has occurred among all industrial sectors from manufacturing to professional services. It is often suggested that office-oriented employment is more suburb oriented than manufacturing industry. Industrial centers have also remained highly economically segregated, so that many suburban office parks are separate from manufacturing industrial parks. To assess these premises, we examine the distribution of employment by nine industrial sectors within Los Angeles County using the location quotient technique.

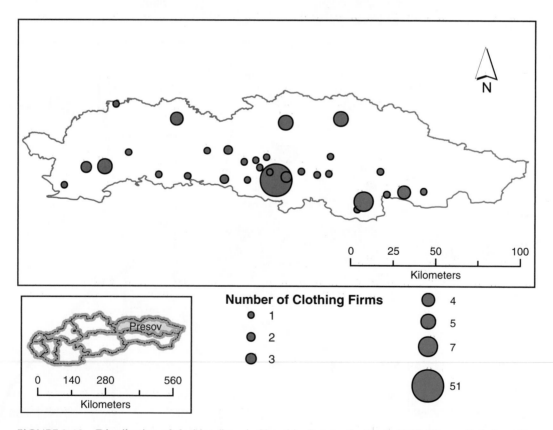

FIGURE 9.12 Distribution of clothing firms in Slovakia. Source: Smith, A. 2003. "Power Relations, Industrial Clusters, and Regional Transformations: Pan-European Integration and Outward Processing in the Slovak Clothing Industry." *Economic Geography* 79:17–40. Figure 2. Page 26. Map showing the number of clothing firms in Slovakia.

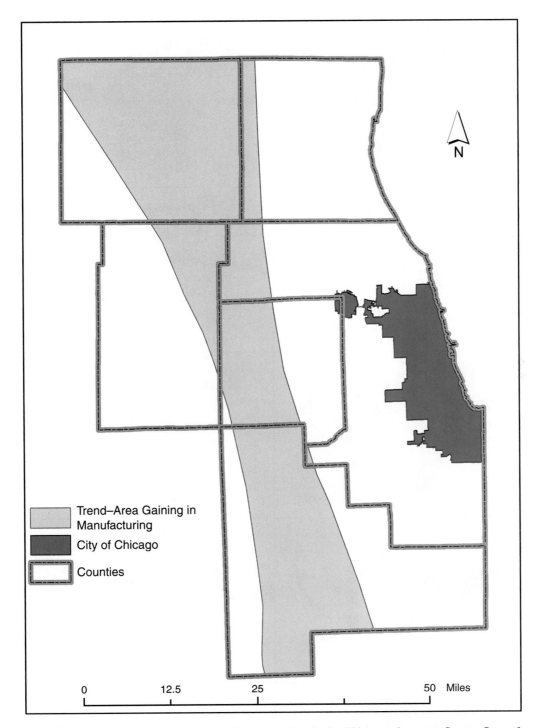

FIGURE 9.13 Pattern of manufacturing firm relocations in the Chicago urban area. Source: Stumpf, M.D. 1996. The Intra-metropolitan relocation of manufacturing firms in the Chicago region, 1987–1992. Master Thesis. Dept. of Geography, Northern Illinois University, Fig. on Trend surface.

9.3.1 Location Quotient

There are several methods for measuring an area's industrial specialization. The method we employ in the chapter's GIS exercise is known as a *location quotient,* which expresses the share of employment in a given industry in

a specified subarea as a percentage of the share of employment in the same industry within the larger area.

If a location quotient for an area is more than 100, the area is considered specialized in that activity. To illustrate, Aguilar (2002) used location quotients (LQs) to show the degree of manufacturing specialization among

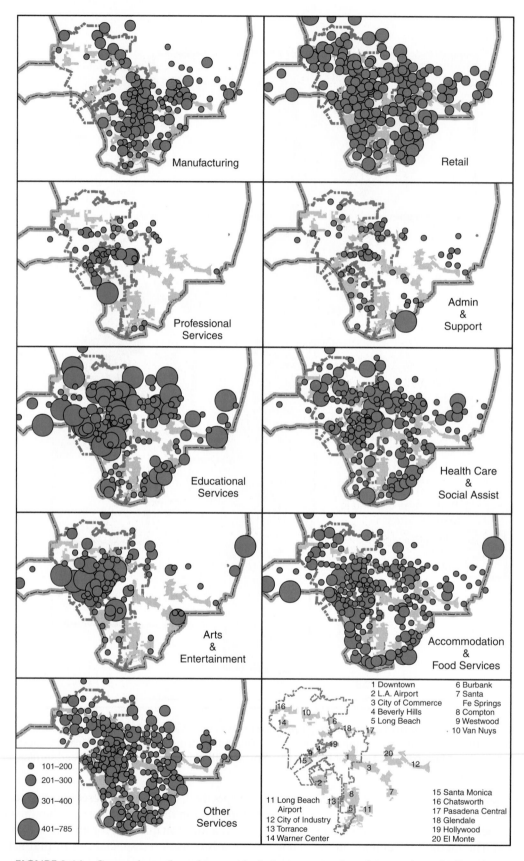

FIGURE 9.14   Comparison of employment by industry sector, location quotients for Los Angeles urban area, 1997.

municipalities in the Central Region of Mexico. The results, partially shown in Figure 9.15, indicate that of the 531 municipalities in Mexico's Central Region, 40 percent registered manufacturing LQs greater than 100. Only 27 percent of these were in metropolitan or urban areas, whereas 73 percent were rural (most outside the map area). The northern part of the Mexico City metropolitan area, shown on the map, was the urban area that benefited the most at attracting manufacturing activity. Other urban areas to gain are the transport corridors

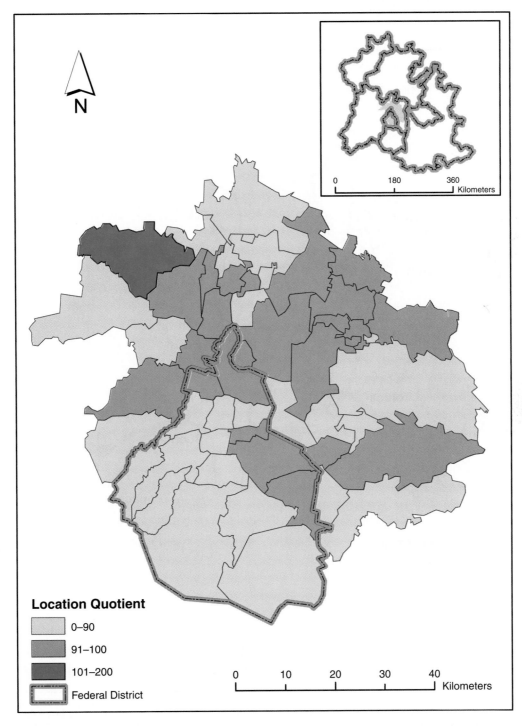

FIGURE 9.15 Manufacturing location quotients for Mexico City metropolitan area. Source: Aguilar, Adrian G. 2002. Megaurbanization and Industrial Relocation in Mexico's Central Region. Figure 4 on P. 666. *Urban Geography* 23:649–673.

connecting metropolitan areas with urban centers. Good examples are the Mexico City-Queretaro and Mexico City-Puebla highways, where cargo vehicle flows register 10,000 and 19,000 per day, respectively (see Figure 9.16). Some rural areas were recipients of simple assembly tasks in textiles and other labor-intensive manufacturing, due to a large, inexpensive labor pool. Aguilar's (2002) study

is just one example of how useful location quotients are for examining industrial specialization within cities or urban regions.

In the GIS exercise of this chapter the subareas are ZIP codes, whereas the larger area is Los Angeles County and we compute the location quotient in the following manner:

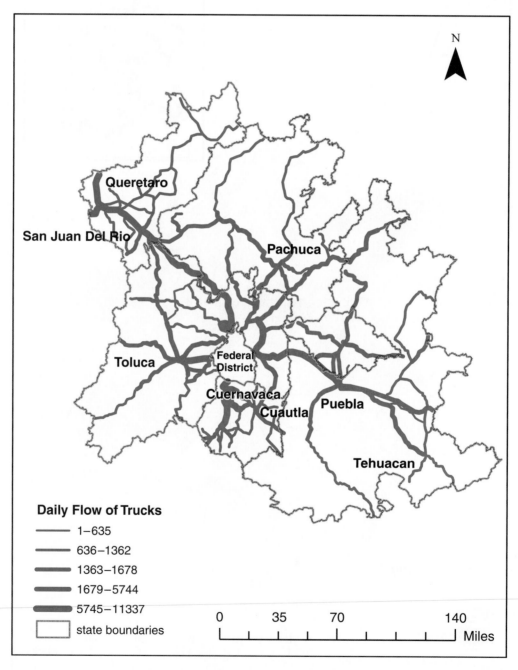

FIGURE 9.16 Cargo vehicle flows in Mexico City megalopolitan region. Source: Aguilar, Adrian G. 2002. Megaurbanization and Industrial Relocation in Mexico's Central Region. Figure 3 on P. 662. *Urban Geography* 23:649–673.

$$LQ = ((E_{ij}/E_j)/(E_i/E_t)) * 100$$

where E_{ij} = employment in subarea j in sector i;
E_j = total employment in subarea j;
E_i = county employment in sector i;
E_t = total county employment

Example of ZIP code X:

	ZIP Code X	County X
Professional services	8.7%	5.5%
Manufacturing	17.6%	22.8%

To determine ZIP code X's LQ in professional services and manufacturing relative to the county, we plug the percentages employed into the following equation:

LQ for professional services (8.7/5.5) * 100 = 158

LQ for manufacturing (17.6/22.8) * 100 = 77

An LQ greater than 100 indicates specialization in the category, and less than 100 indicates that the ZIP code is not specialized in that category. In this example, the ZIP code is specialized in professional services and not specialized in manufacturing.

9.3.2 A Comparison of Professional Services and Manufacturing Spatial Patterns

The professional services map indicates that a heavy concentration of professional services has formed in an archlike pattern both within and outside the city of Los Angeles (Figure 9.14). Starting with the end containing the largest concentration of services to the other, the top spine of the archlike pattern runs east from downtown Los Angeles, through Hollywood, into Beverly Hills, Century City, and West Hollywood, (see Figure 9.17). The second largest concentration of services within the professional group is occurring between Beverly Hills–Century City–West Hollywood and Westwood–West Los Angeles and ends in Santa Monica. Wilshire Boulevard is the major street connecting these centers (see Figure 9.18). Several miles to the north of this same formation is a lesser concentration of services than the first, but one that should be indicated as a heavier concentration of professional services than typical. This concentration forms a linear pattern.

A second major spine of professional services runs from Westwood-West Los Angeles southeast alongside the 405 Freeway to the Los Angeles Airport area located in El Segundo. There are only minor concentrations of professional services in southern Los Angeles.

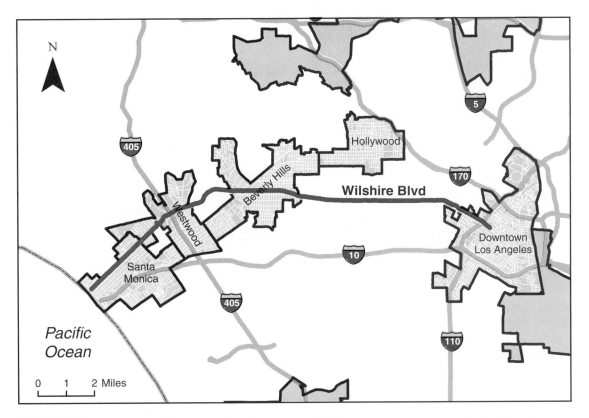

FIGURE 9.17 Job centers in Los Angeles urban area with high concentrations of professional services.

FIGURE 9.18 Wilshire Boulevard business area.

Although the map on manufacturing also shows heavier concentrations, including in downtown Los Angeles, it does not show that these activities string together to create any solid formation of manufacturing over a significant distance, like professional services. The area northwest and southwest of downtown Los Angeles shows a heavier grouping of manufacturing than do areas north and southeast of the downtown area. Heavier manufacturing areas exist further south. A larger, heavier area of concentration exists outside the city, east of the city limits.

Professional services and manufacturing complement each other. Where one dominates, the other has small concentrations, with the exception being downtown Los Angeles itself. Here, both services exist even though it looks as if they do in different sections of the downtown area: Professional services appear higher in the northern end of the downtown area, whereas manufacturing dominates more toward the center and in the southeast portion of the downtown.

Confirming these patterns, Forstall and Greene (1997) found that manufacturing is relatively important in the Commerce-Vernon, Compton, Los Angeles Airport, and Santa Fe Springs employment concentrations and relatively unimportant in the Beverly Hills and Westwood concentrations. Professional services bulk largest in Downtown, Beverly Hills, and Westwood. Professional services and manufacturing are the two highest job concentrations in Downtown.

In the lab exercise, you will calculate LQs throughout the Los Angeles metropolitan area for nine major industry categories, including manufacturing, retail trade, professional/scientific/technical services, and others. After mapping these distributions, you will examine and compare in an essay the distributions of LQ for two of the industry categories and also compare those to the spatial distribution of major job centers. The exercise lets

you apply the concept of LQ to understand the industrial dimensions of a major city.

9.4 INDUSTRY LOCATION TRENDS ACROSS CITIES

Much has been written about the location of manufacturing activities, but as already noted, the economic structure of American cities has undergone and is still undergoing a major transformation characterized by less emphasis on manufacturing and more emphasis on service employment. Noyelle and Stanback (1984) examined this transformation in an urban context by classifying 140 of the largest metropolitan areas according to industrial specialization. A review of their functional classification of cities will help elucidate these important transformations.

In their study, each metropolitan area was placed into a type based on its industrial specialization. The authors used LQs to determine the relative levels of specialization, and then a cluster technique to break out those groups of areas that were most similar. The procedure resulted in nine categories of metropolitan areas (see Table 9.2). The 140 metropolitan areas by type are shown in Figure 9.19. The following is a brief definition of each of the nine types:

1. *Nodal Metros* Metro areas like New York, Chicago, and Atlanta that are specialized in providing distributive and corporate complex services.
2. *Functional Nodal Metros* Metro areas like Detroit, Pittsburgh, and San Jose that are older manufacturing centers where major corporations carry out headquarters and other nonproduction activities.

TABLE 9.2 Noyelle and Stanback's functional typology of metropolitan areas

Type of U.S. metropolitan area	Number in 1980
Nodal	39
Functional Nodal	24
Government-Education	15
Education-Manufacturing	5
Residential	3
Resort-Retirement	9
Manufacturing	25
Industrial-Military	13
Mining-Industrial	7

Source: Noyelle, T., and Stanback, T. (1984) *The Economic Transformation of American Cities.* Toronto, Canada: Rowman &Littlefield.

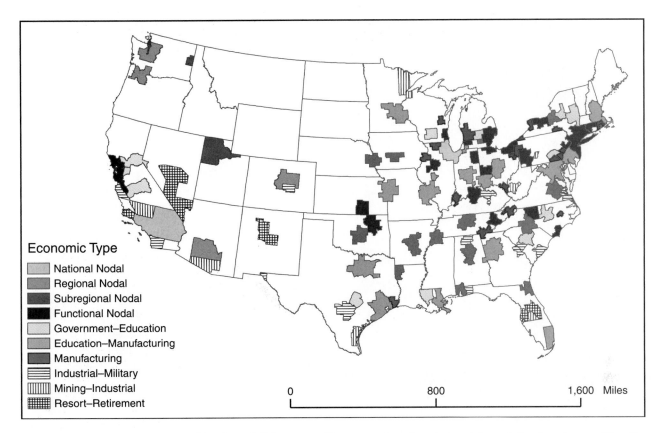

FIGURE 9.19   A functional classification of U.S. metropolitan areas based on industrial specialization. Source: Noyelle, T., and Stanback, T. (1984) *The Economic Transformation of American Cities*. Toronto, Canada: Rowman & Littlefield.

3. *Government-education Metros*   Metro areas like Washington, DC, Austin, Madison, and many other metro areas where government and education comingle (10 on the list are state capitals). These metro areas show a significant emphasis on producer services (see Austin in Figure 9.20).

4. *Education-manufacturing Metros*   Metro areas like New Haven, South Bend, and Ann Arbor, which are both old industrial centers and university towns.

5. *Residential Metros*   Metro areas like Nassau–Suffolk and Long Branch–Asbury Park, which are both extensions of New York City, and Anaheim, which is an extension of Los Angeles. These metros are now consolidated with their respective largest neighboring metro and they have above-average levels of retail and consumer service employment.

6. *Resort-retirement Metros*   Metro areas like Ft. Lauderdale, Orlando, West Palm Beach, and Las Vegas. These places score high in retail and consumer services but also show a strong presence of finance and corporate services. Amenities make these attractive places for companies offering such services.

7. *Manufacturing Metros*   Metro areas like Buffalo, Flint, Greenville, or Erie where manufacturing employment is still strong, if not dominant.

8. *Industrial-military Metros*   Metro areas like San Diego, San Antonio, or Newport News, which house military installations.

9. *Mining-industrial Metros*   Metro areas like Tucson, Duluth, and Johnstown have an overwhelming proportion of their workforce employed in mining or resource-oriented employment.

This typology of metro areas helps frame the discussion about the types of places being impacted by broader changes taking place in the economy. The difference between growth and development was an important component of Noyelle and Stanback's (1984) study because they attempted to identify linkages between urban growth and development. The elements of the changing economic structure, such as the growth of services, the rise of white-collar work, and a decrease in production employment, were viewed as development rather than just growth and decline. Rather, these changes reflect responses to forces at

FIGURE 9.20   Austin, Texas: A government-education metro. Source: SuperStock, Inc.

work in the economy, four of which were singled out by Noyelle and Stanback:

1. The increasing size of markets.
2. Changes in transportation and technology.
3. Increased importance of public-sector and nonprofit activities.
4. The rise of very large corporations.

These forces were critical in the development of the system of cities represented by the nine types of metropolitan areas. Since the early 20th century, traditional manufacturing was diminished across the largest metropolitan areas and was replaced by the development of high-level services. The economic base of many cities has been transformed from goods-production centers to service-production centers.

A recent application of the Noyelle and Stanback (1984) classification of cities revealed that metros at the bottom of the functional hierarchy, such as manufacturing centers, had the highest rates of low-income employment whereas functional nodal metros exhibited less low-income employment (Elliott, 1999). The findings on low-income employment across the urban hierarchy imply that vast inequalities are being created among places as a result of the forces identified by Noyelle and Stanback's pioneering work.

9.4.1 High-Technology Industry

Although manufacturing activity has been declining in importance for most large cities, high-technology industry has been a notable exception. Metropolitan area economic development agencies know this, too, as many are obsessed with developing strategies to lure or maintain high-tech industry for their local areas. To identify the strength of a metropolitan area's high-tech industry, urban scholars examine the area's workforce engaged in science and technology occupations. One recent study applied such an occupational definition of high-tech industry and produced a set of metropolitan rankings based on total increases in high-tech jobs in the 1990s, which contrasted with previous rankings that considered high percentages of high-tech workers (Chapple, Markusen, Schrock, Yamamoto, & Yu, 2004). Their results ranked larger and older industrial areas such as Chicago, New York, and Detroit higher than often-listed areas such as Austin, Raleigh-Durham, and Phoenix. They concluded that areas like Chicago have a large number of high-tech jobs but they are also diversified and as a result they are much less vulnerable to downturns in the economy (see Figure 9.21).

High technology is seen as a driver of economic development partly because it is composed of highly educated,

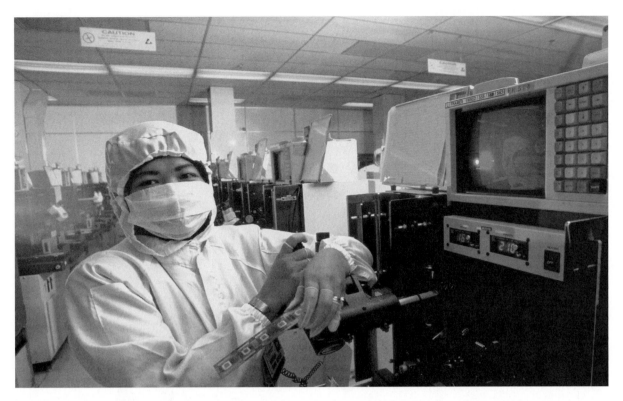

FIGURE 9.21   High-tech worker at Motorola Plant in Chicago. Source: Getty Images/Time Life Pictures.

skilled workers. Florida (2002) referred to this group of highly educated workers as "talent" and suggested that the economic geography of talent is strongly associated with high-technology industry location (see Figure 9.22). Moreover, he found that talent and high-technology industry work independently and together to generate higher regional incomes. He concluded that it is advantageous for metropolitan areas to develop their talent class as an intermediate step to attracting high-technology industry.

9.5 MAQUILADORA INDUSTRY ON THE MEXICO–U.S. BORDER: A UNIQUE MANUFACTURING REGION

The *maquiladora* industry, located mostly at the Mexico–U.S. border, represents an important and successful international manufacturing arrangement (see Figure 9.23). It illustrates many of the concepts raised in this chapter. *Maquiladora* industry is coproduction of manufactured goods, where most of the components are transported from the United States or other advanced countries to Mexico, assembly is done by low-cost labor in Mexico, and assembled products are shipped for sale in U.S. markets. The legal apparatus consists of a U.S. or sometimes a Japanese or Korean company, working with a Mexican counterpart firm that it controls. Usually the top executives and managers at the Mexican plant are American,

often commuting from residences on the U.S. side, whereas most of the middle- and lower level employees are Mexican and live on the Mexican side.

Today, many products familiar to U.S. consumers have largely been assembled in *maquiladora* factories. These include refrigerators, batteries, TV sets, windshield wiper blades, airplane components, auto parts, and more traditional products such as textiles and furniture (Stoddard, 1987). An example is Trico Components, the fifth-largest Mexican *maquila* plant in 1996 (Butler, Pick, & Hettrick, 2001; Expansión, 1997). It manufactures windshield wipers and wiper blades and its market includes the major U.S. automakers as well as many third-party firms. In 1996, it exported $1.96 billion in products and employed 2,974 workers. Trico follows the traditional pattern of twin plants near each other on opposite sides of the border. Trico's Brownsville manufacturing plant produces nearly all of the components for its Mexican plant in Matamoros. In particular, the U.S. plant does metallurgy, casting, stamping, and manufacture of wiper blade arms. It also is the distribution center for shipping wipers and blades worldwide. Every morning 16 large semitrailer trucks load Trico components at the Brownsville Trico plant. They carry these components nine miles across the border to the Trico Componentes plant located in a modern industrial park in Matamoros. There, the components are assembled into wiper blades and

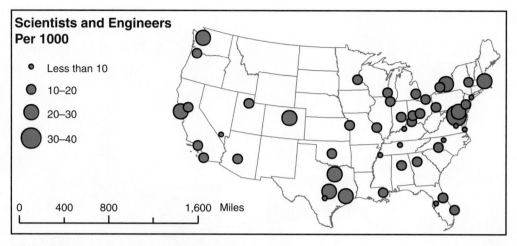

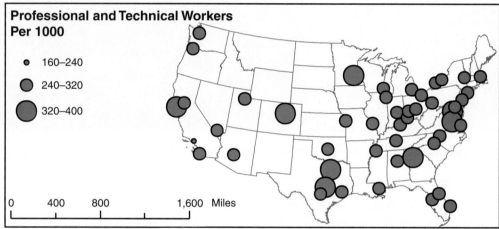

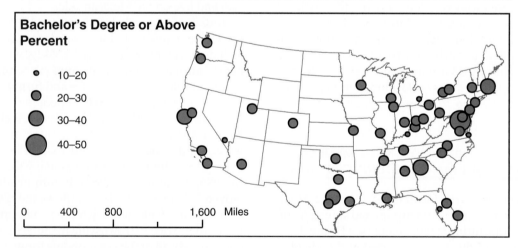

FIGURE 9.22   The geography of talent. Source: Florida, Richard. 2002. The economic geography of talent. Figure 2 on P. 747. *Annals of the Association of American Geographers*, 92:743–755.

blade motor assemblies. At the end of each day, the trucks carry finished products back to the U.S. Trico plant. The major global auto companies then come to the U.S. plant every few days to pick up batches of new products. The whole turnaround from raw materials to finished product and delivery takes only about five days.

Trico is remarkable in dominating the U.S. automotive market for wiper blades. The twin border plants manufacture 62 million blades, arms, wiper motor assemblies, and other items per year. Trico has 80 percent of the U.S. market for blades and arms and 45 percent of the market on motor assemblies. Its customers include Ford, Chrysler,

FIGURE 9.23 *Maquiladora* workers at Pebac plant near the Mexican border. Source: Joel Sartore.

ITT Automotive (which sells to GM and Toyota), CME (which sells to Honda), Nissan, and ASMO (which sells to Toyota and Chrysler), as well as retailers Pep Boys, Carquest, Atlas, and Wheels.

The *maquiladora* industry started in the mid-1960s, and grew exponentially until 2001, when a recession began to hit in the United States and competition grew from Asian *maquiladora*-like production, especially in China ("Wasting Away," 2003). Nevertheless, after three decades of growth and recent decline, the *maquiladora* sector, vis-à-vis the Mexican economy, represents one of Mexico's largest export sectors—behind petroleum and equivalent to tourism (Butler et al., 2001).

As seen in Figure 9.24, the growth in the number of *maquiladora* workers has been huge. *Maquiladora* employment peaked at nearly 1.4 million in 2001, before falling by around a quarter million as a result of the U.S. recession and increasing global competition ("Wasting Away," 2003). The Mexican states with the largest *maquiladora* industry are located predominantly at the northern border, but also in several states with large urban areas in the center of the country, as well as the southeastern state of Yucatan (see Figure 9.25). The employment in these states is given in Table 9.3 for 1998, a

year in which there were about 1 million *maquiladora* employees.

Rapid *maquiladora* growth has been accompanied by parallel growth of the Mexican cities that house the industry, notably Tijuana, Mexicali, Ciudad Juarez, Reynosa, and Matamoros. *Maquiladora* industry has been the principal engine driving the growth of these cities, with the secondary factor being their proximity to the United States. Each of the Mexican border cities has a U.S. sister city, located just across the border. For instance, San Diego is across from Tijuana and El Paso across from Ciudad Juarez. The U.S. part of coproduction, consisting of such factors as distribution, warehousing, and product development, tends to be located in the U.S. sister city.

The *maquiladora* industry illustrates the industrial location concept of economic-base theory and agglomeration economies. The influx of workers also demonstrates the migration application of the gravity model from Chapter 7. In addition, the Mexican labor costs are about one eighth the cost of equivalent labor on the U.S. side of the border. Although transport costs from *maquiladora* plants to U.S. consumption locations are greater than they would be for U.S.-based production, this is far offset by the labor savings available in Mexico.

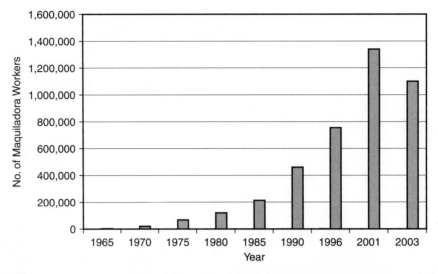

	1965	1970	1975	1980	1985	1990	1996	2001	2003
Maquiladora Enterprises	12	120	454	620	760	1,920	2,411		
Maquiladora Employment	3,000	20,327	67,241	119,546	211,969	460,258	754,858	1,340,000	1,100,000

FIGURE 9.24   Growth of *maquiladora* employment, 1965–2003.

FIGURE 9.25   Automotive *maquiladora* worker at a plant in Mexico City. Source: AP Wide World Photos.

Economic-base theory is demonstrated because the *maquiladora's* rapid growth stimulated concomitant rapid population and economic growth of Mexican border cities and surrounding areas, including ancillary services in those cities. The case also illustrates agglomeration economies because *maquiladora* firms producing similar products are usually spatially grouped together to benefit from common supplies and producer services. This is shown by the spatial proximity of major TV set *maquiladoras* in Tijuana, and by the more traditional grouping of textile *maquiladoras* in Matamoros.

The gravity model of migration introduced in Chapter 7 is demonstrated because the growing and prosperous economies of the Mexican border cities have been a major pull factor for migrants from central Mexico, whereas distance of migration has been small for lower level *maquiladora* workers, mostly young women migrating from nearby rural areas of the border region.

Several geopolitical factors have also influenced the *maquiladora* industry. The North American Free Trade Agreement (NAFTA), approved in 1994, gradually eliminated special Mexican tax incentives on *maquiladora* production, also giving trade preference to U.S. versus Asian *maquiladora* firms. The overall effects on this industry have been mixed.

A more formidable and global challenge to the Mexican *maquiladora* industry, however, is the rise of very

TABLE 9.3 Mexican maquiladora employment, 1998

State	Maquiladora employment	Percent of national total
Border States		
Chihuahua	256,930	26.3
Baja California	207,801	21.3
Tamaulipas	149,689	15.3
Coahuila	88,296	9.1
Sonora	87,438	9.0
Nuevo Leon	43,448	4.5
Nonborder States		
Durango	23,207	2.4
Jalisco	20,646	2.1
Aguascalientes	18,305	1.9
Puebla	17,413	1.8
Yucatan	13,187	1.4
State of Mexico/D.F.	10,832	1.1
Others	38,188	3.9
Border States Subtotal	833,602	85.5
Total	975,380	100.0

Source: Secretaria de Comercio, 1998.

cheap coproduction in China and other Asian nations that is positioned for the U.S. market. A number of *maquiladora* plants have been closed and moved to China in the last several years. The Mexican *maquiladora* assembly worker receives about $2 per hour, whereas the Chinese worker earns only 50 to 80 cents per hour ("Wasting Away," 2003). It is unknown how far this Asian offshore trend will go, and what its ultimate effect will be on Mexico and on consumers in the United States.

9.6 CONSUMER SERVICES AND CITIES

Two types of service employment can be identified: consumer services and producer services. Consumer services deal with the final consumption of goods and services and therefore are principally related to retail industry. Producer services, on the other hand, relate to intermediate demand or aid in the production of other goods and services. These services aid in the production process and might include, for instance, legal services for a financial corporation or a manufacturing firm. Producer services are the topic of the final section. Retailing is an important industrial component of the city, not only because a large share of a metropolitan area's workforce is engaged in retail but also because it shapes many of the daily activity patterns of consumers. Research on the geography of retailing concerns both the location and forms that retail clusters take within metropolitan areas, but it also addresses consumer behavior with respect to shopping travel patterns.

9.6.1 Retail Location Patterns Within Cities

Retail employment in the United States increased by 15 percent from 1992 to 1997, when there were 21 million retail workers (U.S. Census Bureau, 2000). In addition, retail sales between 1992 and 1997 increased by 34 percent. These two trends speak to the importance of retailing in cities as places of both employment and consumption. It raises the important question of where workers and consumers go within the city for this retail activity.

Prior to the automobile's influence on city form, the Central Business District (CBD) was the dominant location of retail activities. With the increased suburbanization of population, large-scale retailing soon followed. Similar to the suburbanization of population, retailing oriented itself along transportation routes, often in radial patterns. There is always a lag between when people arrive in an area and when retail services are offered. In many respects, the latter is an extension of central place theory, already discussed in Chapter 5, although the city displays a much more complex pattern of hierarchal centers because of widely varying population densities and socioeconomic characteristics.

Central cities remained strong throughout the 1950s and 1960s and as a result, retailing in the CBD and along corridors extending outward from the CBD also remained viable. Nevertheless, this would prove to be short-lived as shopping centers began to develop rapidly across metropolitan areas. The regional mall, one of the largest types of shopping center, first appeared in the 1950s, although there were prototypes dating back to the 1920s (Hartshorn, 1992). The Northgate Center built in 1950 in Seattle was the first to meet today's definition of a regional mall, as it included a major department store anchor. Southdale Shopping Center in Minneapolis (1956) was the first regional mall to be fully enclosed (Hartshorn, 1992). Many variations of regional malls would follow after the 1950s including mixed-use centers, megamalls, and theme centers (see Figure 9.26).

Shopping centers in the United States today number approximately 44,000 and many have been renovated or

FIGURE 9.26 Eaton's Mall in Toronto.

are under renovation; fewer are newly constructed (Soriano, 1998). Shopping centers face competition from a number of other shopping formats. A relatively new phenomenon introduced to the metropolitan landscape is known as a power center. There is no standard definition of a power center but it normally implies an agglomeration of big-box retailers (Hahn, 2000). Big-box retailers include category killers offering a wide selection in a special category at a low price such as Home Depot, Circuit City, and The Sports Authority; discounters such as Wal-Mart and Target; and warehouse clubs such as Sam's Club (Hahn, 2000). Power centers often locate near regional shopping centers but do not compete directly with them; instead they compete with local community shopping centers (Hahn, 2000). Power centers

consume a lot of space in terms of building and parking area and their popularity has resulted in increased travel time for consumers as they bypass local community centers for better bargains (Hahn, 2000).

A more recent newcomer to the metropolitan retail landscape is the lifestyle center, which is anchorless and offers upscale retailing and fine dining to consumers in a convenient, open-air layout (Hazel, 2003). The concept attempts to create a hub of activity by including a mix of uses that might include residential development, office space, hotels, churches, or municipal facilities (Mattson-Teig, 2004). The focus is on the consumers and their community rather than a particular tenant of the center. The number of these lifestyle centers has nearly doubled since 1997 and there are now nearly 60 (see Table 9.4). Lifestyle

TABLE 9.4   Lifestyle centers in the United States, 2002, built since 1990

Year opened	Name	City, state	Total gross leasable area, in square feet	Total stores*
2002	Eastwood Towne Center	Lansing, Mich.	393,000	45
2002	Village of Rochester Hills	Rochester Hills, Mich.	375,000	40
2002	The Grove	Los Angeles	575,000	44
2002	Fountain Walk	Novi, Mich.	737,134	13
2002	Geneva Commons	Geneva, Ill.	418,673	61
2001	Aspen Grove	Littleton, Colo.	243,900	50
2001	The Summit Louisville	Louisville, Ky.	367,500	58
2001	The Shoppes at Brinton Lakes	Glen Mills, Pa.	153,000	14
2001	Old Mill District at River Bend	Bend, Ore.	150,000	23
2001	Jefferson Pointe	Fort Wayne, N.J.	562,000	60
2000	Deer Park Town Center	Deer Park, Ill.	506,000	60
2000	Rookwood Commons	Norwood, Ohio	326,462	45
2000	Palladium at CityPlace	West Palm Beach, Fla.	600,000	75
2000	Centro Ybor	Tampa, Fla.	313,899	25
1999	SouthPointe Pavilions	Lincoln, Neb.	500,000	42
1999	Mt. Pleasant Towne Center	Mt. Pleasant, S.C.	426,748	62
1999	Southlake Town Square	Southlake, Texas	201,000	70
1999	The Avenue at East Cobb	Marietta, Ga.	225,000	52
1999	The Shops at Sunset Place	South Miami, Fla.	506,000	60
1999	The Avenue of the Peninsula	Rolling Hills Estates, Calif.	374,000	60
1999	Premier Centre	Mandaville, La.	275,026	22
1998	Denver Pavillions	Denver, Colo.	350,000	40
1998	Gardens on El Paseo	Palm Desert, Calif.	200,000	50
1998	The Shops at Riverwoods	Provo, Utah	192,000	35
1998	The Commons at Calabasas	Calabasas, Calif.	198,388	35
1997	Alamo Quarry Marketplace	San Antonio, Texas	585,000	60
1997	Huebner Oaks Shopping Center	San Antonio, Texas	380,000	60
1997	The Summit Birmingham	Birmingham, Ala.	760,000	72
1997	Redmond Town Center	Redmond, Wash.	569,000	100
1997	University Village	Seattle, Wash.	400,000	93
1996	Town Center Plaza	Leawood, Kan.	700,000	95
1996	The Promenade at Westlake	Thousand Oaks, Calif.	201,563	28
1992	Town Square Wheaton	Wheaton, Ill.	179,000	75
1991	Mizner Park	Boca Raton, Fla.	236,000	47
1990	Reston Town Center	Reston, Va.	250,000	50
1990	Bradley Fair Shopping Center	Wichita, Kan.	220,000	35
1990	CocoWalk	Coconut Grove, Fla.	163,000	40

*Store count includes restaurants
Note: updated as of 2/2003
Source: Hazel, D. 2003. Lifestyle centers look good, but are they earning their keep? Shopping Centers Today (ICSC monthly magazine cover story), May 2003.

centers include upscale retailers and are located in affluent areas. A recent addition to the 2004 list of life style centers was Victoria Gardens, which is also the largest at 1.3 million square feet of space located in Rancho Cucamonga, an affluent suburb of Los Angeles. The introduction of the lifestyle center is indicative of the extreme competition that exists in the retail industry where retailers are frequently looking for new twists in their product delivery (see Figure 9.27). Wrigley and Lowe (2002) observed a similar trend of new retail formats in the United Kingdom such as warehouse clubs and factory outlet centers.

9.6.2 The Consumer City Hypothesis

The *consumer city hypothesis* offers a provocative viewpoint on consumer services and cities that deemphasizes the role of production in the growth of cities and emphasizes the role of consumption (Clark, 2003). Advocates of this hypothesis go so far as to suggest, for cities, that "trying to keep manufacturing is probably useless and because of the negative amenities related to manufacturing possibly even harmful" (Glaser, Kolko, & Saiz,

2001, p. 5). They suggest cities should attract people with high human capital or talent because this group will have the highest income-earning potential and thus the highest demand for consumer products and services. They noted four critical urban amenities for the future growth and economic health of cities:

1. The presence of a rich variety of consumer goods and services. Cities that have more restaurants and offer plenty of entertainment have grown the fastest.
2. Aesthetics and physical setting. Physical amenities such as favorable climate, rolling terrain, higher ground, and proximity to water are all attractive to consumers.
3. Good public services. Good schools and low crime rates are important for attracting a highly educated workforce.
4. Transportation infrastructure that is good for ease of getting around, as consumers avoid areas with high transport costs.

One phenomenon often cited in support of the consumer city hypothesis is reverse commuting from dense

FIGURE 9.27   Geneva commons lifestyle center in suburban Chicago.

TABLE 9.5   Commuting patterns-US metropolitan areas

	Daily commutes (millions)			Annualized growth rate	
	1960	1980	1990	1960–1980	1980–1990
City–city	18.8	20.9	24.3	0.52%	1.52%
City–suburb	2.0	2.6	5.9	3.65%	3.46%
City–other	0.6	1.2	1.9	3.63%	4.70%
Suburb–city	6.6	12.7	15.2	3.34%	1.81%
Suburb–suburb	11.3	25.3	35.4	4.09%	3.42%
Suburb–other	1.1	3.7	6.8	6.22%	6.27%
Total	40.5	68	89.5	2.62%	2.79%

downtowns, by which some residents choose to reside in the center even though their jobs might be in the suburbs (see Table 9.5). These residents choose to stay in the central city where rents are higher because it is convenient to consume high-quality consumer goods and services.

Clark (2003) extended the theory by differentiating between natural amenities and constructed amenities. Natural amenities have been considered in past studies as a major determinant of growth (Mueller-Wille, 1990). Constructed amenities include opera houses, juice bars, museums, and coffee shops. An example, seen in Figure 9.28, of an elaborate constructed amenity is Experience Music

Project in Seattle (*http://www.emplive.com*). Perhaps what is most radical about this approach is that it turns classical retail geography around, in that it has always been suggested that once a critical mass of consumers is reached in an area, a service such as a coffee shop will be offered. Here, however, it is suggested that people move toward amenities like a coffee shop. The detail of the argument is as follows:

This reasoning may seem simple, but it reverses the "traditional economic determinism" which suggests that as individuals (and cities) grow more affluent, they consume more

FIGURE 9.28   Experience Music Project, Seattle's Rock and Roll Museum.

luxury goods, like meals in fine restaurants. In this traditional view, individuals with more income cause restaurants to emerge. But if this is broadly applicable over an individual's life course, to apply the same logic to a city is mistaken, powerfully illustrating overextended "methodological individualism". Why? Because discrete individuals move in and out of cities all the time, yet urban amenities like opera or lakefronts change more slowly, and thus drive location decisions of individuals. This is especially true for talented and younger persons who change jobs frequently—and even the average American changed jobs about every four years in the 1990s. (Clark, 2003, pp. 2–3).

The two different views might actually be complementary where a critical mass is reached and the coffee shop is provided but the coming of the coffee shop encourages additional growth. Constructed amenities have been found to drive growth where young people abound, whereas natural amenities explain the growth of areas of elderly residents (Clark, 2003).

9.7 U.S.–MEXICO TWIN CITIES CONSUMER INTERACTION ON THE BORDER

An international example of a consumer city is the consumer interaction for the U.S.–Mexico sister cities that were discussed earlier regarding *maquiladoras*. Both sides of the border see advantages to shopping consumption on the other side. Shoppers often drive but sometimes walk to the other side. Walking has been even more popular since September 11, 2001, because there are now much longer lines of cars to get across the border from Mexico to the United States due to tightened security. The large number of border pedestrian shoppers is illustrated in Table 9.6, which indicates that for all the border cities, 45 million pedestrians crossed into the United States in 1997 (U.S. Customs Service, 1997).

A short case example gives the feeling of this shopping in the quad cities of Brownsville and McAllen, Texas, and the Mexican sister cities of Matamoros and Reynosa, Mexico, all located at the easternmost part of the border not far from the Gulf of Mexico (see Figure 9.25). The U.S. side of this region is referred to as "the Valley," and has a strong cultural influence from Mexico and Mexican-Americans. The Valley is heavily Hispanic and has experienced a large increase in overall population in the past 10 years. Brownsville and Matamoros had 2000 populations of 139,722 and 516,001, respectively, whereas the population of McAllen, Texas in 2000 was 106,414, and Reynosa across the border had 403,718 residents.

On the Mexican side, the region's economy is powered by *maquiladoras*. However, there are high unemployment and poverty rates on the U.S. side, as well as historically low levels of education, although the state of Texas is investing heavily in remedies at the university

TABLE 9.6 Pedestrian border crossings incoming from Mexico to the U.S., 1997

State and border crossing	Number of incoming pedestrians
California	
Calexico	8,167,540
Andrade	1,360,393
Otay Mesa	621,517
San Ysidro	8,476,225
Tecate	272,484
Arizona	
Douglas	599,082
Lukeville	76,274
Naco	71,839
Nogales	4,643,538
Sasabe	3,097
San Luis	2,220,799
New Mexico	
Columbus	119,418
Texas	
Brownsville	3,726,740
Del Rio	262,717
Eagle Pass	529,897
El Paso	4,542,646
Fabens	14,737
Hildago	2,429,241
Laredo	5,427,815
Presidio	11,890
Progreso	1,164,483
Rio Grande City	85,919
Roma	443,949
Total	45,272,240

Source: U.S. Customs Service, Office of Field Operations.

level. Quite a lot of agriculture is also present, the best known being citrus production.

The Valley offers tourism and leisure as well. For instance, it is a retirement mecca, and South Padre Island near Brownsville is an ever-popular destination, especially for spring break vacationers. "Winter Texans" are a special feature; typically they are older Midwesterners who spend the winter in south Texas. Many of these Winter Texans live in the RV and trailer parks that are scattered up and down the Valley; there are more than 500 of them with a total of 68,000 spaces (Valley Partnership, 2004). The Winter Texans provide consumer benefits to local merchants, especially in Brownsville and Matamoros (Valley Partnership, 2004), as well as shopping and providing consumer benefits on the Mexican side.

The following excerpt from an interview with a former resident of the Valley gives a further feeling of the sister city consumer atmosphere:

> Residents of the Valley do not say they are "going to Mexico." They'll either say they are going to Reynosa or Matamoros, or they will simply say they are "going across." I encountered

someone from Laredo using this term also. When I lived in Mission from 1989–1990, there were people going across for prescription drugs, which were cheaper in Mexico. I am sure there are more people doing that now as prescription drug prices have increased. I mentioned how US residents go across to shop and that Mexicans do the same thing. Saturday is a very busy day in Downtown McAllen. However, on Sundays the downtown is a virtual ghost town (at least when I was there). The stores weren't open Sundays. There are the Winter Texans, typically Midwesterners who spend the winter in South Texas. Mission, the city where I lived for a year-and-a-half, has many trailer/mobile home parks. They are not the stereotypical trashy trailer parks conveyed through the popular media. These are well-maintained and serve as home for the Winter Texans. (Pinnau, 2003).

This brief example demonstrates the idea of "consumer city" in a binational context. Even in a poor border region with low levels of education, consumption is growing in importance, and the culture favors international exchange of consumption especially through the large factor of border shopping.

9.8 THE GEOGRAPHY OF PRODUCER SERVICES

Research on the geography of producer services is not as well established as manufacturing location theory and the research into consumer services, in part because it has risen in prominence in recent decades as manufacturing has declined in relative importance. Coffey (2000) presented several other reasons for why there is relatively little written on the topic:

1. The modern researcher had to contend with the prevailing paradigms that viewed all services as nonproductive activities and that those remaining after manufacturing had been accounted for.

2. Study of producer services had long been confined to the limited framework of central place studies, so it became conceptually stagnant for several decades.

3. Because of the early emphasis on central place studies, more emphasis was placed on consumer services to the exclusion of the more dynamic producer services.

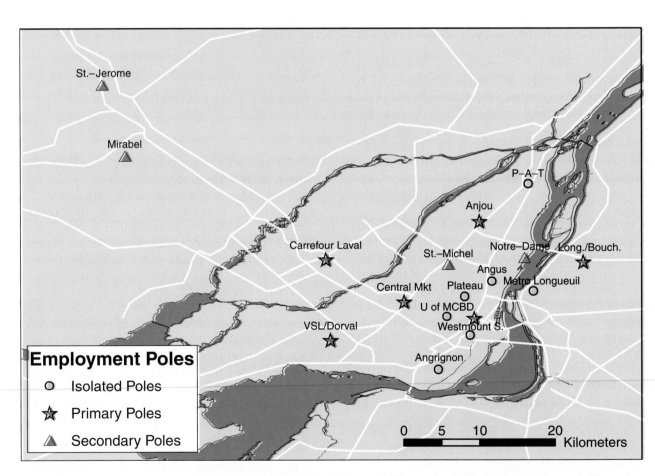

FIGURE 9.29 Montreal employment poles. Source: Coffey, W.J. and G. Shearrmur. 2002. Agglomeration and Dispersion of High-order Service Employment in the Montreal Metropolitan Region, 1981–96. Urban Studies 39:359–378. Page 365. Figure 1. "Employment poles in the Montreal CMA, 1981–96."

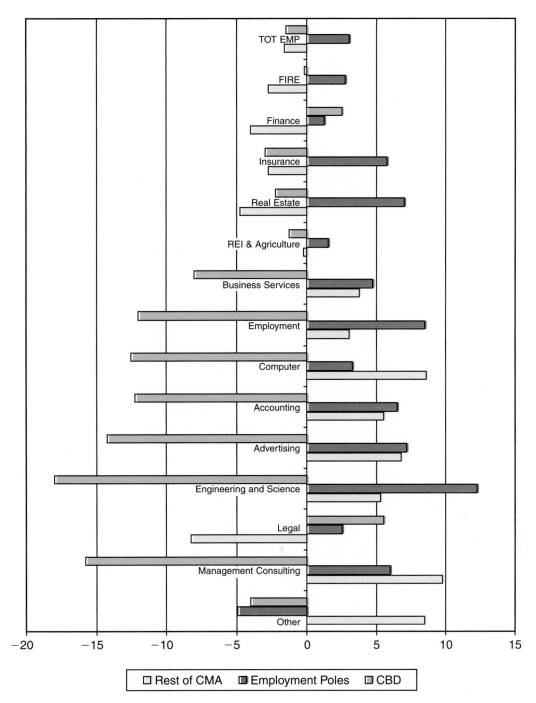

FIGURE 9.30 Changes in the composition of Montreal service employment, 1981–1996. Source: Coffey, W.J. and G. Shearrmur. 2002. Agglomeration and Dispersion of High-order Service Employ-ment in the Montreal Metropolitan Region, 1981–96. *Urban Studies* 39:359–378. Page 371. Figure 4. "Changes in share of CMA employment, Montreal, 1981–96."

Note: FIRE = finance, insurance, and real estate. TOT EMP = total employment.

Services represent the largest share of jobs in the Unit-ed States, 38 percent as measured from the 1997 Economic Census (U.S. Census Bureau, 2000). The service sector is also the fastest growing as demonstrated by the addition of 6.4 million new jobs, or more than half of all new jobs between 1992 and 1997 (U.S. Census Bureau, 2000). A

large proportion of this growth was in business services, or what many urban researchers consider to be the heart of producer services. *Producer services* or *business services* represent inputs into the production process of busi-nesses and organizations as opposed to services provid-ed to households or individuals. These typically include activities such as computer services, accounting and book-keeping services, advertising services, and legal services. Some researchers have broadened the definition of producer services to include finance, insurance, and real estate along with business services (Coffey, 2000).

Nevertheless, the task of differentiating between consumer services and producer services from census classifications might not be that straightforward. One solution to separate the two types of services is to construct an economic input–output model to better estimate the producer–service component of employment (Wernerheim & Sharpe, 2001).

Eventually, researchers recognized the importance of producer services in the role of the city. According to Coffey (2000), although still framed in the economic-base context, much knowledge was gained on producer services from the 1980s to the mid-1990s, as they were shown to have high export value and thus seen as a way to promote regional economic growth. Coffey (2000) made note of a surge in research on the decentralization of producer services from downtowns to edge cities.

A study of Atlanta found that suburban downtowns in Atlanta overshadow its CBD in producer services (Fujii & Hartshorn, 1995). Another analysis of Atlanta with more recent data noted an increase of 200,000 jobs in business and professional services between 1982 and 1997 in the metropolitan area overall, but the central city percentage dropped by 20 percent (Gong & Wheeler, 2002). Most of the increase occurred in the northern suburbs and could be explained by proximity to a well-educated professional class and highway access. A similar pattern occurred for Montreal (see Figure 9.29), where the CBD has declined in relative importance to edge cities with respect to the agglomeration of producer services (Coffey & Shearmur, 2002). Their principal argument for this shift to the suburbs is that it reflects an increasing specialization in certain producer services and the subsequent expansion of other high-level services, which the CBD cannot possibly absorb (see Figure 9.30). Thus the map illustrates the location of these suburban poles relative to the downtown.

SUMMARY

This chapter has examined industrial location with respect to cities. It covered basic concepts and principles, and then moved to the major types of industries in cities. It reviewed the earliest theories on manufacturing location and examined the processes accounting for current-day patterns of manufacturing patterns within cities. The chapter then focused on intrametropolitan location of other economic activities and developed an urban issues section on Los Angeles. Finally, the chapter turned to the service economy and addressed consumer services including urban retailing and producer services separately.

REFERENCES

Aguilar, A. G. (2002) "Megaurbanization and industrial relocation in Mexico's central region." *Urban Geography* 23:649–673.

Blackley, P. R., and Greytak, D. (1986) "Comparative advantage and industrial location: An intrametropolitan evaluation." *Urban Studies* 23:221–230.

Butler, E., Pick, J. B., and Hettrick, W. J. (2001) *Mexico and Mexico City in the World Economy.* Boulder, CO: Westview.

Cervero, R. (2001) "Efficient urbanization: Economic performance and the shape of the metropolis." *Urban Studies* 38:1651–1671.

Chapple, K., Markusen, A., Schrock, G., Yamamoto, D., and Yu, P. (2004) "Gauging metropolitan 'High Tech' and 'I-Tech' activity." *Economic Development Quarterly* 18:10–29.

Clark, T. (2003) "Urban amenities: Lakes, opera, and juice bars: Do they drive development?" in *The City as an Entertainment Machine*, pp. 103–104. New York: JAI Press/Elsevier.

Coffey, W. J. (2000) "The geographies of producer services." *Urban Geography* 21:170–183.

Coffey, W. J., and Shearmur, G. (2002) "Agglomeration and dispersion of high-order service employment in the Montreal metropolitan region, 1981–96." *Urban Studies* 39:359–378.

Elliott, J. R. (1999) "Putting global cities in their place: Urban hierarchy and low-income employment during the post-war era." *Urban Geography* 20:95–115.

Expansión. (1997) "TLCY Maquiladoras." *Expansión*, October 8.

Florida, R. (2002) "The economic geography of talent." *Annals of the Association of American Geographers* 92:743–755.

Forstall, R. L., and Greene, R. P. (1997) "Defining job concentrations: The Los Angeles case." *Urban Geography* 18:705–739.

Fujii, T., and Hartshorn, T. A. (1995) "The changing metropolitan structure of Atlanta, GA: Locations of functions and regional structure in a multinucleated urban area." *Urban Geography* 16:680–707.

Glaser, E., Kolko, J., and Saiz, A. (2001) "Consumer city." *Journal of Economic Geography* 1:27–50.

Giuliano, G., and Small, K. (1991) "Subcenters in the Los Angeles region." *Regional Science and Urban Economics* 21:163–182.

Gong, H., and Wheeler, J. O. (2002) "The location and suburbanization of business and professional services in the Atlanta metropolitan area." *Growth and Change* 33:341–369.

Hahn, B. (2000) "Power centers: A new retail format in the United States of America." *Journal of Retailing and Consumer Services* 7:223–231.

Hartshorn, T. (1992) *Interpreting the City: An Urban Geography.* New York: Wiley.

Hazel, D. (2003) "Lifestyle centers look good, but are they earning their keep?" *Shopping Centers Today* (ICSC monthly magazine cover story), May.

Hinojosa, R., and Pigozzi, B. (1988) "Economic base and input-output multipliers: An empirical linkage." *Regional Science Perspectives* 18:3–13.

Hum, T. (2003) "Mapping global production in New York City's garment industry: The role of Sunset Park, Brooklyn's immigrant economy." *Economic Development Quarterly* 17:294–309.

Kenworthy, J., and Laube, F. (1999) *An International Sourcebook of Automobile Dependence in Cities: 1960–1990.* Boulder, CO: University Press of Colorado.

Leone, R. A., and Struyk, R. (1975) "The incubator hypothesis: Evidence from five MSAs." *Urban Studies* 13:325–331.

Liu, W. Dicken, P., and Yeung, H. W. C. (2004) "New information and communication technologies and local clustering of firms: A case study of the Xingwang industrial park in Beijing." *Urban Geography* 25:390–407.

Mattson-Teig, B. (2004) "Lifestyle retail: A new generation of retail centers is aimed at meeting changing consumer and community needs." *Urban Land* February:48–53.

Mota, J. Q., and Castro, L. M. (2004) "Industrial agglomerations as localized networks: The case of Portuguese injection mould industry." *Environment and Planning A* 36:263–278.

Mueller-Wille, C. (1990) *Natural Landscape Amenities and Suburban Growth: Metropolitan Chicago, 1970–1980.* Chicago, IL: University of Chicago.

Mulligan, G., and Kim, H. (1991) "Sectoral-level employment multipliers in small urban settlements: A comparison of five models." *Urban Geography* 12:240–259.

Noyelle, T., and Stanback, T. (1984) *The Economic Transformation of American Cities.* Toronto, Canada: Rowman & Littlefield.

Pinnau, M. (2003) Personal communication, April.

Rantisi, N. M. (2002) "The competitive foundations of localized learning and innovation: The case of women's garment production in New York City." *Economic Geography* 78:441–462.

Rigby, D. L., and Essletzbichler, J. (2002) "Agglomeration economies and productivity differences in US cities." *Journal of Economic Geography* 2:407–432.

Scott, A. J. (1982) "Locational patterns and dynamics of industrial activity in the modern metropolis." *Urban Studies* 19:111–142.

Scott, A. J. (2002a) "Competitive dynamics of Southern California's clothing industry: The widening global connection and its local ramifications." *Urban Studies* 39:1287–1306.

Scott, A. J. (2002b) "Industrial urbanism in late-twentieth-century southern California," in *From Chicago to LA:* *Making Sense of Urban Theory*, ed. by M. J. Dear, pp. 163–179. Beverly Hills, CA: Sage.

Smith, A. (2003) "Power relations, industrial clusters, and regional transformations: Pan-European integration and outward processing in the Slovak clothing industry." *Economic Geography* 79:17–40.

Soriano, V. V. (1998) "The shopping center industry through 1997." *ICSC Research Quarterly* 5:9–10.

Stoddard, E. R. (1987) *Maquila: Assembly Plants in Northern Mexico.* El Paso, TX: Texas Western Press.

Stumpf, M. D. (1996) *The Intra-Metropolitan Relocation of Manufacturing Firms in the Chicago Region, 1987–1992.* Master's thesis. Department of Geography, Northern Illinois University.

Stutz, F. P., and Warf, B. (2005) *The World Economy.* Upper Saddle River, NJ: Prentice Hall.

U.S. Census Bureau. (2000) *The 1997 Economic Census.* Washington, DC: U.S. Census Bureau.

U.S Customs Service. (1997) *Office of Field Operations.* Washington, DC: U.S. Customs Service.

Valley Partnership. (2004) "Valley Economy." Available at *http://www.valleychamber.com/visitor-guide/economy.shtml.*

Waddell, P., and Shukla, V. (1993) "Manufacturing location in a polycentric urban area: A study in the composition and attractiveness of employment subcenters." *Urban Geography* 14:277–296.

"Wasting away: Despite SARS, Mexico is still losing export ground to China." (2003) *Business Week* June:42–44.

Webber, M. J. (1984) *Industrial location.* Beverly Hills, CA: Sage.

Weber, A. (1929) *Theory of the Location of Industries.* Chicago, IL: University of Chicago Press.

Weimer, A. M., and Hoyt, H. (1939) *Principles of Urban Real Estate.* New York: Ronald Press.

Wernerheim, C. M., and Sharpe, C. A. (2001) "The potential bias in producer service employment estimates: The case of the Canadian space economy." *Urban Studies* 38:563–591.

Wrigley, N., and Lowe, M. (2002) *Reading Retail: A Geographical Perspective on Retailing and Consumption Spaces.* London: Arnold.

EXERCISE

Exercise Description

To examine clustering of industrial activity of various types for Los Angeles County using location quotient analysis. Urban geographers have become increasingly interested in the dispersion of employment away from the central core area of cities to suburban employment subcenters. The shift of employment to the suburbs has occurred among all industrial sectors from manufacturing to professional services. It is often suggested that office-oriented industry is more suburban oriented than manufacturing industry. Industrial centers have also remained highly segregated, so that we often refer to suburban office parks separate from manufacturing industrial parks. In this project we examine the distribution of employment by nine industrial sectors within Los Angeles County to see if the patterns that emerge are consistent with the industrial geography literature.

Course Concepts Presented

Industrial location, location quotients, and location theory.

GIS Concepts Utilized

GIS overlay and thematic mapping through graduated symbols.

Skills Required

ArcGIS Version 9.x

GIS Platform

ArcGIS Version 9.x

Estimated Time Required

1 hour

Exercise CD-ROM Location

ArcGIS_9.x\Chapter_9\

Filenames Required

La_zip_emp.shp
La_emp_cntr.shp
Lacity.shp
Lacnty.shp

Background Information

Location Quotient

There are several methods for measuring an area's industrial specialization. The method we use in this project is known as a location quotient (LQ). The LQ expresses the share of employment in a given industry in a specified subarea as a percentage of the share of employment in the same industry within the larger area. In this exercise the subareas are ZIP codes and the larger area is the county.

Example of ZIP code X:

	ZIP Code X	County X
Professional services	8.7%	5.5%
Manufacturing	17.6%	22.8%

To determine ZIP code X's specialization in professional services and manufacturing relative to the county, we plug the percentage employed into the following equation:

$$\text{Professional services } (8.7/5.5) * 100 = 158$$
$$\text{Manufacturing }\quad (17.6/22.8) * 100 = 77$$

An LQ greater than 100 indicates specialization and less than 100 indicates that the ZIP code is not specialized in that category. In this example, the ZIP code is specialized in professional services and not in manufacturing.

The data for this exercise are from the 1997 Economic Census for Los Angeles County, California. To perform the necessary calculations for the LQ, the county total employment by sector is required. These figures are shown in the following table.

NAICS Code	Description	Employment	Sector employment divided by total county employment
31 (31–33)	Manufacturing	622,302	.280
44	Retail trade	343,656	.155
54	Professional, scientific, and technical services (taxable)	346,290	.156
56	Administration and support and waste management and remediation services	290,946	.131
61	Educational services (taxable)	10,869	.005
62	Health care and social assistance (taxable)	196,543	.088
71	Arts, entertainment, and recreation (taxable)	58,793	.026
72	Accommodation and food service	267,406	.120
81	Other services (taxable)	86,614	.039

Readings

(1) Forstall, R. L., and Greene, R. P. (1997) "Defining job concentrations: The Los Angeles case." *Urban Geography* 18:705–739.

(2) Greene, R. P., and Pick, J. B. *Exploring the Urban Community: A GIS Approach*, Chapter 9.

(3) Scott, A. J. (2002) "Competitive dynamics of Southern California's clothing industry: The widening global connection and its local ramifications." *Urban Studies* 39:1287–1306.

(4) Scott, A. J. (2002) "Industrial urbanism in late-twentieth-century southern California," in *From Chicago to LA: Making Sense of Urban Theory*, ed. by M. J. Dear, pp. 163–179. Beverly Hills, CA: Sage.

(5) Stanback, T. (1991) *The New Suburbanization—Challenge to the Central City*. Boulder, CO: Westview.

(6) Waddell, P., and Shukla, V. (1993) "Manufacturing in a polycentric urban area: A study in the composition and attractiveness of employment subcenters." *Urban Geography* 14:277–296.

Exercise Procedure Flowchart

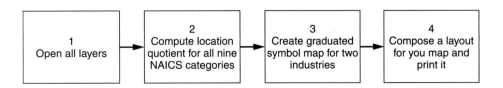

| 1 Open all layers | → | 2 Compute location quotient for all nine NAICS categories | → | 3 Create graduated symbol map for two industries | → | 4 Compose a layout for you map and print it |

Exercise Procedure

Part 1

1. Start ArcMap with Start Using ArcMap With A New Empty Map selected. Click OK.

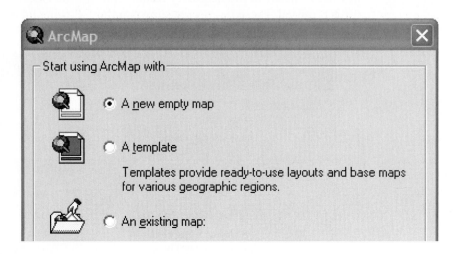

2. Click Add Data and navigate to the exercise data folder.

Note: *If you do not see the folder where you placed the exercise data, you will need to make a new connection to the root directory that holds the data.*

3. Add the following layers from the Chapter 9 folder:

> La_zip_emp.shp (Zip Code Points with Reliable 1997 Employment Data)
> La_emp_cntr.shp (Employment Centers over 50,000 Employees)
> Lacity.shp (L.A. City Boundary)
> Lacnty.shp (L.A. County Boundary)

You can add all four at once by clicking one of the file names, then clicking the others while holding down the Ctrl key on the keyboard, and clicking Add.

4. Keep the drawing order as shown above and right-click on Lacity.shp and Lacnty.shp, select Properties, double-click on the symbol, and make the polygons hollow with city and county line symbols. Right-click La_emp_cntr and select Zoom To Layer.

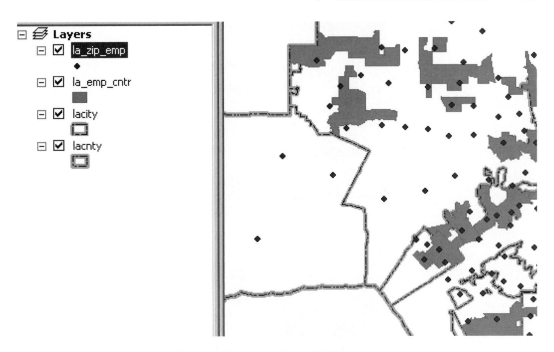

5. Right-click La_zip_emp and select Open Attribute Table.

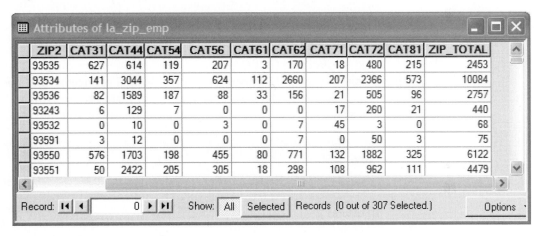

Note the fields for ZIP code, employment by nine industry types (CAT fields), and total employment.

6. Click Options and add a field for the LQ for manufacturing employment. Assign the name loc31, type to float, 16 for precision, and 0 for scale.

7. Right-click the loc31 field and calculate (.280 in denominator is found in table in the Background Information earlier in this exercise):

$$((([CAT31]/[ZIP\_TOTAL])/(.280))*100$$

8. Repeat steps 6 and 7 for the remaining industry types by substituting cat31 and loc31 with the correct industry number. Also substitute the denominator .280 with the appropriate number found in the table in the background section.

9. Make a graduated symbol map for two industries of your choice using the location quotient fields. For instance, to map the LQ for manufacturing, right-click La_zip_emp and select Properties. On the Symbology tab, under Quantities, select Graduated Symbol, set the Value field to loc31, change classes to 4, click classify, manually change the range to that shown and click OK, and click Apply and OK.

Note: *When you change a range, you only need to type in the higher value of the range. For example, for 0–100, you only need to type 100.*

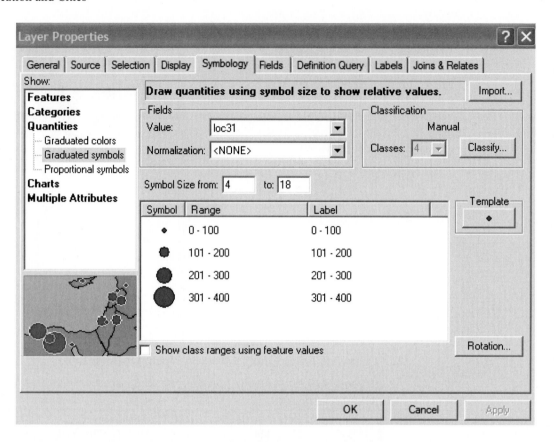

10. To show a second industry (e.g., professional services), right-click La_zip_emp and click
    Copy. Then click the Insert tab and select Data Frame. Now right-click New Data Frame
    and select Paste Layer(s). Copy and paste the three other layers from the above data frame

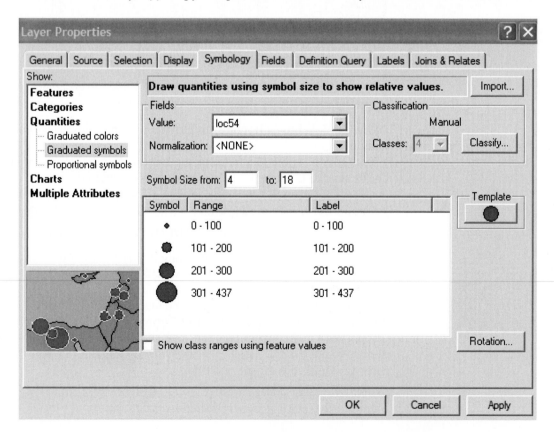

to the new data frame in the same drawing order. Do not forget to zoom to the map extent of the employment centers by right-clicking La_emp_cntr and Zoom to Layer. Now right-click La_zip_emp, click Properties, and change the LQ to loc54 and change the ranges to those shown below and Apply and OK.

11. Click View and click Layout View. Compose a layout that shows your two chosen industries.

Note: *You can size the two map windows by clicking and holding on corners of the windows. You can also zoom and position the maps in the windows.*

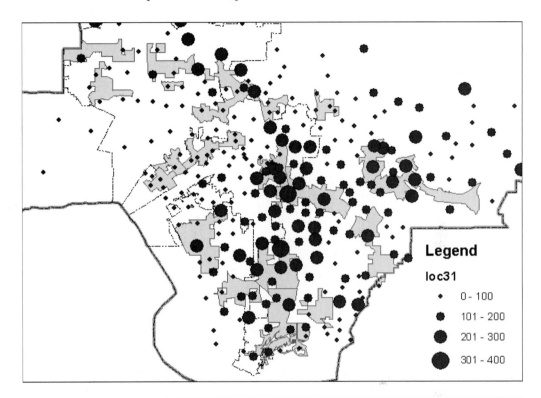

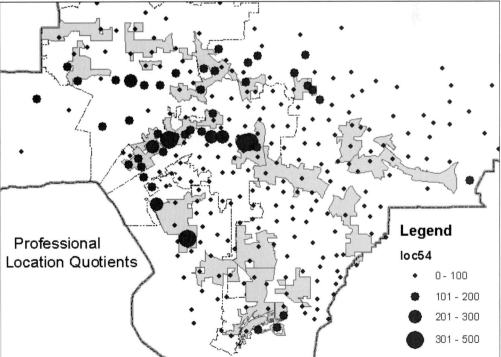

Comparison of Manufacturing and Professional Employment.

Part 2: Essay

Write a short essay describing the location of industry types within Los Angeles County. Include maps showing variation of two or more industrial specializations within the county. Explain the shapes and differences in the employment patterns. What are the implications of the patterns for the Los Angeles economy and its businesses? Refer to the readings listed in the Background Information section. Do the patterns match what the literature suggests? Also, the employment center layer, derived from Forstall and Greene (1997), has a name field that references the name of the job center (all job centers with 50,000 workers or more in 1990). These might provide additional information on the locations of industry types.

10

Urban Core and Edge City Contrasts

The physical pattern of urban development has been a focus of urban geography for a long time. *Urban morphology* is the subdiscipline of urban geography that specifically examines the physical layout of cities. It is a more detailed study of the physical layout of cities than the urban structure models presented in Chapter 4. In contemporary times, urban morphologists and other urban scholars have been preoccupied with the increasing rate at which development has spread over vast areas, often along highways, a characteristic often referred to as *urban sprawl*. The rise of urban sprawl has been blamed for a growing number of urban ills such as traffic congestion, the demise of the town center, reduction in access to employment, loss of sense of community belonging, and the loss of prime farmland, to name just a few. Because the topic of sprawl involves the process of development spreading out over space, geographic information systems (GIS) techniques are being used by a variety of federal, state, and local governments to help them understand the pace and pattern of this phenomenon as well as a means for developing alternative planning scenarios.

A root cause of sprawl is the decentralization of people and jobs from urban cores to the suburbs, a topic that has especially occupied the attention of urban geographers in the last half-century. Both housing and employment growth and redistribution have occurred in suburban areas to such an extent that many metropolitan areas are now *polycentric*— they have multiple downtowns. One concomitant of this trend is that between 1993 and 1996, employment growth in suburban areas exceeded growth in downtowns in all but 20 percent of the largest metropolitan areas in the United States (Brennan & Hill, 1999), and more than 80 percent of the new housing construction in the United States took place in the suburbs between 1986 and 1998 (Hoffman, 1999).

This chapter considers the geography of sprawl, its root causes, and its multiple forms. First, the relative trends of central city and suburban change are reviewed. We then consider the topic of urban sprawl with a case study analysis of the Chicago region, which is linked to the GIS exercise at the end of the chapter. Another section examines the metropolis of Sydney, Australia, which falls between two extreme international models of sprawl. The impact of transportation on the form of urban sprawl is also analyzed in the context of a four-stage model of influence. A comparison of Chicago and Los Angeles shows the relationship of population growth to transportation developments from 1960 to the present.

Following the discussion of sprawl, the concept of edge cities, including their delineation and measurement, is introduced. An edge city is a center of employment in the outskirts of the urban area, associated with highways. Because edge cities represent alternative employment hubs to the main downtown, they play an important role in the future direction and growth of sprawl. The planning concept of "smart growth" is introduced. The chapter finishes by looking at an example of government planning response to sprawl—Portland's growth boundary.

10.1 CENTRAL CITIES

In the United States, in the second quarter of the 20th century, the populations of large central cities began to spill beyond their borders in large numbers. The 1970 Census was the first in which the number of residents of suburbs surpassed the number of residents of central cities. Since 1910 the proportion of metropolitan area population living in central cities has declined every decade, although it leveled off somewhat in the 1980s.

In many respects, the 1980s represented a departure from the big-city declines experienced directly after World War II. In many big Northeastern and Midwestern cities, perennial population declines slowed in the 1980s from the losses experienced in the 1950s, 1960s, and 1970s (see Figure 10.1). For instance, loss in St. Louis's central city accelerated from the 1950s to the 1960s and again in the 1970s, when it reached its greatest-ever decennial loss of 169,435 persons. However, in the 1980s and 1990s, St. Louis's central city population decline slowed significantly, amounting to 56,116 and 48,496, respectively. In the 1990s, these perennially declining cities had losses ranging from a low of 1,675 in Newark, New Jersey, to a high of 84,860 in Baltimore. Baltimore's 1990s loss was twice that of the 1980s but smaller than the loss of 119,046 it experienced in the 1970s.

On the other hand, many cities in the South and West have continued to grow from 1950 through the 1980s. In the 1990s, those hemmed in on their edges by other municipalities or natural features have leveled off or grown slowly, whereas those that are not inhibited from annexing (e.g., Phoenix and San Antonio) have continued to grow at a fast pace (see Figure 10.2). Annexing can be lowered or halted because of physical limitations, limited political interest by the city, and lack of interest or resistance by its target cities or unincorporated areas. For example, Atlanta is not completely hemmed in, but it has not undertaken significant annexation since 1952. Many cities in Texas and Oklahoma have extensive boundaries because they want to prevent population clusters from incorporating and blocking their future expansion. Another factor that might be influencing annexation growth is the relative strength of the metropolitan government. If it is strong, there might be less pressure to annex, such as has been seen in Miami.

The 1990s saw the relative decline of the central city persist in some large metropolitan areas but not in others (see Table 10.1). For instance, the cities of Detroit and Philadelphia experienced changes of −7.5 percent and −4.3 percent, respectively, whereas their overall metropolitan growth rates were 5.1 percent and 4.7 percent. This differential in growth rate between central city and overall metropolitan area indicates that the declining trend of central cities has continued in both Philadelphia and Detroit. On the other hand, the city of Chicago expanded in population between 1990 and 2000 for the first time since the 1950 Census. Nevertheless, such population turnarounds do not necessarily return the central

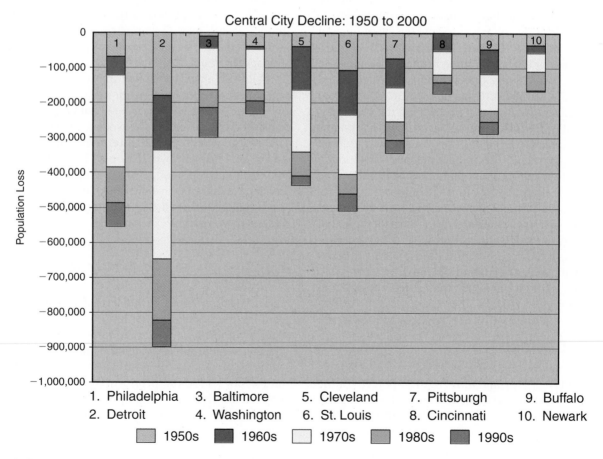

FIGURE 10.1   Central-city decline, 1950–2000, for selected large central cities.

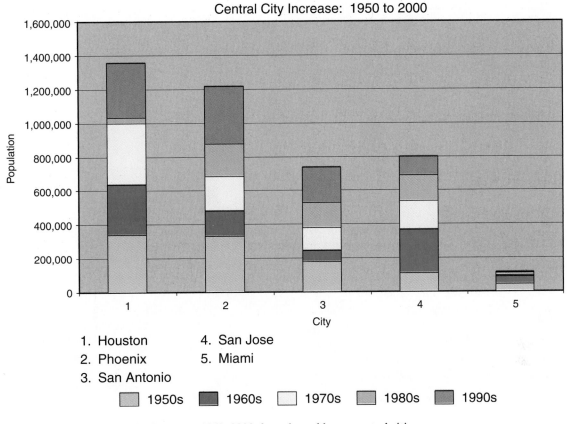

FIGURE 10.2   Central-city increase, 1950–2000, for selected large central cities.

TABLE 10.1   Population change for the 10 largest metropolitan areas and cities: 1990–2000

10 Largest metropolitan areas	Population, April 1, 1990	Population, April 1, 2000	Percent change, 1990 to 2000
New York-Newark-Bridgeport, NY-NJ-CT-PA Combined Statistical Area	19,710,239	21,361,797	8.4
Los Angeles-Long Beach-Riverside, CA Combined Statistical Area	14,531,529	16,373,645	12.7
Chicago-Naperville-Michigan City, IL-IN-WI Combined Statistical Area	8,385,397	9,312,255	11.1
Washington-Baltimore-Northern Virginia, DC-MD-VA-WV Combined Statistical Area	6,665,228	7,538,385	13.1
San Jose-San Francisco-Oakland, CA Combined Statistical Area	6,290,008	7,092,596	12.8
Philadelphia-Camden-Vineland, PA-NJ-DE-MD Combined Statistical Area	5,573,521	5,833,585	4.7
Boston-Worcester-Manchester, MA-NH Combined Statistical Area	5,348,894	5,715,698	6.9
Detroit-Warren-Flint, MI Combined Statistical Area	5,095,695	5,357,538	5.1
Dallas-Fort Worth, TX Combined Statistical Area	4,138,010	5,346,119	29.2
Miami-Fort Lauderdale-Miami Beach, FL Metropolitan Statistical Area	4,056,100	5,007,564	23.5
10 Largest cities			
New York	7,322,564	8,008,278	9.4
Los Angeles	3,485,398	3,694,820	6.0
Chicago	2,783,726	2,896,016	4.0
Houston	1,630,553	1,953,631	19.8
Philadelphia	1,585,577	1,517,550	−4.3
Phoenix	983,403	1,321,045	34.3
San Diego	1,110,549	1,223,400	10.2
Dallas	1,006,877	1,188,580	18.0
San Antonio	935,933	1,144,646	22.3
Detroit	1,027,974	951,270	−7.5

Source: U.S. Censuses of 1990 and 2000.

FIGURE 10.3   Seattle skyline.

city back to the days when it represented the principal location for metropolitan residents, as symbolically demonstrated by the vibrancy of the center of Seattle (see Figure 10.3). The Chicago metropolitan area population increased in the 1990s by 11.2 percent, which is much higher than the city's growth rate of 4.0 percent, due in part to the continued growth of metropolitan Chicago's edge communities.

The ramifications of the large central-city decline are significant. In extreme cases, as population decentralizes from America's metropolitan central-city areas to their suburban edges, urban infrastructure in the central city becomes underutilized or even abandoned. The abandonment of the urban core by the middle class has even been linked to a rising urban underclass, as illustrated in Chapter 8. A later section shows that many owners of industry and businesses in the inner city are also relocating to the suburbs and edge cities, taking with them valuable urban resources and helping contribute to a cycle of central-city decline. When capital and investment are redistributed from the urban core to the suburban periphery, the metropolitan settlement fabric is redefined:

> The replacement for an obsolete building has almost certainly been in another block, another part of the metropolis, or another region of the country. Cost and sheer mechanical

inconvenience, as well as changing locational advantages, have usually discouraged replacement of obsolete structures on site ... In this process, the nation is not simply replacing an inventory of buildings. Because of changed locational advantages, it is also replacing the major part of the fabric three generations have taken for granted as the bedrock geographic pattern of American settlement. (Borchert, 1991, pp. 233–235)

As Borchert (1991) so aptly described, much of our urban infrastructure, heretofore taken for granted, could be lost as locational advantages shift outward. In the next section, we turn to the topic of urban sprawl, one of the consequences of a tendency in many cities for population to redistribute from the core to the edge.

10.2 URBAN SPRAWL

Urban growth is not new; however, a number of urban studies scholars have become concerned that the quickening rate of this expansion is inefficient and unsustainable. Whyte (1956) referred to this process early on as *urban sprawl* and the term has continued to be used widely through the current day. Urban sprawl, shown in Figure 10.4, was advanced to a national policy concern in the 1980s and continues as such today. For instance,

FIGURE 10.4   Urban sprawl in Las Vegas. Source: © Lester Lefkowitz/CORBIS.

Al Gore included the issue as part of his Democratic platform in the 2000 election by advocating incentives to stimulate construction within central cities (Kriz, 1999). There are countless examples of local and regional initiatives that have arisen in recent years to curtail sprawl as well. These initiatives include taxpayers giving approval to their local governments to purchase private land for open space, the creation of purchase of development rights (PDR) programs, and the creation of various regional and state-level land use controls. An example, Portland's growth boundary, is covered at the end of the chapter, and these other initiatives are more fully examined in Chapter 12.

The name Levittown has become associated with urban sprawl because it was implemented early in the post–World War II cycle of suburbanization and is well recognized as having perhaps the greatest impact on the design of American suburbs in this period. William Levitt (see Figure 10.5) built the first Levittown in 1947 on former Long Island potato fields 25 miles east of Manhattan. This automobile-oriented, mass-produced subdivision was extremely attractive to returning veterans and their growing families. Eventually growing to 17,400 houses and 82,000 residents, Levittown became the largest housing development built by a single developer (Jackson, 1985). Levittown was especially successful because it

represented all the ideals of the American dream, a single-family house with a yard at low cost:

> The typical Cape Cod was down-to-earth and unpretentious; the intention was not to stir the imagination, but to provide the best shelter at the least price. Each dwelling included a twelve-by-sixteen-foot living room with a fireplace, one bath, and two bedrooms (about 750 square feet), with easy expansion possibilities upstairs in the unfinished attic or outward into the yard. Most importantly, the floor plan was practical and well designed, with the kitchen moved to the front of the house near the entrance so that mothers could watch their children from kitchen windows and do their washing and cooking with a minimum of movement. Similarly, the living room was placed in the rear and given a picture window overlooking the back yard. This early Levitt house was as basic to post–World War II suburban development as the Model T had been to the automobile. In each case, the actual design features were less important than the fact that they were mass-produced and thus priced within the reach of the middle class. (Jackson, 1985, pp. 235–236)

Two more Levittowns were built, Levittown, Pennsylvania, in the 1950s in Bucks County near Philadelphia (see Figure 10.6) and Levittown, New Jersey, in the 1960s also outside Philadelphia. To the critics of this new suburban style the Levittown developments became

FIGURE 10.5   Bill Levitt. Source: Time.Com.

FIGURE 10.6   Levittown, Pennsylvania. Source: AP Wide World Photos.

FIGURE 10.7   Levittown houses: Bland but affordable. Source: Corbis/Bettmann.

synonymous with mass-produced, bland housing (see Figure 10.7).

Critics claim that the shift of population from city to suburb coincided with the expansion of low-density *single-use* developments (i.e., subdivisions with single-family housing only) with *leap-frog* tendencies (i.e., by-passing less desirable parcels closer in), and isolated strips of commercial development that are served only by automobiles. The problems associated with this form of development included the following:

- Loss of open space including some prime farmland (Platt, 1991).

- Disruption of natural habitats (Acevedo & Buchanan, 2000).

- Traffic congestion as a result of the increased separation of commercial land uses from residential land uses (Downs, 1992).

- Diminished air quality as a result of increased vehicle-miles traveled, as evidenced by the number of urban ozone alerts issued in major metropolitan areas (National Research Council [NRC], 1995).

- Widening socioeconomic disparities as a result of selective migration from the inner cities to the suburbs (Powell, 1998).

Mieszkowski and Mills (1993) identified two groupings of theories designed to explain the causes of suburbanization. Urban theorists and transportation scholars favor the *natural evolution theory*. This theory, which draws from the work of Alonso (1964), holds that residential location choices are determined by minimizing commuting costs to the central business district (CBD). The theory recognizes that central areas are developed first. As land close to the center becomes filled in, development moves to more open tracts of land, likely to be in the suburbs. As new housing is built at the periphery, persons in higher income groups who can afford larger and more modern housing tend to move outward to this area. As a trade-off for the longer commute, the home-owner is rewarded with lower housing prices and a larger lot farther from the CBD.

The second grouping of theories (Fiscal-Social Problems Approach), in contrast, emphasizes fiscal and social stresses of the central city as the main impetus for suburbanization. Ironically, these stresses are a result of suburbanization itself, because as the middle class leaves the inner city for the periphery it leaves behind a decaying area that the parent city finds more difficult to maintain with a falling tax base. The desire to leave these declining neighborhoods grows, resulting in a repeating cycle of poorer school districts, and rising racial tensions, crime, congestion, and environmental degradation. The middle-class movers arriving in affluent suburbs might actually lower the per-capita income there.

Some urban historians suggest that the federal government has contributed to inducing sprawl, through the mortgage insurance programs of the Federal Housing Administration (FHA), begun in 1934, and the Veterans Administration (VA), begun in 1944. These programs provided the funds that allowed many Americans to afford home loans (Hanchett, 2000). However, recent data do not show a bias in mortgage loans toward the suburbs. Almost half (46 percent) of FHA's single-family loans in 1996 were made for homes located in central cities (General Accounting Office, 1999).

Even more important to the federal government's leading role in contributing to sprawl was the advent of the 1956 Federal Interstate Highway Act. The interstate highway system eventually reached 42,500 miles with the federal government paying 90 percent of the cost (Jackson, 1985, p. 249). Although now almost 100 percent complete, most recent expansion of federal highway capacity is taking place at the urban fringe, which often is expanding in population (NRC, 1995, p. 194). Not only did these highways make it convenient for people to leave the city, but a further profound impact was to make it more advantageous for industry to locate outside the city toward suburban superhighway interchanges for superior logistical access.

It is important to note that not all urban scholars are critical of suburbia. For instance, Richardson and Gordon (2004) contended that many of the arguments of the anti-sprawl coalition are flawed. For instance, they contended on the issue of sprawl's impact on traffic congestion that:

> Most people live and work in the suburbs, and most commuting is suburb to suburb. Moreover, our highly mobile population relocates to economize on trips in response to congestion, resulting in the "commuting paradox," where increasing route congestion is compatible with stable or shorter travel times. Average commuting speeds keep rising. The U.S. Department of Transportation reports that average speeds were 28.0 m.p.h. in 1983, 32.3 m.p.h. in 1990, and 33.6 m.p.h. in 1995. Distances are increasing, but trip times have changed little because more trips take place on less congested suburban roads. Where there truly is congestion, it could easily be ameliorated by road-user pricing, where motorists pay to drive on freeways at peak times of the day. Suburbanization is a traffic safety valve, more of a solution than a problem. In fact, the worst traffic and

dirtiest air are found in the highest density areas. (Gordon & Richardson, 2000, p. 68)

Others have taken issue with Gordon and Richardson on decreasing travel times in the suburbs:

> Even on their own terms, G & R's macro comparisons are suspect. In one paper after another, they have argued that decentralization of firms and households raises average travel speeds enough to compensate for longer trips. Recent evidence suggests otherwise. Average commute times became worse during the 1980s in 35 of the 39 metropolitan areas with more than one million people. By the end of the decade, average commute times were significantly greater in the suburbs than in central cities. (Ewing, 1997, p. 113)

Horner (2004), in a recent review of the commuting literature, saw no signs of congestion abating any time soon. Nevertheless, the question of changes to suburban commuting remains unresolved. It would be closer to solution if more data were available. The true answer might vary from one metropolitan area to another.

10.3 TRENDS IN URBAN DEVELOPMENT RATES

Although there is debate among researchers over the significance and impacts of sprawl, there is much evidence to suggest that sprawl is continuing unabated at faster rates than in the past. One source of information on the amount of urban and developed land is the National Resources Inventory (NRI), which, until recently, was conducted by the Natural Resource Conservation Service of the U.S. Department of Agriculture every five years on a large sample of land units throughout the United States. More recently the agency turned to an annual NRI. When those results are combined with the previous results, it estimates that the nation's 73 million acres of urban and built-up land in 1982 grew to 106 million acres in 2001, an increase of 33 million acres, or a 45 percent increase, over the 19-year period (see Figure 10.8).

What is especially noteworthy in the NRI data is the rising annual rate of urban development. On the basis of the 19-year period, U.S. residents added an average of 1.7 million acres of urban land every year between 1982 and 2001 (see Table 10.2). The rate was especially high

TABLE 10.2 Rate of annual increase of urban land, 1982–2001

(Millions of acres)			
1982 to 1987	1987 to 1992	1992 to 1997	1997 to 2001
1.2	1.5	2.2	2.2
Source: 2001 National Resource Inventory, NRCS, 2003.			

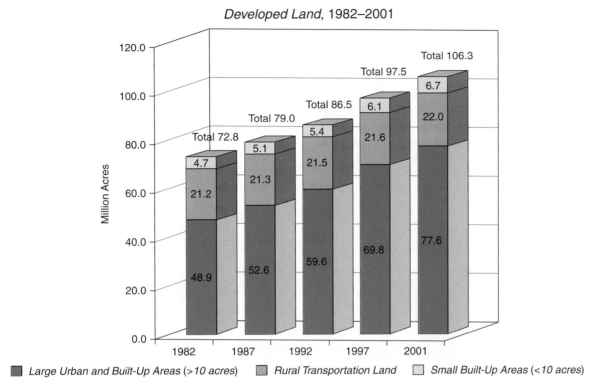

FIGURE 10.8   Developed land in the United States, 1982–2001. Source: 2001 Annual National Resource Inventory, NRCS of the USDA, July 2003.

during the last 9-year observation period, when Americans added 2.2 million acres per year of new urban land, which provides credibility to the argument that the form of urban development today is more land consuming than during previous development periods. Similarly, Lopez and Hynes (2003) developed a sprawl index and found a similar increase in land consumption. To demonstrate the variability of views on this issue, the next section, "Analyzing an Urban Issue," examines the topic at length. One aspect of this urban issue, the forever changing urban fringe, is the study objective of this chapter's GIS laboratory exercise.

10.4 ANALYZING AN URBAN ISSUE: URBAN SPRAWL, ITS MEASUREMENT, AND ITS IMPACTS

During the early 1970s, Clawson compared the patterns of urban development in Britain and the United States and made the following observation about the American pattern:

> The characteristic development form of the postwar United States has been a series of separate subdivisions, with intervening tracts of open land, spreading away in all directions from the larger urban areas. (Clawson & Hall, 1973, p. 130)

Clawson's observations not only capture the essence of the term sprawl, but they describe a key feature of urban areas known as the urban–rural fringe. Urban sprawl is defined as dispersed, auto-dependent development outside compact urban centers (Vermont Forum on Sprawl, 2003). It is typified by low-density settlement, often in erratic building patterns, located along highways in formerly rural areas, and consuming excessive amounts of land. As the name implies, the *urban–rural fringe* is where the zone of urban land is ending and rural land beginning, although no single line separates the two. More recently the urban–rural fringe was described as a peri-metropolitan bow wave:

> The bow wave is a standing wave that always remains immediately in front of the bow of a ship moving through water. The urban–rural fringe, then, is the bow wave of the built-up area of the city, the zone of intensively cultivated, high-priced agricultural land that remains immediately in front of the expanding urban edge. (Hart, 1991, p. 36)

Although Hart was using the bow-wave analogy to argue that the high price of agricultural land in this zone was due to its location rather than the quality of the soil, the analogy has broader significance, as it shows the urban–rural fringe as highly dynamic, and not capable of being delineated precisely, as a "snapshot" of the fringe at one point in time is hard to achieve. This exercise utilizes this bow-wave idea, but instead of using it for agricultural land, it applies the concept to the expanding and developing metropolitan growth rings of Chicago during the period from 1960 to 2000.

Evidence supporting the bow-wave analogy can be seen in recent growth trends for Chicago (see Figure 10.9). Similar to Hart's (1991) case study of agricultural production occurring in rings leading outward from New York City, rings (10 miles in width) were generated out from downtown Chicago. Census tracts were identified as growth ones for the period between 1960 and 1980 or

between 1980 and 2000 and were included as long as they had a base year population greater than zero and increased in population. For each ring, the proportion of the ring's total population in these growing census tracts was then computed (see Table 10.3). Considering the first time period, 1960 to 1980, the ring that contained the greatest share of residents living in growing tracts was Ring 3 (Figure 10.9), at nearly 45 percent. We can place the center of the late 1970s to early 1980s fastest growth ring somewhere in central DuPage County, about 30 miles due west of downtown Chicago.

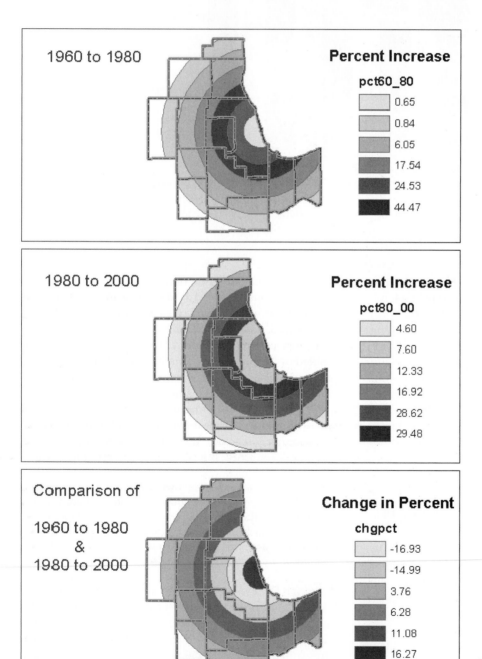

FIGURE 10.9 Growth rings around downtown Chicago, 1960–2000.

TABLE 10.3 Chicago ring analysis of population increase for
two time periods

	1960 to 1980		1980 to 2000		Difference
					Change in percent
	Total	Percent	Total	Percent	
Ring 1	132,448	6.45%	303,417	16.99%	+10.54
Ring 2	505,229	24.58%	136,482	7.64%	−16.94
Ring 3	915,004	44.52%	529,055	29.63%	−14.89
Ring 4	360,550	17.54%	513,073	28.74%	+11.20
Ring 5	124,404	6.05%	220,785	12.36%	+6.31
Ring 6	17,271	0.84%	82,406	4.61%	+3.77

Note: Percent derived by dividing total growth in ring by total growth in region: 2,054,906 total regional growth between 1960 and 1980. 1,785,218 total regional growth between 1980 and 2000.

The pattern that emerged in the second growth period, 1980 to 2000, is more complex, mainly because the inner core's share of residents living in growing tracts increased from less than 1 percent between 1960 and 1980 to 17 percent. In this exercise, the inner core refers to the innermost growth circle, referred to in the tables as Ring 1. Much of this increase actually occurred in the 1990s, during which time the city of Chicago experienced its first reversal of population decline since the 1950 Census. Another trend observed in the data is that the fourth growth ring had increased its share of total metropolitan growth from 19 percent in the first growth period to 29 percent by the second growth period. Meanwhile, the third ring, although accounting for a 29 percent share of growth, was significantly down from its earlier 44 percent. The full story is told in the third map of the sequence where the fourth outer ring, 30 to 40 miles from downtown, was the latest locus for the region's fastest growth. This fourth ring, centered in eastern Kane County, was bounded to the east by two growth rings declining in significance. The true urban fringe or bow wave is probably in Ring 5 or Ring 6, just to the west of the high-growth Rings 3 and 4, where the share of population living in growing tracts is also increasing but where densities are lower, and the prevalence of half-acre lots or larger is consistent with the densities of the urban–rural fringe in growth Ring 6 (Heimlich & Anderson, 2001).

The urban–rural fringe as peri-metropolitan bow wave raises a number of heated issues and debates. Key among these is the impact of urban expansion on farmland and especially the loss of prime farmland. Greene (1997) studied portions of Ring 4 and Ring 5, as defined in the preceding section, and found that as they were urbanizing, their agricultural land was reduced from 73 percent to 49 percent of the rings' total areas between 1975 and 1990 (see Figure 10.10). Moreover, the agricultural land that was lost was overwhelmingly prime farmland.

Like many other large Midwestern cities, Chicago, on its western side, adjoins the largest belt of prime farmland in the nation.

The belief, however, that urban development is consuming an increasing share of the nation's best farmland due to an acceleration of population decentralization has been challenged. In a national study of 135 U.S. counties that grew rapidly between 1970 and 1980, Vesterby and Heimlich (1991) showed little change in marginal rates of urban land consumption between 1960 and the early 1980s. A more recent study of U.S. cropland trends from 1949 to 1997 concluded that only a small fraction of the nation's cropland has been lost to suburban development (Hart, 2001).

Meanwhile, others have drawn attention to the quality and location of the replacement lands that are offsetting losses in prime farmland (Platt, 1991). When cropland located on naturally productive soils is lost to urban development, it is often replaced by cropland located on nonprime farmland, with multiple negative effects on the environment. Greene and Stager (2001) found that the approximately 11 million acres of cropland converted to urban land between 1982 and 1997 matched an equal amount of rangeland converting to cropland. In other words, the rangeland-to-cropland conversions equate to replacement lands for cropland lost to urban encroachment. Approximately one third of the cropland lost to urban development involved prime farmland, whereas the new cropland converted from rangeland was more likely to be nonprime. It also was likely to be irrigated as it was located principally in the arid western United States (see Figure 10.11).

The prime farmland that was converted to urban and built-up land occurred in clusters located in northeastern Illinois, eastern Pennsylvania, southern Michigan, and around the urban areas of Ohio and North Carolina. Other clusters of converted prime farmland are seen in the central valley of California, central Texas, south-central Florida, and near Phoenix and Tucson in Arizona. The study also showed a doubling in the percentage of irrigation for the rangeland converted to cropland during the analysis period, calling into question the sustainability of this new cropland given competing demands for water in these arid regions.

Another concern is that by only focusing on land transitions within the urban–rural fringe we are missing millions of acres of open land in exurban areas being converted to low-density human settlement. By measuring housing density at a small geographic unit, the census block group, Theobald (2001) was able to reveal the fine-grained details of this emerging exurban settlement system (see Figure 10.12). The darker patterns on the map look familiar because the housing density cutoff for urban results in a spatial pattern that parallels the Census

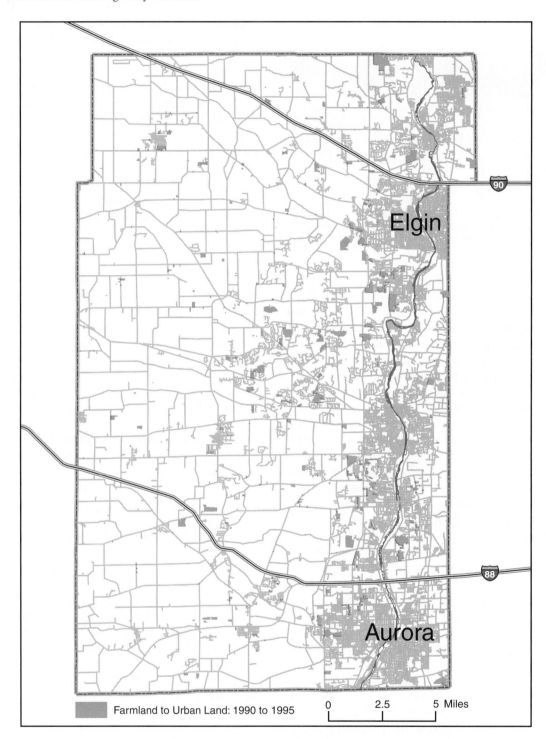

FIGURE 10.10 Prime farmland lost to sprawl on Chicago fringe between 1990 and 1995.

Bureau's own urbanized area (UA) definition. However, by utilizing housing density, the maps reveal zones of second homes, such as vacation areas, and ranchette-style developments in the West that would be somewhat less distinguishable in the mapping of population density.

Light gray rings of exurban development form rings around large metropolitan areas beyond the urban fringe of those metropolitan areas.

Perhaps the most noteworthy pattern in Theobald's (2001) analysis is the amount of new exurban development

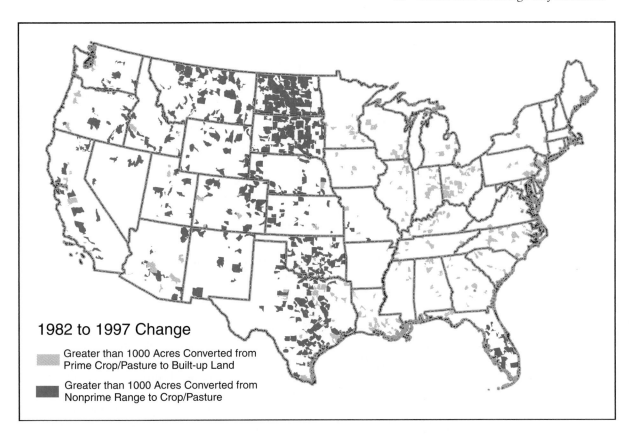

FIGURE 10.11   Cropland and rangeland conversions in the United States, 1982–1997. Source: Greene, R.P. and J. Stager, 2001. Rangeland to cropland conversions as replacement land for prime farmland lost to urban development. *The Social Science Journal*, Vol. 38, pp. 543–555. Fig. 2 on p. 551.

that occurred between 1960 and 2000. Most of the new exurban development is fill-in between urban areas, particularly in the Northeast, eastern Midwest, and South, but some of it is widely scattered west of the 100th meridian and east of the Pacific Coast, including in some remote areas. The Las Vegas cluster represents an unusual circumstance, as it was the fastest growing area in the United States during the 1990s (see Figure 10.13). The Las Vegas map shows the census blocks of 1,000 persons per square mile for the 1990 and 2000 censuses. The largest buildup occurred in the periphery of a sector stretching northwest from the Las Vegas core. In addition, a sizable area was built up and developed to the south of the Las Vegas core, principally in the fast-growing suburb of Henderson. Much further northwest and unconnected to the other clusters but still within the Las Vegas orbit is Pahrump, considered a "retirement hot spot" by the *Atlas of the New West* (Riebsame, 1997).

The ramifications of these fringe and exurban patterns are clear: More open land than originally thought is being encroached on. Although not all of the encroached land is being removed from farmland use, the options for its alternative uses are being reduced by its use for a low-density network of human settlement.

To summarize, studies are largely in agreement over the extent of urban sprawl, although there is much disagreement over how to measure it and interpret its significance. The bow-wave analogy for the urban–rural fringe seems realistic as the outward push of urbanized areas always increases the value of the adjacent land in the path of urban expansion. The latter was demonstrated for Chicago, where the outward advancement of the urban–rural fringe even occurred while the core of the region had regained some population. As noted earlier, the NRI data indicate that the rate of land consumption has increased in the latest period of expansion. This increasing rate of land consumption has raised concerns over excessive loss of prime farmland, in the wake of urbanized-area expansions. One study concluded that the amount of farmland lost to suburban development was minimal. Another study suggested it was more significant when considering that much of the "replacement" land for high-quality farmland does not have satisfactory potential for agriculture. Finally, land development in exurban areas far removed from the conventional urban–rural fringe has not been well studied. However, these developments away from the fringe are growing.

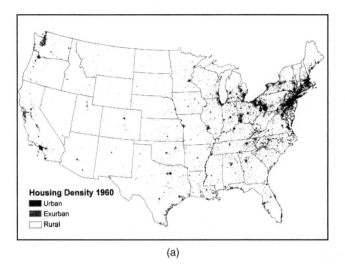

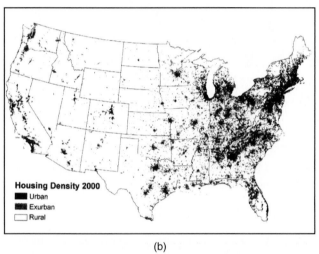

(a)

(b)

FIGURE 10.12 Housing densities in the United States, 1960 and 2000, showing exurban development. Source: Theobald, DM., 2001. "Land-Use Dynamics Beyond the American Urban Fringe." *Geographical Review*, 91:544–564. Figure "Housing Density 1960 and 2000."

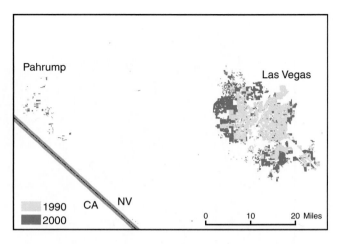

FIGURE 10.13 Exurban development in Las Vegas and Pahrump, Nevada, 1990 and 2000.

10.5 SPRAWL OUTSIDE THE UNITED STATES: THE CASE OF SYDNEY

Forsyth (1999) examined the process of planning for the development of residential housing in Sydney, Australia. Forsyth argued that Australia is between two extreme models of urban sprawl, one represented by the United States, which is largely developer-driven, and the other Great Britain, which is more government-led and centralized. Forsyth examined the Rouse Hill development, a planned community that adjoins Sydney. Government plays a different role in the development of housing property in Australia than it does in the United States. In Australia, government plans for the number of houses in a new development and takes the lead role in its detailed plans.

Although the government in Australia plays a more important role in the development process than it does in the United States, the issues are similar. Forsyth (1999) identified four major interest groups. The first are the *expansionists,* who have a strong and evocative image of an egalitarian society, inextricably linked to low-density suburbia. For *developers,* the second group, the image of the city is motivated by personal gain; they see their job as producing housing, which they can sell at a profit. A third group, the *scientific environmentalists,* favor the environment and have a regional view of the metropolis, as opposed to local environmentalists who are local in their view, similar to NIMBYs ("Not In My Back Yard") in the United States. Finally, there were the *consolidationists,* who favor compact, mixed-use development as a way to help solve the environmental, social, and financing problems of suburban developments.

10.6 TRANSPORTATION AND URBAN SPRAWL

One issue on which there is very little disagreement is the role that transportation has played in the formation of sprawl. Cities before the Industrial Revolution were predominantly very compact and dense because their geographic extents were controlled by the distances people could walk, although a few cities were larger. As a result, urban sprawl was nonexistent in its present form. In the early part of the 20th century, large North American cities grew along radial transit lines, particularly those for rail and early automobile transport, which led to concentrated growth of enterprises, housing, and other development along these routes. Especially since the development of expressways in the 1960s, development has been more spread out because of the ease of movement facilitated by the automobile. As a result, urban sprawl has

FIGURE 10.14   Automobile commuting in Minneapolis.

become almost synonymous with automobile-oriented development. Automobile commuting in Minneapolis is shown in Figure 10.14.

It is no coincidence, then, that the opponents of sprawl often favor an alternative form of planned community known as a *transit-oriented development* (TOD). In a

TOD, there is a rail station that is accessible to the residents, so they can rely on rail rather than depending on a car. The station is surrounded by high-density development, and successively lower density development moving outward (*TDM Encyclopedia,* 2003). The diameter of one TOD is usually a quarter- to a half-mile, with stations spaced a half-mile to a mile apart. Residents are able to walk or cycle between the station and their homes (*TDM Encyclopedia,* 2003).

Using housing construction statistics from 1889 to 1960, Adams (1970) developed a four-stage model to illustrate the impact of transport technology on Midwestern city structure (see Figure 10.15). Adams first laid down the essential foundation for the model's processes, as follows. Housing densities tend to conform to roughly concentric patterns reflecting the construction period in which the units were built. On the other hand, the pattern of socioeconomic status tends to evolve in an axial or sector pattern.

In Stage 1, the *walking/horsecar era* before the 1880s, primitive horse-drawn transport systems and the distances that people could walk limited the physical expansion of cities. Lacking a transit network, the territory of movement conformed to a roughly concentric circle pattern. Immigrants arrived in large numbers during the 19th century, leading to overcrowding because the jobs were located in the central city. Thus, with the transit limitations prior to 1880 the larger American cities were both highly compact and overcrowded. Housing consisted primarily of multiple-family dwellings:

In 1880 the core of Chicago consisted of six-story buildings. The full impact of the elevator and the electric streetcar,

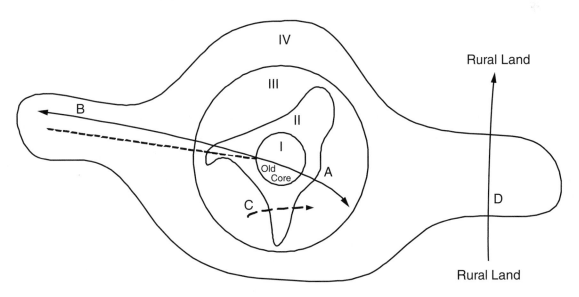

FIGURE 10.15   The Adams four-stage model of development and transport. Source: Adams, John S. 1970. "Residential Structure of Midwestern Cities." Annals of the Association of American Geographers 60:37–62. Figure 15. p. 56. "Transport technology stages and their impact on urban form."

which were introduced on a wide scale in the 1880s, was not felt in the location and design of new residential construction until the next cycle of feverish building activity. Four- and five-story walkup apartment buildings, visible for mile after mile in places like Harlem, Brooklyn, South Chicago, and the north side and south side of St. Louis, represent the transportation and housing technology of the period in which they were built. (Adams, 1970, p. 49)

Although horse car lines existed at the time, they were of limited importance as most people walked until the 1880s.

The residential fabric of cities changed radically with the introduction of Stage 2, the *electric streetcar,* which lent its name to the era that lasted from the 1880s to World War I. Streetcar rails encouraged elongated corridors of residential development, especially where land speculators purchased land in advance, sometimes with knowledge or coercion of where the tracks would be laid (Warner, 1962). Elongation was also encouraged by commuter stops along the interurban rail lines, such as that connecting Chicago and Milwaukee and many others in the East and Midwest. The housing type changed from multiple-family to single-family and two-family housing, and as shown in Figure 10.16, the urban form started to take on a star shape.

During the 1920s, or the beginning of Stage 3, the *recreational auto era,* the growing use of automobiles caused a dramatic increase in the supply of land without a proportionate increase in demand, so land in many areas became plentiful and relatively cheap. Adams (1970) referred to the 1920s as "recreational" because the widespread use of cars for commuting did not take place until the late 1930s. In some larger cities, such as Chicago and New York, the suburban railroads expanded and became an important transit feature. This led to the pre-1945 stereotype image of a suburbanite rail commuter, mostly male and affluent, who each workday left to work in the city center and returned to families in the suburbs. At the same time, urban transport networks matured and many cities became more star-shaped as a result.

The *freeway era,* post–World War II, saw the widespread use of the automobile for both recreation and commuting (Stage 4). The 1956 Interstate Highway Act was the most important event for the freeway era. It brought about a cataclysmic change in the spatial pattern of cities. Fast highways encouraged the very rapid expansion of the urban area. Using simple math based on the area of a circle versus its radius, Risse and Heyman (2000) calculated that the automobile increased the area of acceptable housing opportunities reachable from a given employment center by a factor of 625, due to the accessibility that the automobile network offered. The rapid expansion from the more prevalent, dispersed employment centers led to rapid growth in urban land use.

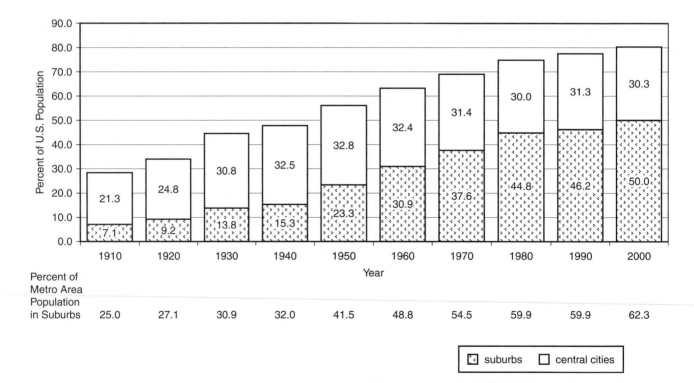

FIGURE 10.16   Percentage of total U.S. population living in metropolitan areas and their central cities and suburbs, 1910–2000. Source: U.S. Census 2002.

The building of thousands of miles of interstate highways and the flexibility the expanded highway networks offered to individuals resulted in a more circular form of urban development, rather than a star-shaped form. People could drive directly into work or they could live farther than the walking distance to the train station or streetcar line. This contrasted with the early stages of automobile usage, when most of the expansion had been solely residential in nature. At the start of the freeway era, most manufacturers were reluctant to abandon the railroad corridors that cut across suburban rings, as the large-sized tractor-trailer truck was decades away from being an effective and efficient short-distance hauler of goods (Wheeler, 1998).

This expansion of the urban area in the freeway era occurred in three waves. The first wave's urban growth was through the construction of more residential units farther from the city core. This was fueled by the substantial pent-up demand for housing, which had built up during the 1930s economic depression followed by World War II. When the war was over in 1945, millions of returning veterans wanted housing, much of which was being built in the suburbs (Randall, 2000).

This was followed by the second and third waves of development that had a more profound impact on the morphology of the metropolitan region. The second wave consisted of the outward movement of major retail business to the suburbs and the third wave was the outward spread of employment. Vance (1991) noted that the suburbanization of retail significantly reduced the remaining support for commuting rail lines. As seen in Figure 10.16, the percentage of U.S. population located in the suburbs grew from 7 percent in 1910 to half in 2000, and the percentage of metro area population in suburbs versus central cities increased from one quarter in 1910 to 62.3 percent in 2000 (Schneider, 1992). Today, approximately two-thirds of U.S. metro area residents are in suburbs.

10.7 EDGE CITIES

Population decentralization from the core to the periphery began early in the last century. It has a much longer history than the same process for industry and businesses. As described earlier, the process has occurred in waves; the first of which was residential dispersion. Commercial and industrial activities decentralizing in large volume in the decades after World War II represent the second wave, and the growth of offices in the suburbs in the last quarter of the 20th century represents the third wave (Hartshorn & Muller, 1989). Office clusters in suburbs have been called a variety of names, including suburban employment center, suburban nucleation, suburban downtown, and edge city.

The term that seems to have become the most widely accepted is *edge city*, coined by Garreau (1991). A

Washington Post journalist, Garreau was perhaps better at popularizing the suburban employment center concept than his academic contemporaries. Garreau first introduced his ideas through a series of news articles and then with the publication of his book *Edge City* in 1991.

An *edge city* (Garreau, 1991) has the following characteristics:

- Five million square feet or more of leasable office space.
- 600,000 square feet or more of leasable retail space.
- More jobs than bedrooms.
- Perceived by the population as one place.
- Not considered part of a "city" as recently as 30 years ago.

Garreau also published lists of edge cities for many urban areas around the country. For example, his list for Phoenix included Uptown/Central Avenue, the Camelback Corridor, Scottsdale, and Tempe/I-10 (see Figure 10.17). Scottsdale is shown in Figure 10.18. According to Garreau, an edge city looks and feels like a new city.

In this text, edge city is defined as a center of employment on the outskirts of the urban area, often in the suburbs, having office and retail space and perceived as one place, usually located at a highway interchange (based on Garreau, 1991). Edge cities take on many shapes and sizes, sometimes forming linear series or corridors, for example, the one outside Chicago from Schaumburg through Hoffman Estates along the Northwest Tollway outside Chicago. This remarkable corridor of edge cities includes O'Hare Airport, Woodfield Mall, and many office parks. Its status was elevated in 1991 when Sears moved its corporate headquarters and 5,000 jobs to Hoffman Estates from the company's former headquarters in the 110-story Sears Tower in downtown Chicago.

Many people are attracted to edge cities because of their convenience, but also because there is much to do in an edge city. Besides being a place where people work, edge cities provide many shopping and entertainment opportunities. In an edge city, regional malls might be intermixed with commercial corridors and upscale hotels, often present for out-of-town visitors who are on business in the edge city or even for those who live in the region who are looking for a weekend escape. Many of these characteristics sound much like the ones ascribed to big downtowns because in fact edge cities are providing something of a downtown atmosphere.

An earlier investigation of the processes underway in creating the edge cities is the multiple nuclei model. As we saw in Chapter 4, this model had growth and development occurring around multiple centers of economic activity in a metropolitan area, rather than a single one. The classic hypothetical representation showed a

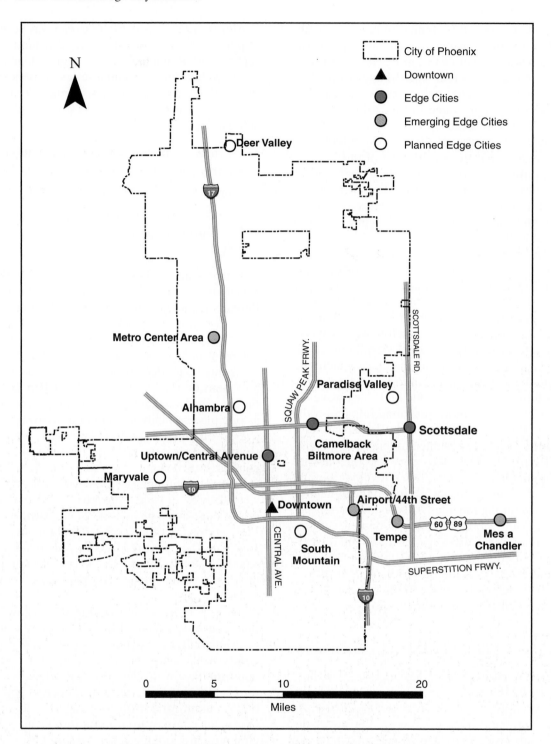

FIGURE 10.17   Edge cities in the Phoenix metropolitan area. Source: Garreau, J., 1991. *Edge City: Life on the New Frontier.* New York, NY: Doubleday. Map of Phoenix p. 181.

metropolis with a CBD, a business center(s), and industrial suburb(s), all this observed in the 1940s (Harris & Ullman, 1945). Although the type of outlying center envisioned by early researchers was different, the idea of an outlying business activity center is the same.

Urban geographers in the 1970s, 1980s, and 1990s were very active in monitoring both the evolving forms and impacts of these outlying activity centers. Baerwald (1978) provided precise delimitation of an evolving edge city outside Minneapolis through the use of aerial photographs

Their cutoff to identify an employment center was a zone with at least 10 jobs per acre, i.e. 6,400 jobs per square mile (Giuliano & Small, 1991).

Another useful measure to delimit edge cities is the ratio of employment to resident workers (E/R ratio), reflecting the balance of workers and jobs. The ratio is defined as follows:

$$\text{E/R ratio} = \frac{number\_of\_workers\_working\_in\_area}{number\_of\_workers\_residing\_in\_area}$$

The ratio varies depending on the net commuting at the local level. An E/R of 1.0 or greater implies that a metropolitan zone has more jobs than it has resident workers, in other words, a net commuter inflow into the zone. An E/R below 1.0 indicates that the metropolitan zone has fewer jobs than resident workers, that is, net outflow of workers from the area.

Edge cities can be identified in terms of places with high E/R ratios, which correspond to the third and fourth parts of Garreau's (1991) definition. In a study of the Los Angeles combined metropolitan statistical area (CMSA), employment centers were defined on the basis of census tracts with an E/R ratio above 1.25 (Forstall & Greene, 1997). A map showing E/R ratio values (see Figure 10.19) reveals some high E/R tracts (E/R over 5) located on the metropolitan periphery. These tracts are identified as constituting edge cities by the limited criterion of this measure. An advantage of the E/R ratio is that it emphasizes in-commuting to peripheral job centers. It is a good measure to sense important job centers like airports having a lot of land that are often missed with the job-density measure. However, it does not include parts 1, 2, and 5 of Garreau's definition, because it does not include leasable office and retail space or lack of prior perception of the job center as "city." Those and other criteria could be added in a more intensive study.

Using a different approach, Fujii and Hartshorn (1995) isolated characteristics normally associated with a downtown area to delineate suburban employment centers in Atlanta. The importance of all these studies is to monitor the trend of the decentralization of employment from the core to the edges of metropolitan areas. However, comparisons from one city to another are difficult to make because the studies employ a variety of methods to define suburban employment centers.

Nevertheless, two studies allow us to compare two metropolitan areas, Los Angeles (Forstall & Greene, 1997) and Chicago (Greene & Forstall, 2001). In these studies, the same data sources and methods were used to define 120 job centers for Los Angeles and 121 for Chicago (Figure 10.20). Consider, for instance, the important question of how decentralized employment is in these two metropolitan areas. One part of the answer is that the two downtowns are still the largest employment centers

FIGURE 10.18   Scottsdale, Arizona: Fashion Square Mall. Source: Alan Keohane © Dorling Kindersley.

and other data. Muller (1981) conducted a broad and comprehensive analysis of the trends of suburban nucleation. Erickson (1986) modeled the process of multinucleation and found, among other things, that its spatial distribution was becoming more random. Cervero (1989) examined a number of traffic-congestion-related issues for the 50 largest suburban employment centers of the time. Knox (1991) linked the trend of increasing flexibility in the spatial organization of economic production to these emerging edge cities using Washington, DC, as a case study.

Subsequent to the appearance of the edge city terminology, studies have attempted to develop better empirical methods for defining and comparing edge city growth through time (Forstall & Greene, 1997; Fujii & Hartshorn, 1995; Giuliano & Small, 1991).

Giuliano and Small (1991) delimited suburban employment centers in Los Angeles using a job-density criterion:

$$\text{Job density} = \frac{number\_of\_jobs\_in\_area}{square\_miles\_of\_area}$$

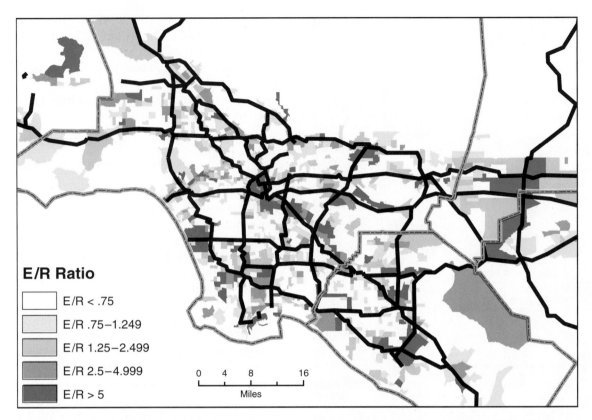

FIGURE 10.19   Employment/Residence (E/R) ratios for Los Angeles relative to Freeways, 1990. Source: Forstall, R.L. and R.P. Greene, 1997. Defining Job Concentrations: The Los Angeles Case. *Urban Geography*, Vol. 18, 705–739. Fig 3b p. 717.

in their respective metropolitan areas, although Chicago's downtown is much bigger (see Figure 10.21). A second part of the answer is that the two areas differ in the proportion of the region's workers that work in downtown. For Los Angeles, the proportion of the region's workers working in the traditional downtown was only 5.6 percent, whereas downtown Chicago included 14.8 percent of its region's workers. Alternatively, the extent of traditional downtown dominance is seen by comparing the gap between the downtown and the next largest employment center. For Chicago, it was almost 500,000 workers, but for Los Angeles only 82,000 workers. Although both metropolitan areas are polycentric (i.e., having multiple downtowns), these studies show that jobs are more dispersed among the multiple downtowns in Los Angeles than they are in Chicago.

There are many implications of the growing number of edge cities. Edge cities contribute to employment opportunities both for employers and employees. Employers can locate enterprises at locations with less clogged transportation and often better logistical routings. In the Internet era, communications at a peripheral location can often function as well as anywhere else in the metropolis. Usually, rental prices are lower in edge cities than the

downtown. Commuting to an edge city might reduce commuting time compared to traveling to the downtown, although a study in the Netherlands found commute times higher in polycentric systems (Schwanen, Dieleman, & Dijst, 2003).

The environmental quality of work life is likely to be better in the edge city than in the center because pollution sources and impacts, loss of biota, and disruption to natural cycles tend to be higher in the city center. Other working and living stresses might be reduced, such as noise and crime. The weaknesses of edge cities relate to the sophistication and complexity of business and society. The range of edge cities services is reduced versus the CBD. The cultural and educational centers of the metropolis are less likely to be close. These pluses and minuses are in flux over time, because the composition of metropolitan centers and suburbs is changing dynamically. Also, there is significant variety in edge cities, so it is hard to generalize, and practical costs and benefits might relate more to a particular edge city relative to its particular metro area.

Edge cities play an important role in the future direction and growth of sprawl, as they represent competing employment hubs to the long-established downtowns of metropolitan areas. As edge cities continue to grow and

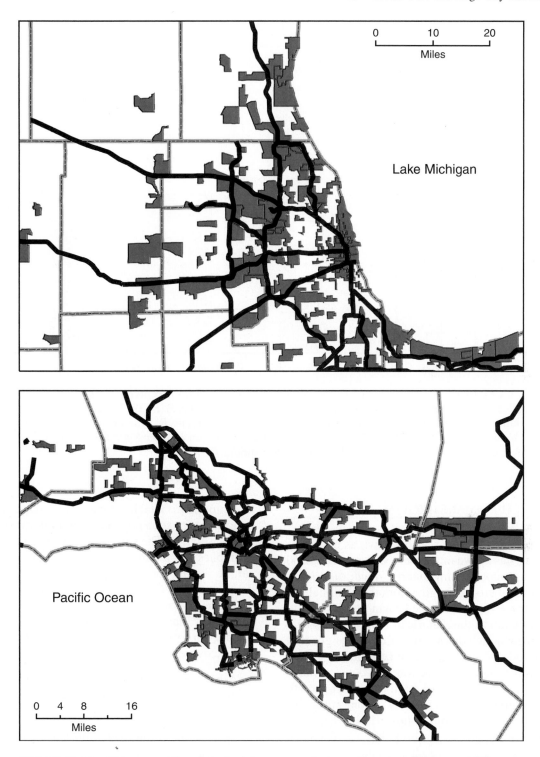

FIGURE 10.20   Chicago and Los Angeles job centers relative to Interstate Highways, 1990.

expand, the urban–rural fringe has the potential for expanding much further outward than it did in previous cycles of growth, when the downtown dominated. This is occurring as an individual figures his or her acceptable commuting range not to the downtown, but to an edge city where he or she has a job. The last section looks at "smart growth" and one example of the government planning consequences of urban sprawl and edge cities, namely, Portland, Oregon's growth boundary. Other planning consequences are discussed in Chapter 12.

Job Center Sizes

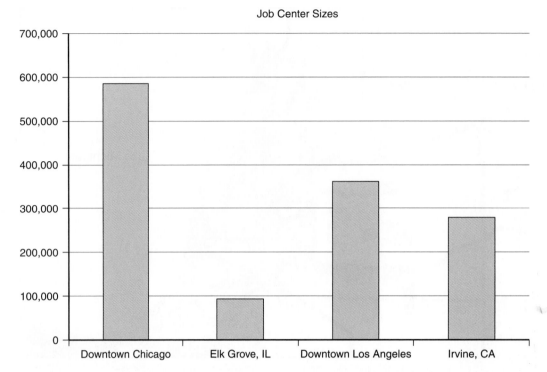

FIGURE 10.21   Relative sizes of largest job center to next largest job center, Chicago and Los Angeles metropolitan areas, 1990. Source: Forstall, R. L., and Greene, R. P. (1997) "Defining job concentrations: The Los Angeles case." *Urban Geography* 18 and Greene, R. P., and Forstall, R. L. (2001) "Urban expansion of Chicago: The late post-suburban period," in *Hamburg and Its Sister Cities*. ed. by J. Lafrenz. Hamburg, Germany: University of Hamburg:705–739.

10.8 SMART GROWTH

The smart growth movement, also known as neotraditional planning, has grown immensely, with more than 60 public interest groups forming *Smart Growth America,* a diverse coalition of groups ranging from the National Association of Home Builders to the Urban Land Institute and the Sierra Club to the American Planning Association. Smart growth includes many planning designs that are contrary to how most suburban communities have developed, such as building more compact developments that incorporate mixed uses. In addition, smart growth promotes development that is friendly to commuters, including the provision of public transportation access. It often includes architectural features that are absent in many suburbs, such as front porches to encourage social interaction and gridiron streets as opposed to the more common cul-de-sac design. Smart-growth initiatives often include the protection of agricultural and environmentally sensitive lands by setting them aside as open space.

Smart-growth initiatives also address the sociological underpinnings of community that were often ignored by postwar suburbia (Whyte, 1956). The spatial layout of homes, parking, yards, and common spaces in a community is vital in promoting social contacts, helping to account for such key factors as patterns of friendships versus isolation. The smart-growth concept of sustainability implies that it is important to develop communities that can be livable for future generations without seriously damaging an area's ecosystems and natural resources.

To help communities achieve smart growth, the American Planning Association (2002) has developed certain core principles, including five that relate to the topics covered earlier:

- A balanced, multimodal transportation system that plans for increased transportation choice.
- A regional view of community.
- Efficient use of land and infrastructure.
- Central-city vitality.
- A greater mix of uses and housing choices in neighborhoods and communities focused around human-scale, mixed-use centers accessible by multiple transportation modes.

The transportation principle emphasizes pedestrian-friendly development and provision of transportation alternatives to the car, including public transit and bicycles.

FIGURE 10.22   Pedestrian-friendly smart growth in Seattle.

A pedestrian-friendly example of Seattle's smart growth is seen in Figure 10.22. The regional view of the community cannot ignore municipal boundaries, but it attempts to integrate community decision making through the regional planning process. Regarding land use, smart growth encourages high-density development, infill development, redevelopment, and the reuse of existing buildings. Central-city vitality is fostered by luring new development to older neighborhoods. On the fringe, smart growth encourages mixed-use development where schools, shopping, housing, community facilities, and jobs are all linked.

10.9 PORTLAND'S GROWTH BOUNDARY

Portland, Oregon, has received much attention in recent years because of its many smart-growth initiatives. Portland established an urban growth boundary (UGB) in 1985 to protect adjacent, open lands from urban sprawl. All of Oregon's cities are surrounded by UGBs. The UGB is a line drawn on planning maps, delimiting the areas within which the city expects to grow (see Figure 10.23). In addition the UGB legislation established a Portland regional agency, Metro, which oversees the UGB (Metro,

2003). State law requires that the UGB include enough land for 20 years of anticipated residential development, so the line is subject to review and has been moved outward over time, a feature that has resulted in some criticism from interests opposed to growth, who would like to freeze the line's position.

Nevertheless, Portland's UGB has achieved much success in containing sprawl and making Portland a more efficient user of land (Porter, 1997). Metro's responsibilities include regional urban long-range planning; future vision for the metro area; transportation planning; operation of a solid waste disposal system, the Oregon Zoo, the Oregon Convention Center, and other trade and visitor buildings; management of parks and open spaces; fish and wildlife protection; natural disaster planning and response; and regional GIS and databases (Metro, 2003). Many of these include economic incentives that encourage growth to occur within the UGBs (Metro, 2003). The tough growth control measures of Oregon, including UGBs had a setback from a November 2004 ballot initiative that allowed property owners who can prove that UGBs have hurt their investment to seek compensation or an exemption from the rules. The measure is too recent to assess its long-term impact on the success of UGBs, but it caught the Oregon urban planning community by surprise.

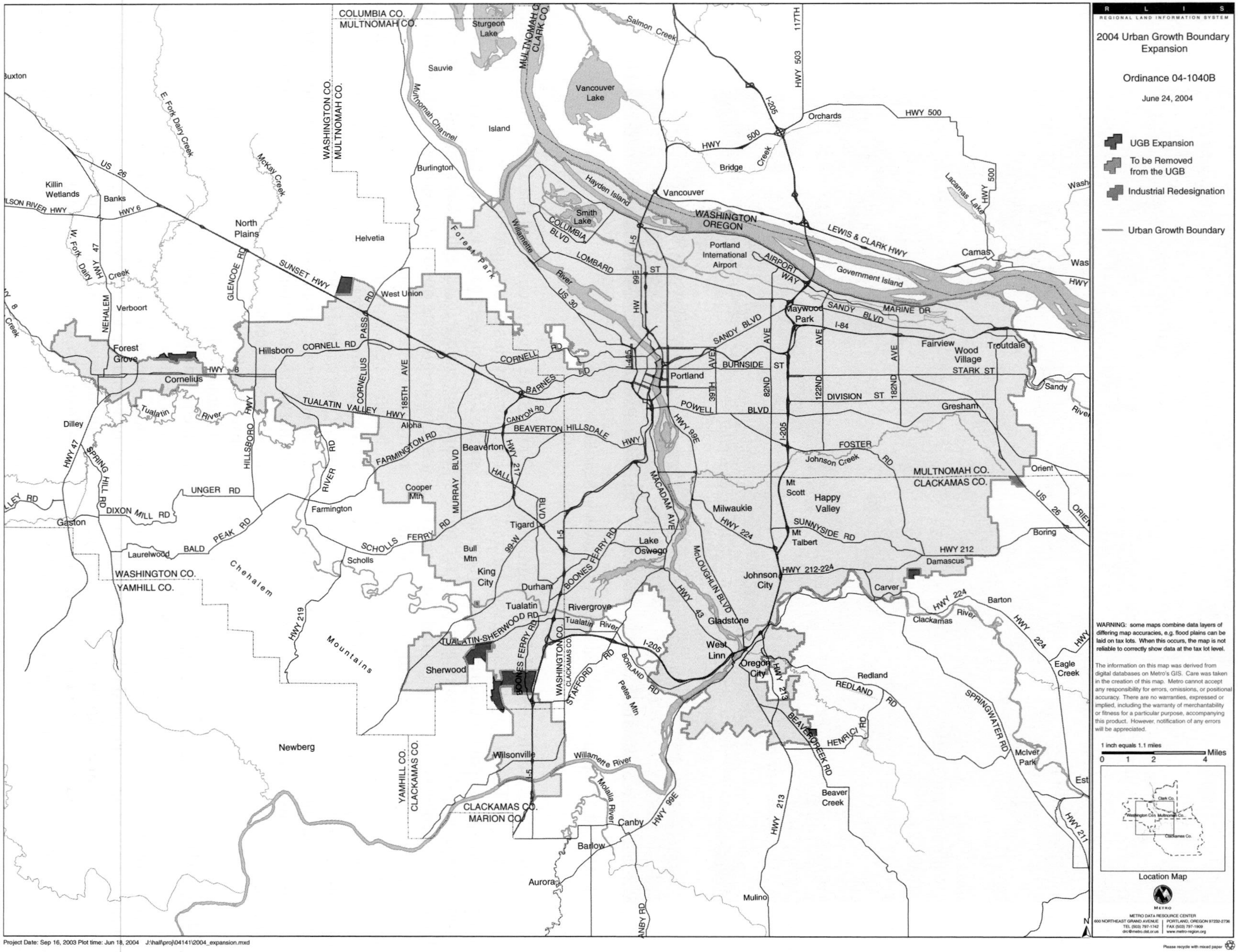

FIGURE 10.23 Portland's urban growth boundary. Source: Metro Data Resource Center, Portland, Oregon (*www.metro-region.org*).

376

SUMMARY

This chapter has covered the key topics in urban geography of suburban growth, urban sprawl, edge cities, and smart growth. By 1970, the number of Americans living in suburbs outnumbered those living in central cities. In the 20th century, central cities steadily declined as a percentage of metro areas, whereas suburbs grew to have a two-thirds share of the overall metropolitan population. One consequence of this redistribution for some central cities is underutilization or abandonment of urban infrastructure built for high population densities. For suburbs, the growth has been largely characterized by urban sprawl, with mainly low-density, single-use development that has strained urban transportation systems. Employment has also been redistributing from the CBD to edge cities, whose locations often are convenient to highway interchanges, thus providing excellent vehicular transport access. Edge cities have expanded the reasonable commuting reaches of metropolitan areas because they lie between the urban fringe and the downtown. Smart-growth policies that are designed to strengthen the central city and make development more land-efficient in the suburbs are being more widely applied throughout the United States. Portland, Oregon, has served as a model metropolitan area for such smart-growth initiatives.

REFERENCES

Acevedo, W., and Buchanan, J. (2000) "Defining the temporal and geographic limits for an urban mapping study." EROS Data Center Web site, *www.edcwww2.cr.usgs.gov*.

Adams, J. S. (1970) "Residential structure of midwestern cities." *Annals of the Association of American Geographers* 60:37–62.

Alonso, W. (1964) *Location and Land Use: Toward a General Theory of Land Rent*. Cambridge, MA: Harvard University Press.

American Planning Association. (2002) *Policy Guide on Smart Growth*. Chicago, IL: Author. Available at *http://www.planning.org/policyguides/smartgrowth.htm*.

Baerwald, T. J. (1978) "The emergence of a new downtown." *Geographical Review* 68:308–318.

Borchert, J. R. (1991) "Futures of American cities," in *Our Changing Cities*, ed. by J. F. Hart, pp. 218–250. Baltimore, MD: Johns Hopkins Press.

Brennan, J., and Hill, E. W. (1999, November) "Where are the jobs: Cities, suburbs, and the competition for employment." Center on Urban & Metropolitan Policy, The Brookings Institute, Survey Series.

Cervero, R. (1989) *America's Suburban Centers: The Land Use Transportation Link*. Boston, MA: Unwin Hyman.

Clawson, M., and Hall, P. (1973) *Planning and Urban Growth: An Anglo-American Comparison*. Baltimore, MA: Johns Hopkins University Press.

Downs, A. (1992) *Stuck in Traffic: Coping with Peak-Hour Traffic Congestion*. Washington, DC: The Brookings Institute.

Erickson, R. A. (1986) "Multinucleation in metropolitan economies." *Annals of the Association of American Geographers* 76:331–346.

Ewing, R. (1997) "Is Los Angeles-style sprawl desirable?" *Journal of the American Planning Association* 63:107–126.

Forstall, R. L., and Greene, R. P. (1997) "Defining job concentrations: The Los Angeles case." *Urban Geography* 18:705–739.

Forsyth, A. (1999) *Constructing Suburbs: Competing Voices in a Debate over Urban Growth*. Amsterdam: Gordon & Breach.

Fujii, T., and Hartshorn, T. A. (1995) "The changing metropolitan structure of Atlanta, Georgia: Locations of functions and regional structure in a multinucleated urban area." *Urban Geography* 16:680–707.

Garreau, J. (1991) *Edge City: Life on the New Frontier*. New York: Doubleday.

General Accounting Office (1999) "Extent of federal influence on "urban sprawl" is unclear." GAO/RCED-99–87. Washington, DC: Author.

Giuliano, G., and Small, K. A. (1991) "Subcenters in the Los Angeles region." *Regional Science and Urban Economics* 21:163–182.

Gordon, P., and Richardson, H. W. (2000) "Defending suburban sprawl." *The Public Interest* Spring:65–128.

Greene, R. P. (1997) "The farmland conversion process in a polynucleated metropolis." *Landscape and Urban Planning* 36:291–300.

Greene, R. P., and Forstall, R. L. (2001) "Urban expansion of Chicago: The late post-suburban period," in *Hamburg and Its Sister Cities*, ed. by J. Lafrenz, pp. 483–529. Hamburg, Germany: University of Hamburg.

Greene, R. P., and Stager, J. (2001) "Rangeland to cropland conversions as replacement land for prime farmland lost to urban development." *The Social Science Journal* 38:543–555.

Hanchett, T. W. (2000) "Financing suburbia: Prudential insurance and the post World War II transformation of the American city." *Journal of Urban History* 26:312–328.

Harris, C. D., and Ullman, E. L. (1945) "The nature of cities." *Annals of the American Academy of Political and Social Science* 242:7–17.

Hart, J. F. (1991) "The perimetropolitan bow wave." *Geographical Review* 81:35–51.

Hart, J. F. (2001) "Half a century of cropland change." *Geographical Review* 91:525–543.

Hartshorn, T. A., and Muller, P. O. (1989) "Suburban downtowns and the transformation of Atlanta's business landscape." *Urban Geography* 10:375–395.

Heimlich, R. E., and Anderson, W. D. (2001) *Development at the Urban Fringe and Beyond: Impacts on Agriculture and Rural Land*. Agricultural Economic Report No. 803. Washington, DC: U.S. Department of Agriculture, Economic Research Service.

Hoffman, A. V. (1999) *Housing Heats Up: Home Building Patterns in Metropolitan Areas*. Harvard University Joint Center for Housing Studies, The Brookings Institute, December.

Horner, M. W. (2004) "Spatial dimensions of urban commuting: A review of major issues and their implications for future geographic research." *The Professional Geographer* 56:160–173.

Jackson, K. (1985) *Crabgrass Frontier*. New York: Oxford University Press.

Knox, P. L. (1991) "The restless urban landscape: Economic and sociocultural change and the transformation of metropolitan Washington, D.C." *Annals of the Association of American Geographers* 81:181–209.

Kriz, M. (1999) "The politics of sprawl." *National Journal* 31:332.

Lopez, R., and Hynes, H. P. (2003) "Sprawl in the 1990s: Measurement, distribution, and trends." *Urban Affairs Review* 38:325–355.

Metro. (2003) *Metro of Portland, Oregon.* Available at *http://www.metro-region.org.*

Mieszkowski, P., and Mills, E. (1993) "The causes of metropolitan suburbanization." *Journal of Economic Perspectives* 7(3):135–147.

Muller, P. O. (1981) *Contemporary Suburban America*. Englewood Cliffs, NJ: Prentice Hall.

National Research Council. (1995) *Expanding Metropolitan Highways: Implications for Air Quality and Energy Use.* Transportation Research Board, Special Report 245.

Platt, R. H. (1991) *Land Use Control: Geography, Law, and Public Policy*. Englewood Cliffs, NJ: Prentice Hall.

Porter, D. (1997) *Managing Growth in America's Communities*. Washington, DC: Island Press.

Powell, J. A. (1998) "Race and space: What really drives metropolitan growth." *Brookings Review* 16:20–23.

Randall, G. C. (2000) *America's Original GI Town: Park Forest, Illinois*. Baltimore, MD: The Johns Hopkins University Press.

Richardson, H. W., and Gordon P. (2004) "US population and employment trends and sprawl issues," in *Urban Sprawl in Western Europe and the United States*, ed. by H. W. Richardson and C. H. C. Bae, pp. 217–235. Burlington, VT: Ashgate.

Riebsame, W. (1997) *Atlas of the New West: Portrait of a Changing Region.* New York: Norton.

Risse, E. M., and Heyman, I. M. (2000) *The Shape of the Future*. Fairfax, VA: SYNERGY/planning, Inc.

Schneider, W. (1992) "The suburban century begins." *The Atlantic Monthly* July:33–44.

Schwanen, T., Dieleman, F. M., and Dijist, M. (2003) "Car use in Netherlands daily urban systems: Does polycentrism result in lower commute times?" *Urban Geography* 24:410–430.

TDM Encyclopedia. (2003) "Transit oriented development: Using public transit to create more accessible and livable neighborhoods." British Columbia, Canada: Victoria Transport Policy Institute. Available at *http://www.vtpi.org.*

Theobald, D. M. (2001) "Land-use dynamics beyond the American urban fringe." *Geographical Review* 91:544–564.

U.S. Department of Agriculture. (2003) *2001 Annual National Resources Inventory*. Washington, DC: United States Department of Agriculture. Available 2004 at *http://www.nrcs.usda.gov/technical/land/nri01/nri01dev.html.*

Vance, J. E. Jr. (1991) "Human mobility and the shaping of cities," in *Our Changing Cities*, ed. by J. F. Hart, pp. 67–85. Baltimore, MD: Johns Hopkins Press.

Vermont Forum on Sprawl. (2003) *What Is Sprawl?* Burlington, VT: Author. Available at *http://www.vtsprawl.org.*

Vesterby, M., and Heimlich, R. E. (1991) "Land use and demographic change: Results from fast growing counties." *Land Economics* 67:279–291.

Warner, S. B. (1962) *Streetcar Suburbs: The Process of Growth in Boston, 1870–1900*. Cambridge, MA: Harvard University Press.

Wheeler, J. O. (1998) *Economic Geography (3rd Edition)*. New York: Wiley.

Whyte, W. H. (1956) *The Organization Man*. Garden City, NY: Doubleday.

EXERCISE

Exercise Description

The goal is to measure the ever changing and outward expansion of the Chicago urban fringe in the period from 1960 to 2000.

Course Concepts Presented

Urban fringe and urban sprawl.

GIS Concepts Utilized

GIS overlay procedures.

Skills Required

ArcGIS Version 9.x

GIS Platform

ArcGIS Version 9.x

Estimated Time Required

1 hour

Exercise CD-ROM Location

ArcGIS_9.x\Chapter_10\

Filenames Required

Ch_dwntwn = point for location of downtown
Cnty13 = county boundaries for Chicago metro
Ch_60_80 = census tracts with positive population increase between 1960 and 1980 and population greater than 0 in base year (some areas not tracted in base year)
Ch_80_00 = census tracts with positive population increase between 1980 and 2000 and population greater than 0 in base year (some areas not tracted in base year)

Background Information

The Peri-Metropolitan Bow Wave

In this project, we accept the idea of the urban fringe as a peri-metropolitan bow wave:

> The bow wave is a standing wave that always remains immediately in front of the bow of a ship moving through water. The urban–rural fringe, then, is the bow wave of the built-up area of the city, the zone of intensively cultivated, high-priced agricultural land that remains immediately in front of the expanding urban edge. (Hart, 1991, p. 36)

You are asked to create 10-mile rings outward from the center of downtown Chicago and to measure each ring's proportion of population living in growing census tracts for two time periods, 1960 to 1980 and 1980 to 2000. You are then asked to determine whether or not the results confirm that the Chicago urban–rural fringe is analogous to a bow wave.

Readings

(1) Greene, R. P. (1997) "The farmland conversion process in a polynucleated metropolis." *Landscape and Urban Planning* 36:291–300.
(2) Greene, R. P., and Pick, J. B. *Exploring the Urban Community: A GIS Approach,* Chapter 10.
(3) Hart, J. F. (1991) "The perimetropolitan bow wave." *Geographical Review* 81:35–51.

Exercise Procedure Flowchart

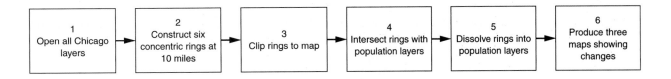

Exercise Procedure

Part 1: Location of Peri-Metropolitan Bow Wave and Concentric Ring Analysis of Sprawl: 1960 to 1980

1. Start ArcMap with Start Using ArcMap With A New Empty Map selected. Click OK.

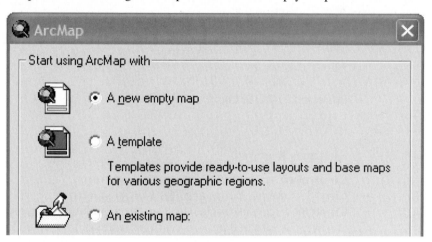

2. Click Add Data and navigate to the exercise data folder.

Note: *If you do not see the folder where you placed the exercise data, you will need to make a new connection to the root directory that holds the data.*

3. Add Ch_dwntwn, Cnty13, Ch_60_80, and Ch_80_00 from the Chapter 10 folder. You can add all four at once by clicking one of the file names, clicking the others while holding down the Ctrl key on the keyboard, and clicking Add.

4. Click ArcToolBox and select Analysis Tools, Proximity, and double click Multiple Ring Buffer.

5. Create six 10-mile rings around downtown Chicago. In the Multiple Ring Buffer window, enter Ch_dwntwn.shp (full path, which might be different on your machine) as the Input Features and Ch_rings.shp (full path) as the Output Features class. Note that Ch_rings.shp does not exist yet, and this step creates it. Enter 10 into the Distances box and click the plus sign to the lower right. Repeat by entering 20 through 60. Enter Miles as the Buffer Unit. Please note you may need to toggle down with the right-hand slide bar to enter this. Click OK. Close the resulting processing box when completed.

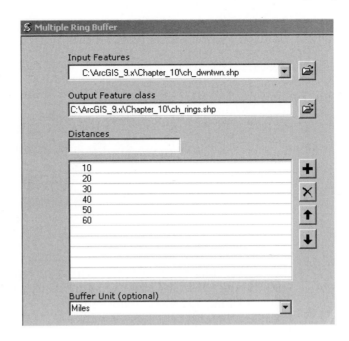

6. When finished, you will have a new layer of rings called ch_rings.

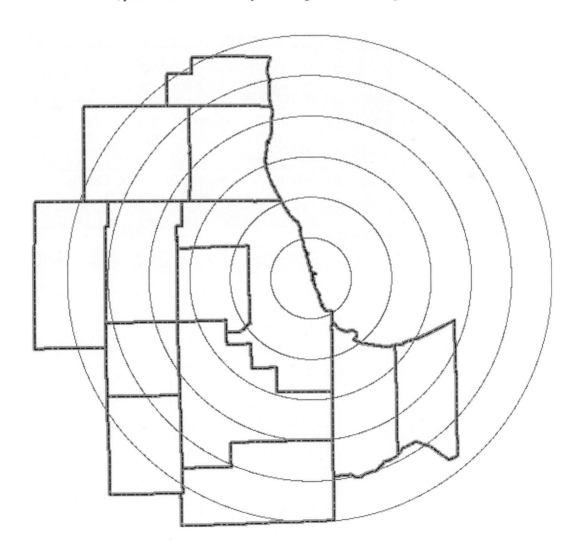

7. The rings extend beyond the study area, so we clip them by the county file. Click ArcToolbox, and select Analysis Tools, Extract, and Clip.

8. In the Clip window, enter Ch_rings.shp as the Input Features, Cnty13.shp as the Clip Features, and Ch_rings_Clip.shp as the Output Features Class to create. Click OK. Close the resulting processing box when completed.

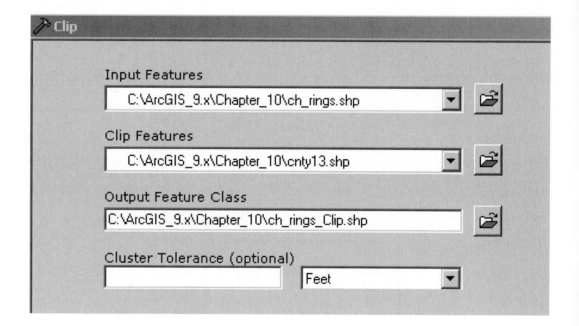

9. When finished, turn off ch_rings and you will see the new layer of rings clipped to the study area boundary.

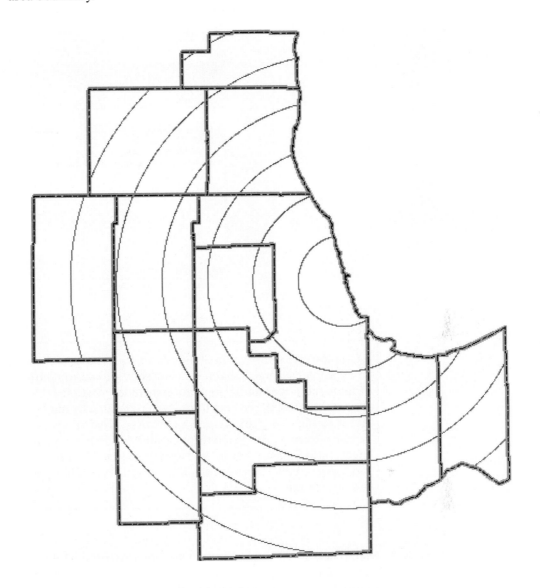

You will now perform a GIS intersection of Ch_60_80 and Ch_rings_clip.

10. Click ArcToolbox, then select Analysis Tools, Overlay, and Intersect.

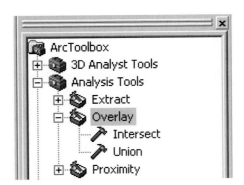

11. For the Input Features, click the folder button to the right and select Ch_rings_Clip.shp. Click the plus sign to the lower right. Repeat for Ch_60_80.shp as the second Input Features. Enter Ch_ring_6080.shp as the Output Feature Class to create. Click OK. Close the resulting processing box when completed.

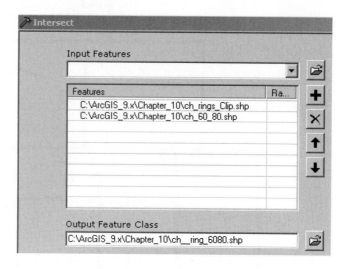

After the GIS intersection of the census tracts layer with ch_rings_clp, the output layer includes census tracts that have been split into multiple tracts. In addition, the new attribute table containing population includes multiple records and population is grossly overcounted. We correct for this overcounting by multiplying population in the final layer by the new_area divided by the orig_area. This treats the tracts as though population is evenly distributed throughout the original tract, an assumption that might not necessarily be true but is necessary to make. To implement this correction you first add new_area to our resulting intersected layer table, and then compute the corrected population.

12. Right-click ch_ring_6080 and select Open Attribute Table. Click Options, and click Add Field, call it new_area as type float with a precision of 16 and a scale of 5.
13. Right-click on new_area and select Calculate Values (answer Yes if you are asked to continue), click Load, and browse to the return_Area function in the Chapter 10 folder.
14. Select the file and click Open. Click OK to perform the calculation.
15. Add a field called final_pop, as float, precision 16, and scale of 0.
16. Right-click final_pop, select Calculate, clear the Advanced check box, and calculate as follows:

$$[chg60\_80] * ([new\_area]/[orig\_area])$$

17. Close the table.
18. Click ArcToolbox, then select Data Management Tools, Generalization, and Dissolve.

19. Click the folder browser button and select Ch_ring_6080.shp as the Input Features layer and Ch6080_dis.shp as the Output Feature Class to create. Select Distance as the dissolve field. You will then scroll to the bottom of the Dissolve dialog box and add final_pop as a field to compute a sum of during the dissolve process. To do this, enter final_pop in the Statistics Field box and click on the plus sign. Click OK. Close the resulting processing box when completed.

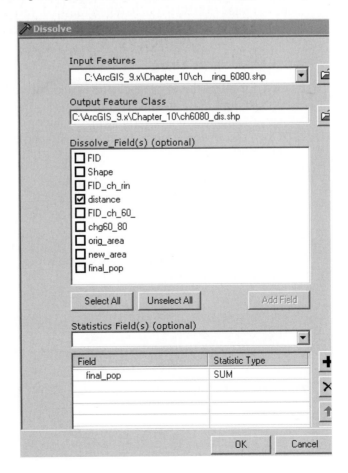

20. When finished, turn off all layers except ch6080_dis and identify each ring (identify icon) and verify that the population increase in growing census tracts by ring is correct. You should have the following results shown in the total column under "1960 to 1980":

| | 1960 to 1980 | | 1980 to 2000 | | Difference |
	Total	Percent	Total	Percent	Change in percent
Ring 1:	132,448	6.45%	303,417	16.99%	+10.54
Ring 2:	505,229	24.58%	136,482	7.64%	−16.94
Ring 3:	915,004	44.52%	529,055	29.63%	−14.89
Ring 4:	360,550	17.54%	513,073	28.74%	+11.20
Ring 5:	124,404	6.05%	220,785	12.36%	+6.31
Ring 6:	17,271	0.84%	82,406	4.61%	+3.77

Note: Percent derived by dividing total growth in ring by total growth in region:

2,054,906 total regional growth between 1960 and 1980.

1,785,218 total regional growth between 1980 and 2000.

Part 2: Location of Peri-Metropolitan Bow Wave and Concentric Ring Analysis of Sprawl: 1980 to 2000

Repeat steps 10 through 20 for the 1980 to 2000 period (use ch_80_00 as an input layer in the intersection step and be careful to give unique file names for the output layers that you create). Change the numerator in the equation in step 16 to [chg80_00]. For step 19, use appropriate file names to refer to 1980 to 2000. When finished, verify that the data in the remainder total fields of the above table under column B (1980 to 2000) and column C (Difference) are correct.

Part 3: Mapping the Change

21. We only need the ch_ring_clip layer and cnty13 layer, so right-click the other layers and select Remove. You should only have two layers remaining, ch_ring_clip and cnty13. Turn off cnty13 and have only ch_ring_clip turned on.
22. Open the attribute table for ch_ring_clip and add these three fields to the table by clicking Options:

> Pct60_80, float, precision 16, scale 2
> Pct80_00, float, precision 16, scale 2
> Chgpct, float, precision 16, scale 2

23. Close the table.
24. Turn on editing tools (right-click gray at top menu bar), click Editor, and click Start Editing. Set ch_ring_clip as the target.

25. Click the Edit tool to the right of the word Editor.

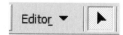

26. Click the innermost ring (Ring 1) to select it.

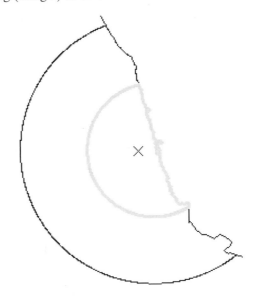

27. Click the Attributes tool on the Editor toolbar.

28. Refer back to the table and enter the percentage values reading across the table for each ring.

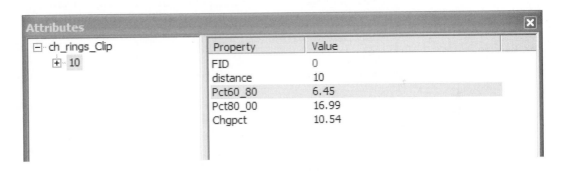

29. Keep the Attributes dialog box open, click the next outer ring, and enter the percentage values reading across the table for Ring 2.
30. Repeat this procedure for the remaining rings.
31. When finished, click Editor, click Stop Editing, and answer Yes to save your edits.
32. Right-click ch_rings_Clip and select Open Attribute Table. Your table should like the following:

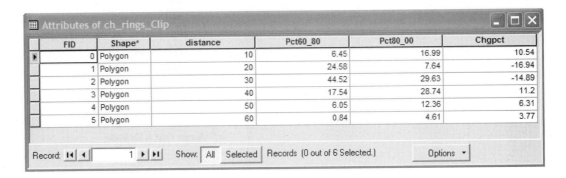

FID	Shape*	distance	Pct60_80	Pct80_00	Chgpct
0	Polygon	10	6.45	16.99	10.54
1	Polygon	20	24.58	7.64	-16.94
2	Polygon	30	44.52	29.63	-14.89
3	Polygon	40	17.54	28.74	11.2
4	Polygon	50	6.05	12.36	6.31
5	Polygon	60	0.84	4.61	3.77

Record: ◄◄ ◄   1   ► ►◄   Show: All Selected   Records (0 out of 6 Selected.)   Options ▾

33. Close the table.
34. Right-click ch_rings_Clip and select Properties. Click the Symbology tab.
35. Select Categories and select Unique Values. In the Value Field drop-down list, select Pct60_80, clear the <All Other Values> check box, click Add All Values, highlight all the values shown, and click OK.

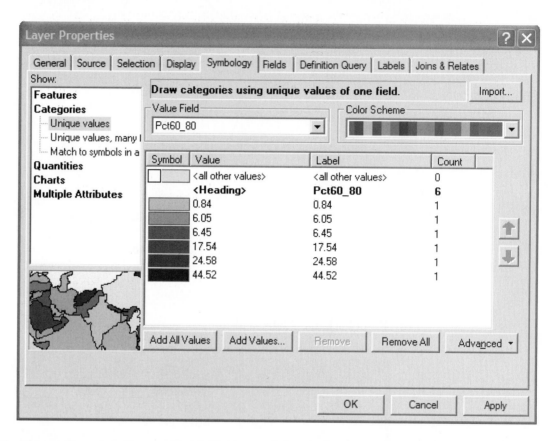

36. Change the colors if you wish by clicking each box individually and selecting colors of your choice (gradations from light to dark are best to correspond to low values ranging to high values). Click Apply and then click OK to see your results.
37. Click Insert on the top menu bar and click Data Frame. Repeat by clicking Insert and clicking Data Frame again. You should have the following:

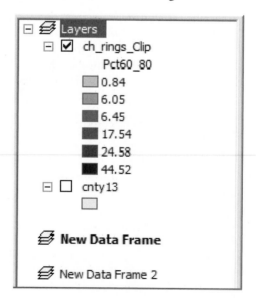

38. Right-click the Layers data frame and select Activate. Now the original data frame is active. Select the ch_rings_Clip layer and hold down the Shift key, then select cnty13. Now that they are both selected, right-click ch_rings_Clip and select Copy.
39. Right-click New Data Frame and select Paste Layer(s).
40. Right-click New Data Frame 2 and select Paste Layer(s).
41. Now all three data frames have the same layers.

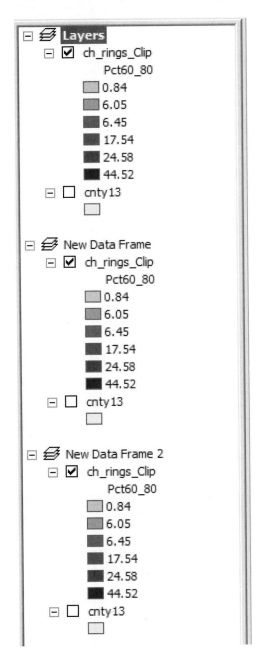

42. Right-click New Data Frame and select Activate.
43. Right-click ch_rings_Clip within New Data Frame and select Properties. In the Properties dialog box, under Fields, select Pct80_00. Click Add All Values, change the colors to be consistent with the Layers data frame, click Apply, and then click OK to see your results.
44. Right-click New Data Frame 2 and select Activate.
45. Right-click ch_rings_Clip within New Data Frame 2 and select Properties. In the Properties dialog box, under Fields, select Chgpct. Click Add All Values, change the colors to be consistent with the previous two data frames, click Apply, and then click OK to see your results.

46. For each of the data frames, right-click cnty13 in the table of contents and select Properties. Click the Symbology tab and then the Symbol box.

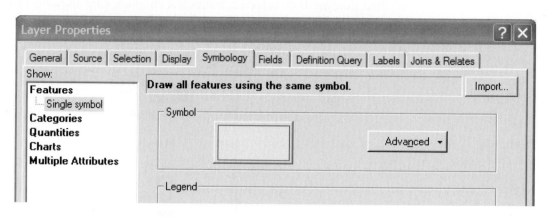

47. In the Symbol Selector dialog box, click the Hollow box, then click Properties.

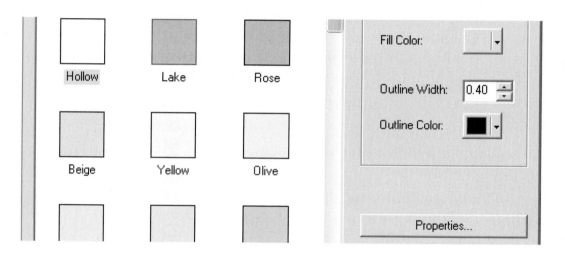

48. In the Symbology Property Editor dialog box, click Outline.

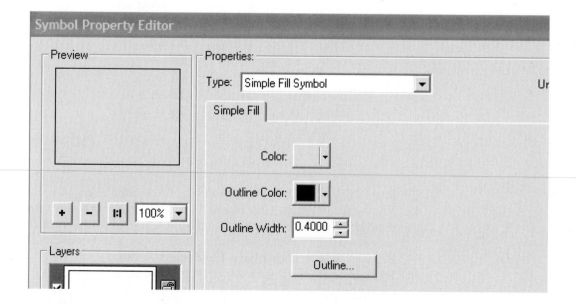

49. In the Symbol Selector dialog box, scroll down and choose the symbol for county boundaries. Click OK on each box as you return to the map.

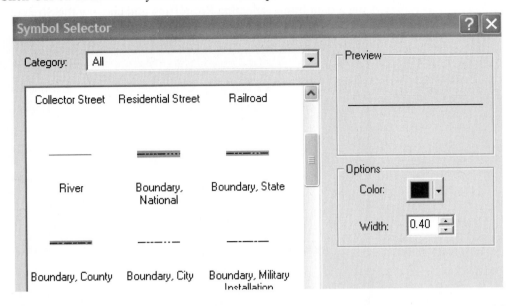

50. Click the cnty13 layer name in the table of contents and drag the layer to the top position above ch_rings_clip. Make sure the cnty13 layer is now turned on.
51. Once all of the data frames have the same format with county boundaries on top and turned on, you can select Layout View from the View menu to compose a map for printing.
52. Within the Layout View you can see multiple data frames at a time. All 3 data frames are there; you just need to reposition them in the following format with associated legend and text. Note that the legends are ordered by the magnitude of values, not by ring order.

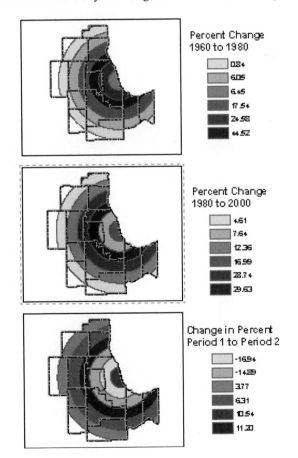

53. To achieve a similar format to the one just shown, you can standardize the *x, y* position of the lower left corner of each data frame and assign a standard height and width by right-clicking a data frame, selecting Properties, and clicking the Size And Position tab.

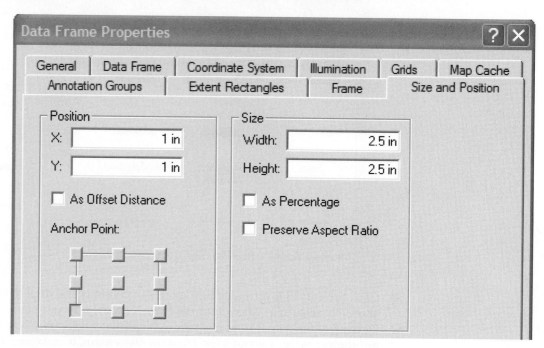

54. In the preceding example for New Data Frame 2, the size included a width of 2.5 inches and a height of 2.5 inches, the *x* position was set to 1 inch, and the *y* position to 1 inch. Note that the anchor point that is depressed is the lower left; this is the *x* and *y* position you are specifying on that corner of the page. The New Data Frame should include the same width and height as well as the same *x* position. However, its *y* position should be set to 4. The Layers data frame *y* position should be set to 7.

55. Make the Layers data frame active and click Insert to insert a legend. In the first dialog box, make sure that the only legend item is ch_rings_clip. To remove cnty13, select it in the right column and use the arrow to move it to the left column.

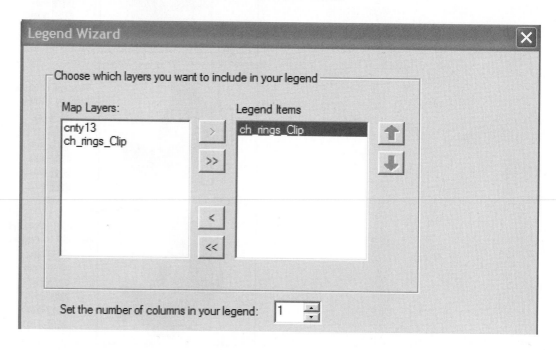

56. Click Next, erase the word Legend, and replace it with:

    Percent Change
    1960 to 1980

57. Click Next in the next several dialog boxes and click Finish.
58. Now double-click the selected legend. Click the Items tab, and in the right column double-click ch_rings_Clip, click Properties, click the General tab, and clear the Show Layer Name and Show Heading check boxes. Click Apply and then click OK. Click OK as you exit back. Reposition the legend to a suitable spot.
59. Repeat 55 through 58 for New Data Frame. The legend should read:

    Percent Change
    1980 to 2000

60. Repeat 55 through 58 for the Data Frame 2. The legend should read:

    Change in Percent
    Period 1 to Period 2

61. Click File and select Print to print the final map (alternatively you could click File and select Export Map to export it to one of your favorite image formats).
62. Exit ArcMap.

Part 4: Essay

In your essay, review the readings, including the relevant Chapter 10 sections on sprawl and the urban fringe. Discuss the expansion of Chicago 1960 to 2000. What were the major ring trends over the forty years and why did they occur? Discuss the strengths and weaknesses of the method employed to monitor outward expansion of the urban fringe and sprawl for the Chicago region. Explain how the three maps can be used in support of the hypothesis that the urban fringe is analogous to a bow wave.

11

Environmental Problems

The city comprises parts of the natural and human environment. It contains biota and ecosystems; consumes energy, water, and food; is a part of man-made food chains; emits wastes; and faces challenges to manage these flows and processes. This chapter focuses on the role of environmental problems in the city. That is a large role because worldwide almost half of the population lives in urban areas, and a higher proportion of man-made environmental problems occurs in cities.

The role of urban geography is to elucidate the spatial aspects of urban environmental flows, for example, solid-waste locations, transport of pollutants, water supply, and land conversion on the metropolitan fringe. The chapter examines the major concepts and processes of the urban environment. The earth's natural cycles include the carbon cycle, water cycle, and energy flows. It presents the natural ecology concepts of species, habitat, and ecosystem. Humans have intervened in all of these cycles and systems. The important pollution types are examined.

The chapter considers population growth as the fundamental cause of urban environmental problems. Chapter 7 presented population principles for urban geography, emphasizing that urban population is growing on a worldwide basis, both by natural increase and net migration from rural to urban areas. The urban growth is uneven because rural-to-urban migration varies a lot between nations and urban fertility also varies, although generally less than that country's rural fertility. At the periphery of cities, agricultural land is at risk of conversion into urban uses. An example is given of citrus farming versus development in the city of Redlands, California.

Added to the effects of population growth is a long-term trend for per-capita increase in the consumption of energy and products, as well as more damaging environmental impacts from consumption. This can be understood by the model known as IPAT with PC (Kates, 2000).

IPAT is an abbreviation for impacts, population, affluence, and technology (Ehrlich and Holdren, 1972), while PC stands for population and consumption (Kates, 2000). The IPAT/PC model is given by: $I = P \times C/P \times I/C$, where I is impacts on the environment, C is consumption, and P is population. Significantly, the consumption per-capita ratio in this model increased by about five times in the 20th century (Kates, 2000) and is continuing to grow. The ratio of impacts per consumption has also grown. For example, more carbon dioxide per consumer has been released into the atmosphere during the past century. More research is needed to understand the numbers for this equation worldwide and among countries (Kates, 2000).

Water, essential for plant and animal life on earth, follows a water cycle. Water is evaporated and transpired by plants into the atmosphere, where it forms clouds and is precipitated back to earth. Once on the earth's surface, it flows through surface streams and groundwater in the earth back to lakes and oceans where it starts the cycle again. Other precipitation is taken up by plants, photosynthesized and transpired, also starting the cycle again. The city is subject to this water cycle, through which it must satisfy its need for adequate water supply and quality. Many cities, mostly in the developing world, suffer from water scarcity and contamination, as well as from oversupply leading to urban flooding.

Other environmental aspects are urban wastes and their management, energy resources, and air pollution. The "Analyzing an Urban Issue" section examines Mexico City's air pollution, sometimes regarded as the worst in the world.

A key question for urban geography is where, within the cities, the negative environmental impacts are felt. Unfortunately, poor, deprived, and, often, minority populations absorb most of these impacts. This imbalance and the need to rectify it is referred to as *environmental justice.* The complex issues and causes of these inequalities can stem from conscious policy or accidents of history. A case study is presented for the crisis over solid waste disposal in New York City that exemplifies the complexities.

Smart growth and new urbanism are current planning approaches that attempt to plan and develop cities and urban areas more sensibly and in better concert with nature. Successes are best achieved through cooperation between business and government.

11.1 CONCEPTS AND PROCESSES OF THE URBAN ENVIRONMENT

This section presents basic concepts of ecological, population, water, energy, and geochemical cycles that underpin life on earth, including human life in the city. These cycles are modified by the city's high concentrations of economic activity, built infrastructure, and human population. Finally, the complexities of urban environmental processes can be modeled by complex simulation models, which are useful tools.

The earth's energy is derived originally from the sun. The sun's energy that arrives at the earth can be utilized directly by plants for photosynthesis; can be deflected back out of the atmosphere; can serve to warm the planet; or can be captured by solar energy facilities to generate heat and electricity. The other major forms of energy are *fossil fuel energy,* which is energy extracted from decomposed organic compounds stored in the earth such as coal and petroleum; *geothermal energy,* derived from hot water in the earth; *hydro energy,* coming from harnessing the flow of water; and *nuclear energy,* from atomic processes of uranium and other elements and compounds. Most of this energy originally came from solar energy hitting the earth. In the natural environment, energy cycles are essential because all plants and animals require energy. Man-made energy systems vastly increase the amounts of certain forms of energy, such as nuclear and fossil, but bring with them environmental and economic costs.

Energy is measured in watts, which represents the power produced by a current of 1 ampere across a potential difference of 1 volt *or* 1/746 horsepower. A kilowatt is 1,000 watts; a megawatt is 1 million watts; and a gigawatt is 1 billion watts. Worldwide, the total amount of energy consumed by humans on an annual basis in 1998 was 111.7×10^6 gigawatt hours or an average of 0.189 gigawatt hours per capita (United Nations Development Programme, 2000). Of this, about 60 percent is consumed by people in urban areas.

There is a finite amount of water on the earth and in its atmosphere. As seen in Figure 11.1, water moves through a *water cycle.* It leaves the earth and enters the atmosphere through *evaporation* and *transpiration* from photosynthesis occurring in vegetation and falls back to the earth's land and ocean surfaces through *precipitation.* Some water runs off into oceans and other water bodies, and some *infiltrates* into the soil, where it might be absorbed by plants or reach the groundwater, also called the zone of saturation, which stores large amounts in *aquifers* or *reservoirs. Groundwater* is water inside the earth that is supplied by the infiltration of surface water through the soil into underground aquifers or reservoirs. From the groundwater, it comes again to the surface through natural springs and man-made wells. The man-made interventions in this cycle include surface water diversions, dams, flood-control channels, canals, pipelines, and reservoirs, many of them essential for contemporary societies. However, these facilities might damage ecosystems, including leading to urban flooding, climate change, and loss of natural waterways. The latter occurs when the natural courses of rivers and streams are altered through damming and water diversions. New York City's water reservoir in Central Park is shown in Figure 11.2.

Water supply is commonly measured in *acre-feet.* One acre-foot is the amount of water volume in an area of 1 acre by a depth of 1 foot.

Mexico City represents an urban example of the importance of the water cycle. Underlying the metropolitan area is an immense aquifer. Today, the metropolitan area's 18.1 million residents receive 1.6 million acre-feet of water per year, with 67 percent of their water supply coming from the aquifer, 2 percent from surface sources inside the metro area, and 31 percent pumped up to Mexico City from the surrounding Lerma and Cutzamala basins (Ezcurra, Mazari-Hiriart, Pianty, & Guillermo, 1999; Grafton, 2001). The immediate challenge is obtaining enough water supply to keep up with the population growth because the aquifer has a large but limited capacity. Another short-term problem is the hazard of spillage of contaminated wastes that do not receive sewage treatment and are transported out of the city in surface canals (Ezcurra et al., 1999). More underground sewage drainage canals are being built but their leakage is even more dangerous because they leak more deeply in the soil and might endanger the aquifer over time. The serious longer term problem is the potential contamination of the aquifer (Ezcurra et al., 1999; Mazari & Mackay, 1993; Mazari-Hiriart, De la Torre, Mazari Menzer, & Exequiel, 2001). The challenge for the metropolitan governments is to improve the management of water supply and sewage disposal and treatment and to recognize much more than it has the natural water cycle of the Mexico City basin.

There are many chemical cycles, in which chemicals circulate from the atmosphere to the earth's land surface to the subsurface, from which they eventually return

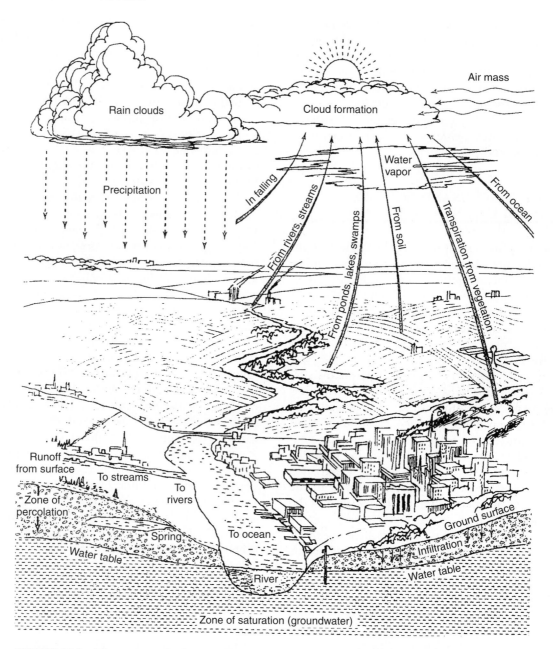

FIGURE 11.1   The water cycle. Source: Chiras, Daniel D. D., John P. Reganold, J. P., and Owen., O. S. (2002). *Natural Resource Conservation: Management for a Sustainable Future*. Upper Saddle Reiver, New Jersey: Prentice Hall.

through underground or surface flows to the ocean, and back to the atmosphere. The most important ones for urban geography are the carbon and nitrogen cycles (see Figures 11.3 and 11.4). In the *carbon cycle,* carbon dioxide, respired from animals and microorganisms, is absorbed during photosynthesis by plants and other organisms in the form of carbon dioxide. When plants die, carbon dioxide is processed as part of the decomposition

and recycled into the soil, where over a long period, some of it might become embedded as fossil fuel. The fossil fuels eventually can be mined for fuel use in factories and transport, which leads to carbon dioxide return to the atmosphere. Other plant remains might end up in ocean sediments or be absorbed by phytoplankton in the ocean.

The nitrogen cycle is shown on the left in Figure 11.4. Nitrogen dioxide in the atmosphere is absorbed by

FIGURE 11.2   Water reservoir in Central Park, New York City. Source: Peter Arnold, Inc.

nitrogen-fixing bacteria in the soil, which produce nitrates. Soil nitrates also stem the actions of denitrifying bacteria on decomposing dead organic matter. From the soil, nitrates can undergo chemical reactions and become atmospheric nitrogen dioxide. Other nitrates are incorporated into fertilizer, as well as being utilized in factory processes, which eventually return nitrogen to the atmosphere. Among the major man-made interventions into these two cycles and others are atmospheric emissions by automotive vehicles and industrial plants of carbon monoxide, nitrogen oxides, and sulfur dioxide. Their contribution to urban air pollution is covered later.

An impact of concern from these cycles is the production of elevated amounts of carbon dioxide, which is implicated as a major cause of global warming. The upward trend in carbon dioxide is seen in Figure 11.5, which plots, over the past 150 years, the world's carbon dioxide concentration and population, and world industrial production over the past 75 years (Genske, 2003). The earth's carbon dioxide concentration grew from 300 parts per million (ppm) in 1850 to 375 ppm in 2000, a 25 percent expansion in a century and a half. It is evident that the carbon dioxide concentration is growing at a rate somewhat

slower than industrial production but faster than population growth. The buildup of carbon dioxide and other greenhouse gases in the high atmosphere serves to retain more of the earth's heat. Figure 11.6 shows carbon dioxide being emitted into the atmosphere by an industrial plant.

Many scientists ascribe this as the cause of global land temperature rise of 1 degree Fahrenheit from 1940 to 2000, as measured by the U.S. National Climatic Data Center (Wright & Nebel, 2002). However, global temperature has fluctuated up and down in long cycles of heating and cooling, so the global temperature rise was also 1 degree Fahrenheit in the more extended period measured from 1855 to 2000 (Wright & Nebel, 2002). The degree of future global warming is subject to debate, but one important series of projections was performed in the U.S. National Assessment, a part of the U.S. Global Change Research Program in 2000. The program commissioned five models from leading climate change centers that projected global temperature increases from 2000 to 2100 ranging between 3 and 8.5 degrees Fahrenheit (U.S. Global Change Research Program, 2000). Even the lowest 21st-century rise would be threefold more than during the past century

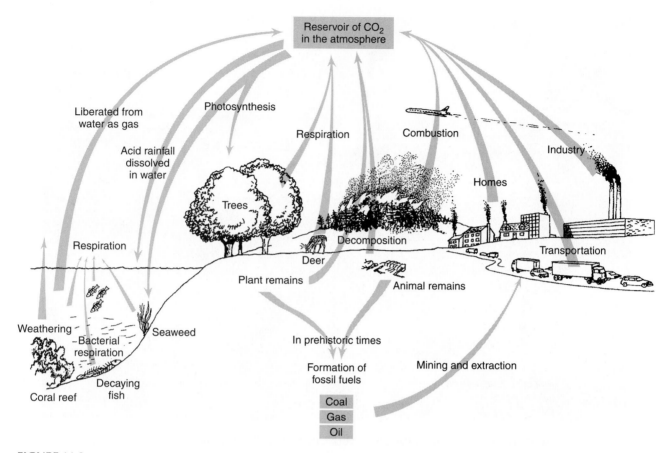

FIGURE 11.3   The carbon cycle. Source: Chiras, Daniel D. D., John P. Reganold, J. P., and Owen., O. S. (2002). *Natural Resource Conservation: Management for a Sustainable Future*. Upper Saddle Reiver, New Jersey: Prentice Hall.

and a half. The potential consequences include partial melting of glaciers and polar ice caps, rising sea levels with intrusions on coastal areas including major coastal port cities, climate changes, and shifts in the habitat ranges for many species (Wright & Nebel, 2002).

An urban example involving the nitrogen and carbon cycles would be the hypothetical addition of a new semiconductor manufacturing plant in Albuquerque, New Mexico. Semiconductor plants consume a lot of energy, so the new plant would add to the demand on the New Mexico power grid, which is powered 75 percent by coal and natural gas plants, the other 25 percent by nuclear plants (PNM, 2003). The new energy requirements would add to carbon dioxide, sulfur dioxide, and nitrogen oxide emissions of the fossil plants, as well as to demand for cooling water from all the plants. The carbon dioxide would contribute marginally to global warming. All three substances would enter the carbon and nitrogen cycles shown in Figures 11.3 and 11.4 and have a wide variety of potential soil, oceanic, and biotic effects. For instance, it would contribute carbon and nitrogen to plants and organisms that would later lead through their decomposition to carbon compounds and nitrates in the soil. The

semiconductor plant would also produce toxic substances that would have to be disposed of in waste sites, so as not to be carried in other cycles not discussed.

Plants and animals are part of biological systems that have long-term natural interactions. All organisms are classified into different *species,* which are defined as groupings of animals, plants, or microorganisms of a single type. Single type implies similarity of appearance and the ability to produce fertile offspring. Estimates are that the earth contains between 3.6 million and 13.6 million species, although only 1.75 million have been identified and classified. Species have developed through the process of evolution, and could be subject to genetic changes over time, as well as extinction. Species occupy particular land or aquatic areas, known as habitats. A habitat depends on food sources, environmental factors, behavior including mating, and other factors. Taken together, groups of plants and animals in a region constitute an ecosystem, which is defined as a grouping of animals, plants, and other organisms interacting with the environment and with each other so that the group of species perpetuates itself. Examples of ecosystems include tropical rainforest, tundra, or desert.

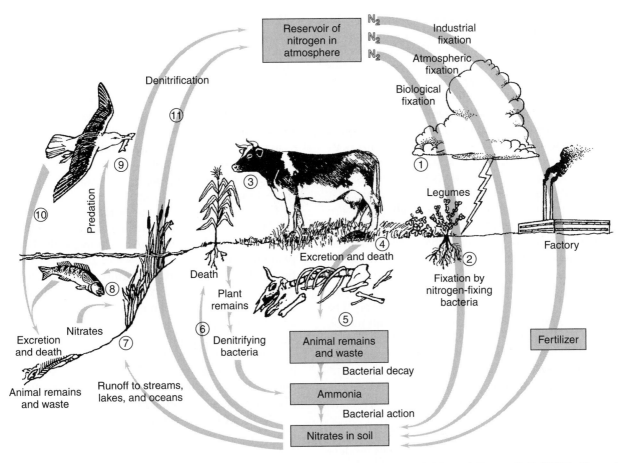

FIGURE 11.4   The nitrogen cycle. Source: Chiras, Daniel D. D., John P. Reganold, J. P., and Owen., O. S. (2002). *Natural Resource Conservation: Management for a Sustainable Future.* Upper Saddle Reiver, New Jersey: Prentice Hall.

The scientific discipline of ecology studies organisms in the environment, systems that govern their interactions, their population growth and decline, and recycling and energetics of biota. The discipline is essential to understanding urban environments. The city also has natural biota and ecosystems—natural cycles and processes, often modified, which support its plant, animal, and microbial life. The city depends on inflows of food and other natural substances from local and more distant ecosystems, and also releases an outflow of waste products that impact ecosystems. An example is acid rain, which stems primarily from industrial outputs in urban areas but impacts broad areas and their biota.

Figure 11.7, modified from Folke (1999), shows the interactions of the global natural ecosystem on the left and the global human system on the right. The global ecosystem can be divided geographically into regional ecosystems such as those in the temperate region and the tropical region. A regional ecosystem can be further divided into local ecosystems (e.g., a forest ecosystem in Oregon). The global ecosystem is impacted by the hydrological, energy, chemical, and other cycles, as well as by

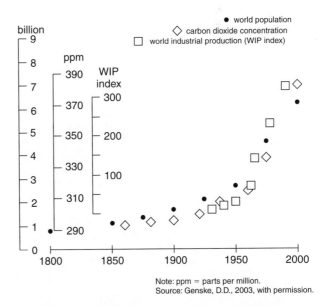

Note: ppm = parts per million.
Source: Genske, D.D., 2003, with permission.

FIGURE 11.5   World levels of carbon dioxide, population, and industrial production, 1850–2000. Source: Genske, Dieter D. D. (2003). *Urban Land: Degradation, Investigation, Remediation.* Berlin, Germany: Springer-Verlag.

weather and ocean circulation. The global human system, on the right, includes the global economy and societies such as nations and groups of nations. These large human units receive environmental inputs into their economies that include energy and resources, and produce outputs. Among the inputs to the global human system are food, plant and animal products, energy, chemicals, and nonrenewable natural resources. Its outputs are processed foods, purified water, wastes, agricultural products, fertilizers, irrigation water, pollutants, nuclear wastes, and various other items that impact the natural system. Societies have environmental policies, knowledge, and institutions that influence the inputs and outputs. At a smaller scale are local human communities, such as provinces, counties, tribal groups, and metropolitan areas, which also have individual environmental inputs and outputs that influence local ecological policies, knowledge, and institutions.

All told, this is a dynamic set of systems that form highly complex interactions and flows. A major portion of the global human system, perhaps more than half, is urban. Hence, understanding the components of this complex world model is important for urban geography.

For several decades, research groups have worked at establishing dynamic models of the global environmental system. Among the well-known models were the systems dynamics models that originated in the late 1960s at the Massachusetts Institute of Technology (MIT; Forrester, 1969, 1971) and the Meadows models

(Meadows & Meadows, 1972; Meadows, Meadows, & Randers, 1992). Like any computer simulation, model predictions are dependent on a complex set of assumptions including demographic and economic magnitudes and growth rates, population age structure, magnitudes and consumption rates of natural resources, and food consumption rates. The systems dynamics models commonly gave predictions that extended for 100 or more years. The accuracy of such models in predicting out many decades is low, but the models are, nevertheless, good tools for understanding dynamic process interactions under given conditions.

In long-range city planning, some cities have employed similar types of models for the local metropolitan area and its surrounding region. An example is the Blue Ring plan for Seattle, which is a vision of 100 years to preserve the quality of life (City of Seattle, 2002a, 2002b; Olson, 2003). The former Green Ring was established in 1903 by the famous planners Frederick Law Olmsted and John Charles Olmsted. It established loops of wooded parkways and parks and other green open spaces. During the subsequent 100 years, this long-term vision worked fairly well because the many projects, developers, and builders adhered to the long-term plan. The Blue Ring plans for the further enhancement of open spaces, with emphasis on the downtown area, which is where many U.S. cities lack open space (City of Seattle, 2002a, 2002b). The plan and land-use model was worked out between

FIGURE 11.6 Emission of carbon dioxide from industrial plant. Source: Black Star.

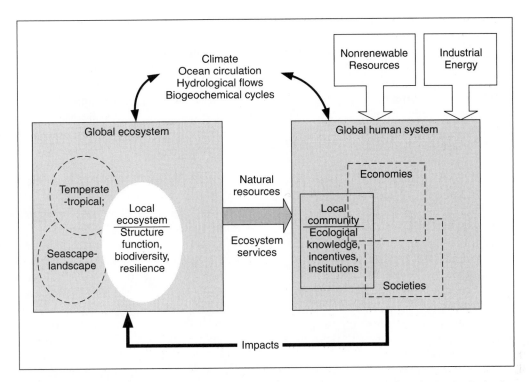

FIGURE 11.7 Combined system of humans and nature. Source: Folke, C. (1999) "Ecological principles," in *Handbook of Environmental and Resource Economics*, ed. by J. C. J. M. van den Bergh, pp. 895–911. Cheltenham, UK: Edward Elgar.

2000 and 2003 and involved many stakeholders, including inputs from developers, neighborhood groups, and public interest groups and boards. The planning was done by CityDesign, an office in the city government (Olson, 2003). A serious challenge is the high land values and lack of available land plots for parks in the downtown. However, the planners discovered that street right of ways constitute about 40 percent of the project's land.

The developers of the plan walked and bicycled around many neglected land parcels to add insights and refinements to the planning. The Blue Ring plan is intended to set a long-term vision for use of open spaces, prior to the occurrence of many infrastructure and development projects anticipated over the next decade and the next 100 years, such as redeveloping the downtown waterfront, improving transport, and revitalizing infrastructure and neighborhoods. The plan is an example of an environmental model developed by a city that will be applied for a very long-range strategic purpose (City of Seattle, 2002a, 2002b).

11.2 LAND USE

This section explores land use in the urban environment. Urban land is allocated to industrial, commercial, residential, park, and other uses. There are many forms of land ownership and regulation throughout the world. Land is often regulated by city, county, and other governmental authorities. For instance, zoning in the United States is done mostly by local municipalities and somewhat by counties. However, states and the federal government might sometimes designate their own uses for state or federal lands within metropolitan areas, even superseding local zoning. In developing world cities, land zoning is often less strictly imposed (Pick & Butler, 1997). Although land ownership is usually observed worldwide, in some cities, such as Mexico City and Sao Paulo, squatters can often lay claim to land just by occupancy.

Land use must be viewed as dynamically changing. What is the present land ownership and zoning of the land parcel where you are now reading? Think what it might have been a century or two centuries ago. Over time, urban land can undergo development, be in active use, become neglected, and even be abandoned, especially in older areas of large cities, which in the United States are often near the city center. Abandoned, deteriorating industrial sites called *brownfields* are eyesores and often sources of contamination, but can be regenerated at a cost into relatively clean and habitable *greenfields*. Besides industrial land degradation, other causes of urban land deterioration are use in disposal of wastes, resource extraction, natural disaster, abandonment, and warfare, which is illustrated in Figure 11.8 by the devastation of Hiroshima after the atomic bomb was dropped on August 6, 1945. At the edge of growing cities,

FIGURE 11.8   Hiroshima, Japan, after the U.S. atomic bombing of August 6, 1945. Source: Corbis/Bettmann.

there is pressure to convert rural or agricultural land into urban land ready to be developed. This agricultural land conversion process has advantages and disadvantages.

Land cover is a static representation of the land surface. For instance, an aerial or satellite photo shows land cover, as does a geographic information system (GIS) map layer of land use, such as was done for some of the GIS layers in the lab exercise in Chapter 10. However, when GIS analysis techniques are applied to a layer, the meaning shifts from static land cover to active land use.

In the United States, residential land has standards for suitable habitation. However, beyond the minimum thresholds there are benefits referred to as *amenities.* They include such things as water frontage, scenic views, convenience to parks or cultural sites, trees, and home gardens. *Disamenities,* features perceived as adverse, include noise, susceptibility to flooding, and adjacency to industrial pollution, waste-disposal sites, power plants,

and other toxic areas. Amenities favor the comfort of residents and higher land values, whereas disamenities have the opposite effects. There are cultural differences in the perceptions of what constitutes a residential amenity or disamenity. For example, although oceanside location is regarded as a big plus in the United States and Europe, it is viewed as a hazard in Accra, Ghana, because of occasionally inclement weather conditions, reducing real estate value (Meyer, 1999).

There are interactions between residential land-use patterns and the natural environment. The design and layout of residential land can affect air pollution, natural habitats, and hydrological systems, and emphasize or restrict open spaces (Meyer, 1999). The trade-offs become very evident for land being altered by urban sprawl at the periphery of cities. At the periphery, changes in land layout depend heavily on the amount of and locations for automobile transport access, extent of agricultural land consumption, and need for access to water.

11.2.1 Brownfields and Greenfields

A brownfield is an abandoned, neglected, or underutilized commercial land site for which redevelopment is retarded by a perception, not always accurate, of environmental degradation or contamination. Examples of brownfields are abandoned factories, transport facilities, and service stations; neglected waste-disposal sites; vacant warehouses; and uninhabited run-down buildings. By contrast, a greenfield is defined as a land site with natural ecological functions such as groundwater renewal and carbon cycling that has an ample level of biodiversity. The territory of brownfields and greenfields is small compared to overall urban area, but they are significant as locations for particular types of urban development. The U.S. Conference of Mayors has estimated that there are 21,000 brownfield sites in the 210 largest U.S. metropolitan areas (Sierra Club, 2003). Many sites are contaminated, some at toxic levels. In fact, some brownfields appear on the U.S. list of Superfund sites, places designated by the U.S. government as the most toxic in the nation. Brownfields are often located in urban core areas that were formerly industrial, but are sometimes on the periphery where there are abandoned mines and industrial plants (Genske, 2003). For instance, in Britain, abandoned mines are located in small and medium-sized towns northwest of Birmingham, and in the former East Germany there are many abandoned mines filled with debris that have not been cleaned up. A brownfield with a derelict tank in the Ruhr District of Germany is shown in Figure 11.9.

Commonly a brownfield lacks people or organizations willing to take responsibility to clean it up. Further, the abandonment might attract crime, require social services, and reduce the value of nearby property. A goal for brownfields is to clean them up through redevelopment and convert them into beneficial uses. An example of brownfield rehabilitation was an abandoned gas station in Trenton, New Jersey (see Figure 11.10), which was renewed to become a new firehouse (see Figure 11.11). The costly process involved not only surface demolition and building, but also subsurface cleanup of residual ground contamination. The U.S. Environmental Protection Agency (EPA) points to the benefits of redevelopment, including the following:

• Economic improvement of the urban core.

• Avoidance of urban sprawl by emphasis on redeveloping the center.

• More centralized and compact industrial facilities, which in turn lowers the need for pollution monitoring.

• Reuse and recycling of abandoned properties.

• Change of brownfields into greenfields.

FIGURE 11.9 Brownfield with derelict tank in Ruhr district, Germany. Source: Genske, Dieter D.D. (2004).

A number of steps are being taken in different countries to revitalize brownfields. One of the strongest national mandates is in Britain, where the government has passed legislation that requires that by 2015, 60 percent of new development must take place in brownfields. Progress is being made, but it is too early to know the outcome. In the United States, an exemplary city in cleaning up brownfields is Pittsburgh. In 1960, the city was among the most polluted in the United States. As the city's coal and steel industries were largely eliminated, many zones were cleaned up and revitalized. As seen at the top of Figure 11.12 for 1947, the section of the city center bordering the Monongahela and Allegheny rivers contained transport and industrial facilities that were eventually abandoned. By 1990, this section was renewed with a park, statuary, new bridges, and other useful features and amenities (see Figure 11.13).

FIGURE 11.10   Brownfield before rehabilitation: abandoned gas station, Trenton, New Jersey. Source: City of Trenton.

FIGURE 11.11   Rehabilitated brownfield from Figure 11.10: Firehouse, Trenton, New Jersey. Source: City of Trenton.

FIGURE 11.12 Downtown Pittsburgh in the 1930s. Source: Corbis/Bettmann.

It is possible to examine brownfields on a smaller scale. The EPA identified two buildings on Grand Street in Hoboken, New Jersey, as so contaminated with mercury, asbestos, and other substances that they were put on the U.S. Superfund list (EPA, 1997). They were put through the process of revitalization by the following steps: mercury decontamination and recovery, asbestos abatement, demolition of the two buildings on the site and disposal of any contaminants, disposal of contaminated soils, and groundwater monitoring for any recurring contamination.

11.2.2 Agriculture to Subdivisions: Urban Land Conversion

As cities expand in concert with worldwide urbanization trends, agricultural and other land at the urban periphery is converted to urban land. The conversion process for the United States was introduced in Chapter 10. The U.S. is converting land to urban uses at a much more rapid rate versus its population when compared with two European countries, Germany and Switzerland (see Figure 11.14). Switzerland's low relative level of rural land conversion reflects strict governmental norms and regulations, and a public appreciation of its mountainous environment. Germany is in between in its intensity of land conversion. Although it also has norms and regulations for conversion, its rural areas have partly yielded to growth pressures (Genske, 2003).

In the United States, the Soil Conservation Service estimates that a net of 900,000 acres yearly were converted from agricultural to urban land in the 1980s. Most of the agricultural land being converted, in terms of the value of its agricultural production, is located near cities. Farmers near cities, faced with higher land values, have to produce higher value crops such as citrus, vegetables, fruit, and flowers, to make a profit. The huge losses in this prime farmland are being partly replaced by marginal agricultural land located mostly in the nonmetropolitan counties of the arid West (Greene & Harlin, 1995).

The area nationally at risk of agricultural land loss is large. On the basis of the Census in 1990, 202 of the

FIGURE 11.13 Downtown Pittsburgh in the early 21st century. Source: The Image Works.

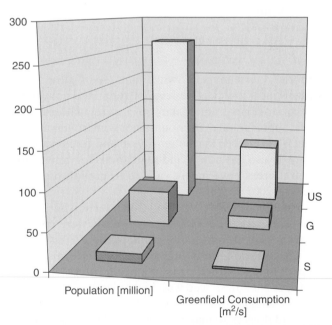

FIGURE 11.14 Estimated land conversion for Switzerland (S), Germany (G), and United States (US), in cubic meters per second. Source: Genske, Dieter D. D. (2003). *Urban Land: Degradation, Investigation, Remediation.* Berlin, Germany: Springer-Verlag.

approximately 2,000 metropolitan and metropolitan-adjacent counties out of the 3,069 U.S. counties were considered at high risk of losing valuable farmland to urban encroachment (Greene & Harlin, 1995). They constituted 6 percent of all counties, but accounted for 25 percent of agricultural sales. They averaged 18 percent population growth during the 1980s. A second set of 109 agricultural counties, considered at risk of losing farmland and with significant high-value farmland, grew by 17 percent during the 1980s. Both sets of counties at risk grew much faster than the U.S. county average of 4.1 percent (Greene & Harlin, 1995). The high population growth in these counties is leading to more and more loss of the prime farmland through urban encroachment.

Unlike Switzerland, the United States has no national land-use policy. Rather, controls for land use are approved and enforced at the state and local levels. This leads to inconsistencies across the country. However, some states have moved quickly to fill in the gap. Vermont, Florida, and California have enacted statewide land-use controls that replace some of the local controls. They are encouraged by generally positive support from residents in the most impacted urban fringes (Greene & Harlin, 1995).

11.2.3 Redlands, California: A Case of Conflicts over Agricultural Land Conversion

Redlands, California, is a city of 64,000 located east of the city of Los Angeles and just to the east of San Bernardino, bordering the Los Angeles combined statistical area of 16.4 million persons (see inset of Figure 11.15). Redlands had a strong tradition of citrus growing that went back to Spanish missionary times in the late 18th century. The town underwent development in the late 1900s as a vacation resort. In the 1950s and 1960s, in Redlands as in many other places in southern California, developers emphasized buying up farmland and attempted to convert it for development (Ackerman, 1999). As sprawl moved closer and the post-World War II economy boomed, citrus farming began to be threatened. As seen in Figure 11.16, citrus production in Redlands and San Bernardino reached a peak of 51,000 acres in 1945.

Huge conversions of citrus land to housing occurred in the second half of the century. Many owners of citrus orchards report that they have anticipated eventual conversion of their lands to housing by intentionally not caring for their land, that is, by not spraying or fertilizing,

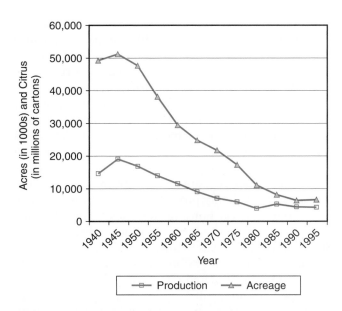

FIGURE 11.16   Citrus production and acreage, Redlands, California, 1940–1995. Source: Ackerman, W. V. (1999) "Growth control versus the growth machine in Redlands, California: Conflict in urban land use." *Urban Geography* 20(2):146–167.

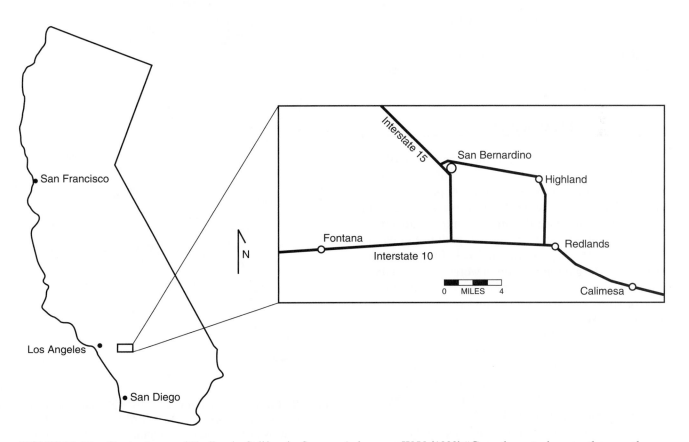

FIGURE 11.15   Context map of Redlands, California. Source: Ackerman, W. V. (1999) "Growth control versus the growth machine in Redlands, California: Conflict in urban land use." *Urban Geography* 20(2):146–167.

not replacing dead trees, and neglecting irrigation and frost protection (Ackerman, 1999). The spatial sequence of conversion over 40 years is seen in Figure 11.17—the solid encirclement of the city of Redlands by citrus groves in 1949 changed to peripheral citrus patches by 1990.

Political opposition to developers converting the citrus orchards (seen in Figure 11.18) was centered on slow-growth and no-growth groups beginning in the late 1970s (Ackerman, 1999). In 1978, Proposition R was approved by statewide voters to reduce growth and insure environmental values. It guaranteed a citizen voice in future growth and life quality, protected citrus groves and open space, put caps on housing densities, reduced urban sprawl, and lowered traffic congestion (Ackerman, 1999; Sullivan, 2003). This was followed by additional statewide Measures N and U in 1987 and 1997, which tightened up restrictions and directed city planning to conform to "managed development."

Presently, there continue to be mixed feelings about the slow growth and development restrictions. The public is generally in favor of slow growth, but most citrus landowners hope to eventually sell their groves to developers. They are intentionally putting their land in a "wait state" of partial neglect. Developers are highly constrained from conversion because of the added fees from the city to buy and develop citrus properties (Ackerman, 1999). They have either moved their focus to other localities or are contemplating lawsuits. In summary, this case highlights that agricultural land conversion is a complex and often lengthy process. Economics, geography, agriculture, and politics all come into play.

11.3 WATER RESOURCES

Water resources are essential to the functioning of any city. Besides residential supply, water is utilized for commercial and industrial uses, recreation, parks, cooling water for power plants, and other purposes. The amount of water withdrawn each day in the United States from surface and ground sources was 1,505 gallons per capita in 1995 (Solley, Pierce, & Perlman, 1998). In urban areas, an estimated 260 gallons per capita were consumed daily for public use and 109 gallons per capita for daily industry use other than power generation (Solley et al., 1998). There are many challenges and problems of water resources for cities, including water quality, supply, and management. An associated problem covered in the next section is urban flooding, that is, coping with too much water. This section first reviews in more detail the urban water cycle and how it relates to land uses in the city, and next turns to water quality and pollution.

The water supply cycle was shown in Figure 11.1. However, that simple diagram did not allow for water pollution.

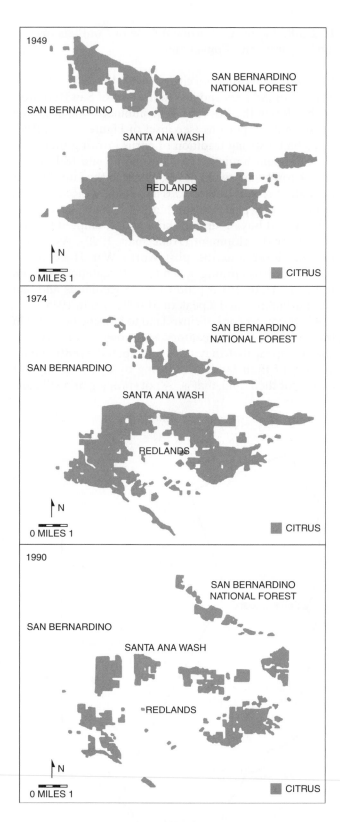

FIGURE 11.17 Distribution of citrus in Redlands, 1949, 1974, and 1990. Source: Ackerman, W. V. (1999) "Growth control versus the growth machine in Redlands, California: Conflict in urban land use." *Urban Geography* 20(2):146–167.

FIGURE 11.18 Citrus agriculture impacted by development, Redlands, California.

A more complete picture of the water cycle in a typical urban system (see Figure 11.19) reveals many sources of water pollution. The figure is at the scale of a city and its surroundings, similar to Figure 11.1. The water supply and pollution sources are shown in the figure on the left, whereas those for agriculture are on the right. The sources of water for the city are stream flow, rainfall, and groundwater. Potential pollution sources occur at each step along the way. The water being evaporated into the atmosphere contains pollutants from urban air pollution, industry, and power plants. When this water returns locally in the form of precipitation, it might have pollution, including acid rain. The city's sewage treatment plant releases nutrients and dissolved oxygen into the river body, which is likely to increase biological contamination of the river. Pollutants are being released into the groundwater through a variety of causes, including leakage from landfills and sewer lines, pollutants generated by vehicles that infiltrate into the soil, storm water infiltration, and outfall leakage from a sewage treatment plant. The agricultural uses might affect the water flows for metropolitan areas because agriculture is often intensive around the urban fringes.

This urban water system is highly complex, with many diverse flows and pollution sources. Nature can interfere randomly with parts of the system, such as the weather and soil erosion processes. It is difficult to model because major parts of it are underground, in water bodies, and in the atmosphere.

The most important water pollution impacts in cities on human health and disease and on biota are summarized in Table 11.1 (Marsh & Grossa, 2002). This table reveals the real threats of water pollution for cities. An underlying factor in urban water pollution problems is the sheer size of city populations, especially in the largest urban areas such as New York and Los Angeles. Twenty million people produce directly or indirectly a huge amount of waste each day, small amounts of which find their way into the consumption portion of the water system. Also, water pollutants tend to be more concentrated in fresh water, which comprises less than 1 percent of the all water flow, fresh and saline.

Groundwater is often a major supply source for cities. For instance, Mexico City is principally dependent on water stored in its underground aquifer. Cities need to

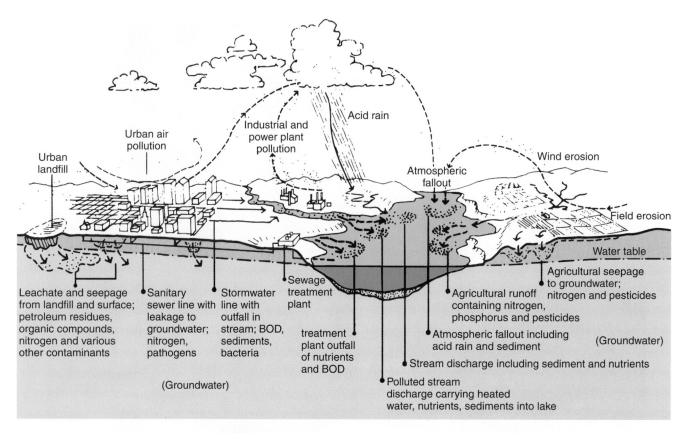

FIGURE 11.19   Water cycle in a typical urban system. Source: Marsh, W. M., and Grossa, J., Jr. (2002) *Environmental Geography: Science, Land Use, and Earth Systems.* New York: Wiley.

be concerned about the potential pollution of their groundwater supply. Solid waste eventually decomposes into a liquid waste called leachate, which infiltrates into the groundwater and can be an insidious pollutant. Likewise, hazardous and toxic wastes can infiltrate into the groundwater. When groundwater becomes polluted, it is often a long-term problem to restore it because direct intervention is not possible inside the earth. Cities need to make a conscious effort to protect their groundwater supplies over the long term, which might involve measures such as protecting the location of wellheads from underground pollutants.

11.3.1 Water Supply

Another key issue in urban water resources is *water supply,* the provision of sufficient water for consumption by residents and companies. Water supply comes mainly from freshwater bodies such as rivers, streams, and lakes, but can also be provided by man-made irrigation canals and by rainfall collected in high-precipitation areas. A more modern technological source is *desalinization,* or the desalting of seawater. Table 11.2 shows public water consumption for several nations projected to 2000, from a UNESCO-sponsored study (Shiklomanov, 1999). Italy,

Japan, and the United States have higher levels of per-capita public water consumption in the range of 40 to 75 gallons per day. At the lower range are developing nations such as China, South Africa, and Brazil, in the range of 7 to 18 gallons per day. Other countries mostly moderate in economic development, with Canada as an exception, are in a middle consumption range from 19 to 74 gallons per day. The lowest consumption of 8 gallons per day is for Brazil, a developing country. The implication is that metropolitan-area water demands are associated with level of country development.

Four considerations are crucial in planning for future water supply (Vaux, 2002). First, the current *water availability* needs to be thoroughly understood. In the United States, this especially applies to western states and counties facing the limits of water supply (Vaux, 2002), whereas it is less urgent in some other states. For instance, there is increasing struggle in urban southern California and Las Vegas, Nevada, to obtain water supply. By contrast, in the wetter southeastern part of the nation, Atlanta, Georgia (Vaux, 2002) also has a less urgent water supply problem.

A second key factor in urban water planning is the extent of *population growth.* In the United States, population continues to increase. As was seen in Chapter 1, most U.S. metropolitan areas grew moderately in the 1990s.

TABLE 11.1 Water pollution types by disease category

Pollutant category	Health impact on humans or biota
Oxygen-demanding wastes	Sewage and organic waste decomposing in water bodies absorbs oxygen, which impacts aquatic biota.
Plant nutrients	They stimulate certain aquatic plants, which could clutter up water bodies and impact other aquatic life.
Sediments	Sediments run off from terrestrial urban and agricultural uses. They cloud and clutter up water bodies and affect biota on the bottom.
Disease-causing organisms	Viruses, bacteria, worms, and other microorganisms mostly from human waters find their way into the water supply and can cause a variety of illnesses.
Toxic minerals and inorganic compounds	These are heavy metals, asbestos, and other products that can sometimes be carcinogenic and also harm biota.
Synthetic organic compounds	They are man-made, issuing from industrial processes. They might cause cancer and other diseases.
Radioactive wastes	They are emitted from nuclear plants, defense and military facilities, and specialized health and other industry processes. They can cause cancer and other diseases.
Thermal discharge	Heated cooling water from power plants and industrial processes is released into water bodies. This hot water stimulates some organisms and inhibits others.

Source: Marsh, W. M., and Grossa, J., Jr. (2002) *Environmental Geography: Science, Land Use, and Earth Systems.* New York: Wiley

TABLE 11.2 Water consumption by country, 2000

Brazil	173.8	18.5	7.7
Canada	30.2	10.8	25.9
China	1275	301	17.1
Egypt	68.8	38	40.0
India	1017	396	28.2
Israel	6	2.03	24.5
Italy	56.5	56.6	72.5
Japan	125	88.6	51.3
Mexico	102	47.2	33.5
South Africa	46.6	8.5	13.2
U.S.	275	166	43.7

Note: The 2000 figures are projected from 1995 data.

Source: Shiklomanov, I. A. (1999) *World Resources and Their Use.* St. Petersburg, Russia: State Hydrological Institute. Available at *http://espejo.unesco.org.uy*

In metropolitan areas where the current water supply is somewhat constricted, such as Los Angeles, San Diego, Phoenix, and Las Vegas, the population factor is especially pressing. From Chapter 1, recall that the Sunbelt has grown more than 70 percent since 1970 and is substantially located in the urban areas in Texas, the Southwest, and southern California, large parts of which are arid. Other semiarid or arid areas of the world, such as the Middle East, face even worse problems in this century because of rapidly growing population and severely limited water supply (Roudi-Fahimi, Creel, & De Souza, 2002). The Middle East and North Africa have 6.3 percent of the world's population but only 1.3 percent of its fresh water supply, so water is critical for many of its countries (Roudi-Fahimi et al., 2002).

The third factor for urban water planning in the United States and in some other nations is *old-fashioned institutional arrangements* (Vaux, 2002; White, 2000). Many American water arrangements at the national, state, and local levels were established in the 19th century.

For instance, state water laws in the West, known as *prior appropriation* laws, were designed to give long-term rights to farmers, who historically needed firm assurance of water supply to take the risks involved in agriculture. Today many believe the laws require more flexibility to deal with emerging water scarcities (Vaux, 2002). At the federal level in the United States, the problem is a "slicing" of federal responsibility for water regulation among multiple agencies, which often do not collaborate. As seen in Table 11.3, quite a few large and bureaucratic federal agencies have partial responsibility for water. Some interagency communication problems improved somewhat between 1966 and 1982, when the federal Water Resources Council acted as a coordinating body, including with state agencies (Vaux, 2002). At the state level, some states such as California have emphasized for decades the need for large systems of transport, delivery, and storage of water. Figure 11.20 shows the huge California water

TABLE 11.3 United States federal water agencies and their authority

Federal water agency	Authority
Corps of Engineers	Flood control and navigation
Bureau of Reclamation	Irrigation and power
Federal Power Commission	Power, including cooling water and water wastes
U.S. Forest Service	Forest management, including water aspects
Soil Conservation Service	Soil conservation and its water aspects

Source: White, 2000.

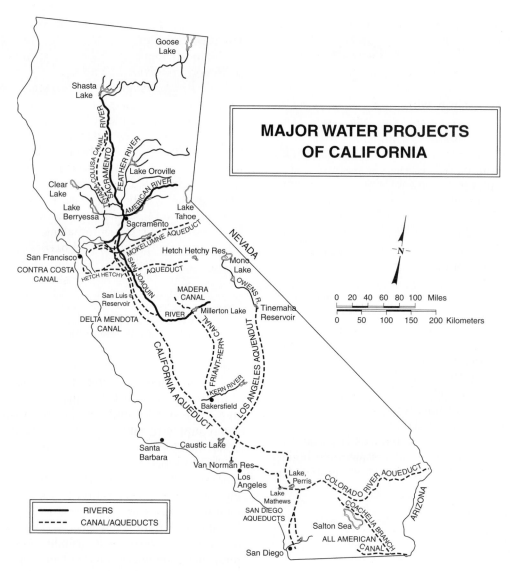

FIGURE 11.20   California water system. Source: Hundley, N., Jr. (2001) *The Great Thirst: California and Water, A History* (revised ed.). Berkeley: University of California Press.

system, which includes the huge California Aqueduct, other aqueducts and canals, rivers and lakes that serve as reservoirs, and agricultural sump of the Salton Sea. Most of the state's major metropolitan areas are served by this water system, including Sacramento, San Francisco, Bakersfield, Los Angeles, and San Diego, as well as the state's richest agricultural areas of the Central Valley and parts of Imperial Valley (Hundley, 2001). Started in 1960, the California Aqueduct extends for 450 miles.

At the metropolitan level, there are local water districts and private firms that provide the vast bulk of domestic water (White, 2000). In some metropolitan areas, such as Los Angeles, the local water districts are largely professional, experienced, and working in concert with each other (Grafton, 2001). A challenge to local water

districts is not focusing only on supplying water to the consumer at the demanded amount, quality, and prices, while ignoring wastewater disposal and other impacts on ecosystems (White, 2000). Unfortunately, the federal government has not actively encouraged balanced planning by communities. For instance, there have long been federal water quality standards to guarantee drinking water safety, but there are no analogous standards to establish a threshold for wastewater disposal impacts (White, 2000).

Two small urban communities in the Shenandoah Valley in Virginia are a good example of local planning to serve both human consumption and ecosystems. Their water plans include ecosystem impacts. This type of planning can be forward looking, so that it can "help guide the location and quality of further urban and suburban development" (White, 2000, p. 35).

A historical multipurpose program is the Tennessee Valley Authority (TVA), which began in 1933 and is still in operation today. The TVA took a coordinated approach on water and energy with a focus on the entire Tennessee River Basin, considering both consumer demand and ecosystem impacts. It has responded to changes in its political environment including ramping up power production for World War II industries; building some nuclear plants, then canceling most; and becoming a very efficient producer in the last two decades (TVA, 2004).

Institutional arrangements present in developing nations have strengths and weaknesses. For instance, Mexico City lacks a coherent water development policy. Water supervision is done by strong-willed federal agencies that have limited expertise and often ignore each other (Grafton, 2001), so there is no coherent metropolitan plan. The water production agencies are mining water resources without conserving or protecting them. The environmental programs that do exist have been highly politically charged (Grafton, 2001). This might change in today's political climate in Mexico.

At the same time, governments in some countries are encouraging advanced, balanced designs to conserve and recycle water, as well as to support biota and ecosystems. Figure 11.21 shows the design for a German public building where rainwater from the roof and other runoff water from the building are stored and utilized as gray water to

FIGURE 11.21 Design to conserve water in a German public building. Source: Genske, Dieter D. D. (2003). *Urban Land: Degradation, Investigation, Remediation.* Berlin, Germany: Springer-Verlag.

water gardens, flush toilets, and clean the building's solar panels (Genske, 2003).

The fourth key factor in planning water supply is the movement in advanced nations toward preserving *environmental amenities*. New urbanism, discussed later in the chapter, stresses in part compact growth to preserve natural zones within and adjacent to metropolitan areas. However, that approach depends on a supply of water suitable for maintaining substantial biota in natural and agricultural settings, which exceeds the usual level for metropolitan uses. Hence, these ecosystem services require substantial additional water (Vaux, 2002). The usual fix in urban areas is to construct water impoundments, dams, and holding ponds to balance out water scarcity, but the downside are physical barriers and other interference with ecosystems. One adverse outcome of this fix tends to be the sharp diminishment of wetlands in or near cities (Vaux, 2002).

11.4 URBAN FLOODING

This issue is an extension of the problems of water resources. Although flooding occurs in unperturbed natural environments, it is common in urban areas. In an unperturbed area, high precipitation tends to be absorbed into the ground through the process of *infiltration* down through soil layers. This alleviates the water overload when there is a lot of rain. However, cities replace soil layers with hardscape surfaces. Consequently, natural infiltration is greatly reduced, and instead, the rainwater or snowmelt runs along the surface into streams, riverbeds, and man-made channels, with the risk of eventually overflowing them and spilling out onto a floodplain. As seen in Figure 11.22 at the top, the unperturbed watershed has a lot of forest cover, which helps in evenly bringing water to the surface, where it infiltrates downward and eventually ends up in groundwater or in the river system. In the same area with urban development (Figure 11.22 at the bottom), the rainfall is more likely to run off and overflow the channel, potentially leading to a flood. There is reduced soil infiltration due to hardscape surfaces and less groundwater resupply. Another exacerbating circumstance is that the sediment and debris washed along with the runoff into the channel fills in the channel and constricts it. Flood engineers can try to compensate by straightening out channels to increase flow. However, this is usually disruptive to the environment and it also often transfers the high water flow downstream, where the impacts could be just as damaging.

A developing flood situation can be tracked by a *hydrograph,* a graph showing stream or channel water level versus elapsed time under specific climatic conditions. The example in Figure 11.23 shows two scenarios of

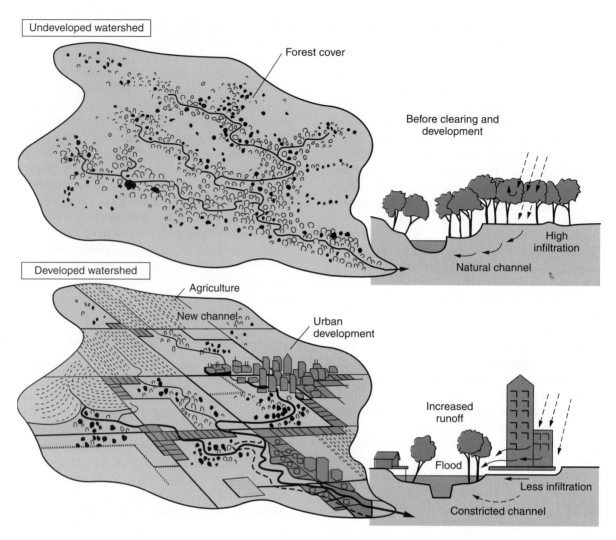

FIGURE 11.22 Runoff and streamflow in undeveloped and developed watersheds. Source: Marsh, W. M., and Grossa, J., Jr. (2002) *Environmental Geography: Science, Land Use, and Earth Systems*. New York: Wiley.

level of stream flow, one for the natural setting before urban development and the second after urban development. In the natural setting, there is delay of one day, after which a moderate amount of water has run off or infiltrated into the streambed, leading to a gradual rise in water level to a peak that is still below flood level. With urban development, after only a half-day there is a sharp rise, as much as five times the natural rise, in channel water level to considerably above flood level. It is clear that hydrographs monitoring all the major waterways of a metropolitan area can help government understand and manage the complexity of flood threats.

As an example, the great Mississippi River flood of 1993 covered 9.3 million acres and caused 70,000 people to become homeless, as well as resulting in several deaths. The impact of the substantial rainfall was magnified because of man-made planning mistakes. A large

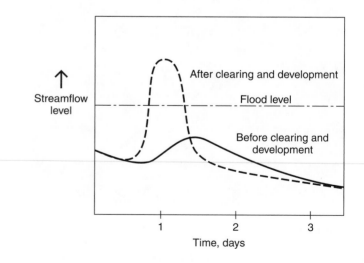

FIGURE 11.23 Urban hydrograph for watersheds in Figure 11.22.

agricultural area adjacent to the Mississippi upstream of the flood had been "enhanced" with extensive man-made channels and piping to increase irrigation. Downstream near St. Louis, the river had been narrowing for more than a century because of construction of various river-related facilities. From 1949 to 1969, the river width at this point had narrowed from 4,300 feet to 1,900 feet (Marsh & Grossa, 2002). As the rains increased, large volumes of water flowed rapidly from the upstream agricultural areas into the river and then to the constricted downstream part, where it overflowed man-made levees and overspilled onto the large floodplain. Disaster ensued. In retrospect, one mitigating measure would have been to construct large storage facilities for storm water. More important would have been better upstream irrigation planning to reduce potentially huge flow.

There are a number of dangers in the longer term for urban flooding. One is that as population grows in the United States, there is more temptation for human settlement on floodplains. Settlement there takes place because the local land-use controls over it are usually minimal. In fact, people are often unaware that they are moving onto a floodplain. Some local governments have been overly optimistic about engineering flood solutions.

Some of the natural habitat of floodplains could be threatened by certain engineering solutions. The Federal Emergency Management Agency (FEMA) has authority over flood emergencies, but it is focused on saving lives and preventing human material losses, rather than preserving ecosystems (White, 2000). In 100 years of changes in federal policy, the trend has been away from a balanced approach to managing the complex relationships of humans, biota, agriculture, and water on floodplains. Other nations face similar challenges for urban floodplain management. Among the most advanced nations in planning for floods is the very susceptible Netherlands. Some cities in developing nations have had severe problems; for instance, Caracas, Venezuela, suffered grave flood losses in the 1990s because of short-sighted planning for mountain runoff.

An example of an extreme urban flooding incident was the December 26, 2004 tsunami that killed over 165,000 persons, mostly in small- to medium-sized cities, largely in Indonesia, Sri Lanka, India, and Thailand—Asian countries bordering the Indian Ocean. A tsunami is a disastrous ocean wave that is usually caused by a sub-oceanic earthquake with magnitude of 6.5 or higher on the Richter scale. The 2004 tsunami's epicenter was located to the east of the Indonesian island of Sumatra. Entire

---

Case Study: Wildfire as an Urban Environmental Hazard: San Bernardino County in 2003

A serious urban threat comes from uncontrolled wildfires. From October 25 to 31, 2003, a devastating series of wildfires occurred in six major forested locations in southern California, ranging from the Simi Valley north of Los Angeles to eastern San Diego County. The total devastation was 700,000 acres burned, 3,346 homes destroyed, and 20 persons killed including one firefighter (Emergency Management, Inc., 2004). Some of the fires are speculated to have been started by arsonists, although evidence and proof is notoriously difficult to establish, as so much was burned (Gorman, Pugmire, & Henry, 2003). This case focuses on the mountain fires in San Bernardino County, the Grand Prix and Old Fires, shown in Figure 11.24. Those fires ended with 162,000 acres burned, 902 homes destroyed, and 4 deaths.

The San Bernardino fire problem stemmed from a number of natural and man-made reasons. One was lack of prior fires in this area, so the forest stands had become too thick. Additionally, a southern California drought had dried and stressed the trees. However, the problems were compounded by the bark beetle, a species endemic to the region, which has exploded in numbers because of the drought-stressed situation of the trees. This allowed the beetles to bore into the bark of pines more easily to

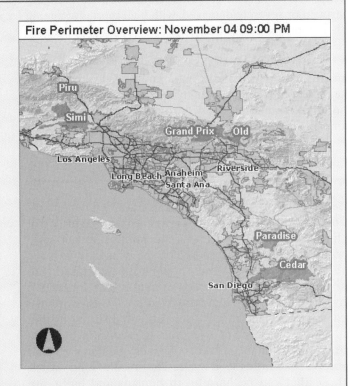

FIGURE 11.24 San Bernardino fires, November 4, 2003.

reach their objective, the moist inner core (Sahugun, 2004). Estimates are that, at the time of the 2003 fires, about one third of the trees in the San Bernardino Mountains were dead or near death from a combination of drought and bark beetle infestation (Sahugun, 2004).

Misguided urban development contributed to another side of the problem. The issue stems from rapid population growth and a history of easy approvals for developers (Martin, 2004). From 1990 to 2000, Riverside and San Bernardino Counties, on the eastern side of the Los Angeles metropolitan area, termed the "Inland Empire," grew by 25 percent, and growth rates are predicted to continue to increase over the next 20 years. The mountain area of San Bernardino has a population of 60,000, growing at a rate equal to the region. Responding to population growth, home developers gained approval for more and more mountain projects with increasingly high densities and in some cases, ones too tightly packed near forest stands (Tempest, 2003).

There were a number of unfortunate outcomes of this disaster. One involved the need for firefighters to use bulldozers to carve firebreaks around threatened mountain cities, such as Big Bear Lake. On the one hand, this was crucial to saving the city from fire destruction, but in the long term, it is potentially threatening to animal species in the area. This is a delicate balancing act between reducing imminent dangers to cities and not causing irreparable long-term harm to nature. Another outcome was huge property losses, estimated at $3 billion, of which $1.35 billion was for the San Bernardino fires (Reich, 2003). The only California natural disaster surpassing this loss was the San Francisco earthquake of 1906, which cost $5.7 billion in inflation-adjusted dollars.

The lessons learned have been applied by the San Bernardino County Board of Supervisors, which in April 2004 approved strict regulations of future home building in the mountain areas, in spite of strong protests by developers (Martin, 2004). Another lesson was how valuable GIS was as a tool in mitigating the fires (ESRI, 2004). It was utilized for allocation of equipment and resources, mapping of the fire incident perimeters, 3-D mapping of fire situations, developing response plans, displaying public information, implementing public safety measures, and conducting law enforcement, including maps of evacuations, monitoring, and arson investigation. Many aspects of this case carry over to other types of natural disasters in urban areas including floods, tsunamis, hurricanes, and earthquakes.

---

oceanside cities, such as Banda Aceh, Indonesia, were destroyed or severely damaged, with high mortality among their populations. It will take decades for some of the urban areas to recover and be rebuilt.

In summary, urban flooding is due largely to manmade alterations that disrupt natural soil absorption systems. Flooding can be planned for and alleviated. Modern solutions should also take into account environmental impacts on the natural systems of floodplains.

11.5 SOLID WASTES

Solid wastes are the solid materials discarded by humans and institutions. This section examines the magnitude of the problem in urban areas, the composition of solid wastes, how they are disposed of, and what pathways the residuals take. Solid waste should not be confused with *sewage,* which is waste solids or liquids that are carried away by a sewer system. *Municipal solid wastes* are discarded materials from households and institutions that are largely disposed of in landfills. A *landfill* is a managed area, usually on the outskirts of a metropolitan area, that receives and stores solid wastes. Detrimental solid wastes, those that can be harmful to humans and biota, are termed *toxic wastes.* Examples are strong acids, petroleum products, and nuclear materials. The federal government has tried to clean up the worst toxic sites through the Superfund, created by the U.S. Congress in 1980. Presently there are 2,000 Superfund solid-waste sites located throughout the nation, but concentrated most heavily in the northeast Boston-to-Washington corridor, northern Washington state, and northern California, areas with numerous defense-related facilities. However, the Superfund has suffered from lack of allocated funds, and so far only dozens of sites have been cleaned up.

In the United States, each resident produces an average of 44 tons of municipal solid wastes per year (Miller, 2003). The annual U.S. production of municipal solid waste is 78 million tons, with 1.14 billion tons of industrial wastes (Miller, 2003). There are 9 billion tons of solid waste generated yearly from mining and petroleum production (Miller, 2003). As seen in Figure 11.25, on a per-capita basis in 1994, this is about five times more than the per-capita production of solid wastes in Europe (Genske, 2003). In Paris, the per-capita production of solid wastes is a third that of the average U.S. resident. Solid wastes are considerably lower per capita in developing nations, at a level of 2 percent to 3 percent of the U.S. level. It is worrisome that these consumption patterns increased rapidly for all these categories in the 20th century and continue to do so. Although there are more than 40,000 possible substances that comprise solid wastes, typical European solid wastes (see Table 11.4) consist mostly of paper, compostable material, and plastics.

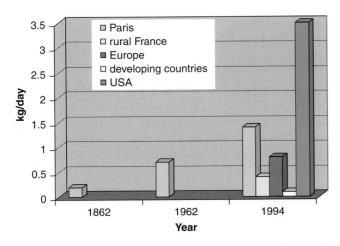

FIGURE 11.25 Average daily household waste production in Paris, rural France, Europe, developing countries, and the United States. Source: Genske, Dieter D. D. (2003). *Urban Land: Degradation, Investigation, Remediation*. Berlin, Germany: Springer-Verlag.

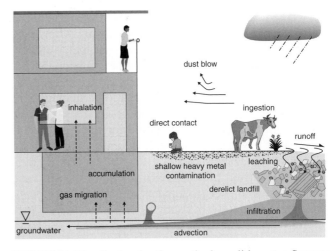

FIGURE 11.26 Contamination paths for solid wastes. Source: Genske, Dieter D. D. (2003). *Urban Land: Degradation, Investigation, Remediation*. Berlin, Germany: Springer-Verlag.

In the United States, huge solid-waste outputs are disposed of in landfills (73 percent), by burning (14 percent), and by recycling (13 percent). The wastes eventually decompose into leachate, a foul-smelling liquid. Part of the management challenge is to retain the leachate in the landfill so it does not pollute the soil, or worse, the groundwater (Genske, 2003).

Solid wastes can be carried by runoff into water bodies or get into the air through mismanaged burning. The dangers from improperly buried and contained solid wastes, seen in Figure 11.26, are direct contact by humans, ingestion by animals with potential harm later to humans, air pollution, runoff into water bodies, infiltration into deep groundwater, and movement of waste-generated gas from groundwater into buildings (Genske, 2003).

TABLE 11.4 Average composition of municipal solid wastes by weight in Switzerland in 1990s.

Solid waste component	Percentage
Paper	29
Compostable waste	22
Plastics	15
Ceramics, tiles	8
Natural products (e.g., wood, bones, meat)	8
Composite materials (e.g., appliances, etc.)	6
Packing material	4
Glass	3
Metals (e.g., iron, etc.)	3
Textiles	2
Total	100

Source: Genske, D. D. (2003) *Urban Land: Degradation, Investigation, Remediation*. Berlin, Germany: Springer-Verlag.

Recycling is an efficient solution to many solid-waste problems. Worldwide the United Nations estimates that about 35 percent of solid wastes are recycled. The recycling process starts at the home site, where the wastes are separated by type. Then the different types are transported to the recycling center. Here, each type is converted into a usable product. Energy savings from using recycled versus new products can be huge, often 75 percent or more. For aluminum cans, the savings is more than 90 percent (Marsh & Grossa, 2002).

Solid wastes cause controversy in cities. They produce odors and pollution, and reduce land values. They might have health detriments, such as spread of infectious disease and susceptibility to cancer, and psychological impacts due to fear and anxiety about their presence. They often lead to political disputes stemming from NIMBYism ("not in my backyard").

Love Canal, a 15-acre housing development located in Niagara Falls, New York, illustrates a toxic-waste contamination issue. Starting in the 1970s, residents noticed odors and chemicals leaking into houses, as well as a high incidence of disease symptoms such as birth defects, cancer, kidney defects, and skin disorders (SUNY Buffalo, 1998). Although there were many complaints to the local government, there was no response. The contamination contributed to local economic problems, as well as impacting the residents and the adjacent ecosystems. Finally, links were discovered between the residents' problems and industrial contamination of the water, air, and soil. Love Canal was eventually designated as a federal disaster area in 1978, and residents were ordered by President Jimmy Carter to evacuate (SUNY Buffalo, 1998). Two hundred and thirty-nine families were relocated (SUNY Buffalo, 1998). It was discovered that the former solid-waste landfill on which the Love Canal development was located had earlier received 20,000 tons of toxic chemicals, mainly from the Hooker Chemical Company. After 1978, the site underwent many years of cleanup, at a total government repair cost now estimated at $250 million (Marsh & Grossa, 2002). By 1998, although the toxic area was covered with a plastic liner and declared permanently off limits, the rest of the neighborhood was renamed Black Creek Village and made available for new settlement (CNN, 1998). Nearly all of the 239 homes were renovated and sold. The newly habitable area passed federal tests and was resettled (CNN, 1998). In summary, this case demonstrates the severe impacts of a toxic-waste problem on a community, but shows that focused and costly government action can result in remediation.

11.6 ENERGY

The modern city relies on energy for heating and cooling; electricity for buildings, homes, commerce, and industry; transportation; and public services. This dependence can be seen dramatically during electrical crises, when cities grind to a near standstill. This happened during the New York blackout in July 1977, and again in the Northeastern and Great Lakes states as well as Canada in the summer of 2003. In wartime, electrical crises can occur resulting from the conflict linked to problems existing beforehand—for instance, Baghdad's electrical outages during the Iraq War in spring of 2003.

Energy has a geographical aspect, in that it is produced in particular locations and transferred to other locations

for consumption. Its wastes and impacts are felt in particular geographical areas. GIS is a useful tool for analyzing urban energy flows.

The world is consuming energy in ever larger quantities. For instance between 1970 and 2000, worldwide energy consumption grew by 33 percent (International Energy Agency, 2001). It is projected to expand by another 50 percent between 2000 and 2020 (International Energy Agency, 2001). The dominant energy forms are oil, natural gas, coal, and nuclear. A plot of past and projected world primary energy demand from 1971 to 2020 reveals that use of all forms of energy have grown and will continue to do so, except for nuclear, which is projected to peak around 2010 and then decline (see Figure 11.27). The continued rapid growth in oil and natural gas usage portends increasing dependence of the United States and other advanced nations on nonrenewable energy sources and high reliance on nations rich in fossil energy. Overall, by 2020, the world's energy demand is forecast to double from 1971, at a rate of increase of 1.4 percent, which is slightly more the rate of population growth (International Energy Agency, 2001; Population Reference Bureau, 2002).

A model of energy supply and consumption for the United States (U.S. Energy Information Administration, 2004), shown in Figure 11.28, portrays energy flows in 1999. It is expressed in terms of *Quads*, which are quadrillions of British Thermal Units (BTUs). A *BTU* is the energy needed to increase the temperature of 1 pound of water by 1 degree Fahrenheit. In 1999, the United States was supplied with 100.42 Quads and consumed 96.6 Quads

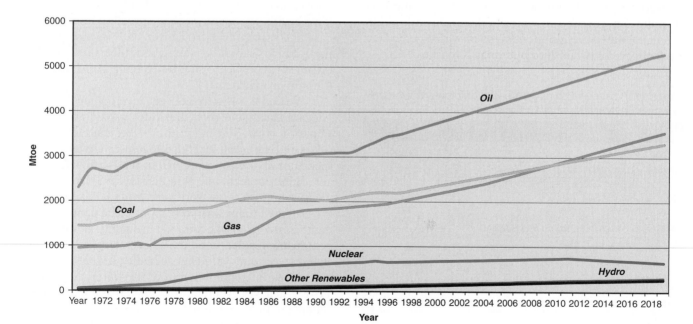

FIGURE 11.27 World primary energy demand by fuel, 1971–2020. Mtoe = million tons of oil equivalent. Source: International Energy Agency. (2001) *World Energy Outlook*. Paris, France: Organization for International Development.

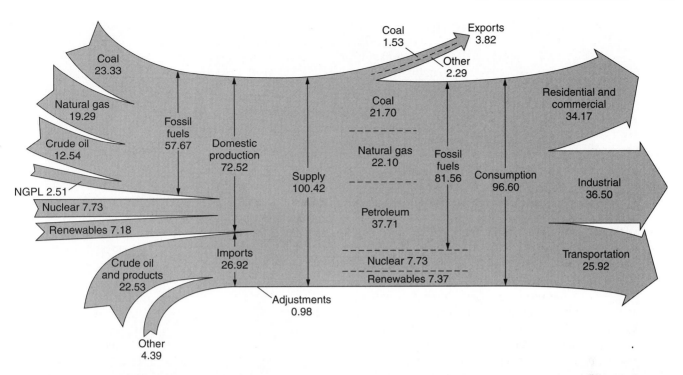

FIGURE 11.28 U.S. energy flow in 1999: Supply and consumption. Source: Hinrichs, R. A. and Merlin, K. (2002) *Energy: Its Use and the Environment* (3rd edition). Brooks/Cole.

(U.S. Energy Information Administration, 2004). The extra energy of 3.82 Quads was exported to other nations, mostly as coal. The high dependency of the United States on imported energy is seen by the import of 26.92 Quads, or 27 percent of supply. This is especially high for crude oil, for which two thirds is imported, mostly from Venezuela, Mexico, Nigeria, and the Middle East. Consumption is dominated by fossil fuels (84 percent), followed by nuclear (8 percent), and renewable energy (8 percent).

This consumption picture is similar to the world data in Figure 11.27, except that, for the United States, the proportion of nuclear power consumption is lower. That difference is because the nuclear energy program in the United States has been greatly slowed for the past two decades, while it has continued to grow and even dominate in some other nations, such as France.

For the United States, the end uses of energy are evenly split among industrial (38 percent), residential and commercial (35 percent), and transport (27 percent). About half of the transport sector use of energy is due to automobile transport.

Energy in cities can be consumed as heat, electricity, or combustion. The electricity portion of overall energy has been rising more quickly than heat, because modern consumer and industrial products and appliances utilize more and more electricity. This includes use for information systems, air conditioning, refrigerators, dryers, and electric motors (Marsh & Grossa, 2002).

The major forms of energy have quite different origins, supply sources, and operation.

Fossil Fuels

Fossil fuels are based on energy that arrived on the earth from the Sun tens of millions of years ago and was photosynthesized to form plants. Plants decompose over time and under certain conditions are deposited in the earth in gaseous, liquid, and solid states. This form of energy is mostly transported or piped, often over long distances, from extraction sites to power plants, where it is converted to electricity or heat. Oil and especially natural gas are less environmentally polluting than coal, so their use is growing more rapidly than coal (Figure 11.27).

Nuclear

Nuclear energy depends on extraction and supply of uranium, often from other countries. Nuclear power plants are very large and very expensive. Their energy production capacities are often 2,000 to 4,000 megawatts, about two to five times the output of a fossil plant. A nuclear energy plant located in an urban area is seen in Figure 11.29. Nuclear technology is restricted by security provisions in some parts of the world. It has the low but ever-present potential for causing severe environmental damage, with the worst instance having been the Chernobyl disaster in the Ukraine, from which tens of thousands of persons

FIGURE 11.29   Nuclear energy plant in an urban area. Source: © W. Cody/CORBIS.

have suffered serious health problems, not to mention damage to biota.

A major problem associated with nuclear energy is waste disposal. In the United States today, wastes are largely kept at nuclear sites, awaiting the opening of a permanent federal underground nuclear disposal site, which has been approved by Congress near Las Vegas, Nevada, although strongly opposed by that state. However, there are potential contamination dangers from transport of the wastes to the site. On the positive side, nuclear energy plants have low environmental impacts in their day-to-day operations, with cooling-water consumption being their major impact.

Renewable Energy

Renewables include hydro, solar, geothermal, wind, and tidal energy. *Hydroelectric energy* comes from the mechanical energy of water flow, which turns power plant turbines to generate electricity. Hydro requires damming of rivers, which could severely damage the natural environment. Otherwise, its operations are minimally intrusive. However, hydro plants are often located far away from major urban centers, so energy must be gridded over long distances. This results in grid losses of electricity and land consumption from transmission corridors. For instance, Mexico City is mostly powered by hydro energy sent from plants located many hundreds of miles to its south in the Pacific Ocean state of Chiapas.

Solar energy comes both in the form of solar power plants and individual units attached to homes and buildings, which are utilized for heating. Solar power plants are low polluting but have the adverse impact of consuming large land areas with solar collectors. Both forms are especially effective in zones that receive high solar radiation such as Arizona, southern California, and northern Mexico.

Geothermal energy consists of energy production from hot water, steam, and gases generated inside the earth, which are brought to the surface and utilized to generate power for heating. Exploitable geothermal energy tends to occur along tectonic plate edges, for example, in Japan, the Philippines, Indonesia, and the U.S. Pacific Coast (northern California, and near the U.S.–Mexico border). Its energy plants need to be located near the energy sources because very hot water cools rapidly when transported. So far, geothermal is a minor source of energy worldwide. The plants and pipelines have technical problems, because the water is often salty and corrosive. Geothermal's environmental impacts are not negligible, and include noise, smell, thermal impacts on water, air pollution, and land consumption.

Wind energy is one of the environmentally soundest forms of energy, and although minor, has been growing rapidly. It has the major disadvantages that wind sources are erratic and its facilities have land and visual impacts.

Tidal energy is utilized in special regions such as Japan that have flat areas of ocean shelf with substantial movement of tides. It has low economies of scale and its technology is expensive. It is a form of energy that is fixed to particular power plant locations, which are unlikely to be convenient for major urban areas.

Energy is essential for the functioning of cities, and is tied intimately to the urban economy. Its environmental impacts, wherever it is located, need to be determined on humans and the environment. This is especially true in or near cities, where substantial human populations

FIGURE 11.30 Cleveland smoke stack.

could be affected. An example is seen for Cleveland in Figure 11.30. It has not been studied by geographers as much as some other environmental factors such as water and toxic substances, partly because geographically referenced energy data are hard to come by. However, because it is so influential on manufacturing, transportation, commerce, and information technology, it constitutes an important part of urban geography.

11.7 AIR POLLUTION

Urban air pollution is closely related to energy, industry, and transportation. This section presents the major forms of air pollution; its prevalence, dispersion, and transport; its impacts; and strategies to prevent and control it, including monitoring. The next section analyzes the urban issue of the air pollution in Mexico City, leading up to the lab exercise.

Air pollution derives from energy production of different types for vehicular transport, industry, lighting, heating, and domestic uses. It is linked, as are other environmental hazards, to expanding population.

The major categories of air pollutants are the following:

- *Sulfur dioxide and nitrogen oxides (SO_2 and NO and NO_2, often called NO_x).* They are released through combustion of fossil fuels in stationary sources. In developing countries, they are emitted through burning of firewood and coal. Nitrogen oxides are also released from automobiles and other vehicles.

- *Carbon monoxide (CO).*   It is released primarily from motor-vehicle engines.
- *Ozone (O_3).*   It is a photochemical oxidant. This means that it is formed from chemical oxidation reactions that involve nitrogen oxides and volatile organic compounds, which are emitted from transport, industry, and energy distribution. Ozone is most prevalent in hot, smoggy climates that are full of traffic (e.g., Los Angeles).
- *Suspended particulates.*   They include both natural particulates emitted from the soil and ecological processes, and man-made particulates from motor vehicles, industry, rubbish, and energy combustion.
- *Trace air pollutants.*   They are a variety of toxic pollutants, beyond the standard ones, that are present in trace amounts. Some examples are carbon tetrachloride, benzene, formaldehyde, and methyl chloride (Miller, 2003). Hard to measure because of their slight concentrations, they can be carcinogenic and damaging to health (United Nations, 1992).

The common pollutants in developed countries differ somewhat from those in developing ones. Advanced nations tend to have more pollutants that stem from motor vehicles, especially nitrogen oxide, carbon monoxide, ozone, and particulates, whereas developing nations have a wider range of sources including more particulates. Within developing countries, the mix might vary; for instance, Latin America tends to have more vehicle-based pollutants, whereas Africa has very few.

The concentration levels of the six major air pollutants are shown for the United States in year 1999 in Figure 11.31. They include carbon monoxide, nitrogen dioxide, ozone, particulate matter, sulfur dioxide, and lead (EPA, 2004; Environmental Defense, 2004). This chart shows the areas of the country that have reached attainment of EPA standards for the six pollutants (light), counties for which part of the county is in nonattainment of standards (medium shade), and counties entirely in nonattainment (dark shade). It is clear that air pollution is following the location of major metropolitan areas. All of the nation's largest metropolitan areas are in nonattainment. The largest contiguous area of nonattainment is in and near the populous Boston-to-Washington corridor. The cause is its high concentration of vehicles, fossil power plants, and manufacturing industry. The second wide area of high nonattainment is in the urban parts of central and southern California. The elevated air pollution exposures of greater Los Angeles, Chicago, and Houston–Galveston are reflected in the 10 counties with the most person-days exceeding national ambient air quality standards (NAAQS) for six criteria air pollutants, as seen in Table 11.5 (Environmental Defense, 2004; EPA, 2004). All told, more than 100 million Americans are living in urban counties in nonattainment of federal air pollution standards (Environmental Defense, 2004; EPA, 2004).

Air pollution is dispersed, trapped, and transported through a combination of topographic, atmospheric and

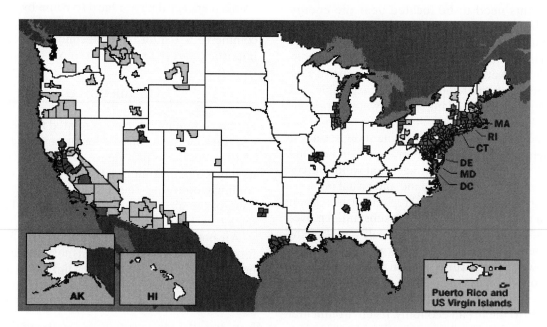

FIGURE 11.31   Concentration levels of the six most common air pollutants in the United States.
Source: Environmental Defense, 2004. Note: The grey shadings are explained in the text.

TABLE 11.5 Ten counties with the most person-days in excess of national ambient air quality standards for six criteria air pollutants, 2002.

Rank	County	Associated major urban area	Person-days in excess of NAAQS
1	Los Angeles, CA	Los Angeles	2,588,936,328
2	San Bernardino, CA	San Bernardino	815,416,328
3	Cook, IL	Chicago	510,863,165
4	Harris, TX	Houston	508,658,865
5	Riverside, CA	Riverside	458,760,240
6	Fresno, CA	Fresno	322,167,952
7	Kern, CA	Bakersfield	266,518,656
8	Maricopa, AZ	Phoenix	194,928,684
9	Tarrant, TX	Fort Worth	140,517,672
10	Sacramento, CA	Sacramento	114,847,216

Note: The criteria air standards are carbon monoxide, lead, nitrogen dioxide, ozone, particulate matter, and sulfur dioxide.

Source: EPA (2002) *Solid Waste Management: A Local Challenge with Global Impacts*. EPA 530-F-02-206d. Washington, DC: Author; Environmental Defense, 2004.

TABLE 11.6 Air pollution impacts

The health impacts are from direct inhalation and also by indirect routes including contaminated drinking water, contaminated food, and transfer from skin. Impacts are common on respiratory and cardiovascular systems.

EFFECTS ON HUMAN HEALTH

Pulmonary function affected by higher sulphur dioxide and particulates. Also nitrogen dioxide and ozone.

Ozone causes irritation to eyes, nose, and throat.

CO_2 can displace oxygen in the blood. This can have heart and neurobiology results.

Lead "inhibits hemoglobin synthesis in red blood cells in bonemarrow, impairs liver and kidney function, and causes neurological damage." Younger and more elderly are more susceptible

EFFECTS ON THE ENVIRONMENT

Acid deposition "Sulphur and nitrogen oxides are the principal precursors of acidic deposition." Long-range transport of sulphur and nitrogen oxides and acidic products cause the acidification and adverse impact on aquatic and terrestrial ecosystems.

Forests. Sulphur dioxide, nitrogen dioxide, and ozone can cause crop loss and damage to forests. Sulphur dioxide can damage buildings and art works.

Source: United Nations, 2002.

climatic conditions. Hills and mountains can serve as barriers that trap pollutants. Mexico City is a dramatic example. In a situation of *thermal inversion,* present often in Los Angeles, the normal temperature gradient is reversed, so temperature rises with altitude. In this case, the pollutants become trapped near their sources, close to the ground, and do not disperse. In an *urban heat island,* heat emitted from the city rises up, and cold air comes in underneath to replace it, drawing in polluted air from the surrounding regions (United Nations, 1992). On a large scale, *street canyons* can form along city streets bounded by high buildings. The canyon traps pollutants and prevents their escape (United Nations, 1992). Pollutants not only impact the immediate metropolitan area, but for large cities with great pollution, the pollutants can be transported as an air pollution plume over great distances and register effects hundreds of miles away.

There are many health impacts of air pollution on humans and natural organisms that are summarized in Table 11.6. It is evident that the damages can be insidious, especially for more vulnerable young and elderly people.

Cities, regions, states, and the federal government are trying to combat this damaging problem in a number of ways. They have set emissions standards that have been useful in benchmarking and monitoring. Many urban governments have set up air pollution monitoring networks. These are a series of measuring stations networked together that continually record pollution levels. Cities

can regulate air pollution by banning industrial plants, restricting or banning vehicular emissions, and requiring scrubbers, exhaust equipment, and other cleaning equipment on plants and vehicles. Additionally, urban governments can require energy conservation, which reduces the root cause.

Over the past two decades, Los Angeles has been a success story in reducing pollution. The Los Angeles Air Pollution Control Board (LAAPCB) and Southern California Air Quality Management District (SCAQMD) were established in 1947. Stringent vehicular and industry emission standards were put in place starting in the 1970s by the SCAQMD. They made major progress in fighting sources of air pollution, including restricting petroleum refineries in Long Beach, a steel plant, and lead smelters located inland. Consequently, nitrogen oxides fell from 450 ppm in 1987 to 325 ppm in 2000, and carbon monoxide was lowered during this period from 1,800 ppm to 1,150 ppm. The progress in halving the peak ozone concentration from 1970 to 1996 is seen in Figure 11.32. This example demonstrates that local governments can get their act together and concertedly attack environmental problems. Los Angeles still has a long way to go (Table 11.5), but the current situation would have been much worse without the mitigation achievements.

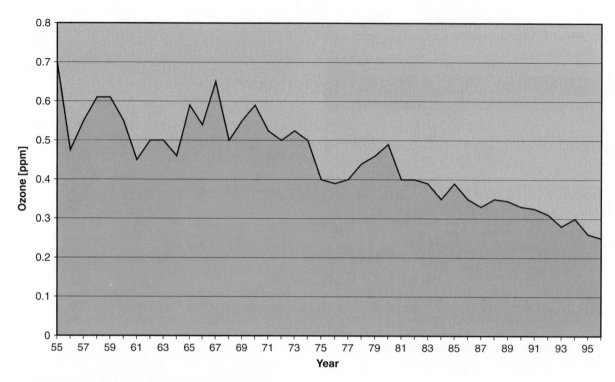

FIGURE 11.32   Peak ozone levels in Los Angeles, Orange, and parts of Riverside and San Bernardino counties, 1955–1996. Source: South Coast Air Quality Management Board, 1997.

11.8 ANALYZING AN URBAN ISSUE: AIR POLLUTION IN MEXICO CITY

Mexico City has the worst air pollution of the world's megacities, according to a comprehensive study by the United Nations (1992). Twenty of the world's 24 megacities were studied, and all had serious air pollution. The health impacts identified included higher disease rates, death, reduced lung function, higher cardiovascular disease, and neurobehavioral problems. There also were serious impacts on the material structures of these cities, including artworks and buildings. Ecosystems and their biota were disturbed. The United Nations pointed to the air pollution problem in megacities as likely to grow because the megacity populations were projected to grow. A small portrait of Mexico City's air pollution appears in Figure 11.33.

Mexico City presents many unique aspects for urban air pollution. It sits in a high-altitude valley at 7,350 feet, surrounded on the east, south, and southwest by high volcanoes and mountains. The wind patterns blow during the day primarily from the northeast to the southwest. At night the patterns tend to reverse, and flow mostly from the southwest to the northeast, although some cross-currents go to the southeast. The spatial distribution of Mexico City's air pollution is shown in Figure 11.34.

The metropolitan area is subject to thermal inversions for eight months of the year from October through May,

so air pollution can be trapped in inversion layers for several consecutive days. Besides the monthly variations, air pollution tends to vary diurnally and weekly. It is lower on weekends, when vehicular traffic and industry are much reduced.

The metropolitan area has a huge vehicular fleet of more than 14 million vehicles, including more than 7.5 million automobiles. It has more than 35,000 factories, of which 4,000 emphasize combustion and transformative processes that generate air pollution emissions.

The major air pollutants in Mexico City and their sources are as follows:

- *Carbon monoxide.*   It is largely vehicular in origin and is concentrated in the Federal District center that has the heaviest traffic. Until 1990, air pollution controls on vehicles were relaxed. Starting in 1993, more stringent vehicle exhaust controls were put in place, as well as restrictions of vehicles to use on only four or five workdays per week.

- *Hydrocarbons.*   They come both from motor vehicles, concentrated in the center of the megacity, and industrial plant emissions, located all over the city but mostly in the near northwest.

- *Nitrogen oxides.*   They are expected also in the more central areas of the city having higher vehicular traffic. Over three quarters of these emissions in the city come from motor vehicles (United Nations, 1992).

FIGURE 11.33   Child impacted by air pollution, Mexico City.

- *Particulates.*   They come mainly from numerous industrial sources. The heaviest industrial production area is the near northwest factory zone. Besides industry, other sources of particulates are from blowing dust, biological decomposition processes, motor vehicles, combustion, and rubbish.

- *Sulfur dioxide.*   It is geographically the most widely distributed air pollutant, and tends to be in all sections of the megacity, except the old central zones. It grew somewhat from 1986 to 1992, before stricter industry controls were put in place, and has moderated since 1992. The main sources are industry and power plants, which tend to be located in the near northwest part of the megacity.

There has been more cooperation on air pollution in the past decade between the city government of the Federal District and the government of the surrounding State of Mexico, which includes the outer ring of greater Mexico City, to set pollution control programs and other strategies. Accomplishments include the closure of one large fossil-fuelled power plant inside the city and the switch of vehicular gasoline to unleaded fuel. Among the long-term goals of these programs are to reduce the sulfur content of fuel oil and diesel oil; retrofit buses, vans, and trucks to use natural gas; increase limitations on carbon monoxide emissions from vehicles; and equip all automobiles with catalytic converters.

In analyzing this urban issue in the lab, you will be conducting spatial analysis of the distribution of five major air pollutants (carbon monoxide, hydrocarbons, nitrogen dioxide, particulates, and sulfur dioxide), comparing them to wind patterns at two daily time points, and justifying the patterns.

11.9 ENVIRONMENTAL JUSTICE AND PRESERVING THE NATURAL ENVIRONMENT

This part of the chapter looks at social and political responses to urban environmental problems. It introduces the concept of *environmental justice* and why it is important to cities. It also considers the environmental aspects of new paradigms in urban design under the names of smart growth and new urbanism that recognize many of the issues in this chapter. The full treatment of both environmental justice and the new paradigms takes place in the concluding chapter of this text, including a full case study that illustrates the planning and ethical complexity of the issues involved.

Environmental justice is defined as the fair treatment and significant involvement of all people, including those of different races, colors, national origin, and income, relative to the development, implementation, and enforcement of environmental projects, policies, and laws (modified from Bullard, 2003). Many research studies

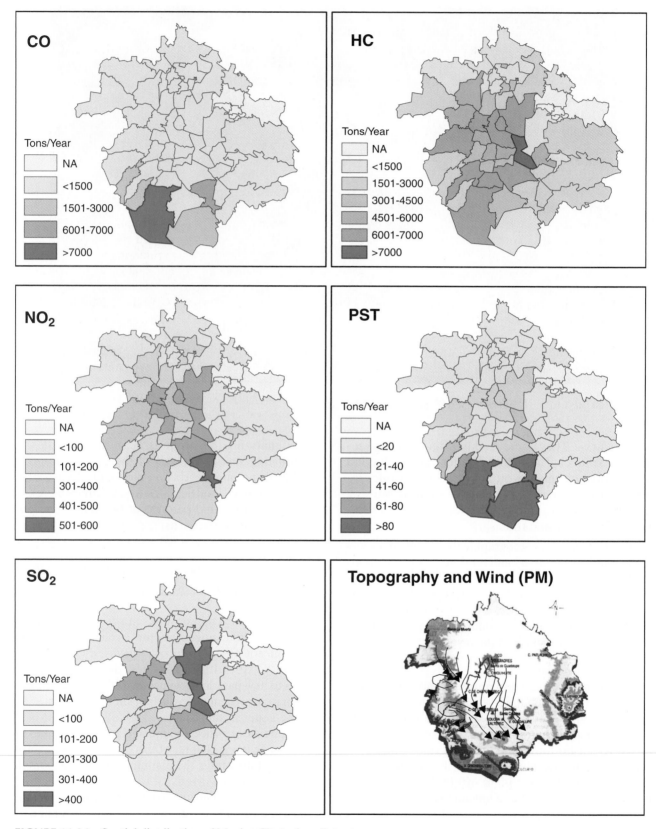

FIGURE 11.34　Spatial distribution of Mexico City's air pollutants.

CO = carbon monoxide　　HC = hydrocarbons
NO$_2$ = nitrogen dioxide　　PST = particulates
SO$_2$ = sulfur dioxide

have shown sharp inequalities in adverse environmental impacts with weightings toward disadvantaged groups (Bullard, 2000; Pullido, 2000). This has been shown fairly consistently in many areas of the United States and for a variety of environmental impacts. It was perhaps initially uncovered in the 1987 study of Los Angeles commissioned by the United Church of Christ (Bullard, 2003). This careful statistical study confirmed that race was more highly associated with toxic waste sites than any other characteristic, such as land cost, property values, and income. Another pioneering environmental justice study was that of the U.S. government's Agency for Toxic Substances and Disease Registry (cited in Bullard, 2003). The study concluded that lead poisoning disproportionately affected poverty-stricken, inner-city children. By ethnicity, it affected mostly Black population. This was exacerbated because Black children in the inner city were highly segregated. Following these studies, there have been many studies all over the country that have largely corroborated the associations of pollution and social deprivation.

Another example of environmental justice is a long-term dispute involving solid-waste disposal in New York City. The problem has been the huge amounts of solid waste generated by the megacity without adequate facilities to accept it (Gandy, 2002). Much disposal over the past 75 years has depended on the Fresh Kills landfill located on Staten Island. It has become stacked up to a level of 500 feet—more than the giant pyramids of Egypt—leading to activism and protests by residents of Staten Island and finally to the city's commitment to its closure in 2001 (Gandy, 2002). This presented New York City planners with difficult choices. Among the alternatives historically considered were ocean dumping, recycling, alterative landfills, incineration, and export of wastes to distant communities for disposal. Ocean dumping was ruled out by court cases in the 1930s, and recycling on this scale has been too expensive (Gandy, 2002). There are no alternative landfills nearby, so the latter two alternatives have been the ones considered by the city.

Incineration was seriously pushed in the 1980s and early 1990s through efforts by the city to gain approval for huge incineration plants to be located on the former Brooklyn Navy Yard (Gandy, 2002). However, the major ethnic constituencies nearby in the Greenpoint-Williamsburg section of Brooklyn reacted with a solid wall of protest. The two major opposition groups were Latino and Hassidic Jewish. Although they had traditionally had very bad relations, the planned incinerators caused these communities to surprisingly unite on an issue. Protests through actions included a major united march of protest by Latinos and Hassidics across the Williamsburg Bridge in January 1993. In the end, these groups prevailed in overturning the city's plans. Currently, the city is looking to external communities to receive the waste at a cost.

This brief case, which is fully described elsewhere (Gandy, 2002), underscores a number of environmental justice issues. There is the question of which communities are responsible to receive environmental damages and at what cost. What are the responsibilities of urban entities to absorb their own waste and pollution within their own territory or trade waste disposal for money payments to other communities? How much detailed information should be gathered and provided to the general public of urban areas about environmental damages and trade-offs?

The body of work on environmental justice gives us pause that the solutions to urban environmental problems might not rest only with engineering and regulatory controls. There might be another dimension to environmental problems that involves the greater society and its inequities (Pulido, 2000; Bullard, 2000; Bullard, 2003).

Another dimension of the solution of urban environmental problems are the new paradigms of smart growth and new urbanism. They call for a different kind of planning and development of cities than what took place in the second half of the 20th century. Among the elements are compact development; attention to the conservation of environmental zones and farmland; redevelopment of deteriorated central-city areas, such as some brownfields, through an approach called "infill"; mixed land uses rather than sharply separated single land uses; and planning that includes transportation from the start. They are mentioned here because they include environmental sustenance and improvement as central goals. Rather than adopting a "master builder" approach that more concrete and higher buildings are better, they seek a more even balance between natural systems and human systems. These approaches, including a complex planning case study of Irvine Ranch in southern California, are fully explored in the final chapter.

SUMMARY

This chapter examined urban environmental science and problem solving as a part of the urban system. It covered basic concepts and principles, and then moved to the major types of environmental problems, including land use, water, flooding, fire, solid waste, energy, and air pollution. Each of those has complex relationships with the city, as well as policy and regulatory issues. The chapter focused somewhat more deeply on analyzing the urban air pollution of Mexico City, the topic of this chapter's lab. Finally, the chapter turned to environmental justice, and the planning, social, and ethical aspects of the urban environment, issues that are explored more in the final chapter.

REFERENCES

Ackerman, W. V. (1999) "Growth control versus the growth machine in Redlands, California: Conflict in urban land use." *Urban Geography* 20(2):146–167.

Bullard, R. (2000) *Dumping in Dixie: Race, Class, and Environmental Quality*. Boulder, CO: Westview Press.

Bullard, R. (2003) *Environmental Justice: Strategies for Creating Healthy and Sustainable Communities*. Macon, GA: Mercer University. Virtual lecture available 2004 at *www.law.mercer.edu*

Chiras, D. D., Reganold, J. P., and Owen, O. S. (2002) *Natural Resource Conservation: Management for a Sustainable Future*. Upper Saddle River, NJ: Prentice Hall.

City of Seattle. (2002a) *The Blue Ring: Connecting Places. 100-Year Vision*. Seattle, WA: Author.

City of Seattle. (2002b) *The Blue Ring: Connecting Places. The Next Decade*. Seattle, WA: Author.

CNN. (1998) "Despite Toxic History, Residents Return to Love Canal." Posted on August 7, 1998, at CNN.com.

Ehrlich, P. and Holdren, J. 1972. "Review of the closing circle." *Environment* April, pp. 24–39.

Emergency Management, Inc. (2004) *Wildfires in Southern California*. Los Angeles, CA: Author. Available at *http://www.emergency-management.net*.

Environmental Defense (2004) *Scorecard: The Pollution Information Site*. New York, NY: Environmental Defense. Available 2004 at *www.scorecard.org*.

EPA. (1997) "Grand Street Mercury." Record of Decision System. EPA ID: NJ00013227733. Available at *http://www.epa.gov/superfund/sites/rodsites/0204030*.

EPA. (2002) *Solid Waste Management: A Local Challenge with Global Impacts*. EPA 530-F-02-206d. Washington, DC: Author.

EPA. (2004) *National Ambient Air Quality Standards (NAAQS)*. Washington, DC: Environmental Protection Agency.

ESRI, Inc. (2004) "GIS helps response to southern California fires." *ArcNews Online*. Winter 2003–2004. Available at *http://www.esri.com*.

Ezcurra, E., Mazari-Hiriart, M., Pianty, I., and Adrián Guillermo, A. (1999) *The Basin of Mexico: Critical Environmental Issues and Sustainability*. Tokyo, Japan: United National University Press.

Folke, C. (1999) "Ecological principles," in *Handbook of Environmental and Resource Economics*, ed. by J. C. J. M. van den Bergh, pp. 895–911. Cheltenham, UK: Edward Elgar.

Forrester, J. W. (1969) *Urban Dynamics*. Cambridge, MA: MIT Press.

Forrester, J. W. (1971) *World Dynamics*. Cambridge, MA: Wright-Allen Press.

Gandy, M. (2002) *Concrete and Clay: Reworking Nature in New York City*. Cambridge, MA: MIT Press.

Genske, D. D. (2003) *Urban Land: Degradation, Investigation, Remediation*. Berlin, Germany: Springer-Verlag.

Gorman, A., Pugmire, L., and Henry, W. (2003) "Southern California fires: Arson cases tough to prove, even tougher to prosecute." *Los Angeles Times* November 2:A29.

Grafton III, W. D. (2001) *Comparative Research of Groundwater Issues in Mexico City and Los Angeles: Some Possible Solutions*. Unpublished masters thesis, Program in Environmental Studies. Los Angeles, CA: University of Southern California.

Greene, R. P., and Harlin, J. M. (1995) "Threat to high market value agricultural lands from urban encroachment: A national and regional perspective." *The Social Science Journal* 32:137–155.

Hinrichs, R. A. and Merlin, K. (2002) *Energy: Its Use and the Environment* (3rd edition). Fortworth, TX: Harcourt College Publishers.

Hundley, N., Jr. (2001) *The Great Thirst: California and Water, A History* (revised ed.). Berkeley, CA: University of California Press.

International Energy Agency. (2001) *World Energy Outlook*. Paris, France: Organization for International Development.

Kates, R. W. (2000) "Population and consumption: What we know and what we need to know." *Environment* 42(3):10–19.

Marsh, W. M., and Grossa, J., Jr. (2002) *Environmental Geography: Science, Land Use, and Earth Systems*. New York: Wiley.

Martin, H. (2004) "Growth debate follows fires." *Los Angeles Times* April 5:B1.

Mazari, M., and Mackay, D. M. (1993) "Potential groundwater contamination by organic compounds in the Mexico city metropolitan area." *Environmental Science Technology* 27:794–802.

Mazari-Hiriart, Marisa, De la Torre, L., Mazari Menzer, M., and Exequiel E. (2001) "Ciudad de México: Dependiente de sus recursos hídricos." *Ciudades* 51:42–51.

Meadows, D. H., and Meadows, D. (eds.). (1972) *The Limits to Growth: A Report for the Club of Rome's Project on the Predicament of Mankind* (2nd edition). New York: Universe Books.

Meadows, D. H., Meadows, D. L., and Randers, J. (1992) *Beyond the Limits: Confronting Global Collapse, Envisioning a Sustainable Future*. White River Junction, VT: Chelsea Green.

Meyer, W. B. (1999) "Land use and environmental quality," in *Handbook of Environmental and Resource Economics*. ed. by J. C. J. M. van den Bergh, pp. 551–559. Cheltenham, UK: Edward Elgar.

Miller, G. T. (2003) *Environmental Science: Working with the Earth*. Pacific Grove, CA: Brooks Cole.

Olson, S. (2003) "With this ring: A newly connected Seattle." *Seattle Post-Intelligencer*, May 12, p. E1. Available on-line at *http://seattlepi.nwsource.com*.

Pick, J. B., and Butler, E. W. (1997) *Mexico Megacity*. Boulder, CO: Westview Press.

PNM (2003) *Corporate Website of PNM*. Albuquerque, NM: PNM. Available at *www.pnm.com*.

Population Reference Bureau. (2002) *World Population Data Sheet*. Washington, DC: Author.

Pulido, L. (2000) "Rethinking environmental racism: White privilege and urban development in southern California." *Annals of the Association of American Geographers*, 90(1):12–40.

Reich, K. (2003) "Fire Insurance payouts could reach $3 billion." *Los Angeles Times* November 18:B6.

Roudi-Fahimi, F., Creel, L., and De Souza, R.-M. (2002) "Finding the balance: Population and water scarcity in the middle east and north Africa." *Middle East, North Africa Policy Brief.* Washington, DC: Population Reference Bureau.

Sahugun, L. (2004) "Dead pines pose risk of another 'mega-fire'". *Los Angeles Times* April 24:A1.

Shiklomanov, I. A. (1999) *World Resources and Their Use.* St. Petersburg, Russia: State Hydrological Institute. Available at *http://espejo.unesco.org.uy.*

Sierra Club. (2003) "Brownfields." Available at *http://www. sierraclub.org.*

Solley, W. B., Pierce, R. R., and Perlman, H. A. (1998) "Estimated use of water in the United States in 1995." *U.S. Geological Survey Circular 1200.* Denver, CO: U.S. Geological Survey.

Sullivan, S. (2003) "Redlands has retained the charm of days past." *Los Angeles Times* April 27:K16.

SUNY Buffalo. (1998) "Background on the Love Canal." *Love Canal Collection, University Archives.* Buffalo: State University of New York at Buffalo.

Tempest, R. (2003) "Southern California fires: Homes should never have been built." *Los Angeles Times* October 31:A21.

TVA (2004). *A short history of TVA: From the New Deal to a new century.* Knoxville, Tennessee: Tennessee Valley Authority. Available 2004 at *www.tva.gov.*

United Nations. (1992) *Urban Air Pollution in Megacities of the World.* Report of the United Nations Environment Programme and World Health Organization. Oxford, UK: Blackwell.

United Nations Development Programme. (2000) *Energy and the Challenge of Sustainability.* New York: Author.

U.S. Energy Information Administration. (2004) *Energy Glossary.* Washington, DC: U.S. Energy Information Administration. Available 2004 at *www.eia.doe.gov.*

U.S. Global Change Research Program. (2000) *Climate Change Impacts on the United States: The Potential Consequences of Climate Variability and Change.* New York: Cambridge University Press.

Vaux, H., Jr. (2002) "A U.S. water research agenda for the twenty first century." *Environment* 44(4):32–43.

White, G. F. (2000) "Water science and technology." *Environment* 42(1):30–38.

Wright, R. T., and Nebel, B. J. (2002) *Environmental Science: Toward a Sustainable Future* (8th edition). Upper Saddle River, NJ: Prentice Hall.

EXERCISE

Exercise Description

Air pollution in Mexico City and the effect of topography and wind.

Course Concepts Presented

The concepts addressed by this exercise include the quality of the urban environment as assessed through emissions mapping.

GIS Concepts Utilized

Environmental analysis, thematic mapping, spatial overlay.

Skills Required

ArcGIS Version 9.x

GIS Platform

ArcGIS Version 9.x

Estimated Time Required

1 hour

Exercise CD-ROM Location

ArcGIS9.x\Chapter_11\

Filenames Required

mx_mun03_final2.shp
Arrow_p87.shp
Arrow_p88.shp
Rectifyp24map.tif

Background Information

Air pollution in megacities has become a critical urban policy issue. Mexico City provides a good case study because its air pollution is influenced greatly by its unique topography and microclimates.

The emissions that are to be mapped are described by level of emission. The five emissions that are used in this exercise and their categories are described next.

Carbon Monoxide (CO)

A colorless and odorless gas that is harmful to humans when inhaled, carbon monoxide inhibits the absorption of oxygen by the blood. For more information on the description, sources, and health effects of carbon monoxide, see the Environmental Protection Agency (EPA) Web site at *http://www.epa.gov/air/urbanair/co/what1.html*. For the purposes of this exercise, the data are classified into six categories. The range of values for each category is shown in the following table.

Carbon Monoxide (CO) field name: CO_CAT

Category	Emission level (tons/year)
1	<1500
2	1501–3000
3	3001–4500
4	4501–6000
5	6001–7000
6	>7000

Hydrocarbons (HC)

Hydrocarbons are carbon-based gases (i.e., methane and benzene) emitted from automobiles and industry. For more information on the description, sources, and health effects of hydrocarbons, see the EPA Web site at *http://www.epa.gov/otaq/invntory/overview/pollutants/hydrocarbons.htm*. For the purposes of this exercise, the data are classified into six categories. The range of values for each category is shown in the following table.

Hydrocarbons (HC) field name: HC_CAT

Category	Emission level (tons/year)
1	<1500
2	1501–3000
3	3001–4500
4	4501–6000
5	6001–7000
6	>7000

Nitrogen Dioxide (NO_2)

This pollutant is produced mainly by motor vehicles and power plants that burn fossil fuels. For more information on the description, sources, and health effects of nitrogen dioxide see the EPA Web site at *http://www.epa.gov/air/urbanair/nox/what.html*. For the purposes of this exercise, the data are classified into six categories. The range of values for each category is shown in the following table.

Nitrogen Dioxide NO_2 field name: NO2_CAT

Category	Emission level (tons/year)
1	<100
2	101–200
3	201–300
4	301–400
5	401–500
6	501–600

Particulates ($\geq 10\ \mu$)

These consist of solid and liquid airborne particles of a wide range of sizes. For more information on the description, sources, and health effects of particulates ($\geq 10\ \mu$) see the EPA Web site at *http://www.epa.gov/air/urbanair/pm/what1.html*. For the purposes of this exercise, the data are classified into five categories. The range of values for each category is shown in the following table.

Particulates ($\geq 10\ \mu$) field name: PST_CAT

Category	Emission level (tons/year)
1	<20
2	21–40
3	41–60
4	61–80
5	>80

Sulfur Dioxide (SO_2)

Sulfur dioxide is a corrosive gas emitted in the burning of fossil fuels. For more information on the description, sources, and health effects of sulfur dioxide see the EPA Web site at *http://www.epa.gov/air/urbanair/so2/what1.html*. For the purposes of this exercise, the data are classified into five categories. The range of values for each category is shown in the following table.

Sulfur Dioxide (SO_2) field name: SO2_CAT

Category	Emission level (tons/year)
1	<100
2	101–200
3	201–300
4	301–400
5	>400

Readings

(1)   Greene, R. P., and Pick, J. B. *Exploring the Urban Community: A GIS Approach,* Chapter 11.

(2)   World Health Organization (1994) "Air pollution in the world's megacities." *Environment* 36(2):4–13, 25–37.

Exercise Procedure Flowchart

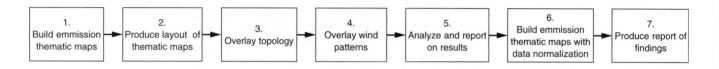

Exercise Procedure

Part 1: Examine the Pollutant Patterns

1.   Start ArcMap with Start Using ArcMap With A New Empty Map selected. Click OK.

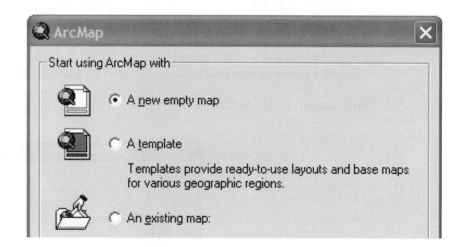

2.   Right-click Layers, then select Properties. Click General, and change name to CO, then click OK.

3.   Click Add Data and navigate to the exercise data folder.

Note: *If you do not see the folder where you placed the exercise data, you will need to make a new connection to the root directory that holds the data.*

4.   Add mx_mun03_final2 from the Chapter 11 folder by selecting it and clicking Add.

5.   Right-click mx_mun03_final2 and select Properties. Click the Symbology tab.

6.   Select Categories and Unique Values and a light-to-dark color ramp. Change the Value Field to co_cat (carbon monoxide) and click Add All Values. Clear the check box next to the symbol for <all other values>. Note: The emission values for all of the emission categories were described previously. A value of 0 indicates that there were no data available, which only affects 6 municipios.

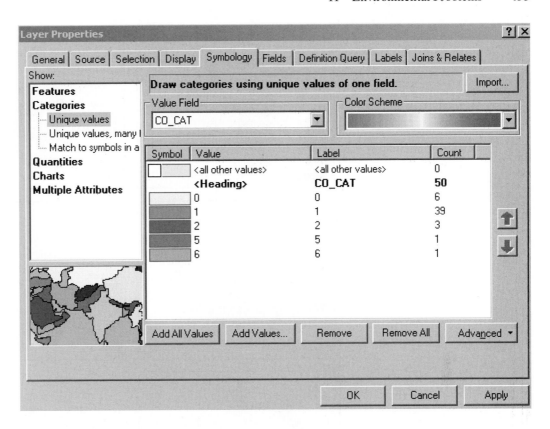

7. Click Insert and then click Data Frame. Repeat steps 2 through 6 for HC (hydrocarbons), NO_2 (nitrogen dioxide), PST (particulates), and SO_2 (sulfur dioxide). Note: In steps 2 and 3 the CO will be replaced by the emission code (HC, NO, PST, and SO). Also, the corresponding fields for mapping (step 6) are: hc_cat for HC, no2_cat for NO_2, pst_cat for PST, and so2_cat for SO_2.

8. Create a layout view of the five types of emissions with appropriate titles.

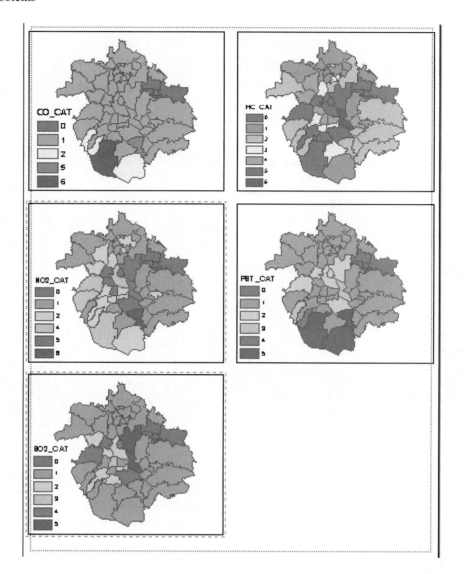

9. To get this placement, rearrange the five data frames, select one in the lefthand menu, right-click it, select Properties, and choose Size and Position. Set all widths to 4.0 and all heights to 3.25. For the X, Y positions use the following rule:
   a. X for left column is 0.125 and for right column it is 4.375.
   b. Y is 7.5 for top row, 4.0 for middle row, and 0.5 for bottom row (assumes a lower left anchor point).

10. To insert the correct legend select a frame and click Legend. Click Next, remove the word Legend, and replace it with the pollutant name. In the next dialog box, continue to click Next and Finish. Resize and reposition.

    Note: *You can clean up the legend title appearance by double clicking on the legend, double clicking on the legend item, and choosing the appropriate option for appearance.*

Part 2: Topography and Wind Analysis

11. Toggle back to the Data View and click Insert, then click Data Frame. Add the following layers to the new data frame: Rectifyp24map.tif, arrow_p87, and arrow_p88.
    a. Rectify24map.tif (image showing topography and Volcanoes)
    b. Arrow_p87.shp (lines indicating 4 PM Wind Direction)
    c. Arrow_p88.shp (lines indicating 9 AM Wind Direction)

12. Click on the Line Symbol for Arrow_p87.shp. In the Symbol Selector dialog box, scroll toward the bottom and select Arrow At End. Click OK, click Apply, and click OK.

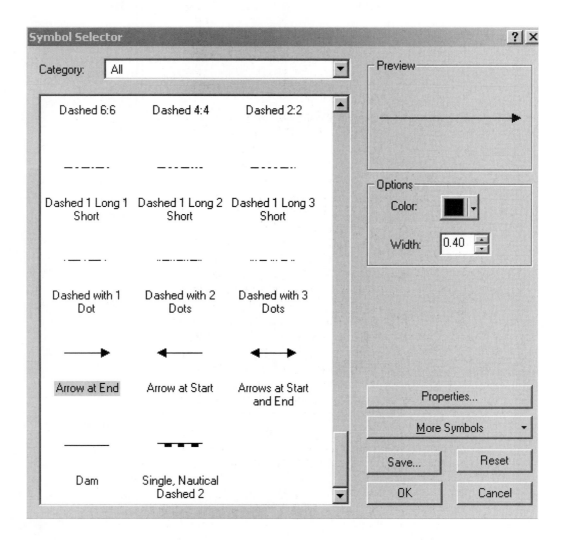

13. Repeat step 12 for Arrow_p88.shp.
14. Toggle back to the Layout frame and reposition the final data frame so that it is in the lower right.

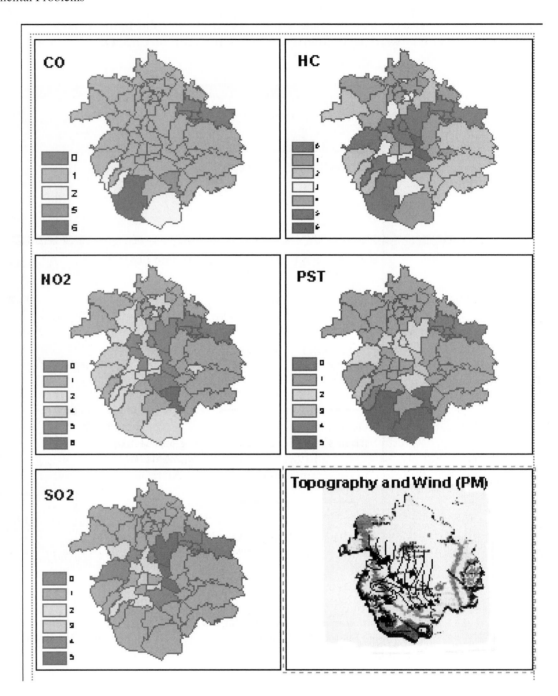

Part 3: Essay

Write an essay discussing to what extent the topography and wind patterns help explain the geographic concentration of each of the pollutants. Explain the spatial distributions of pollutants from the types of sources. How do you explain the locations of the sources? Are there some pollutant distributions that show a stronger relationship to the wind patterns and topography? If so, why might this be the case? What information from the World Health Organization (1994) reading or other more recent publications might help to explain these patterns?

12

Urban and Regional Planning

Cities have been planned since ancient times. Some cities centuries ago planned and implemented projects that stand out today. Contemporary planning confronts some of the same urban problems, as well as new ones, but can apply many modern methods, concepts, and technologies. Urban planning is one of the most important aspects of the study of cities. It is, after all, the way that design, order, integration, and redevelopment can occur. Many of the topics already covered bear on urban planning, including urban economic development, urban structure, population, neighborhoods, poverty zones, transportation, industrial location, urban sprawl, edge cities, environmental processes, and geographic information systems (GIS).

A geographical perspective to city and regional planning is essential. Geography informs planning on spatial proximities, flows, scale relationships, the arrangement of the natural environment, and many other aspects. This chapter is an appropriate capstone for a text on urban geography because urban planning must be integrative; it cannot separate environment, transportation, and neighborhoods, but must adopt an integrative, multidimensional perspective. Many famous plans, such as England's garden cities, the center of Paris, and Chicago's park system, encompassed multiple dimensions of planning. This is not to say that the integration always has worked or has included all needed elements, but the perspective was broad and multifaceted.

The chapter starts with the London fire of 1666 and the planning reactions resulting from it, including the world's first Building Act. At the same time, planning lessons were learned the hard way, but progress was made. The backdrop shifts to mid-19th-century New York City confronting a serious lack of open space. The solution, Central Park, turned out to be a milestone that gave rise to a series of improvements termed the City Beautiful movement. In the late 19th century, the next generation of urban thinkers discovered a new set of urban issues and problems; their discontent crystallized in the Progressive movement, which adopted a more formal approach to planning, including the first zoning precedents. Zoning continued to develop, becoming a mainstay of city planning.

Modern planning in the United States started in the 1920s and 1930s, and emphasized going beyond a single city to encompass the emerging metropolitan areas and regions. Metropolitan agencies known as councils of governments (COGs) were formed to plan for broad regions.

After World War II, the U.S. economy began an extended population and economic boom. However, that expansion put difficult pressures on many communities, leading to the concept and practice of growth management, under which methods were put forward to plan and administer an expanding city or metropolis, while striking compromises between key stakeholders. Precedent-setting examples from the 1950s and 1960s, discussed in the chapter, were Ramapo, New York, and Petaluma, California.

The "Analyzing an Urban Issue" section concerns the spatial aspects of integrated growth management planning. The exercise introduces a multidimensional spatial approach to analyze the impacts of a plant siting and associated population growth on a fringe watershed in Chicago. The section leads up to this issue by discussing GIS-based planning for the eastern coastal Australian city of Hervey Bay.

The smart growth topic from Chapter 10 is revisited in the context of planning. How can planning be done in the context of smart growth? Does it lead to urban benefits? The case example of Irvine in Orange County, California, demonstrates elements of the three major planning approaches covered in the chapter.

The end of the chapter addresses the planning-related issues of urban governance and politics and technology. Understanding and dealing with these issues are essential to urban planning success. Finally, several examples of the issues of urban planning in the developing world are presented for Mexican cities and for Shanghai, China.

12.1 ANTECEDENTS

The chapter's story of planning starts with London in the 1660s. Much of the history of urban planning is not discussed here, but the reader can access excellent sources for it (Hall, 1998; Levy, 2003). London had been growing in the mid-17th century, from a population of 191,000 at the beginning of the century to about 400,000 by 1660 (adjusted from Chandler, 1987). Although medieval city walls framed the city, London had overgrown their extent. Its structures were mostly of wood, although a building code dating from 1189 had specified that building facades needed to be stone or brick (Platt, 1996). Earlier, in 1580, Queen Elizabeth I had recommended establishing a green buffer zone for three miles outside the city walls, but her proposal had been largely ignored (Hearsey, 1965).

The great London fire occurred in 1666, spreading rapidly throughout the housing stock, fanned on by high winds (Hearsey, 1965). Extending over the city walls, it also burned the city periphery on the west. There was no possibility of containing it because of lack of technology for water pumping. When the fire was quelled, most of the city within the wall was burned (Loftie, 1884). Although 100,000 people became homeless, miraculously, only four people died.

The shock of this episode galvanized the British king and Parliament. King Charles II consulted Christopher Wren, the leading architect of the time, who came up with an ambitious plan for a capital city. However, the practical king did not subscribe to this idea, and instead presented a rebuilding plan having five major points (Platt, 1996):

• Street widths would be proportionate to the street's importance.

• No wood would be used on facades of buildings facing streets.

• A wide and open buffer area would be developed next to the Thames River, which would give firefighters access to water.

• Public nuisance structures, particularly a tannery and brewery, would be moved to the outskirts.

• Property owners who were restricted on public lands from rebuilding by the new design would be compensated.

A blue-ribbon panel of architects and other experts was appointed to recommend the best practices for the rebuilding (Hearsey, 1965). In 1667, the Act for Rebuilding London was approved by Parliament. This act essentially established a building regulatory code. It included four main building types, ranging from small residences to large buildings. The code keyed a particular section of street to the allowable building types. Regulations for new buildings emphasized fire prevention. The code specified exterior materials, ceilings, floors, wall thickness, room height, and the extent of roof overhangs, among other things (Platt, 1996).

The act had the weakness of lack of teeth to administer and enforce it. Nevertheless, it constituted a watershed event in urban planning history. In formalizing the government's regulatory role in a way not done before, it established a precedent for construction regulation in Britain. Many planning lessons from England were incorporated into land-use planning for Philadelphia, Savannah, and other colonial American cities (Levy, 2003).

During the first half of the 19th century, North American cities were growing rapidly as a consequence of high rates of immigration and urbanization, a process discussed in Chapter 4. In spite of the valuable precedents derived from Europe and, particularly, England, pressing problems remained. In the case of New York City, by 1850, rapid growth had magnified the problems of urban poverty and disease (Platt, 1996). This was accentuated by the lack of sufficient open and park space. Consequently, in 1853, the New York State Legislature passed a law that called for the formation of Central Park. The 770-acre site of the park was mostly open country 30 minutes' walking distance from the then edge of the city, which was already 30th Street in Manhattan.

In 1858, Frederick Law Olmsted was appointed the park superintendent. A Yale graduate with a degree in agricultural science, he had not settled into a career pattern nor had he designed a park (Platt, 1996). He proceeded to submit a design for Central Park, along with one of his assistants, Calvert Vaux. To their surprise, the plan was selected. It was full of innovation—such as changes in circulation patterns to allow pedestrians to move around in the park, without being blocked, while bicycles could circulate unrestricted through other routes, and horse-drawn carriages through others. The trick was including many over- and underpasses that allowed different vehicles to circulate simultaneously. At the time, Americans touted the country and rural landscapes as ideal; for example, in art this was expressed by the Hudson River School of landscape painters, who portrayed idyllic rural scenes. Central Park adhered to the ideals of the time by introducing a country landscape right into the metropolis.

The park was a hugely popular and financial success. Although it cost $14 million, it was made up within several years by the tax increases from the improved property values of housing adjacent to the park. In fact, within 10 years of its opening, the yearly increase in property tax for land adjacent to the park was $4 million (Platt, 1996). By 1871, there were more than 10 million visitors each year, meaning that on average, each New York resident made multiple visits. The park succeeded in providing water, meadows, and trees that provided welcome variation from city crowding (Platt, 1996). Another outcome was the identity that Central Park lent to the whole city. Subsequently, other similar parks followed in the city, such as Prospect Park in Brooklyn, also designed by Olmsted and Vaux (Levy, 2003).

Olmsted went on to design or redesign famous parks in a dozen U.S. cities. They included the Emerald Necklace, a connected series of parks circling Boston. Although generally praised, not all of them were successful, as evidenced by the "fenway" area in Boston, which became run down. A fenway is an area that contains fens, which are low lands entirely or partially covered with water that can be drained. Late in his career, Olmsted was the landscape architect for the 1893 Columbian Exposition in Chicago (see Figures 12.1 and 12.2). He designed the monumental Jackson Park, which went along with the Exposition's "White City" building design. Although Olmsted maintained his high standards of landscape design, the overall Exposition design leader was Daniel Burnham, who introduced the idea of functionalism, that is, coordinating the buildings, waste disposal, transportation, water supply, and electrical system together in the plan (Platt, 1996).

The next section examines the City Beautiful movement that Burnham led. It is remarkable that these two landmark planners collaborated for the Exposition at the opposite ends of their careers.

Burnham, the Chicago architect, leveraged the opportunities of the Columbian Exposition, in his first city-wide project in Washington, DC, in 1902 (Levy, 2003). Washington, DC had originally been planned in 1792

FIGURE 12.1   Colombian Exposition of 1893 in Chicago: Reflecting pool. Source: © Bettmann/CORBIS.

FIGURE 12.2 Colombian Exposition of 1893 in Chicago: Visitors in amusement park. Source: Corbis/Bettmann.

by Pierre L'Enfant, a military engineer from France, who had served with the Americans in the Revolution. It was a sweeping French design with broad and grand boulevards and stately buildings housing the principal functions of government. Two buildings, the Capitol and the White House, were placed on slight rises, connected by two straight vistas. From each major building, there was a radiating set of avenues that met in circular intersections, often also located on slight rises. By the mid- to late 19th century, population growth of the city and the accompanying overcrowding occurred. Because the nation was becoming more complex, there was an increase in government functions, requiring additional buildings not in the original plan.

The federal government sensed a serious problem with the increasingly outmoded design and, in 1900, appointed the McMillan Commission, chaired by a senator. The other panel members were Daniel Burnham; Charles McKim, another architect; Frederick Law Olmsted, Jr., landscape architect and the son of Central Park's designer; and sculptor Augustus Saint-Gaudens.

In 1902, the Commission produced the first city and regional plan in the United States. The key elements were the following (Levy, 2003; Platt, 1996):

- The Mall, Washington, DC's main axis, was redesigned and replanted. It would be extended west to include additional area for the future Lincoln Memorial and reflecting pool.

- Public structures, fountains, and gardens were added to the Mall.

- Union Station, designed by Daniel Burnham, was added northwest of the mall. It largely solved the problems of the extensive rail facilities in the city center.

- A less imposing but important landscaping plan was developed by Frederick Olmsted, Jr. Its main feature was a network of parks along the Potomac River and Rock Creek, as well as a parkway extending to Mount Vernon in Virginia. There was the strong influence of Olmsted's father, who had died a year earlier.

The plan elements were largely constructed and have proven robust over the century since. Of course, many adjustments have been made. This success was an early part of the movement known as City Beautiful, referring to efforts to augment the beauty and elegance of American cities. Other cities that Burnham subsequently led in redesigning were Cleveland, San Francisco, Baltimore, and also Manila in the Philippines.

Burnham's plan for Chicago in 1909 was especially important. It was embodied in a document written with architect Edward H. Bennett, entitled *Plan of Chicago* (Levy, 2003). The key components were the following:

- Chicago's lakefront park system was installed.
- There was limited consolidation of railroad terminals.
- The Cook County forest preserve system was established. It was regarded as justified to buffer the city's anticipated rapid population increase.
- A bridge for vehicles was built over the railroad yards on the south side of the downtown Loop area.
- An important central business district thoroughfare, Wacker Drive, was made double-deck.

Other elements envisioned by Burnham were less possible. For instance, he proposed a large civic center next to the harbor, having boulevards radiating in different directions in the French baroque manner, but this was never built. What was built from the Burnham–Bennett plan were those items the city government considered practical and suitable to the growing business city. It formed the basis of planning in Chicago for many decades. Today, elements like the lakefront parts are still crucial to the city's identity. Critics have been largely supportive but have found fault with the plan's absorption with public spaces and lack of concern for private development (Platt, 1996).

Some argue that the visionary city planning by an Olmsted or Burnham is largely absent in the United States today. As we will see later in the chapter, planning was brought increasingly under the regulation of government agencies and the courts, and this expanded bureaucracy is still with U.S. planning today. Consequently, the creativity and vision of these 19th-century planners might not be met. Important problems for students in urban geography to study are what are the best ideals of urban planning and how can the setting of planning encourage them.

12.2 GARDEN CITIES

The Garden City movement in England was coterminous with Beautiful Cities and emphasized towns located in the countryside. These towns were intended to reverse the direction of urban-to-rural migration streams of 19th-century England, which were overcrowding the major cities.

The Garden City concept was devised by town planner Ebenezer Howard, who published the ideas in his 1892 book, *To-morrow: A Peaceful Path to Social Reform*, later retitled *Garden Cities of Tomorrow* (1902). Howard had a varied career as a stenographer, law reporter, social reformer with liberal views, and planner. He proposed garden cities of 32,000 population to be located away from major cities on a land plot of 6,000 acres, of which 1,000 acres would constitute the core section (Levy, 2003). He felt that the size of cities should not be accidental but carefully defined by conscious thought (Platt, 1996).

The Garden City spatial arrangement (see Figure 12.3) consisted of a central park with cultural attractions, surrounded by residential and business sections, and an outlying large green band that is mostly agricultural. The central park area would include a theater, museum, art gallery, hospital, library, concert and lecture hall, hospital, and "crystal palace" shopping arcade (Howard, 1892; Levy, 2003). Surrounding the core was residential housing with gardens located on tree-lined avenues or interior streets. Farther out was an industrial zone, placed to avoid pollution and discourage sweatshops. It was important that the garden city would have its own economy and not be dependent on a major city such as London. The city was compact and surrounded by a green rural ring.

Its governance structure was unique. It would be privately governed as a trust or corporation by a small group of individuals on a nonprofit basis. People oriented toward philanthropy would raise funds in the investment market. The trust would own the city's land and would redirect the rental profits to improving the city on the basis of a master plan (Platt, 1996).

On a regional scale, a group of garden cities would serve to reverse the direction of city-based migration and restore population to the countryside. The idea of town and country is seen in the magnet diagram (see Figure 12.4) created by Howard (1892). "Town" represents big cities with the advantages of economic opportunities, but problems of deteriorated environment and unaffordability, whereas "country" connotes a garden city with a healthy environment and natural beauty, but lack of social diversity, boredom, and a weak civic environment (Howard, 1892). The "people" on the diagram appear to be torn on the choice between the town and country.

In 1903, work started on the first garden city, Letchworth, about 30 miles north of London. It followed fairly well Howard's plan but did not have a central garden or crystal palace. It had rigid land-use controls and was governed by a corporate structure along the lines that Howard proposed. Welwyn Garden City, less strict about

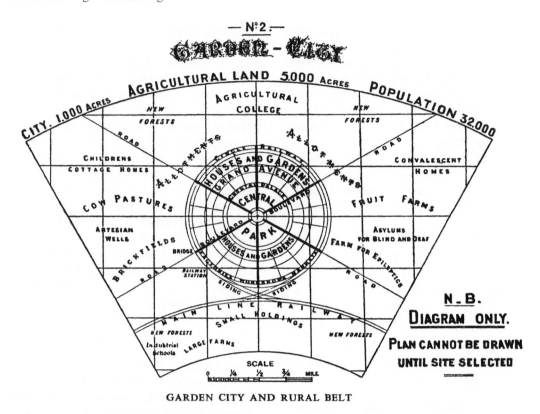

FIGURE 12.3   Ebenezer Howard's diagram for a garden city and surrounding area. Source: Osborn, F.J. (ed.). 1945/1965. *Ebenezer Howard's Garden Cities of To-morrow*. Cambridge: The MIT Press. Figure 6-13. "Howard's diagram of Garden City and its 'rural belt.'"

Howard's plan, was built nearby in 1919. Although the garden cities did not proliferate, after World War II, the New Town concept, patterned somewhat after garden cities, appeared and was more successful, with 14 eventually being developed. The New Towns Act of 1946 called for towns of 30,000 to 140,000 population, not including the surrounding large greenbelt (Levy, 2003; Platt, 1996). The concept was adapted by other countries including Sweden, some in western Europe, the Soviet Union, and the United States.

In the United States, three "greenbelt towns" were designed and funded by the federal government following World War II in Greenbelt, Maryland; Greenhills, Ohio; and Greendale, Wisconsin (Randall, 2000). However, the three communities were never completed and their further expansion was prevented by adjacent development. Perhaps the most successful greenbelt town in the United States was Park Forest, Illinois, done in a collaboration between government and private builders, and involving most of the garden-city ideas (Randall, 2000). Park Forest in the period 1946 to 1952 is seen in Figure 12.5. Its plan responded to a housing shortage following the Depression and World War II and the large number of returning GIs. It was as close to the planning

ideas of Ebenezer Howard as any new town built after World War II (Randall, 2000).

12.3 PROGRESSIVE MOVEMENT AND ZONING

As it unfolded, the City Beautiful movement engendered criticism from its beginning. Critics claimed its beauty features ignored underlying, larger problems of overcrowding, congestion, poverty, and political and business corruption. Those problems stemmed partly from the huge immigration to America in the 19th century that was discussed in Chapter 8 and also from the rampant industrial corruption that muckraker journalists had so poignantly identified.

The spark of the Progressive movement was Benjamin Marsh, a political activist in New York City, who had set the goal of instilling a standard practice of city planning (Platt, 1996). In 1909, Marsh authored the first textbook on urban planning, *Introduction to City Planning* (Marsh, 1909). That same year, the progressives sponsored the first national conference in the United States on modern planning (Cullingworth & Caves, 2003). The tone was practical, that is, to reduce the stated problems through action steps. This movement raised the profile of planning

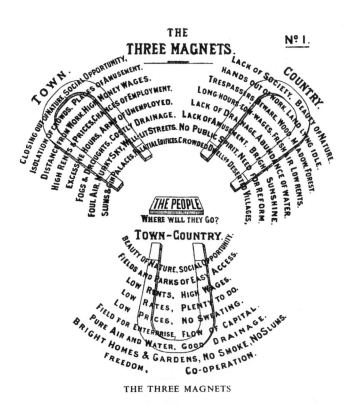

FIGURE 12.4 Ebenezer Howard's "Three Magnets" diagram. Source: Osborn, F.J. (ed.). 1945/1965. *Ebenezer Howard's Garden Cities of To-morrow*. Cambridge: The MIT Press. Figure 6-12.

deficiencies. It underscored the need for planning mechanisms to overcome congestion and other problems and led to the advent of zoning.

Zoning refers to the partitioning of land in a city by ordinance into sections reserved for different purposes. It appeared in the United States in 1916 as the result of a land-use dispute in Manhattan. Merchants in the Garment District were encroaching on the space of Fifth Avenue merchants, including introducing larger buildings. The Fifth Avenue merchants led a coalition to establish a

FIGURE 12.5 Park Forest, Illinois, 1946–1952. Source: Art Institute of Chicago. Photo. "Park Forest, Illinois, 1946–1952.

city zoning ordinance that controlled the building sizes by district; this protected their environment as well as its property values. Besides building bulk, today other types of zoning regulation include land use; population and housing densities; parking requirements, especially for larger buildings; advertising signage; extraction of minerals; nondomestic uses of homes such as for professional offices; and special environmental factors such as wetlands (Platt, 1996). For example, zoning in a city might be based on a weighted combination of housing densities and environmental indicators (Fleissig & Jacobsen, 2002).

The use of zoning spread among municipalities but ran into potential downsides. Although zoning protected certain stakeholders, others such as landowners and developers were hurt by their inability to carry out building projects and realize sales. Legal contention grew, and in 1926, a famous case in the U.S. Supreme Court brought to a head this key issue. In *Village of Euclid v. Ambler Realty Company,* a realty company sought to challenge the zoning restrictions then in place, pointing to its potential losses. The legal issue boiled down to the question of Constitutional police power that enabled zoning versus the 5th Amendment of the Bill of Rights and 14th Amendment of the Constitution, which gave an individual reasonable use of his or her property. The ruling was in favor of the village, so the landowner was not compensated. More specifically, *police power* is the permissible scope of federal or state legislation, taking into account that it might conflict with an individual's right, with respect to health, safety, and welfare of the general public.

Zoning depends on the reasonableness, both in the use of police power by local government and in the use of the land. For cities, it is crucial that a plan be put in place that justifies the zoning options selected; otherwise, a court might not recognize that zoning is reasonable. This accentuates the connection between zoning and planning. Zoning is usually implemented by a planning commission, which is a group of experts appointed by the city council. If disputes occur, they are referred to the board of appeals, a city unit that tries to resolve differences before they go further in the legal arena.

Zoning is done on the basis of three major factors: use of private land, density of structural development per land unit, and the bulk of buildings (Platt, 1996). On the basis of specific criteria, dozens of zoning classes can be arrived at by the city (Levy, 2003). The usual major classes are residential, commercial, and industrial, but the zoning categories could get much more refined. For instance, the residential class might apply if housing unit density is more than three per acre. An example of zoning classes for the town of Fairfield, Connecticut, is given in Table 12.1 (Town of Fairfield, 2003).

Zoning is a constitutional police power right that is delegated from the federal government to the state

TABLE 12.1 Residential zoning classes for the town of Fairfield, Connecticut

Residential zoning district	Permitted use	Minimum lot size
AAA	Single detached dwelling for one family	Two acres
AA	Single detached dwelling for one family	One acre
R-3	Single detached dwelling for one family	20,000 square feet
R-2	Single detached dwelling for one family	14,000 square feet
A	Single detached dwelling for one family	9,375 square feet
B	Single detached dwelling for one family	6,000 square feet
B	Single detached dwelling for two families	9,000 square feet
C	Single detached dwelling for one family	5,000 square feet
C	Single detached dwelling for two families	7,500 square feet
C	Single detached dwelling for three families	10,000 square feet
C	Single detached dwelling for four families	12,500 square feet

Source: Town of Fairfield, 2003.

government and then to local cities. Federal land is exempt from zoning, but state land varies; sometimes it is exempt, whereas at other times, it depends on state law. An interesting question is how municipalities do zoning of their own land; the answer is that they can easily amend their own zoning laws to accomplish zoning of their own land (Platt, 1996).

The practical experience over many decades with zoning has led to a body of experience and certain continuing issues. A legal pattern is that courts, over a long time, have generally tended to favor the zoning ordinances, so long as they are connected to city plans. A controversial part of zoning is the "taking" issue. This issue is the extent to which zoning regulations can reduce the value of a property without compensation to the owner (Levy, 2003). It is backed up in the 5th Amendment, which states, "nor shall private property be taken for public use without just compensation." Over the past 75 years, many disputes and law cases have evolved around taking, with no certain outcome on either side. It depends on the reasonableness of the zoning, connection to planning, and extent of anticipated losses.

Governments have misused zoning to accomplish objectives not keyed to city planning. For instance, the NIMBY ("not in my backyard") attitude implies that a certain set of stakeholders or the city as a whole wants to exclude certain parties it considers threatening. However, a municipality is biased if it uses zoning to satisfy NIMBY complaints. The point is that the issue of NIMBY needs to be brought to a head in ways other than zoning. In *fiscal zoning,* the city uses zoning as a tool principally to increase its revenues, again misusing the instrument. Zoning has also been misdirected for purely political ends.

The long-term trend in zoning use is toward greater flexibility. A city can achieve this by frequent review of zoning classification, willingness to amend zoning ordinances, and granting of exceptions, usually in the form of conditional use permits, which means the zoning decision

FIGURE 12.6   Zoning in downtown Chicago.

is put off and made later by the appeals board. Zoning in Chicago is seen in Figure 12.6. Zoning is the standard in the United States and many other countries, but some (e.g., Mexico) utilize alternative approaches, which are discussed later. Zoning has the advantages of aspiring to achieve fairness and completeness, but it also tends to reduce the creativity of urban planning. For example, zoning regulations would likely have halted such creative thinkers as Frederick Olmsted, Sr. and Ebenezer Howard.

12.4 MODERN PLANNING: METROPOLITAN AND REGIONAL PLANNING

In the 1910s through the 1950s, U.S. metropolitan areas were growing and the metropolitan concepts, discussed in Chapter 5, were taking hold. Cities began to expand more into their suburbs, extending past their city limits. From a planning standpoint, it was no longer enough for the city planning commission to focus on what was within the city limits. Neighboring suburbs and cities had to be considered as well. In 1928, the federal government recognized this in the Standard City Planning Enabling Act of the Department of Commerce (Levy, 2003). It recommended, among other things, that a local planning commission must consider in its master plan the areas outside the immediate city territory that "bear relation to the planning of such municipality" (Platt, 1996). Another milestone was the first metropolitan plan, the Plan for Greater New York in 1929, which was sponsored by New York's Regional Plan Association (Platt, 1996).

There were many stakeholders in the metropolitan and regional setting, and many views and opinions on the importance of coordinated planning—an issue still with us today. Inevitably, the courts would be brought into play, which happened with two landmark decisions, both in New Jersey. The Cresskill decisions in 1949 and 1954 involved whether it was optional or mandatory for the borough of Cresskill to exclude land uses, if they were needed for the surrounding region. At this time, Mount Laurel, New Jersey, was a rapidly growing and sprawling 22-mile-square township with mixed developments and agricultural uses. It had zoned its remaining agricultural land for industry purposes, a move intended to lower average property taxes and appeal to existing landowners. However, this meant that newcomers faced higher land prices. Mixed housing, including affordable low-cost housing, was discouraged. The court ruled that the township had to include a range of land offerings at different pricing levels (Platt, 1996). In both court decisions, the municipalities were forced to take a broader metropolitan and regional perspective.

After World War II, there was continuing metropolitan expansion. The federal government, including the U.S. Census Bureau, recognized this in 1949 by establishing standard metropolitan areas, retitled in 1959 as standard metropolitan statistical areas (SMSAs). Reflecting this real-world growth, in 1957, Jean Gottmann published his famous article "Megalopolis, or the Urbanization of the Northeastern Seaboard," which gave a new name to the series of metropolises in the Northeast (Gottmann, 1957). In the decades following the Cresskill decisions, there was an even greater planning thrust toward covering the whole metropolis and region. These developments were implemented by counties collaborating, states encouraging changes, and federal support.

Planning agencies for groups of cities and counties appeared more frequently in the late 1950s, 1960s and 1970s. The federal government supported their development by giving substantial funding through the 1954 Housing Act (Section 701) to form COGs. To be a COG, at least 51 percent of the board of directors must be elected officials of associated cities or counties (Levy, 2003). The federal funding for COGs was plentiful from 1954 to the beginning of the Reagan administration in 1981, when it largely expired, thrusting the COGs into rough times (Simmie, 2001). This was further encouraged by the requirement of the federal Office of Management and Budget in the mid-1960s that all SMSAs establish an arrangement for planning that covered the entire area.

An example of a COG is the San Diego Association of Governments, known as SANDAG, see *www.sandag.cog.ca.us*. It was formed in 1966 on the basis of a joint powers agreement between San Diego County and the State of California. Its members have come from the city governments within San Diego County, and its mission has been to provide area-wide planning and coordination. Starting with a focus on population, economy, and development, it has added many metropolitan planning elements as it went along (SANDAG, 2003), including transportation (1975), criminal justice (1977), technical assistance to its members (1978), regional housing needs (1979), waste management (1990), congestion management (1991), and an advisory relationship with the water agency (1995). Today, SANDAG is a sophisticated metropolitanwide planning unit utilizing census and other data through advanced GIS capabilities. SANDAG is somewhat unusual for these agencies in comprising the area of only one county; most involve two or more.

In California, two other major examples of regional planning COGs are the Southern California Association of Governments (SCAG, see *www.scag.ca.gov*), headquartered in Los Angeles, and the Association of Bay Area Governments (ABAG, see *www.abag.ca.gov*). In Chicago, the Northeastern Illinois Planning Commission (NIPSE, see *www.nipc.org*) is a regional public planning agency that coordinates public planning across six counties and many cities. A SCAG public meeting is shown in Figure 12.7.

Other types of metropolitan and regional planning entities include economic development commissions, transportation districts, and river basin commissions. These are usually consortia of groups of municipalities and counties, with varied funding drawn from federal, state, and local sources (National Association of Regional Councils, 2004). For instance, large metropolitan areas often have special districts for air pollution control and water resources that cover their whole metropolitan area or a broader region. The South Coast Air Quality Management District (SCAQMD, see *www.aqmd.gov*), originally founded in 1957 and consolidated in 1977, performs air pollution planning, regulation, enforcement, monitoring, technology development, and education for 15 million residents in four counties of southern California. River basin commissions such as the Ohio River Commission, coordinate and manage water resources in multistate and multicity regions. For transportation planning, states have often kept control, sometimes coordinating with regional and metropolitan transportation councils

FIGURE 12.7   Public meeting of Southern California Association of Governments. Source: SCAG. 2004. "SCAG 2003–2004 Annual Report." P. 7.

(Simmie, 2001). Finally, master planning by private companies has sometimes played a role for metropolitan areas and regions. In this chapter's case study for Irvine, California, a corporation, the Irvine Company, that had owned the future city's land for about a century, originated the master planning and continued to participate over time.

In summary, metropolitan and regional planning grew during the 20th century, responding to the steady expansion of cities into metropolises and the development of closely spaced metropolitan areas in certain regions. Courts often backed the responsibility to take a broader metropolitan perspective. The federal and state governments encouraged and funded new entities for regional planning.

12.5 GROWTH MANAGEMENT

Growth management, a planning approach introduced in the late 1960s and early 1970s, responded to the decades of post–World War II high fertility and economic growth, including the original baby boom of 1946 to 1964. Growth management was a response of cities to potentially being overwhelmed by the challenges of rapid population increase, but it raised issues of the rights of private developers as well as the potential barriers for low earners to live in communities where real estate prices and rentals were high.

Growth management differs from zoning in that zoning focuses on the situation of individual owners and parties, whereas growth management focuses on stakeholders across the area as a whole. Another contrast is that zoning usually occurs in piecemeal steps independent of possible effects on overall growth, whereas growth management focuses more on overall municipal or metropolitan objectives. As with zoning, growth management has constitutional constraints, as illustrated by the Ramapo and Petaluma cases discussed later.

To implement growth management, a municipality has a number of available control options. The major categories are as follows (modified from Fulton, 1999):

• *Control of population size through housing restrictions.* Restrictions can be established on housing permits, or policies created that favor sites in neglected areas of the central city, referred to as in-fill.

- *Zoning control.* An example is that a city can down-zone certain city zones to industrial uses. Once industry has been established, this puts constraints on developers' ability to up-zone later for residential projects.
- *Control of total floor space.* This uncommon technique involves putting annual total caps on floor space for residential and commercial development. In California, it has been done in Santa Monica, Pasadena, and San Francisco.
- *Infrastructure control.* The city specifies that new development can only be undertaken if it does not significantly impact infrastructure.
- *Political control.* Growth is directly voted up or down by a general ballot of citizens, city council, or board of supervisors. On the ballot, you might be asked to vote for or against growth.
- *Control by a growth management element in the general plan.* The element defines the growth management goals, objectives, measures, and steps.

In growth management situations, there might be multiple agencies involved at different levels of government. For instance, a state transportation agency's policies might constrain growth for a city.

Growth management is subject to modification by the courts. Two famous early cases established many of the precedents (Cullingworth & Caves, 2003). One involved the town of Ramapo, New York, located in Rockland County 35 miles north of Manhattan (Cullingworth, 1997; Levy, 2003) In the late 1960s, the town's population had undergone rapid increase to 76,000 residents. A master plan was put in effect in 1966, which carefully controlled development in concert with the timed availability of city services and facilities. Once services were in place, a special permit could be granted for development. One outcome was that new projects ground to a halt at the periphery, where little infrastructure was in place (Cullingworth, 1997; Fulton, 1999).

The New York Court of Appeals ruled, "Where it is clear that the existing physical and financial resources of the community are inadequate to furnish the essential services and facilities which a substantial increase in population requires, there is a rational basis for phased growth and, hence, the challenged ordinance [i.e., in the town's plan] is not violative of the federal and state constitutions" (cited in Cullingworth & Caves, 2003, p. 156). This step constituted the first court recognition of the validity of phased growth management; subsequently, timing was added by court action as a growth management tool for cities.

Another famous case involved rapid population expansion in the city of Petaluma, located 40 miles north of San Francisco. New housing units jumped in one year; 300 annual units in 1970 grew to 891 in 1971. In 1971, the city approved a temporary stop to development to allow time for adoption of a growth management plan (Fulton, 1999). A year later, a plan was approved calling for a maximum development of 500 units per year. At the same time, the city developed a multicriteria system of points to determine which development projects could be approved. The criteria included "access to existing services which had spare capacity, … excellence of design, … provision of open space, … inclusion of low-cost housing, and … provision of needed public services" (cited in Cullingworth & Caves, 2003, p. 156). Imposing the criteria slowed growth and also led to an 8 to 12 percent proportion of low- and intermediate-cost housing for the 500 units approved (Cullingworth & Caves, 2003).

Developers were upset and brought a lawsuit, which eventually reached the U.S. Court of Appeals. Ruling in favor of the city, the court stated that Petaluma's growth management policy was a proper use of its "police power" (Fulton, 1999). It noted that although the city's plan was somewhat exclusionary, so was any zoning approach. Including some of the low- and moderate-income housing had helped influence the ruling in favor of the city.

Since these landmark decisions, many disputes on growth plans have come before the courts; in most cases, plans have been supported, with the crucial factors being whether there is enough planning to back up the policy and whether its housing provisions are sufficiently inclusive. The point system developed by Petaluma and other cities has been repeated by many communities, although Petaluma itself later dropped it after population pressures subsided. Some court rulings have gone against growth management, especially if planning was weak or policies appeared to be exclusionary.

Some of the continuing debate in growth management is summarized in Table 12.2, modified from Fulton (1999). It is evident that growth management is not simple, since it is connected to many external factors such as state and federal policies, real estate markets, and demographic booms and busts. State policies influence growth management approaches as well. For example, in California, growth management in coastal municipalities is constrained by the California Coastal Commission, an agency with special powers to regulate coastal lands. Oregon has mandated growth management planning for all of its counties. There has been a statewide concurrence that its cities cannot become oversized, as Oregonians perceived happened in California. Furthermore, all of Oregon's 241 cities have adopted urban growth boundaries (see Chapter 10) that must be approved by a state commission. However, Oregon voters in 2004 modified these policies and regulations to allow compensation of aggrieved landowners affected by the restrictions (Barringer, 2004). By contrast, Hawaii, at the state level, has an unusual centralized system of state land-use control. In summary, growth management is not just a local prerogative but is influenced by state policy and other forces.

TABLE 12.2   The continuing debate on growth management

Some of the areas of continuing debate on growth management are the following:

The building industry says that growth management only leads to higher prices. Response. The point here is that real estate markets are complex, and the building industry argument is oversimplified.

Does growth management lead only to a shift in the type and size of developers? Response. A point system for development project applications, such as Petaluma used, favored large-size homes with amenities and the ability to wait it out. In this case, the bias is given to large-scale, well-financed developers.

Are there outcomes of the overemphasis on residential development and underemphasis on commercial/industrial?

Response. The growth management measures historically began by restricting residential development, and not controlling commercial and industrial development. The result may be that communities end up with high-cost housing but lots of jobs. This is seen in the trend for industry today to move into fringe areas.

Do citizen activists show a lack of connection between supporting residential restrictions and at the same time economic growth? Response. The same citizens who want growth management restrictions often also favor economic growth. However, the two are contradictory.

Growth initiatives in municipalities are too limited, because the issues are regional and statewide. Response. If growth management is controlled in one community, that is not a regional solution, as growth will just move elsewhere. Example: San Luis Obispo, California is a "charming" community with carefully controlled growth. However, other parts of San Luis Obispo County have received a large influx of growth. The growth management crisis has only moved, not been solved.

Source: Fulton, W. (1999) *Guide to California Planning* (2nd ed.). Point Arena, CA: Solano Press Books.

12.6 ANALYZING AN URBAN ISSUE: LAND-USE PLANNING

In the lab exercise for this chapter, you apply what-if analysis to the impacts of location of a cell phone plant on land use and population distribution within the Blackberry Watershed west of Aurora, Illinois. The lab involves first computing the overall population growth rate of the watershed on the basis of historical data. Next, a series of what-if environmental and economic assumptions are made as to how the population growth rate will be allocated spatially to the census block groups within the watershed. Finally, the resultant spatial distribution of population is projected into the future and mapped with GIS. This exercise challenges you to use more sophisticated analysis functions of ArcGIS to arrive at an integrated spatial model that outputs the spatial population distribution. The exercise also draws on your conceptual knowledge from this chapter, in particular, growth management, zoning, and environmental aspects of planning.

This type of what-if analysis is sometimes known as sensitivity analysis. What-if refers to the outcome of making assumptions about urban parameters. In the lab, we make the assumption that a plant will be located in the watershed (the "what-if") and examine what the implications are.

Cities, counties, COGs, special planning agencies, and other planning units frequently undertake what-if analysis and problem solving. GIS is a helpful tool to utilize because it goes beyond just the magnitudes of impacts to show their spatial distribution. As we have seen in the last chapter and this one, it is often the locus of impacts that is the key source of local controversy. For instance, if impacts can be located away from cities in an abandoned or unused area, public controversy might be reduced, although it usually is not absent.

Consider the contrasting example of what-if planning applied to population growth impacts in Hervey Bay Township along the east coast of Australia (Pettit, 2002). The township had been impacted by rapid population growth, an increasing number of immigrants, more community diversity, an older age structure, and rising tourism (Gillam & Taylor, 2001). It needed a tool to cope with the impacts of this growth on its environment, land use, and economic development.

It utilized a GIS system consisting of ArcGIS software and a decision-planning add-on software called What If? 1.1. Inputs from federal, state, and local planning agencies included cadastral land parcels; footprints of buildings; road, sewer, and water lines; land use; vegetation cover; national parks and state forest; coastal wetlands; extent of prime agricultural land; open space; and areas of significance to native population (Pettit, 2002). Hervey Bay's City Council in turn indicated what their priority ratings were for suitability factors. The priorities represented users' assumptions about the nature and policies of land use, for example, whether the city should be kept compact or development should be allowed to be scattered outside the city limits in adjacent areas.

Two examples of outcome maps are shown in Figures 12.8 and 12.9. The coverages include land use, soils, slopes, streets, school districts, and others. The first map (Figure 12.8) shows the implications of allowing a growth scenario. With no land-use controls, the growth occurs beyond the city limits and outside of the map window pictured (Pettit, 2002). The city's greenbelts are maintained and the many undeveloped land tracts are left intact on the fringe. In the compact growth scenario (Figure 12.9), the growth takes place within the city limits, particularly filling in many of the green areas and also causing residential development in rural–urban fringe areas. From this GIS, the council determined that the city limits would need to be expanded to achieve the objectives of maintaining

FIGURE 12.8 What-if planning scenario for Hervey Bay, Australia: Growth scenario, with scatter outside of city limits beyond map window. Source: Pettit, Christopher J. "Planning Application Tool - What - If," ArcNews, Fall 2002, available on-line at www.esri.com. Undeveloped land tracts appear in white. Residential land appears in light grey. Parks and public land appear in black.

green spaces and having compact growth within the city limits. This Hervey alternative example should assist you in gaining a broader understanding of the usefulness of integrated GIS-based what-if planning, prior to undertaking the chapter exercise.

For very large metropolitan areas, more complex modeling and simulations are often performed (Batty and Xie, 2005). For instance, a large-scale urban growth model was constructed for the megacity of Lagos, Nigeria (Barredo, Demicheli, Lavalle, Kasanko, & McCormick, 2004). The city was divided into modeling spatial grid cells of 100 by 100 meters, called urban cellular automata. Each cellular automaton depends on the states of the other cells within its neighborhood. Cells have complex transition algorithms that are beyond the scope of this chapter (see Barredo et al., 2004; Xie & Baty, 2005; Batty, Xie, & Sun, 1999). The model has particularly strong land-use factors that could be calibrated through use of time series data. As a developing-world megacity, Lagos experienced very rapid growth in its population and land extent from 1962 to 2000. The model was run and the city's urban spatial distribution projected for 20 years from 2000 to 2020. It projects that

Lagos in 2020 will become a megacity with 27 million persons, encompassing an area of 969 square kilometers, which is more compact than Mexico City today (Barredo et al., 2004). The model projects continued rapid growth of the urban area, although with considerable consolidation and in-fill. The model can be used for metropolitan planning, and is especially sensitive to land-use assumptions. It serves to demonstrate the need for better infrastructure to provide the key needs of water, energy, and communications (Barredo et al., 2004). Although the model does not include sophisticated components for population, economics, and environment, these can be added as a future enhancement (Barredo et al., 2004).

12.7 PLANNING FOR SMART GROWTH

Smart growth, sometimes called new urbanism or neo-traditional planning, was introduced in Chapter 10. This section follows up by focusing on the planning for smart growth, including an example of it applied to in-fill redevelopment in Chattanooga, Tennessee; it also considers

FIGURE 12.9 What-if planning scenario for Hervey Bay, Australia, compact growth. Source: Pettit, Christopher J. "Planning Application Tool - What - If," ArcNews, Fall 2002, available on-line at www.esri.com. Undeveloped land tracts appear in white. Residential land appears in light grey. Parks and public land appear in black.

criticisms of smart growth and compares it to the other planning paradigms of metropolitan and regional planning and growth management.

As mentioned before, smart growth encourages compact development with mixed uses, redevelopment of in-fill areas and brownfields, pedestrian friendliness of walkways (see Figures 12.10 and 12.11), mass transit, high-quality parks connecting neighborhoods, and architectural designs that encourage social interaction and respect the region and local history (Ellis, 2002).

The planning for smart growth emphasizes compactness, environment, and quality. Recent studies indicate it is favored by between 25 and 40 percent of developers in the suburban market (Ellis, 2002). In the inner city, it is being encouraged by the Department of Housing and Urban Development (HUD)'s Hope VI Program, which favors projects that apply smart growth to in-fill redevelopment (Barnett, 2003; Ellis, 2002).

A key issue in planning for smart growth is how to assure the public as well as the courts that this type of

FIGURE 12.10 Wide suburban street with no sidewalks, under assumption of little or no walking. Source: Ellis, Cliff. 2002. "The New Urbanism: Critiques and Rebuttals." *Journal of Urban Design* 7(3). P. 265. Figure 3. "A wide suburban street with no sidewalks."

FIGURE 12.11 Pedestrian pathway in Kentlands, Gaithersburg, Maryland. Source: Ellis, Cliff. 2002. "The New Urbanism: Critiques and Rebuttals." *Journal of Urban Design* 7(3). P. 265. Figure 4. "Pedestrian pathway in Kentlands, Gaithersburg, Maryland."

planning is inclusive of poor and marginalized classes. However, there is nothing inherent in smart-growth principles that points to any particular social class. For instance, pedestrian and bicycle features can be applied to lower class clientele as well as for the middle class. The keys to planning success with smart growth appear to be the following (Ellis, 2002):

- Open up the project for discussion with the community and public, to educate them on smart-growth principles and gain buy-in. Include public meetings, if possible.

- Be flexible on adherence to some, but not necessarily all, of the smart-growth principles. A smart-growth purist might find it much harder to gain public support and financing.

- Use financial planning techniques to balance objectives. One useful method to check the balance of smart-growth goals is the "Smart Scorecard" (Fleissig & Jacobsen, 2002). As seen in Table 12.3, a designer can score a range of weighted smart growth criteria to determine the overall extent of adherence. For each criterion, specific measures can be combined to achieve a rating.

- Consider how low- and intermediate-income housing demand is addressed in a smart growth plan. Include provision of some lower cost options, especially for larger developments and city plans.

- Involve designers and builders who are experienced in smart growth and regard it as a way to achieve worthy overall city planning results, rather than primarily as a means to attain individual architectural recognition.

TABLE 12.3 The smart scorecard: A way to weight smart-growth criteria.
The Smart Scorecard is used by communities to prioritize their scorecard for smart growth success.
The following table gives hypothetical weightings for one city of smart growth measures:

Smart growth criteria	Weighting: percent of total score
Proximity to Existing/Future Development and Infrastructure	15
Mix and Balance of Uses	12
Site Optimization and Compactness	12
Accessibility and Mobility Choices	12 ·
Community Context and Site Design	10
Fine-Grained Block, Pedestrian, and Park Network	10
Environmental Quality	8
Diversity of Housing	8
Reuse and Redevelopment Options	7
Process Collaboration and Predictability of Decisions	6
Total	100

Examples of specific measures for diversity of housing criterion	
Specific measure	**Rating scale**
Variety of building types and styles.	Rated on 1 to 4 scale
Provide a wide range in pricing structure of units that will be sold or leased at least 20 percent of units priced for 80–100 percent of average median income.	Rated on 1 to 4 scale
Vary residential lot sizes.	Rated on 1 to 4 scale
Address need for civic facilities and amenities, such as day care, teen/senior center, cultural facility, etc.	Rated on 1 to 4 scale

Source: Fleissig, W., and Jacobsen, V. (2002) *Smart Scorecard for Development Projects.* San Francisco, CA: Congress for New Urbanism. Available at *http://www.cnu.org.*

One of the most promising areas for smart growth is in-fill redevelopment because it brings new design ideas to a part of the city that is receiving less attention today. This refers to projects to reinvigorate brownfields and other neglected or abandoned areas that might be located in the central city or more outlying areas.

An example is the Eastgate Town Center in Chattanooga, which is shown in Figure 12.12. The enclosed Eastgate Shopping Mall was built in the early 1960s in the then-new suburban neighborhood of Brainerd, located 12 miles from downtown Chattanooga near farmland (Benfield, Terris, & Vorsanger, 2001).

Although it was very popular for almost two decades, Eastgate began to lose appeal in the 1980s. The opening of a new and larger mall nearby in 1986 was the finishing blow, and Eastgate became largely vacant, with only 30 percent occupancy by the early 1990s (Mid-America Regional Council [MARC], 2003). By the time redevelopment commenced in the late 1990s, Eastgate was emptying out and heading toward eventual abandonment.

In the mid-1990s, the new mayor of Chattanooga addressed with the Chattanooga Regional Planning Agency the question of how to restore the largely vacant Eastgate Shopping Mall. The agency recommended that it be redeveloped to include a mixed use of offices, shops, civic organizations, and some residences (Benfield et al., 2002; MARC, 2003). Next, a developer entered the picture and offered to restore Eastgate as a national call center. Other business parties emerged with a variety of proposals.

A series of public meetings, in what is sometimes called a *charrette process,* involved discussions with hundreds of people on the mall design (MARC, 2003: Segedy & Johnson, 2003). A charrette process is a planning technique that emphasizes public inputs through open forums (Segedy and Johnson, 2003). The ensuing design essentially turned the mall inside out (see sequence of design drawings in Figures 12.12, 12.13 and 12.14). Shops on the ground floor were converted to have windows and face outwards, whereas the mall's second floor was converted to offices with windows (MARC, 2003). A public square was added to allow walking and socialization among town center workers and visitors.

The redevelopment project was approved with a price tag of $30 million. Once completed, the renamed Eastgate Town Center attracted diverse clientele, ranging from major corporations to small businesses, a community college, the YMCA of Chattanooga, an ice rink and other recreation facilities, and some shopping. Mass transit access and pedestrian features were included. Restoration was started in 1998, and by 2002, the occupancy rose from 27 to 90 percent. The new Eastgate Town Center has been a success, with many local customers returning after years of staying away (MARC, 2003). In the process, most smart-growth proposals were attained, with only a few, particularly a greenway trail element, not accommodated. In sum, this case study illustrates successful design and implementation of a smart growth project dedicated to city in-fill.

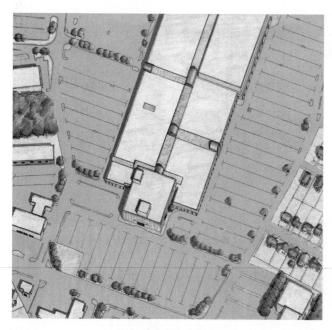

FIGURE 12.12   Eastgate Mall redevelopment plan: Existing conditions. Source: Ellis, Cliff. 2002. "The New Urbanism: Critiques and Rebuttals." *Journal of Urban Design* 7(3). P. 270. Figure 10.

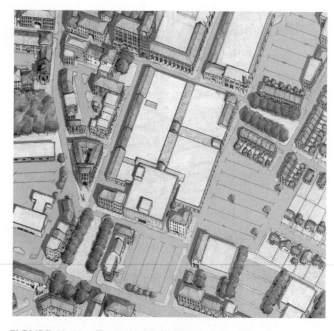

FIGURE 12.13   Eastgate Mall: Changes feasible within two or three decades. Source: Ellis, Cliff. 2002. "The New Urbanism: Critiques and Rebuttals." *Journal of Urban Design* 7(3). P. 270. Figure 11.

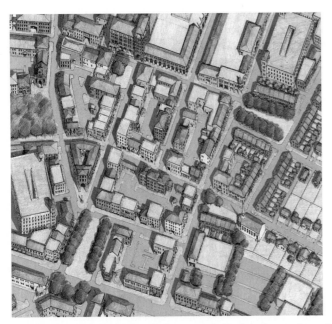

FIGURE 12.14 Eastgate Mall: Long-term transformation of the mall into a town center. Source: Ellis, Cliff. 2002. "The New Urbanism: Critiques and Rebuttals." *Journal of Urban Design* 7(3). P. 270. Figure 12. "Long-term transformation of the mall into a town centre."

Critics of smart growth have doubted whether it can realize its goals. The concerns include its performance relative to compact growth, particularly more mass transit, less traffic congestion, greater inclusiveness, more currency, and higher architectural standards. However, these criticisms have been defended by smart growth advocates (Ellis, 2002). For instance, critics of smart growth argue that because only 5 percent of the U.S. land mass is urbanized, it is not necessary to emphasize compactness. However, the counterargument is that many U.S. metropolitan areas are adjoined by highly productive agricultural land or environmentally sensitive areas, so compactness is justified. Regarding traffic congestion, smart growth has been attacked as increasing suburban congestion on limited traffic arteries (Ellis, 2002). On the other hand, many agree that smart growth has helped traffic patterns in the central city.

Critics of smart growth also doubt that making cities more walkable actually results in people walking more, thereby making claims about health benefits questionable (Seelye, 2003). On the contrary, some reports have shown there are more walkers in compact cities such as high-density, albeit non–smart-growth New York City, as seen in Figure 12.15 (Seelye, 2003). A study by Lawrence

FIGURE 12.15 Walkers in Times Square in New York City. Source: PhotoEdit.

D. Frank of the University of British Columbia, based on 12,000 survey responses, confirmed that "people who live in areas of low building density (i.e., suburbs) tend to weigh more than people in higher-density, mixed-use areas (i.e., cities), even controlling for income, age, sex, and ethnicity" (cited in Seelye, 2003, p. A14). In fact, findings on Atlanta showed that in the highest density areas of the city, 50 percent of White men were overweight and 13 percent obese, compared to 68 percent and 23 percent, respectively, in the lowest density areas. Similar findings were found for White women and Black men, although sampling precluded reliable results for Black women.

Compact, pedestrian-friendly cities such as San Francisco and Boston showed the highest walking prevalence, whereas lower density cities such as Houston and Phoenix did the worst. In sum, the Frank study provides data that counter the walking critique. As smart growth is a new form of urban design and planning, the debate between advocates and critics is likely to continue for some time, just as it has for zoning, growth management, and other approaches discussed in this chapter.

12.8 URBAN PLANNING CASE STUDY: IRVINE, CALIFORNIA

Over the past 50 years, Irvine, California, in Orange County, southeast of Los Angeles, has grown from open ranchland to a city of 164,900 population in 2003 (City of Irvine, 2002b). It has undergone this rapid growth and change through a combination of planning from government and the private sector. No one planning paradigm has been utilized, but a combination of several, especially regional planning and smart growth, as well as environmental planning, which was discussed in Chapter 11. The case shows how city planning might not find a simple solution to a problem in the real world, but might need to draw from a number of approaches and shift over time.

In the 1860s and 1870s, James Irvine, a San Francisco businessman, bought very large southern California ranchlands that had originally been Spanish colonial holdings (see the chronology in Table 12.4). The land contained one of the world's largest orange orchards, as well as extensive farms of walnuts, lima beans, and avocados. The Irvine family held the land until 1977. Since 1983, the Irvine Company has been owned by Orange County and Los Angeles businessman Donald Bren (Forsyth, 2002).

In 1959, the site for the University of California Irvine (UCI) campus was selected and the campus opened in 1965. The design firm of William Pereira and Associates did the master planning of both the UCI campus and the Irvine Ranch housing and commercial developments in the surrounding area. Hence, the first planning was done through a private company. The large size of the property

TABLE 12.4   Irvine ranch time-line

Date	Event
1860s, 1870s	James Irvine I buys ranch, first with partners, who he then buys out
1937	James Irvine Foundation formed
World War II	Military bases constructed on Irvine Ranch
1949	Housing Act
1954	Housing Act
1956	Interstate Highway Act
1957	University of California (U.C.) Irvine site selection study initiated
1959	Presidencies of Irvine Foundation and Company pass to nonfamily members
1959	U.C. Irvine site selected and initial planning of Ranch
Early 1960s	Planning for Irvine
1964	Southern Sector Plan agreed to by Orange County
1965	U.C. Irvine opens
1965	U.S. Department of Housing and Urban Development created
1968	Housing and Urban Development Act—creates Title VI new towns
1969	Taxation Act—affects nonprofits owning for-profit companies
1970	Irvine General Plan accepted by Orange County
1970	Housing and Urban Development Act—creates Title VII new towns
1973	Endangered Species Act
1977	Irvine Ranch Sold
1983	Donald Bren buys out most others, becomes sole owner in 1996
Early 1990s	Habitat protection negotiated

Source: Forsyth, Ann. 2002. "Who Built Irvine? Private Planning and the Federal Government. *Urban Studies*. 39(13). P. 2511. Table 1. "Irvine Rach time-line."

was perceived by the firm as a major opportunity to demonstrate the benefits of private planning.

Orange County planners raised a number of concerns about the developer's planning proposals, including "loss of agricultural land, designs of villages, lack of specific plans for affordable housing or mass transit, traffic projections, and flood control" (Forsyth, 2002, p. 2512). The public reaction to the Irvine Company was and remains mixed, with much dislike engendered by threats to natural areas, loss of farming, and more congestion and pollution (Olin, 1991; Schiesl, 1991). The County was able to reach compromises enough to include a modified Irvine Company plan in its County General Plan in 1970. However, several of those problems have continued to encounter opposition, particularly affordable housing and traffic congestion.

The City of Irvine was not incorporated until 1971. The full original extent of Irvine Ranch was 115,000 acres (see Figure 12.16), which is about one fifth of Orange

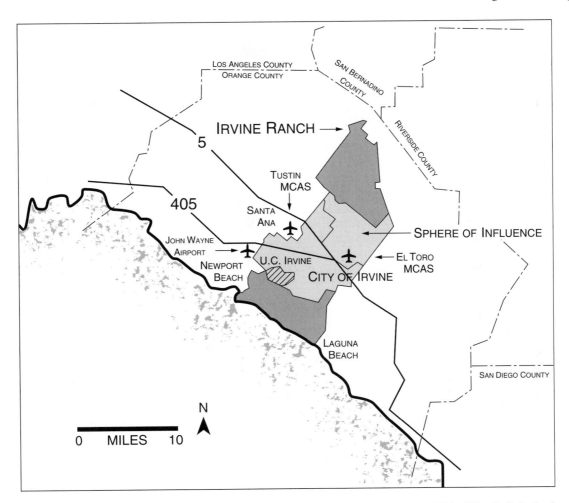

FIGURE 12.16 The Irvine Ranch: Its features and setting. Source: Forsyth, Ann. 2002. "Who Built Irvine? Private Planning and the Federal Government. *Urban Studies*. 39(13). P. 2510. Figure 1. "Irvine Ranch: major features."

County. Also shown is the 53,000 acres of the City of Irvine and the 1,500-acre UCI campus. By contrast, some of the Irvine Ranch land was sold off to form parts of the coastal communities of Newport Beach and Laguna Beach. They were not master planned, and today are upscale communities that largely succeeded through normal city planning, without having a master plan.

Some planners have used the terminology "new community" to describe this type of private–public collaboration on design for large developments, in particular those of more than 2,500 acres (Forsyth, 2002).

From its start, the city of Irvine's plan was set in the context of metropolitan and regional growth. First of all, the rest of the Irvine Ranch property, outside of the city of Irvine in Figure 12.16, was allocated to six other cities (Forsyth, 2002), so a metropolitan perspective was present from the start. The State of California was intimately involved in planning also because of the centerpiece UCI campus (Figure 12.16).

Less obvious was the strong and multifaceted planning involvement of the federal government. As seen in Table 12.5, at each step in the history of the last 50 years, federal government decisions, laws, and policies influenced the local planning. An example was military policies, particularly Cold War defense funding, which channeled large amounts of dollars into Irvine, and the siting of two military bases on the Irvine Ranch. Since the 1980s, federal environmental policies have influenced contentious environmental disputes that have erupted between the Irvine Company and the city (Olin, 1991). The latest interaction involved the alternatives for redevelopment of the former El Toro Marine Station. Defense Department and Marine Corps policies have been crucial factors in deciding on the base's reuse, which has shifted from airport to park and land development.

The environmental planning disputes have centered on tensions between environmentalists and the Irvine Company leadership (Forsyth, 2002; Inman, 1998). The

TABLE 12.5   Effects of federal regulation on the Irvine ranch

The most important federal interventions and impacts were as follows:

1956 Interstate Highway Act, plus associated energy policy. Made auto more attractive and lowered gas prices.

Federal agricultural policy. Did not mandate stop to use of agricultural land for urban development.

New Communities. In the 1960s and 1970s, the federal government promoted "new communities" through certain new programs.

Educational policy. Cold War policies favored higher education.

Mortgage financing. Gave tax breaks including no capital gains for mortgages. At same time, lack of subsidies for low-income housing.

Military policies. Defense funding for Cold War was funneled into mostly Sunbelt. Orange County received defense businesses. Two military bases sited on Irvine Ranch property.

Federal environmental regulation and federal support for environmental infrastructure. From the 1980s onward many federal agencies were involved. Federal interactions for the Irvine Ranch have been both helpful and a hindrance.

Interstate highway locations. The initial impetus came from a federal interstate highway systems project. The California Highway Commission negotiated the location of the Interstate 5 freeway to suit the Irvine Company. The freeway ended up going through the Irvine Ranch, ruining some areas of excellent agricultural land located in the foothills. Later on, the state, county, and a private tollway developer added transport.

Source: Forsyth, A. (2002) "Who built Irvine? Private planning and the federal government." *Urban Studies* 39(13):2507–2530

TABLE 12.6   City of Irvine mission statement and goals

City mission statement
"Our goal is to create and maintain a community where people can live, work, and play in an environment that is safe, vibrant, and aesthetically pleasing. This community promotes the well-being of all people."

City goals
• "Maintain and enhance the physical environment by never letting any visible deterioration take hold in the community.   • Maintain a safe and secure community by assuring the right balance between public safety services and prevention strategies.   • Promote economic prosperity by attracting and retaining businesses and sales tax.   • Promote effective government by assuring that the city organization is flexible, market based, and customer focused in its service delivery."

Source: City of Irvine. (2003a) *City of Irvine Mission Statement*. Irvine, CA: Author. Available at *http://www.cityofirvine.org*.

issue that aroused the most strident reaction was the question of protection of endangered-species habitat. The company opposed federal restrictions on wetlands development but eventually had to give in, despite their own efforts to demonstrate a wetlands solution (Forsyth, 2002). The matter was eventually largely settled by a series of company gifts of large land acreages for wildlife preserves (Forsyth, 2002).

Irvine would have seemed to be a candidate to adopt a growth management approach, but that was not the case. The city grew very rapidly in the 1970s and 1980s, going from its original population of 14,231 in 1971 to 127,200 in 1996. Since then, the city has slowed down to a still rapid growth rate of 3.7 percent yearly. Rather than growth management, the city advocates a mixed-development approach that can be sensed in the city's mission statement (City of Irvine, 2003a) in Table 12.6. Although environmental concerns are evident in its first goal, the city clearly would like to grow its business sectors and tax base, as seen in Goal 3. Another factor has been Irvine's high median housing price, $707,000 in 2004 ("Orange

County Home and Condo Sales," 2004), which puts some damper on the rapid population growth rate.

The smart-growth planning approach for Irvine, although not openly espoused by the city, is evident in some of its planning policies and outcomes (e.g., the business park pictured in Figure 12.17). These include emphases on environmental quality, housing diversity, mixed uses, and mobility (e.g., Irvine has extensive bike paths). Recently, the city has taken a vocal and forceful stance on annexing and redeveloping parts of a huge in-fill site, the El Toro Marine Station (Villanueva, 2003).

This case demonstrates that a community can be planned on the basis of a variety of paradigms, in this case mixing the metropolitan and regional, private-sector planning, and smart-growth approaches. The resultant city of Irvine has been remarkably successful in a short period of time, garnering awards for outstanding educational rankings, accessibility for the disabled, transportation, technology, and home design (City of Irvine, 2002a). The Irvine Company represents an instance of private master planning of a community, blended with strong city government planning (Schiesl, 1991). The inevitable conflicts have been intense, but mostly resolved through both sides giving way on some items (Olin, 1991). However, in many respects, the city has realized smart-growth steps over the past 35 years.

12.9 VIABLE DOWNTOWNS

From a building standpoint, the post–World War II slum-clearance programs tore down deteriorated inner-city housing and replaced it with multistory public housing (Barnett, 2003). These new buildings represented a way to reduce land costs per capita, and also were part of the

FIGURE 12.17   Environmental features of business park, Irvine. Source: PhotoEdit.

general popularity of skyscrapers at the time. However, the high-rise buildings rarely had secured entrances, leading to high crime and vandalism rates (Newman, 1973; Barnett, 2003). Another type of design in this period consisted of rows of buildings barracks-style. Barnett (2003, p. 120) described them as follows: "a long, low row of attached duplex apartments. The apartments were like small separate houses, potentially much more manageable than a tower; but the design of the buildings as utilitarian barracks laid out in rows did not create a comfortable environment. At the time these projects were built, public housing tenants were not allowed to own cars, so there are few internal streets in these projects and wide open spaces between the buildings which became uncared for and unsafe".

Critics have voiced concerns about these planning practices, attacking everything from the stark design of buildings, to lack of protection and encouragement of crime, to loss of the traditional character of neighborhoods (Barnett, 2003). Although the programs of slum clearance largely ended in the 1960s, other problems discussed in this book exacerbated the situation in the center. Low-income housing projects on the Lower East Side of Manhattan are shown in Figure 12.18. The flight from the ghetto meant that middle-class minority members

departed, leaving the remaining population less educated and less employable on the average. Poverty and the underclass received attention as large problems in the late 20th century (Massey & Denton, 1993).

Better central city in-fill designs are being encouraged by the Hope VI program of HUD (New Urban News, 2002; Barnett, 2003). Hope VI was launched in 1993 with

FIGURE 12.18   Low-income housing on Lower East Side of New York City in the early 1990s.

the goal of redeveloping the 86,000 most run-down public housing projects in the nation by demolishing them and rebuilding them into mixed-income developments having both public housing tenants and tenants paying rent at market levels (Popkin et al., 2004). The program supports and recognizes housing developers, designers, and housing authorities for their work "to redevelop failed modernist public housing projects into mixed-income communities." (New Urban News, 2002). An example is Capitol Gateway in Atlanta, a Hope-VI joint redevelopment project of a Philadelphia design firm and the Atlanta Housing Authority, resulting in mixed-income dwellings, low cost apartments, commercial development, and centers for learning and early childhood development (New Urban News, 2002). Although Hope VI has improved the quality of deteriorated housing structures and had a positive effect on many inner city neighborhoods, critics have pointed to sometimes ineffective outcomes, political wrangling with local governments, instances of rebuilding of projects in-place without design innovation, and a significant proportion of original residents whose quality of life has not been improved (Popkin et al., 2004). Hope VI's funding was cut by Congress for fiscal year 2004 and its future is uncertain (Popkin et al., 2004).

What can be done to plan better and achieve viable downtowns? Teitz (1997) advocated the following steps: (1) Develop the local economy, that is, partnering between local government and business. This was encouraged in the Carter administration with the Urban Development Action Grant Program. (2) Community development, which takes a larger view of the city and region, and puts the central city in the context of the whole region. In the Clinton administration, the Empowerment Zones Program emphasized central-city community development. The empowerment zone was introduced in 1994, and focused on empowering communities and people to collaborate on developing a strategic plan in the country's most impoverished areas, with the goal of relieving poverty through new jobs and opportunities. Although hundreds of empowerment zones were created, the success of the program was only moderate. This example is meant to illustrate the potential for community development. In the first recession and its aftermath at the beginning of the 21st century, such programs have been slowed, but some feel they hopefully will pick up.

12.10 URBAN GOVERNANCE AND POLITICS

Cities face many problems that challenge their capability to manage and govern. This chapter has examined the challenges of planning, including such factors as population growth, environment degradation, maintaining appropriate mixtures of land uses, and not embracing growth to the extent of sacrificing quality of life. So far, these have been examined as questions confronting planning agencies. More broadly, the solutions to these and other local and regional issues are addressed through urban governance and politics. *Urban governance* refers to the structures and actions by which a city or metro area manages and provides leadership in its public processes. Because urban governance is usually led by politically elected officials and representatives, urban politics is associated with governance. This section first looks at the broad economic problems facing urban governance and how they can be resolved and then discusses the alternative forms of governance. The importance to urban geography is that much of the functioning of urban areas discussed here depends on, or interrelates with, municipal government actions. Also, many, if not most municipal government activities are spatial, such as municipal service provision, tax collection, and environmental monitoring.

A fundamental challenge for urban governance is how to pay for it. This challenge applies whatever the form of government discussed later: city, metropolitan, or cooperative structures. In a time of economic crisis, in fact, several of these forms of government might be looked at to provide funds. Funding of urban governance comes from a mixture of taxes, fees, borrowing, and revenue sharing involving larger public units such as state and federal government. On the other side of the income statement are the costs, including payrolls, services, welfare, and financial debt service. City and urban governance must balance these inflows and outflows from year to year.

Economic problems arise from the following factors, among others: pressure for pay raises from government workers, demand for more welfare payments, costs related to urban area growth, costs of maintaining and upgrading deteriorating infrastructure, and changes in revenue sharing and grants from state and federal governments. As discussed in Chapters 7 and 8, during the past two decades, many large U.S. cities have experienced high immigration inflow, which puts added pressure on their city services and welfare. At the same time, in the United States, the largest generation ever, the baby boomers, will be retiring in the first two decades of this century, which will increase other pressures on city services, while reducing tax funding.

Examples of urban economic pressures at the breaking point were the New York City fiscal crisis of 1975 and the Orange County, California, bankruptcy of 1994. In both cases, certain costs grew while revenues were not able to keep up. Government officials were not alert and action-minded enough to prevent severe crises.

For New York City, the immediate cost increase was growth in welfare recipients (Knox, 1994). By fall of 1975, the city had a cumulative operating deficit of $2.5 billion. This shortfall was too large to be filled by financial borrowing, and the federal government under President Gerald Ford refused to intervene (Knox, 1994). The city averted looming bankruptcy only by the New York State Legislature stepping in to bail out the city. To achieve a long-term solution, the legislature established the Emergency Financial Control Board, led by the governor and major bank executives, as well as the Municipal Assistance Corporation, a financial entity to locate and arrange borrowing. Eventually, over a number of years, government austerity measures combined with late-arriving federal funding and improved management brought the city back to financial normalcy.

Orange County, which includes the city of Irvine as well as dozens of other cities, was forced into the largest municipal bankruptcy in U.S. history in December 1994. The immediate reason was that its county treasurer had been using county funds for speculative investments instead of following the usually conservative investment approach of the vast majority of municipal governments. Ultimately, the investment losses reached $1.7 billion, forcing the bankruptcy. Besides irresponsibility, an underlying reason for this speculation was California's passage in 1978 of Proposition 13, a citizens' referendum that put a 1-percent upper lid on California residents' property taxes. However, after it went into effect, municipalities did not take quick enough steps to reduce their costs proportionally, so that some developed mounting year-to-year deficits (Public Policy Institute of California, 1998). Other underlying causes of the crisis were political fragmentation and lack of attention to financial oversight by county political leaders and managers (Public Policy Institute of California, 1998). As large urban areas have expanded, often in the Sunbelt, sometimes too many municipalities have been grouped under a single urban governance structure, making it more difficult achieve metropolitan-wide management and controls. In the Orange County case, there were 187 participating municipal governments, school districts, and other agencies (Ryan, 1995). Their coordination became a political problem because numerous political leaders and representatives were involved. Only strong political leadership of the entire urban area can bring such diverse parties together.

It took years of stringency for Orange County to emerge from bankruptcy and get back on a normal pathway. The lesson is that economic distress can lead to desperate measures and cover-up, unless the political leadership takes on its full responsibility. Although not discussed here, other U.S. cities experienced economic pressures in the late 20th century including Boston, Cleveland, Detroit, and Philadelphia. These examples help to understand the challenges of urban governance and politics.

Urban geography is also concerned with what types of city and metropolitan government and political structures exist. The three major types of urban governance are single-tier, two-tier, and cooperative forms (Goldsmith, 2001).

Single-tier governance refers to a single, unified government for an entire urban area. Examples in the United States are San Diego, Phoenix, Houston, Dallas, Oklahoma City, and Kansas City (Goldsmith, 2001). Single-tier government grows largely through annexation. As discussed in Chapter 3, the potential for annexation is often unfavorable. If the urban area is underbounded, having substantial suburban population for which annexation is not supported, single-tier governance becomes unfeasible. The *two-tier* approach involves a government system that is integrated for a whole metropolitan area (Goldsmith, 2001). In the United States, this might be in the form of county government. For instance, Orange County is the metropolitan government for its municipalities. For the Miami area, the metropolitan government is Miami-Dade County. In some cases, the metro government has been formed or reinforced because of crises. For example, Toronto's two-tier metro government was formed as the result of a financial crisis of the municipalities of Toronto (Goldsmith, 2001). In London, England, mounting traffic and infrastructure problems in the 1960s led to formation of a two-tier system, which was abolished two decades later.

A third form of metropolitan government is *cooperative governance*. This consists of a cooperative arrangement of various institutions in the municipalities of a metropolitan area. Usually, cooperative government is the grouping of various entities to manage a particular function, for example, air pollution, regional planning, or finances. An example already mentioned of a multicounty planning organization is SCAG for five southern California counties. In London, after its two-tier system was abolished in 1986, a cooperative agreement was set up between London and its former two-tier municipalities. Sometimes, municipalities form COGs that offer a limited form of metropolitan government. For example, Metro Washington provides limited government for Washington, DC and 15 surrounding counties and cities in Maryland and Virginia (Goldsmith, 2001). Such councils do not replace the full-fledged city or metropolitan government, but they can offer complementary strengths. One evaluation of their pluses and minuses indicated that COGs perform better at providing services than at extracting revenues (Holden, 1964).

City and metropolitan governance varies also along a range from elitist to pluralist. *Elitist* refers to government arrangements that are controlled by a small elite that might consist of leading politicians, leaders of elected representatives, and business leaders. An example of a city that has tended to be more elitist is San Diego (Knox, 1994). At the other extreme is *pluralistic* governance, for which power is spread among many political interest groups. Metro in Portland, Oregon, is an instance of more pluralistic governance. Metro was founded to give voice to many constituencies in forming a vision for the area (Metro, 2004).

Another form of governance is *corporatist,* which involves a symbiotic relationship between government, private, and nonprofit entities such as businesses, community organizations, and labor organizations. Often, for the corporatist approach, the government depends more on professional technocrats to run it. The corporatist approach was encouraged following the fiscal crises in American cities in the 1970s and 1980s (Knox, 1994) because greater efficiencies were being sought. The approach frequently involves *privatization,* which is the outsourcing of formerly government services to profit-making companies. Rising privatization of government services was seen in state governments in the United States, for which private-sector services grew from $27 billion in 1975 to $150 billion in 1992 (Knox, 1994). Similar growth occurred for cities and metropolitan areas. Among the advantages to city and metropolitan governments of privatization are the sharing of financial risks with businesses, higher productivity through reduced costs, and the easing of year-to-year financial jolts. On the other hand, privatization can lead to "low balling" by private contractors, in which a contractor comes in with a low bid to obtain the contract, but with the intent to raise the bid quickly afterward. Also, the city needs to assure that the cost lowering of privatization will be maintained for the long term. Another issue can be low performance by service providers. This can only be offset through the government's continual monitoring and quality control of the providers. There might also be loss of jobs among city workers, many long term, and reduction of jobs that remain in skill and pay levels.

For the corporatist approach, cities have sometimes been entrepreneurial, working in association with businesses to develop large-scale public centers of different types. Although the outcomes are varied, well-known examples of corporatist successes are Quincy Market in Boston, Riverfront in Savannah, Pioneer Square in Seattle, and South Street Seaport I in New York City (Knox, 1994). They offer the combined features of retail businesses, restaurants, museums, concert and performance halls, tourist venues, and high-quality office complexes. Similar entrepreneurial joint projects are the sports complexes, convention centers, and large-scale urban mall developments in many cities.

12.11 TECHNOLOGY

Technology is crucial to urban planning, development, and activities of cities. Rapid developments in computing and technology in the second half of the last century have already significantly influenced cities and certainly will do

so more in the 21st century. One of the consequences of information technology is that the world's cities and their residents are being drawn much closer together. The new technologies influence work, home, and leisure activities. They have also encouraged a trend of telecommuting in cities, meaning that employees can do part of their work away from corporate premises, in their homes or on the road. The new technologies, however, are not affecting all urban residents equally. Instead, some benefit and others are left behind, a division referred to as the *digital divide.* This section focuses on the topics of the urban Internet, telecommuting, and the urban digital divide.

12.11.1 The Internet and the City

The commercial development of the Internet started in 1994, when Netscape Communications Corporation was formed to commercially develop the Mosaic browser, originally created at the University of Illinois. In 10 years, it expanded worldwide, with hundreds of millions of users and backbones circling the globe. Urban geography must examine how this phenomenon affects the city, its infrastructure, work habits, leisure, and communications, as well as to recognize where disparities exist in its use.

Some concepts and definitions are given first. *Real space* refers to physical space, for instance, a car moving on a road. *Virtual space* refers to the messages, communications, and social interactions taking place over the Internet. It does not necessarily have a physical location. However, it is closely related to real space because it depends on the equipment located in real space to support it. In comparing these two concepts of space, Kellerman (2002) divided them into three dimensions: organization, movement, and users. As seen in Table 12.7, the two types of space are quite different (Kellerman, 2002). Among the most important differences are that the content of real space is both physical (e.g., a book) and informational (e.g., contents of the book), whereas virtual space is informational (e.g., message content). They differ in the speed of movement. Whereas traditional communications depended on transport, Internet communications move at the speed of light, so messages can cross the globe in seconds. The differences are important because the virtual space gets away from the coordinate systems and maps that geographers are used to, but might have influences on the city equal to or even greater than traditional location-based entities.

The Internet can be divided into the production of information, its transmission, and its consumption by users (Kellerman, 2002). The production of Internet information takes place in government, businesses, universities, and other institutions that disperse information and engage in electronic commerce. Most of these entities are located in cities. Their Internet communications

TABLE 12.7   Real and virtual spaces

Dimension	Real space	Virtual space
Organization		
1. Content	physical and informational	informational
2. Places	separated	converge with local real ones
3. Form	abstract or real	relational
4. Size	limited	unlimited
5. Construction and maintenance	expensive and heavily controlled	reasonably priced and lightly controlled
6. Space	territory/Euclidean	network/logical
7. Matter	Material/tangible	immaterial/intangible
Movement		
8. Medium	transportation	telecommunications
9. Speed	depends on the mode	speed of light of transport
10. Distance	major constraint	does not matter mostly
11. Time	matters	matters, but events can suspend in time
12. Orientation	matters	does not matter mostly
Users		
13. Identity	defined	independent of identity in real space
14. Experience	bodily	imaginative and metaphorical
15. Interaction	embodied	disembodied
16. Attitude	long-term commitment	uncommitted
17. Language	national-domestic	mainly English-international

Source: Kellerman, Aharon. 2002. *The Internet on Earth: A Geography of Information.* John Wiley and Sons Ltd. P. 35. Table 2.1. "Real and virtual spaces."

go out from cities to regional recipients and to the world at large. The transmission is sent out on high-volume cables, generally fiber-optic, known as backbones, which physically interconnect cities. Figure 4.1 shows the number of Atlanta's Internet backbone connections with other major cities in 2000. The total Internet bandwidth (roughly speaking, transmission flow volume) in U.S. cities serves as good indicator of the amount of production and consumption of Internet-based information. As seen in Table 12.8, in 2000, the most intensive Internet-usage cities were New York, Chicago, Washington, and San Francisco. That is not surprising because the cities are both populous and world centers for particular aspects of information and e-commerce. New York City is a world media, advertising, and finance capital, and Chicago has major universities, research and development, and financial markets. Washington, DC has federal government information producers and users, plus

TABLE 12.8   Top 10 U.S. metropolitan areas based on total bandwidth of the internet backbones that serve them, 1997–2000

1997	1998	1999	2000	Metro area population in 2000 (millions)
Washington, DC	San Francisco	Washington, DC	New York	21.2
Chicago	Chicago	Dallas	Chicago	9.2
San Francisco	Washington, DC	San Francisco	Washington, DC	9.6
New York	Dallas	Atlanta	San Francisco	7.0
Dallas	New York	Chicago	Dallas	5.2
Atlanta	Los Angeles	New York	Atlanta	4.1
Los Angeles	Denver	Los Angeles	Los Angeles	16.4
Denver	Atlanta	Kansas City	Seattle	3.6
Seattle	Seattle	Houston	Denver	2.6
Phoenix	Philadelphia	St. Louis	Kansas City	3.3

Source: Malecki, Edward J. 2002. "The economic Geography of the Internet's Infrastructure." *Economic Geography* 78(4). P. 410. Table 5. "Top ten metropolitan areas in total bandwidth on Internet backbones serving them."; U.S. Census, 2003.

media and technology enterprises, and San Francisco is a world technology center. Many of the world's major technology corporations are headquartered in or near these cities as well, notably IBM in suburban New York, Microsoft in a northern suburb of Seattle, and Oracle in a suburb south of San Francisco. Table 12.8 also indicates that the cities change somewhat in position from year to year, which reflects economic- and technology-related ups and downs. The amount of Internet transmission is not dependent on population, as seen in the right-hand column. Los Angeles, for instance, is in seventh place for Internet volume but second in population. One reason given is that Los Angeles's large entertainment and movie industry had not yet in 2000 become Internet dependent (Kellerman, 2002).

Similar studies for European cities (Malecki, 2002a, 2002b), which measured how connected each one was to European Internet backbones, show London first, followed by Amsterdam, then a tie among Paris, Frankfurt, and Hamburg, followed by a six-way tie of Berlin, Brussels, Dusseldorf, Milan, Munich, and Zurich. London's premier place is not surprising given its importance in finance, media, research and development, and technology.

Internet consumption can be measured by home penetration, which is the percentage of homes using the Internet. The leaders were Portland, Seattle, and San Francisco, with penetration rates from 69.7 to 69.1 percent, followed by Boston, San Diego, and Washington, DC (Kellerman, 2002). Although the high positions of Portland and Seattle might be surprising, these cities have sophisticated users who have a variety of public and private local uses. Portland, for instance, is a leader in integrating Internet and Web uses into community life.

The lessons regarding the Internet in U.S. cities are several. From Chapter 9, there are often strong traditional geographic reasons for similar types of companies and institutions to agglomerate. This helps to account for why so much Internet production and consumption is located in certain cities that also lead in related information industries such as technology, advertising, media, research and development, and education. There is a difference today, however, in that these lead cities are less self-contained, but have huge cross-flows of information going out and coming in through virtual space regionally and worldwide.

12.11.2 Telecommuting

Because medium-sized, large-sized, and megacities all have traffic problems and congestion, it is becoming increasingly difficult for large numbers of commuters to move through a metropolitan area on a daily basis. This is one of the justifications for edge cities (see Chapter 10). Alternatively, settlement patterns can "scatter" into urban fringe developments (Chapter 10).

Telecommuting, or the practice of allowing employees to arrange for work at home or sites other than their corporate office, has been an issue of ongoing interest to researchers and practitioners since it was recognized in 1976 (Nilles, Carlson, Gray, and Hanneman, 1976; Vega, 2003). Many factors have contributed to keeping this issue current, and often controversial. Initially, telecommuting was thought of as an alternative to reduce traffic congestion and pollution in large cities by eliminating the need for employees to commute between their home and their offices (Nilles, Carlson, Gray, & Hanneman, 1976; Vega, 2003). Later on, organizations realized its potential for cost reduction by eliminating needs for office space (Apgar, 1998). Some also noted productivity increases, because of elimination of work distraction and commuting time for employees (Apgar, 1998; Navarrete & Pick, 2003). Highly skilled individuals realized the personal gains of engaging in this working arrangement (e.g., autonomy, financial savings, and reduced stress). Some demanded a telecommuting arrangement as a condition for employment.

The benefits of telecommuting have led many individuals and organizations to adopt it. However, there have also been reports of problems from this alternative work arrangement for individuals and organizations. They include workers' feelings of isolation by staying away from the office, work–family conflicts due to the inability to separate work and family issues, and management's inability to deal with the new mechanisms for controlling and evaluating employees. Similarly, other reports find evidence of a decline in productivity among telecommuters because of their lack of commitment to the organization or to increasing levels of stress brought about by working at home. These adverse outcomes have kept many businesses and individuals from adopting telecommuting (Apgar, 1998; Vega, 2003).

There are several planning points to note here. First, the motivation for telecommuting is that many employees like to work at home. The more distant the commuter is from the workplace, the more likely he or she will be motivated to telecommute. However, there are several losses for the vitality of businesses and organizations. The distance from the workplace might lead to lowered morale and work motivation. The employee's career might suffer from his or her lack of visibility to peers and managers (Zigurs & Qureshi, 2001).

12.11.3 The Urban Digital Divide

Cities are uneven in their use of technology, both within single cities and between groups of cities (Graham, 2002). This unevenness is referred to as the *digital divide*. An example is the San Francisco Bay Area, which has intensive Internet use in the Silicon Valley at its

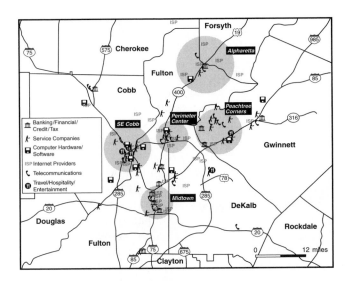

FIGURE 12.19 Locations of high Internet bandwidth in the Atlanta metropolitan area. Source: Walcott, Susan M. and James O. Wheeler. 2001. "Atlanta in the Telecommunications Age: The Fiber-optic Information Network." 22(4). P. 333. Figure 6.

southern flank, in its heavily wired downtown area, and in its university communities such as Berkeley and Stanford. However, Internet use remains very low in disadvantaged parts of the metropolitan area such as parts of Oakland. Referring to Atlanta, as seen in Figure 12.19, there are locations of high Internet bandwidth in medical and technology business clusters and edge cities located to the north of the downtown (Georgia Power Company, cited in Walcott & Wheeler, 2001), but other disadvantaged areas to the south have low levels of Internet connectivity. The differences in technology levels occur also at the county level in the United States, with urban counties tending to have higher levels of technology which is associated with higher levels of education, R& D, and technical/professional workforce (Azari and Pick, 2003, 2005).

The differences are sharply evident in cities in developing countries as well. For example, Bangalore, India, often referred to as India's Silicon Valley, is a world center for programming and software development. Entrepreneurs and scientists have accumulated wealth from this high-tech boom, but at the same time, the city has "an extremely fragmented and polarized urban structure" (Graham, 2002, p. 45). At the high end is the Electronic City area of several hundred acres containing world-class facilities of Texas Instruments, IBM, Motorola, and lead Indian tech firms such as Wipro, which services computers globally (Graham, 2002). Other parts of the city are impoverished and entirely cut off from this high-tech center. This brief example raises the issue of

the urban digital divide. Cities are beginning to grapple with and address it, but many of them have other pressing problems that often push the divide aside in priority. It remains important because of the rapid growth of technology and its impact on the workplace.

12.12 PLANNING CHALLENGES IN THE DEVELOPING WORLD

The last section of this chapter considers planning practices and experiences in other nations besides the United States. Do the approaches differ or are they similar? How do results compare? What are the constraints on achieving good planning outcomes anywhere in the world?

In Mexico, there is a history of urban planning over many centuries. The ancient civilizations of Mexico such as the Mayans and the Aztecs built advanced cities and urban designs for their time. The Spanish colonial settlers in the 16th century originally re-built Mexico City with a careful design modeled on Spanish cities of the era. When the French ruled Mexico for several years in the 1860s, French architects contributed the grand Paseo de la Reforma, modeled after the Champs-Elysees in Paris.

With this impressive history, it is disappointing that planning for Mexican cities today has become a very political process (Herzog, 1990; Ward, 1998). This is explained by the government structure in Mexico, which has at the top a financially well-endowed federal government, below it weaker state government, and below that much weaker municipal government (Herzog, 1990). Although these disparities were somewhat reduced in the mid-1990s by federal legislative changes on the revenue-sharing formulas affecting the three governmental levels, they remain largely in place. The result is that local and regional governments have a limited potential to accomplish planning.

In fact, until the 1970s, there was almost no formal local planning in Mexico (Herzog, 1990). Instead, the federal government stepped in with its national planning of economic sectors, for example, *maquiladora* manufacturing or tourism. It was only in 1976 that the Mexican government recognized local planning prerogatives. However, the results for cities and *municipios* have been weak (Butler, Pick, & Hettrick, 2001; Herzog, 1990; Ward, 1998). Local government planning is limited to construction of small public works and some water and energy services, and the federal government runs planning of the national economy, public housing, tourism, streets and road construction, much energy planning and supply, and environmental controls. States provide much of the primary and secondary education and local economic development. Uneven planning for housing in Tijuana, Mexico, is shown in Figure 12.20.

FIGURE 12.20   Weak planning in Tijuana, Mexico.

At the local level, there are conflicts of interest, pleasing of political constituencies, and some mishandling of funds (Herzog, 1990). Of course, such practices are not unknown in the United States and European countries, but they are relatively more prevalent in Mexico.

What are the hopeful signs? Several large Mexican border cities, most prominently Ciudad Juarez, have formed quasi-government metropolitan planning agencies. In the case of Ciudad Juarez, Mexico's sixth-largest city, the planning agency is known as *Instituto Municipal de Investigación y Planeación* (IMIP). Its charter is to formulate, realize, and evaluate the Urban General Plan and its parts, zoning rules, and other urban standards, in congruence with the state and national plans of urban growth (IMIP, 2003). The agency has nearly 100 employees, only some of whom are planning professionals. IMIP has adopted GIS and makes large databases available to the community. It also has orientation and diffusion programs to inform the people about its programs and to improve the public welfare. It has a large office, library, bookstore, and server-based information system with GIS. In essence, this sounds like an up-to-date American metropolitan planning unit, such as SANDAG. So far, however, the impact of IMIP has been mixed. Some major infrastructure deficits have not yet been remedied, not because of lack of planning, but because of scarcity of municipal funding and, in some cases, political roadblocks.

The challenge for Mexico in the future is to further professionalize its municipal planning, provide teeth to it through enforcement and better legal recourse, improve the funding of municipal planning offices and bodies such as IMIP, reduce political corruption of the planning process, learn from—but not necessarily imitate—the planning knowledge and methods from the United States and other advanced nations, and educate the local electorate and business community about modern urban planning.

There is not sufficient space in this text to cover the many other nations doing urban planning comprehensively, but several other examples are worth noting. The Scandinavian nations have had strong planning, often combined with architecture and urban design. Their cities tend to be well arranged, environmentally friendly, and not overrun with growth problems. Singapore is a noteworthy success in instituting the high-speed Internet throughout the entire city-state (Arun & Yap, 2000; Coe & Yeung 2001), and has set information technology deployment as a high-priority goal. There are concerns, however, about too much government intrusiveness and tight control of the Internet platform (Arun & Yap, 2000).

Finally, in China, Shanghai has remarkably augmented the modernity of this megacity in a space of only about 10 years, with much more ambitious plans in the future to reinforce its role as a leading global city (Yusuf & Wu, 2002). The municipal government committed itself to growth planning through sheer will in setting ambitious strategic objectives and not wavering. In the process, its city leadership has become rather more autonomous from the Chinese central government (Yusuf & Wu, 2002).

The following are some examples of Shanghai's 10 years of accomplishments in planning, strategy, and implementation (Yusuf & Wu, 2002):

- The new Shanghai international airport opened in 1999.
- The huge Economic and Technology Development Zone (ETDZ) of Pudong, across the Huangpu River from the historic center, has grown to become the largest free-trade zone in China, with hundreds of new businesses and buildings.
- Three bridges and two tunnels have been built across the Huangpu River, as well as expressways, a ring road, and two new subway lines.
- Public utilities have been given greater management autonomy.

Perhaps the biggest remaining problem in transforming Shanghai into a global city is the lack of adequate skilled labor. This point goes back to rural-to-urban migration and the Chinese residence registration system as discussed in Chapter 7. The challenge is more than reducing the residency strictures, but also training Shanghai natives as well as immigrants for the highly skilled workforce that is being called for in the new Shanghai global city.

SUMMARY

This final chapter has covered the planning of cities. There is a long and often successful history of planning. In the United States, major recent development approaches have been metropolitan and regional planning, growth management, and smart growth. It is important to understand and appreciate planning as a means to revitalize downtowns and achieve urban open space.

Technology is changing the flows of people and information in cities. The Internet and telecommuting are examples of new work modes that influence work, transportation, physical, and organization environments. Many case examples were discussed in the chapter, including the classical zoning and growth debates in New York City, Ramapo, and Petaluma; growth management in Hervey Bay; mixed planning in Irvine; and recent planning of Shanghai.

The chapter brings together many concepts and methods from the other chapters, stressing that planning is multidimensional and integrative. In concluding, we would like to emphasize the importance of studying urban geography. The world's urban population is continuing to increase rapidly, and the number of large metropolitan areas is growing even more. A geographic perspective is essential to understanding, analyzing, planning, and effecting actions and changes to improve cities. The effectiveness of learning and applying urban geography is enhanced by use of modern technology tools like GIS. GIS is uniquely suited to capture the many spatial layers and dimensions of the contemporary city and provide analytic tools to model its growth and change.

REFERENCES

Apgar, M. (1998) "The alternative workplace: Changing where and how people work." *Harvard Business Review* 76:121–136.

Arun, M., and Yap, M. T. (2000) "Singapore: The development of an intelligent island and social dividends of information technology." *Urban Studies* 37(10):1749–1756.

Azari, R., and Pick, J. B. (2003) "The influence of socioeconomic factors on technological change: The case of high-tech states in the U.S.," in *Current Security Management and Ethical Issues of Information Technology*. ed. by A. Rasool, Hershey, PA: Idea Group Publishing, pp. 187–213.

Azari, R. and Pick, J. B. (2005). "Technology and society: socioeconomic influences on technological sectors for United States counties." *International Journal of Information Management*, in press.

Barredo, J. I., Demicheli, L., Lavalle, C., Kasanko, M., and McCormick, N. (2004) "Modelling future urban scenarios in developing countries: An application case study in Lagos, Nigeria." *Environment and Planning B: Planning and Design* 32:65–84.

Barnett, J. (2003) *Redesigning Cities. Chicago*, IL: Planners Press.

Barringer, F. (2004) "Rule Change in Oregon May Alter the Landscape." *New York Times*, November 26, section A, p. 1.

Batty, M., Xie, Y., and Sun, Z. (1999) "Modeling urban dynamics through GIS-based cellular automata." *Computers, Environments, and Urban Systems* 23(3):205–233.

Benfield, F. K., Terris, J., and Vorsanger, N. (2001) *Solving Sprawl: Models of Smart Growth in Communities Across America.* New York: Natural Resources Defense Council.

Butler, E. W., Pick, J. B., and Hettrick, W. J. (2001) *Mexico and Mexico City in the World Economy.* Boulder, CO: Westview.

Chandler, T. (1987) *4000 Years of Urban Growth.* Lewiston, NY: Edwin Mellen.

City of Irvine. (2002a) *2002 Annual Report.* Irvine, CA: Author.

City of Irvine. (2002b) *History of Irvine.* Irvine, CA: Author. Available at *http://www.cityofirvine.org*.

City of Irvine. (2003a) *City of Irvine Mission Statement.* Irvine, CA: Author. Available at *http://www.cityofirvine.org*.

City of Irvine. (2003b) *Strategic Business Plan, 2001-2006.* Irvine, CA: Author. Available at *http://www.cityofirvine.org*.

Coe, N.M., and Yeung, H.W. (2001) "Grounding global flows: Constructing an e-commerce hub in Singapore," in *Worlds of E-Commerce: Economic, Geographical, and Social Dimensions*. ed. by T. R. Leinbach and S. D. Brunn, pp. 145–166. Chichester, UK: Wiley.

Cullingworth, B., & Caves, R. W. (2003) *Planning in the USA: Policies, Issues, and Processes*. London: Routledge.

Curry, M. R. "The digital individual and the private realm." *Annals of the American Association of Geographers* 87(4): 681–699.

Ellis, C. (2002) "The new urbanism: Critiques and rebuttals." *Journal of Urban Deign* 7(3):261–291.

Fleissig, W., and Jacobsen, V. (2002) *Smart Scorecard for Development Projects*. San Francisco, CA: Congress for New Urbanism. Available at *http://www.cnu.org*.

Forsyth, A. (2002) "Who built Irvine? Private planning and the federal government." *Urban Studies* 39(13):2507–2530.

Fulton, W. (1999) *Guide to California Planning* (2nd ed.). Point Arena, CA: Solano Press Books.

Gillam, E., and Taylor, A. (2001) *Demographic profile for Hervey Bay city*. Hervey Bay, Australia: Planning Information and Forecasting Unit, Department of Local Government and Planning.

Goldsmith, M. (2001) "Urban governance", in *Handbook of Urban Studies*. ed. by R. Paddison, pp. 325–335. London: Sage.

Gottmann, J. (1957) "Megalopolis, or the urbanization of the northeastern seaboard." *Economic Geography* 33:189–200.

Graham, S. (2002) "Bridging urban digital divides? Urban polarisation and information and communications technologies (ICTs)." *Urban Studies* 39(1):33–56.

Hall, P. (1998) *Cities in Civilization: Culture, Innovation, and Urban Order*. London: Weidenfeld & Nicholson.

Hearsey, J. E. (1965) *London and the Great Fire*. London: J. Murray.

Herzog, L. A. (1990) *Where North Meets South: City, Space, and Politics on the U.S.-Mexico Border*. Austin, TX: Center for Mexican American Studies.

Holden. M. (1964) "The governance of the metropolis as a problem in diplomacy." *Journal of Politics* 26:627–647.

Howard, E. (1892) *Garden Cities of To-morrow*. London: Swan Sonnenschein.

IMIP. (2003) Que es el IMIP? Ciudad Juárez, Mexico: Author. Available at *http://www.mexguide.net/imip*.

Inman, B. (1998) "California trends: How Irvine became a southern California oasis." *San Francisco Chronicle* November 15.

Kellerman, A. (2002) *The Internet on Earth: A Geography of Information*. Chichester, UK: Wiley.

Knox, P. L. (1994) *Urbanization: An Introduction to Urban Geography*. Englewood Cliffs, NJ: Prentice Hall.

Levy, J. M. (2003) *Contemporary Urban Planning* (5th ed.). Upper Saddle River, NJ: Prentice Hall.

Loftie, W. J. (1884) *A History of London*. London: E. Stanford.

Marsh, B. C. (1909) *An Introduction of City Planning: Democracy's Challenge in the American City*. New York: Marsh.

Malecki, E. J. (2002a) "The economic geography of the Internet's infrastructure." *Economic Geography* 78(4):399–424.

Malecki, E. J. (2002b) "Hard and soft networks for urban competitiveness." *Urban Studies* 39(5-6):929–945.

Marsh, B. (1909) *An Introduction to City Planning*. New York, NY: Benjamin Marsh. Reprinted 1954 by Arno Press (New York, NY).

Massey, D. S., and Denton, N. A. (1993) *American Apartheid*. Cambridge, MA: Harvard University Press.

Metro (2004). "About Metro." Portland, OR: Metro. Available at *www.metro-region.org*.

Mid-America Regional Council. (2003) "Eastgate town center." Case Studies, Quality Places. Kansas City, MO: Author. Available at *http://www.qualityplaces.marc.org*, July.

National Association of Regional Councils (2004). *NARC: Building Regional Communities*. Washington, DC: National Association of Regional Councils. See *www.narc.org*.

Navarrete, C. J., and Pick, J. B. (2003) "Cross-cultural telecommuting evaluation in Mexico and United States." *The Electronic Journal on Information Systems in Developing Countries* 15(5):1–12.

New Urban News (2002) "Hope VI funds new urban neighborhoods." *New Urban News*, 7(1):9–10. Available at *www.newurbannews.com*.

Newman, O. (1973) *Defensible Space: Crime Prevention through Urban Design*. New York: Macmillan.

Nilles, J. M., Carlson, F. R., Gray, P., and Hanneman, G. G. (1976) *The Telecommunications Transportation Trade-off*. New York: Wiley.

Olin, S. (1991) "Intraclass conflict and the politics of a fragmented region," in *Postsuburban California: The Transformation of Orange County Since World War II*. ed. by R. Kling, S. Olin, and M. Poster, pp. 223–253. Berkeley, CA: University of California Press.

"Orange County home and condo sales for April." (2004) *Los Angeles Times*, Real Estate Section, May 30:K5.

Osborn, F. J. (ed.). (1945/1965) *Ebenezer Howard's Garden Cities of Tomorrow*. Cambridge, MA: MIT Press.

Pettit, C. J. (2002) "Planning Application Tool: What-If," *ArcNews*, Fall. Available at *http://www.esri.com*.

Platt, R. (1996) *Land Use and Society: Geography, Law, and Public Policy*. Washington, DC: Island Press.

Popkin, S. J., Katz, B., Cunningham, M. K., Brown, K. D., Gustafson, J., and Turner, M. A. (2004). *A decade of HOPE VI: Research findings and policy challenges*. Washington, DC: Urban Institute. Available on *www.urban.org*.

Public Policy Institute of California. (1998) "The Orange County bankruptcy: Who's next?" Public Policy Institute of California Research Brief, Issue No. 11, April.

Randall, G. C. (2000) *America's Original GI Town: Park Forest, Illinois*. Baltimore: Johns Hopkins University Press.

Ryan, H. (1995) "The Orange County bankruptcy and California's fiscal crisis." Report available 2004 at *www.howardryan.net/orange.htm*.

Schiesl, M. J. (1991) "Designing the model community: The Irvine company and suburban development, 1950–88," in *Postsuburban California: The Transformation of Orange County Since World War II*. ed. by R. Kling, S. Olin, and M. Poster, pp. 92–141. Berkeley: University of California Press.

SANDAG. (2003) *About SANDAG: History*. San Diego, CA: Author. Available at *http://www.sandag.org*.

Seelye, K. Q. (2003) "Cities made for walking may be fat burners." *New York Times* June 21:A14.

Segedy, J.A., and Johnson, B.E. (2003) *The Neighborhood Charrette Handbook*. Louisville, KY: Sustainable Urban Neighborhoods Program, University of Louisville. Available at *http://www.louisville.edu/org/sun/planning/char.html*.

Simmie, J. (2001) "Planning, power, and conflict," in *Handbook of Urban Studies*. ed. by R. Paddison, pp. 385–401. London: Sage.

Teitz, M. B. (1997) "American planning in the 1990s: Part II, The dilemma of the cities." *Urban Studies* 34(5-6):775–795.

Town of Fairfield. (2003) *Zoning regulations: Town of Fairfield, Connecticut*. Fairfield, CT: Author. Available at *http://www.fairfieldct.org/zoning*.

Vega, Gina (2003) *Managing Teleworkers and Telecommuting Strategies*. New York: Praeger Publishers.

Villanueva, E. (2003, June) "For sale: Vacant Marine air base." *Bdmag.com Online Magazine*. Available at *http://bdmag.com/issues/jun_2003*.

Walcott, S. M., and Wheeler, J. O. (2001) "Atlanta in the telecommunications age: The fiber-optic information network." *Urban Geography* 22(4):316–339.

Ward, P. M. (1998) *Mexico City* (2nd ed.). Chichester, UK: Wiley.

Xie, Y., and Batty, M. (2005) "Integrated Urban Evolutionary Modeling in GeoDynamics," ed. by P. Atkinson, G. Foody, S. Darby, and F. Wu, pp. 273–294. Boca Raton: CRC Press.

Yusuf, S., and Wu, W. (2002) "Pathways to a world city: Shanghai rising in an era of globalisation." *Urban Studies* 39(7): 1213–1240.

Zigurs, I., and Qureshi, S. (2001) "Managing the extended enterprise," in *Information Technology and the Future Enterprise*. ed. by G. W. Dickson and G. DeSanctis, pp. 125–143. Upper Saddle River, NJ: Prentice Hall.

EXERCISE

Exercise Description

An exercise designed to illustrate modeling functions in a GIS through the analysis of population growth impacts in an urban fringe watershed.

Course Concepts Presented

Population growth, forecasting, and smart growth policies.

GIS Concepts Utilized

Spatial analysis tools: buffers, overlays, and generalization.

Skills Required

ArcGIS Version 9.x

GIS Platform

ArcGIS Version 9.x

Estimated Time Required

4 to 6 hours

Exercise CD-ROM Location

ArcGIS_9.x\Chapter_12\

Filenames Required

Blackberry.shp
Flood.shp
land_cover.shp

nwi_poly.shp
str.shp
bb_pop00.shp
plant.shp
41088g34.sid

Background Information

Introduction

The Blackberry watershed is located on the western fringe of Aurora in Kane County, Illinois. Kane County itself is located on the western fringe of the Chicago metropolitan area and in recent years has experienced tremendous growth and development. According to the U.S. Census Bureau's population estimates, the county added 86,648 persons from 1990 to 2000 (a 27 percent increase), which places it among the fastest growing counties in the state.

A telecommunications company, Gpsnet (fictitious), recently announced that it would locate a very large cellular phone plant in the vicinity of Prestbury, Illinois, and Sugar Grove, Illinois, both of which are within the watershed (see graphic below). Gpsnet's announcement was not only a disappointment to economic development officials in Aurora, who tried to lure them to a brownfield site in the city, but to regional planners who had been discouraging additional sprawl within the metropolitan region. The decision also stirred the emotions of farmland preservationists who contended that the population growth associated with the new plant would reduce the rural character of the watershed.

In this project, you need to assess the impact of the new plant on the watershed's resource base and on the growth of the watershed's population to the year 2020.

Blackberry Watershed

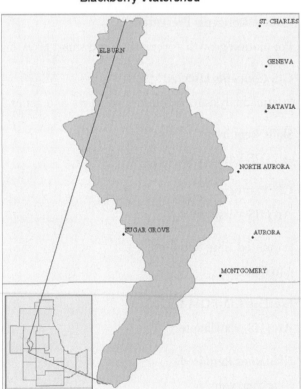

Impact of the New Plant on the Watershed's Population to the Year 2020

We are going to use a simple exponential population forecast model (Krueckeberg & Silvers, 1974). The following formula is used for calculating a growth rate over a given period:

$$P_{t+n} = P_t(1 + r)^n$$

where

$$r = \frac{1}{m}\sum_{t=2}^{d}\frac{P_t - P_{t-1}}{P_{t-1}} = GrowthRate$$

and

P = Population
t = a time index (years)
P_{t+n} = population (n) units of time from (t)
n = number of units of time (in years)
m = number of historical intervals over which average is calculated
d = the date of the latest data being analyzed

Using the following population estimates for Kane County from the U.S. Census Bureau (see URL) illustrates the calculation procedure:
http://www.census.gov/population/www/estimates/countypop.html

Population estimates

1994	1995	1996	1997	1998	1999	2000
346,687	356,995	366,912	367,725	386,103	396,371	404,119

Growth rates

1994–1995	1995–1996	1996–1997	1997–1998	1998–1999	1999–2000	Average
.029	.028	.027	.025	.026	.019	.025

Pop2020 = 2000Pop * (1 + growth rate) ** 20
Pop2020 = 404119 * 1.638616
Pop2020 = 662,755

This example derives a base growth rate (1.64), however, a given area within the county can experience a different growth rate based on a number of characteristics. In this project we assume that the following conditions will influence a block's growth rate.

Distance to Proposed Plant Location

An obvious condition would be proximity to the proposed plant. We assume that workers will want to live close to the new plant, so that they can get to work quickly. Therefore, parcels within one-half mile will receive a higher growth rate.
    Growth rate plus a factor of 1.2

Distance to Major Road

Until the road network improves during this rapid development period, developable land adjacent to major roads leading to the plant will experience faster rates of growth. Parcels within one-half mile of these major roads will receive a higher growth rate.
    Growth rate plus a factor of 1.0

Existing Land Use

The current use of the land will dictate what is available for future growth. Thus, vacant and agricultural land will be able to accommodate the most people. Agricultural and vacant areas will grow at a high rate, whereas parcels composed of residential land will grow at a moderate rate. The movement to protect environmentally sensitive lands will result in a low growth rate for these areas.

Agriculture: Growth rate plus a factor of 0.8
Low-Density Residential: Growth rate plus a factor of 0.2
Environmentally Sensitive Lands: Growth rate minus a factor of 1.2

The values of the growth weights were determined by first establishing 4.64 (1.64 + 3.0 of additional growth) as the maximum possible growth rate for a given block within the watershed. We then allocated the additional growth among the various factors based on our assumptions about the relative importance of each in terms of ability to attract future growth. For instance, consider a block that will grow at the county rate (1.64), is within a half-mile of the proposed plant (1.2), is within a half-mile of a major road leading to the plant (1.0), and is agricultural land (0.8). Adding the growth factors for the latter block yields 4.64 (1.64 + 1.2 + 1.0 + 0.8).

Readings

(1) Greene, R. P., and Pick, J. B. *Exploring the Urban Community: A GIS Approach*, Chapter 12.

Exercise Procedure Flowchart

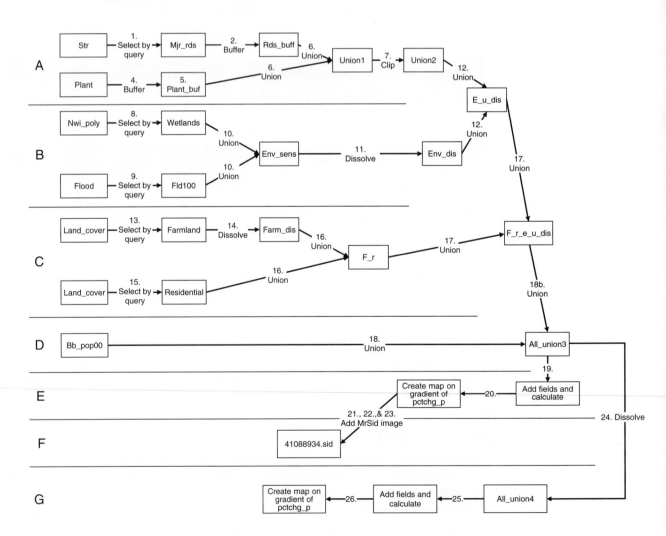

Exercise Procedure

Getting Your Data

1. Start ArcMap with Start Using ArcMap With A New Empty Map selected. Click OK.

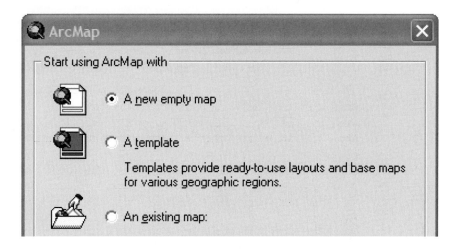

2. Click Add Data and navigate to the exercise data folder.

Note: *If you do not see the folder where you placed the exercise data, you will need to make a new connection to the root directory that holds the data.*

3. Add blackberry, flood, land_cover, nwi_poly, str, bb_pop00, and plant from the Chapter 12 folder. You can add all of these layers at once by clicking on one of the file names, then clicking the others while holding down the Ctrl key on the keyboard and clicking Add.

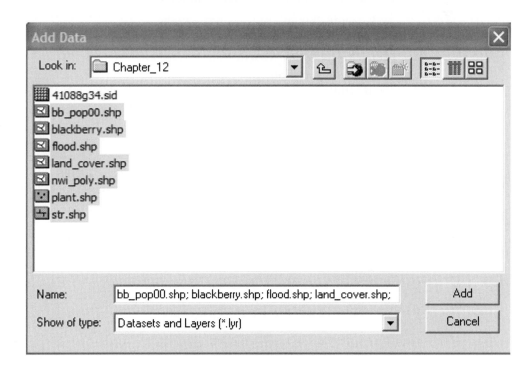

4. Set the data frame properties (map units in meters and display units in feet) by right-clicking on Layers, then selecting Properties on the shortcut menu. Click the General tab and under Units, change the Display setting to Feet. Click Apply and then click OK.

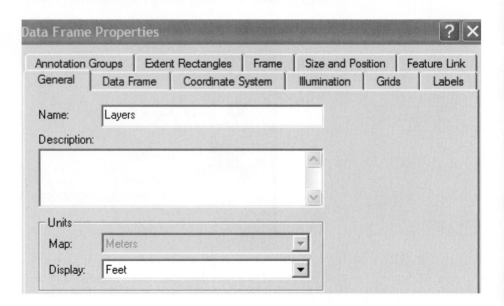

The following steps correspond to the flowchart that has been prepared to show the steps involved in determining the impact of the new plant on the population growth and distribution of the Blackberry watershed (see flowchart, above).

A. Define Major Roads and Plant Buffers

Because the proximity to the plant and major roads affects the population growth, we are setting buffers to define their area of influence.

1. Select the major roads from the *Str* layer:
   a. Select Selection on the menu bar.
   b. Select Select By Attributes.
   c. Set layer to Str.
   d. Enter the following:
      "STREETNAME" in ('Deerpath','Galena','Nelson Lake','Oak','Orchard')

There are two roads by the name of Oak that get added to the selection that are not major roads leading to the plant. In step h below we remove them from the selection.

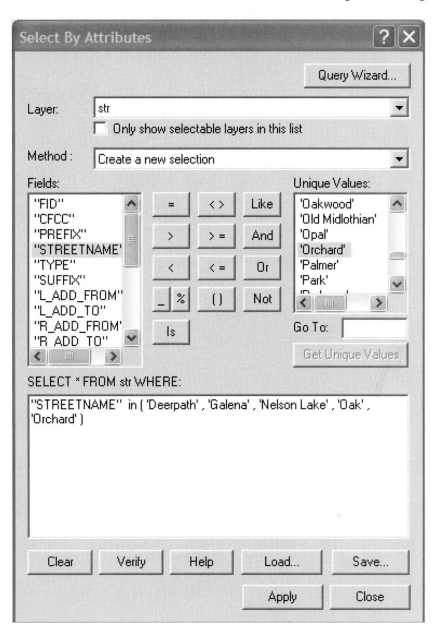

e. To see the full range of streets, click Get Unique Values.
f. Once the query is specified correctly, click Apply, then click Close.
g. Click Selection, click Set Selectable Layers, and clear all check boxes except Str. Click Close.
h. Zoom in on one of the small outliers (these roads are named "Oak" but they are not major roads), either the one in the far northwest or the one in the south. Click the Select Features tool in the Tools Menu and deselect the outlying road segments while holding down the Shift key (if you do not hold down the Shift key, you will lose your legitimate major road selections). Zoom back out and repeat for the second outlying segments. Be sure to deselect all the little pieces of these outliers.
i. With the remaining major roads still selected, right-click Str.
j. Click Data on the drop-down menu, then click Export Data.
k. Change the name from Export_Output.shp to ***mjr_rds.shp***.
l. Click Yes to add it to the map.
m. Click Selection. From the drop-down menu select Clear Selected Features.

2. Buffer the ***Mjr_rds.shp*** layer by a half-mile (2,640 feet):

    a. Click the ArcToolBox button at the top (red button).

    b. Double-click Analysis Tools, Proximity, and Buffer.

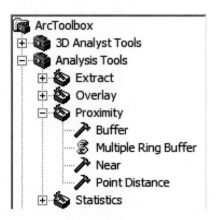

    c. Fill in the Buffer dialog box with mjr_rds as the input, rds_buff as the output, linear unit set to feet with a distance of 2640, and change Dissolve Type to All. Click OK.

> We need to build memory into each of the layers that we create for the final model calculations in section E, so that we can query out each of the growth factors (i.e., within a half-mile of a major road). Thus the following step adds a field to the new buffer and is calculated to a value of one.

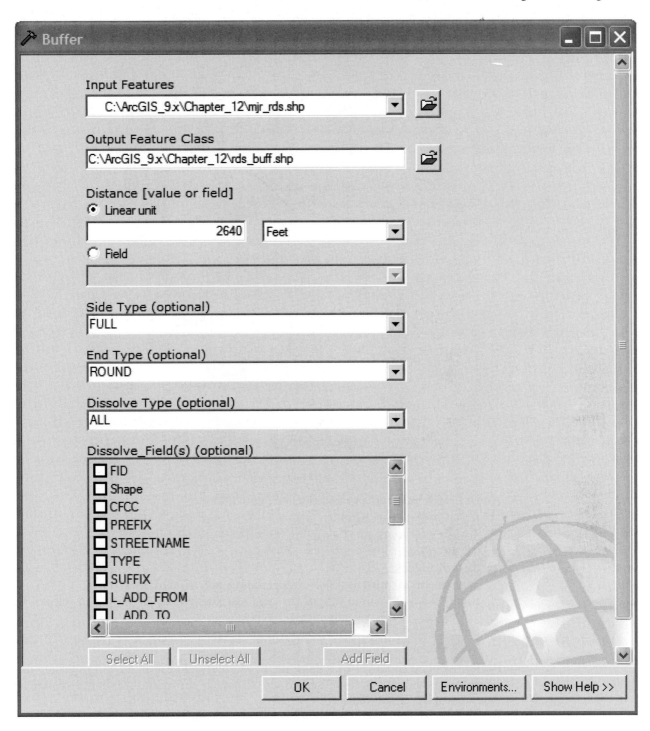

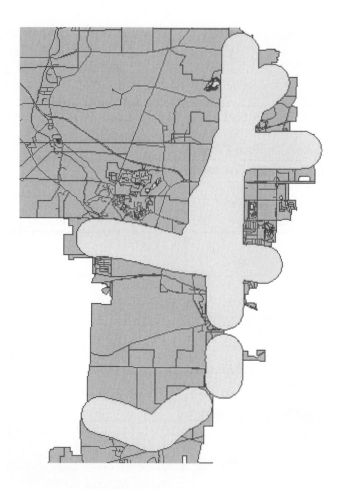

3. Add a field to *rds_buff.shp* and calculate equal to one.
   a. Right-click the *rds_buff.shp* layer.
   b. Select Open Attribute Table.
   c. Click Options.
   d. Click Add Field.
   e. Set name to **in_mjrd** as a short integer and a precision of 0.
   f. Right-click the column title of the field just added (i.e., in_mjrd) and select Calculate Values. Ignore the warning message (click Yes).
   g. Enter a 1 in the large white space under **in_mjrd** =.
   h. Click OK and close the table.
4. Buffer the plant.shp layer by a half-mile (2,640 feet) and call plant_buf.shp using the same procedure as is shown in step 2 (make sure to change the input layer name to plant and call the output layer plant_buf.shp).
5. Add a field to plant_buf.shp called plantin and calculate equal to one using the same procedure as shown in step 3.

We use a GIS overlay procedure known as a union to combine the major roads factor and proximity to the proposed plant into one GIS layer (see illustration).

6. Union the **rds_buff.shp** with the plant_buf.shp to make union1.shp:

   a. Click ArcToolbox, then select Analysis Tools, Overlay, and Union.
   b. Select plant_buf.shp as an input layer and click the plus sign.
   c. Select rds_buff.shp as an input layer and click the plus sign.

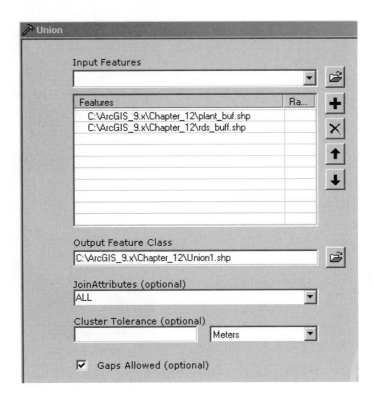

   d. Name the output layer Union1.shp.
   e. Set Join Attributes to All.
   f. Set Cluster Tolerance to Meters.
   g. Click OK.
   h. To illustrate the usefulness of the latter step, try the following:

From the Selection menu, select Select By Attributes. Set the layer to Union1 and then enter the following expression:

$$in\_mjrd = 1 \text{ and } plantin = 1$$

If your previous buffers and union were correct, you should have the following polygon selected (see illustration).

The new union 1 layer extends beyond the watershed limits because the half-mile road buffer included the tips of the roads that intersected the watershed boundary. We use a "cookie cutter" technique known as a clip to remove the areas that extend beyond the watershed (see illustration).

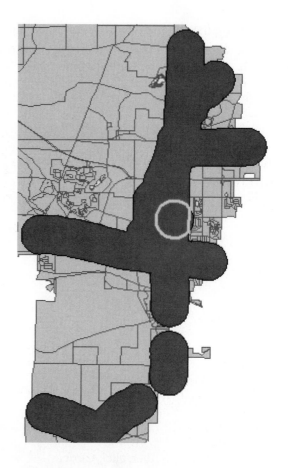

7. Clip the union1.shp layer by the watershed layer called blackberry.shp.
    a. From the Selection menu, select Clear Selected Features.
    b. Click on ArcToolbox, then select Analysis Tools, Extract, and Clip.
    c. Select Union1.shp as the Input Features.

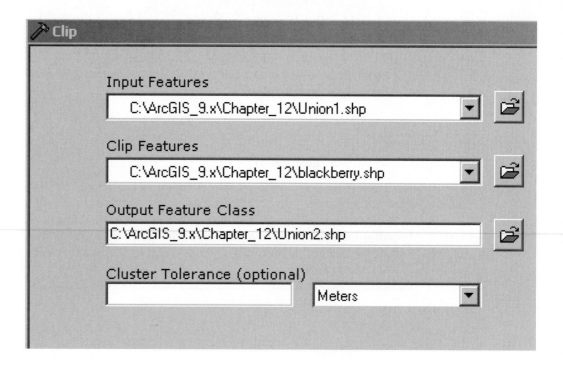

   d. Select Blackberry.shp as the Clip Features.

   e. Name the Output Feature Class Union2.shp.

   f. For Cluster Tolerance, use default Meters.

   g. Click OK.

   h. Right-click plant_buf and select Remove.

   i. Right-click rds_buff and select Remove.

   j. Right-click Union1 and select Remove.

B. Make the Environmentally Sensitive Lands Layer

These next steps identify areas that are environmentally sensitive. These are defined as wetlands and those areas within the 100-year flood zone.

> Our wetlands layer originates from the National Wetlands Inventory (NWI) from the U.S. Fish and Wildlife Service. This data source includes a background polygon that is a non-wetland. In the following step, we need to extract just the legitimate wetlands.

8. Extract wetlands from the nwi_poly.shp layer and call it wetlands.shp:

   a. Click Selection from the menu bar.

   b. Select Select By Attributes from the drop-down menu.

   c. Select nwi_poly for the Layer.

   d. Double-click "CODE" under Fields.

   e. Click (once) on the Not Equal button (i.e., <>).

   f. Double-click 'U' under Get Unique Values. ('U' is the value for the nonwetland background).

   g. Click Apply, then click Close.

   h. Right-click nwi_poly.

   i. Select Data from the shortcut menu.

   j. Select Export Data from the pop-up menu.

   k. Change the Output Shapefile name to wetlands.shp.

   l. Click Yes when asked "Do you want to add this layer to the map?"

   m. Right-click nwi_poly and select Remove Layer.

> The floodplain layer originates from the Federal Emergency Management Agency (FEMA) and includes upland areas that are surrounded by 100 year flood plains (i.e., elevated islands).

9. Extract the 100-year floodplains from the flood.shp layer and call it fld100.shp:

   a. Click Selection from the menu bar.

   b. Select Select By Attributes from the drop-down menu.

   c. Select flood for the Layer.

   d. Double-click "YEAR" under Fields.

   e. Click (once) on the Equal button (i.e., =).

   f. Double-click 100 under Get Unique Values.

   g. Click Apply, then click Close.

   h. Right-click flood.

   i. Select Data from the shortcut menu.

   j. Select Export Data from the pop-up menu.

   k. Change the Output Shapefile name to fld100.shp.

   l. Click Yes when asked "Do you want to add this layer to the map?"

   m. Right-click and select Remove flood.shp.

10. a. Union the wetlands.shp layer with the fld100.shp layer to make env_sens using the union procedure described in step 6.
    b. Remove wetlands.shp and fld100.shp.

> In our model, environmentally sensitive lands are defined as either wetlands or floodplains. Thus we can take advantage of the GIS dissolve procedure to simplify our final model layer.

11. Add a field to **env_sens.shp** called env and calculate equal to one and dissolve:
    a. Use the procedure in step 3 to add the env field and calculate it equal to 1.
    b. Click ArcToolbox.
    c. Click Data Management Tools.
    d. Click Generalization, and click Dissolve.

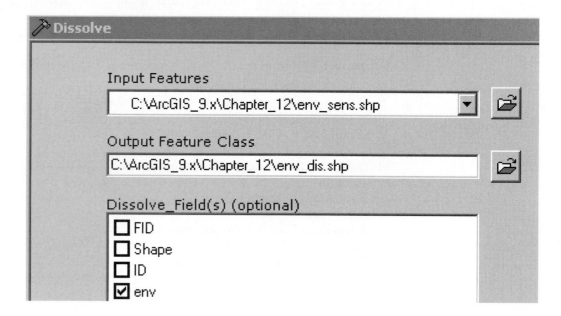

    e. Select env_sens as the Input Features.
    f. Name the output layer env_dis.shp.
    g. Select env as the attribute to dissolve.
    h. We do not use any Statistics Fields.
    i. Click OK.
    j. Right-click env_sens and select Remove.

12. a. Union the env_dis.shp layer with the union2.shp layer to make the e_u_dis.shp layer using the same union procedure shown in previous steps (refer to Step 6).
    b. Right-click union2.shp and select Remove.
    c. Right-click env_dis and select Remove.

Try the following query to see if you have succeeded in the previous steps:
Selection
Select By Attributes
make layer e_u_dis.shp
plantin = 1 and in_mjrd = 1 and env = 1

> *Question:* Which types of polygons get selected?
> *Answer:* Environmentally sensitive lands that are both within a half-mile of the plant and a major road leading to the plant.

C. Make the Cropland and Low-Density Residential Layers

13. Extract cropland from the land_cover.shp layer and call it farmland.shp:
     a. From the Selection menu, select Clear Selected Features.
     b. From the Selection menu, select Select By Attributes.
     c. Select **land_cover** for the Layer.
     d. Double-click COVER2 under Fields.
     e. Click (once) on the Equal button (i.e., =).
     f. Double-click Get Unique Values.
     g. Double-click Cropland.
     h. Click Apply, then click Close.
     i. Right-click land_cover.
     j. Select Data from the shortcut menu.
     k. Select Export Data from the pop-up menu.
     l. Change the Output Shapefile name to farmland.shp.
     m. Click Yes when asked "Do you want to add this layer to the map?"
     n. Click Selection on the menu bar.
     o. Select Clear Selected Features from the drop-down menu.

> Once again, we can simplify our final model layer by reducing the farmland layer utilizing the dissolve function. Currently farmland includes row crops, orchards, and small grains. However, we only need to know that it is cropland. The cover2 field is broad land-use categories and cover1 includes detailed land-use categories.

14. Dissolve ***farmland.shp*** by the cover2 attribute:
     a. Click ArcToolbox.
     b. Click Data Management Tools.
     c. Click Generalization, and then Dissolve.
     d. Select farmland as the Input Layer.
     e. Name the output layer farm_dis.shp.
     f. Select cover2 as the attribute to dissolve.
     g. We do not use any Statistics Fields.
     h. Click OK.
     i. Right-click farmland and select Remove.

15. Extract low-density residential from the land_cover.shp layer and call it Residential.
     a. Click Selection on the menu bar.
     b. Select Select By Attributes from the drop-down menu.
     c. Select land_cover for the Layer.
     d. Double-click "COVER1" under Fields.
     e. Click (once) on the Equal button (i.e., =).
     f. Double-click Low Density Urban under Get Unique Values.
     g. Click Apply, then click Close.
     h. Right-click land_cover.
     i. Select Data from the shortcut menu.
     j. Select Export Data from the pop-up menu.
     k. Change the Output Shapefile name to residential.shp.
     l. Click Yes when asked "Do you want to add this layer to the map?"
     m. Click Selection on the menu bar.
     n. Select Clear Selected Features from the drop-down menu.
     o. Right-click residential then select Open Attribute Table.

p. Right-click COVER2 and select Delete Field (Answer "Yes" to the warning).

Note the latter is a precautionary step because we union residential with farm_dis in the next step and cover2 already exists in farm_dis and is necessary, but it is not necessary for our residential layer.

q. Close the table. Right-click land_cover in the Table of Contents and select Remove.

16. a. Union the farm_dis.shp layer with the residential.shp layer to make the f_r.shp layer using the union procedure shown in previous steps.
b. Remove farm_dis and residential.

17. a. Union the f_r.shp layer with the e_u_dis.shp layer to make the f_r_e_u_dis.shp layer using the union procedure shown in previous steps.

Try the following query to see if you have succeeded in the previous steps:

Selection
Select By Attributes
make layer f_r_e_u_dis.shp
plantin = 1 and in_mjrd = 1 and env = 1 and cover2 = 'cropland'

b. Remove f_r and e_u_dis.

D. Integrate the Census Blocks for the Watershed Area for the Population Analysis

18. From the Selection menu, select Clear Selected Features. Open the attribute table of Bb_pop00.shp. Right-click the pop field and click Statistics.

Note that the watershed had 36,994 people in 2000 (up from 25,754 people in 1990).

We are now ready to union the census blocks layer with f_r_e_u_dis; however, the resulting layer will include census blocks that have been split into multiple blocks. In addition, the new attribute table containing population will include multiple records and population will be grossly overcounted. We correct for this overcounting by multiplying population in the final layer by the new area divided by the original area. This treats the blocks as though population is evenly distributed throughout the original block, an assumption we know is not true but one that is necessary to make.

19. a. Union the bb_pop00.shp layer with f_r_e_u_dis.shp to make all_union3.shp layer using the union procedure shown in previous steps.
b. Remove f_r_e_u_dis.shp and bb_pop00.shp.

A field statistics on the pop field of all_union3.shp will show an exaggeration of the total population (over 200,000 while original was 36,994).

E. Add Fields and Calculate Values

20. Open the attribute table of all_union3.

a. Click Options.
b. Click Add Field.
c. The new field name is new_area, the type is float, Precision is 16, and the scale is 5.
d. Right-click the column name.
e. Select Calculate Values from the shortcut menu.
f. Click Load and navigate to the Chapter 12 folder and load return_area script. Click OK.
g. Add the following additional fields (as float and precision of 16):

Set name to final_pop with a scale of 4.
Set name to plant_rate with a scale of 2.

Set name to road_rate with a scale of 2.
Set name to ag_rate with a scale of 2.
Set name to res_rate with a scale of 2.
Set name to env_rate with a scale of 2.
Set name to grow_rate with a scale of 2.
Set name to pop2020 with a scale of 4.
Set name to pop2020_p with a scale of 4.
Set name to pctchg_p with a scale of 2.

21. Click Selection, select Select By Attributes, set Layer to all_union3, and select orig_area > 0. This is necessary because in the next step we will be dividing by orig_area and there must be no zero values.

22. Open the all_union 3 attribute table and right-click the final_pop field to calculate using the following equation:

$$[pop] * ([new\_area]/[orig\_area])$$

This is used to correct the error from the splitting of the census blocks.

If you do a statistics on final_pop it should be close to the original population.

23. Keep the table open and use the following procedure to calculate the growth rate by the plant.

    a. From the Selection menu, select Clear Selected Features.
    b. In the open attribute table, click Options.
    c. Click Select By Attributes.
    d. plantin = 1
    e. Click Apply.
    f. Click Selected on the bottom of the attribute table.
    g. Right-click plant_rate.
    h. Select Calculate Values from the drop-down menu.
    i. Enter 1.2 in the large box (the growth rate for the plant area).
    j. Click OK.
    k. Clear the selection by clicking All on the bottom of the table and then Options/Clear Selection.

24. Click Show All, click Options, and use the procedure shown in step 23, but query in_majrd for "1" and calculate road_rate $= 1.0$.

    Do not forget to click All, then Options, and Clear Selection after each rate calculation.

25. Use the procedure shown in step 23 but query cover2 for "Cropland" and calculate ag_rate $= 0.8$.

26. Use the procedure shown in step 23 but query cover1 for "Low density urban" and calculate res_rate $= 0.2$.

27. Use the procedure shown in step 23 but query env for "1" and calculate env_rate $= -1.2$.

28. Calculate Grow_rate $= 1.64 + $ ag_rate $+$ env_rate $+$ plant_rate $+$ res_rate $+$ road_rate. Select the fields from the fields menu. Don't type them in.

29. Calculate Pop2020 $= 1.64 * $ final_pop. Select final_pop from the Fields Menu, rather than typing it in.

    A field statistics on this field will show the population in the year 2020 without the new communications plant (60,670).

30. Calculate Pop2020_p $= $ final_pop $* $ grow_rate.

    A field statistics on this field will show the population in the year 2020 with the new communications plant (80,236).

31. Select Final_pop greater than 0 with Select By Attributes.

    Do not clear selection.

32. Calculate Pctchg_p $= (($pop2020_p $- $ final_pop$)/$final_pop$) * 100$.

33. Create a map using a gradient (graduated color map) on pctchg_p of all_union3. Note: This will show the population change as a result of the new plant to the year 2020. First select Clear Selected Features.

F. Examine the Aerial View of the Plant Area

34. From the File menu, select Add Data.
35. Select 41088g34.sid, then click Add.
36. If you wish to see the details of the layers below the aerial photograph, right-click 41088g34.sid, select Properties, click the Display tab, change Transparent to 40%, click Apply, and click OK. Move 41088g34.sid to the top of the layer window.

G. A Challenge Step: Reaggregate Model Results Back up to the Block Level

37. Dissolve all_union3 by the Stfid attribute, carry the sum of final_pop and the sum of pop2020_p, and name the result all_union4.
38. Add a field to all_union4 called pctchg_p.
39. Use the procedure shown in step 20 but query sum_final_pop for "0" and calculate pctchg_p = 0.
40. Click Options and select Switch Selection.
41. Calculate pctchg_p = sum_pop2020_p/sum_final_pop * 100.
42. Create map using a gradient on pctchg_p of all_union4. Note: This will show the population change aggregated to the blocks.

Essay/Report

Write a report discussing the strengths and weaknesses of the above model. Do you consider the spatial procedures appropriate, or would you have changed them? If allotted more time, how would you have improved the model? What other maps would have been helpful to produce for planning purposes? How could the model builder application within ArcGIS help with the testing of alternative scenarios? The final report should be in a standard research format that includes an introduction, description of your study area (exploratory), research methods, analysis and discussion of findings, and conclusions. Your conclusion should also address the land-use planning implications of your analysis.

Index

Figures are indicated by page numbers followed by *f*; tables are indicated by page numbers followed by *t*.

GREENEPICK LICENSE AGREEMENT

Richard P. Greene/James B. Pick
Exploring the Urban Environment through GIS, Lab Exercises Data CD
0-13-146398-5
© 2006 Pearson Education, Inc.
Pearson Prentice Hall
Pearson Education, Inc.
Upper Saddle River, NJ 07458
Pearson Prentice Hall™ is a trademark of Pearson Education, Inc.
Pearson® is a registered trademark of Pearson plc
Prentice Hall® is a registered trademark of Pearson Education, Inc.

YOU SHOULD CAREFULLY READ THE TERMS AND CONDITIONS BEFORE USING THE CD-ROM PACKAGE. USING THIS CD-ROM PACKAGE INDICATES YOUR ACCEPTANCE OF THESE TERMS AND CONDITIONS.

Pearson Education, Inc. provides this program and licenses its use. You assume responsibility for the selection of the program to achieve your intended results and for the installation, use, and results obtained from the program. This license extends only to use of the program in the United States or countries in which the program is marketed by authorized distributors.

LICENSE GRANT

You hereby accept a nonexclusive, nontransferable, permanent license to install and use the program ON A SINGLE COMPUTER at any given time. You may copy the program solely for backup or archival purposes in support of your use of the program on the single computer. You may not modify, translate, disassemble, decompile, or reverse engineer the program, in whole or in part.

TERM

The License is effective until terminated. Pearson Education, Inc. reserves the right to terminate this License automatically if any provision of the License is violated. You may terminate the License at any time. To terminate this License, you must return the program, including documentation, along with a written warranty stating that all copies in your possession have been returned or destroyed.

LIMITED WARRANTY

THE PROGRAM IS PROVIDED "AS IS" WITHOUT WARRANTY OF ANY KIND, EITHER EXPRESSED OR IMPLIED, INCLUDING, BUT NOT LIMITED TO, THE IMPLIED WARRANTIES OR MERCHANTABILITY AND FITNESS FOR A PARTICULAR PURPOSE. THE ENTIRE RISK AS TO THE QUALITY AND PERFORMANCE OF THE PROGRAM IS WITH YOU. SHOULD THE PROGRAM PROVE DEFECTIVE, YOU (AND NOT PEARSON EDUCATION, INC. OR ANY AUTHORIZED DEALER) ASSUME THE ENTIRE COST OF ALL NECESSARY SERVICING, REPAIR, OR CORRECTION. NO ORAL OR WRITTEN INFORMATION OR ADVICE GIVEN BY PEARSON EDUCATION, INC., ITS DEALERS, DISTRIBUTORS, OR AGENTS SHALL CREATE A WARRANTY OR INCREASE THE SCOPE OF THIS WARRANTY. SOME STATES DO NOT ALLOW THE EXCLUSION OF IMPLIED WARRANTIES, SO THE ABOVE EXCLUSION MAY NOT APPLY TO YOU. THIS WARRANTY GIVES YOU SPECIFIC LEGAL RIGHTS, AND YOU MAY ALSO HAVE OTHER LEGAL RIGHTS THAT VARY FROM STATE TO STATE.

Pearson Education, Inc. does not warrant that the functions contained in the program will meet your requirements or that the operation of the program will be uninterrupted or error free. However, Pearson Education, Inc. warrants the CD-ROM(s) on which the program is furnished to be free from defects in material and workmanship under normal use for a period of ninety (90) days from the date of delivery to you as evidenced by a copy of your receipt. The program should not be relied on as the sole basis to solve a problem whose incorrect solution could result in injury to person or property. If the program is employed in such a manner, it is at the user's own risk, and Pearson Education, Inc. explicitly disclaims all liability for such misuse.

LIMITATION OF REMEDIES

Pearson Education, Inc.'s entire liability and your exclusive remedy shall be

1. the replacement of any CD-ROM not meeting Pearson Education, Inc.'s "LIMITED WARRANTY" and that is returned to Pearson Education, Inc., or
2. if Pearson Education, Inc. is unable to deliver a replacement CD-ROM that is free of defects in materials or workmanship, you may terminate this agreement by returning the program.

IN NO EVENT WILL PEARSON EDUCATION, INC. BE LIABLE TO YOU FOR ANY DAMAGES, INCLUDING ANY LOST PROFITS, LOST SAVINGS, OR OTHER INCIDENTAL OR CONSEQUENTIAL DAMAGES ARISING OUT OF THE USE OR INABILITY TO USE SUCH PROGRAM EVEN IF PEARSON EDUCATION, INC. OR AN AUTHORIZED DISTRIBUTOR HAS BEEN ADVISED OF THE POSSIBILITY OF SUCH DAMAGES, OR FOR ANY CLAIM BY ANY OTHER PARTY. SOME STATES DO NOT ALLOW FOR THE LIMITATION OR EXCLUSION OF LIABILITY FOR INCIDENTAL OR CONSEQUENTIAL DAMAGES, SO THE ABOVE LIMITATION OR EXCLUSION MAY NOT APPLY TO YOU.

GENERAL

You may not sublicense, assign, or transfer the license of the program. Any attempt to sublicense, assign, or transfer any of the rights, duties, or obligations hereunder is void. This Agreement will be governed by the laws of the State of New York.

Should you have any questions concerning this Agreement, you may contact Pearson Education, Inc. by writing to

ESM Media Development
Higher Education Division
Pearson Education, Inc.
1 Lake Street
Upper Saddle River, NJ 07458

Should you have any questions concerning technical support, you may write to New Media Production

Higher Education Division
Pearson Education, Inc.
1 Lake Street
Upper Saddle River, NJ 07458

YOU ACKNOWLEDGE THAT YOU HAVE READ THIS AGREEMENT, YOU UNDERSTAND IT, AND AGREE TO BE BOUND BY ITS TERMS AND CONDITIONS. YOU FURTHER AGREE THAT IT IS THE COMPLETE AND EXCLUSIVE STATEMENT OF THE AGREE-MENT BETWEEN US THAT SUPERSEDES ANY PROPOSAL OR PRIOR AGREEMENT, ORAL OR WRITTEN, AND ANY OTHER COMMUNICATIONS BETWEEN US RELATING TO THE SUBJECT MATTER OF THIS AGREEMENT.

SYSTEM REQUIREMENTS

The data contained on this CD are designed for use with ArcGIS Desktop 9.X (ArcView Level) software. This software is supplied separately by ESRI, the Environmental Systems Research Institute – *www.esri.com*.

ESRI Software is designed only for the PC. ArcGIS Desktop 9.X (ArcView Level) is not designed to run on an Apple computer.

- Windows:
Windows 2000/NT/XP
800 MHz Intel Pentium processor – minimum
1.0 GHz Intel Pentium processor – recommended
16 bit Sound Card
256 MB of available RAM – minimum
512 MB of available RAM – recommended
Display – greater then 256 colors
Swap Space – 300 MB minimum
Disk Space – 695 MB
Mouse or other pointing device
8X – CD_ROM

For additional details on ESRI system requirements, please see *www.esri.com*

SUPPORT INFORMATION

If you are having problems with this software, call (800) 677-6337 between 8:00 a.m. and 8:00 p.m. EST, Monday through Friday, and 5:00 p.m. through Midnight EST on Sundays. You can also get support by filling out the web form located at *http://247.prenhall.com/mediaform*

Our technical staff will need to know certain things about your system in order to help us solve your problems more quickly and efficiently. If possible, please be at your computer when you call for support. You should have the following information ready:

- Textbook ISBN
- CD-Rom/Diskette ISBN
- corresponding product and title
- computer make and model
- Operating System (Windows or Macintosh) and Version
- RAM available
- hard disk space available
- Sound card? Yes or No
- printer make and model
- network connection
- detailed description of the problem, including the exact wording of any error messages.

NOTE: Pearson does not support and/or assist with the following:

- third-party software (i.e., Microsoft including Microsoft Office Suite, Apple, Borland, etc.)
- homework assistance
- Textbooks and CD-Rom's purchased and used are not supported and are nonreplaceable. To purchase a new CD-Rom contact Pearson Individual Order Copies at 1-800-282-0693

Support for ArcGIS Desktop 9.X (ArcView Level) ESRI software is available through the ESRI website – *www.esri.com*.